The Epoch of Galaxy Formation

NATO ASI Series

Advanced Science Institutes Series

A Series presenting the results of activities sponsored by the NATO Science Committee, which aims at the dissemination of advanced scientific and technological knowledge, with a view to strengthening links between scientific communities.

The Series is published by an international board of publishers in conjunction with the NATO Scientific Affairs Division

A Life Sciences	Plenum Publishing Corporation
B Physics	London and New York
C Mathematical	Kluwer Academic Publishers
** and Physical Sciences**	Dordrecht, Boston and London
D Behavioural and Social Sciences	
E Applied Sciences	
F Computer and Systems Sciences	Springer-Verlag
G Ecological Sciences	Berlin, Heidelberg, New York, London,
H Cell Biology	Paris and Tokyo

Series C: Mathematical and Physical Sciences - Vol. 264

The Epoch of Galaxy Formation

edited by

Carlos S. Frenk

Richard S. Ellis

Tom Shanks

Physics Department, University of Durham,
Durham, U.K.

Alan F. Heavens

and

John A. Peacock

Royal Observatory, Blackford Hill,
Edinburgh, U.K.

Kluwer Academic Publishers

Dordrecht / Boston / London

Published in cooperation with NATO Scientific Affairs Division

Proceedings of the NATO Advanced Research Workshop on
The Epoch of Galaxy Formation
Durham, U.K.
18–22 July 1988

Library of Congress Cataloging in Publication Data

```
NATO Advanced Research Workshop on the Epoch of Galaxy Formation (1988
  : Durham, England)
    The epoch of galaxy formation / edited by Carlos S. Frenk ... [et
al.].
        p.   cm. -- (NATO ASI series. Series C, Mathematical and
physical sciences ; vol. 264)
    "Proceedings of the NATO Advanced Research Workshop on the Epoch
of Galaxy Formation, Durham, England, July 18-22, 1988."
    Includes indexes.

    1. Galaxies--Evolution--Congresses.   I. Frenk, Carlos S.
II. Title.   III. Series: NATO ASI series.  Series C, Mathematical
and physical sciences ; no. 264.
    QB857.5.E96N38   1989
    523.1'12--dc19
```

ISBN-13: 978-94-010-6902-1   e-ISBN-13: 978-94-009-0919-9
DOI: 10.1007/978-94-009-0919-9

Published by Kluwer Academic Publishers,
P.O. Box 17, 3300 AA Dordrecht, The Netherlands.

Kluwer Academic Publishers incorporates the publishing programmes of
D. Reidel, Martinus Nijhoff, Dr W. Junk and MTP Press.

Sold and distributed in the U.S.A. and Canada
by Kluwer Academic Publishers,
101 Philip Drive, Norwell, MA 02061, U.S.A.

In all other countries, sold and distributed
by Kluwer Academic Publishers Group,
P.O. Box 322, 3300 AH Dordrecht, The Netherlands.

printed on acid free paper

TABLE OF CONTENTS

I. FORMING GALAXIES

II. GAS CLOUDS AT HIGH REDSHIFT

III. QUASARS

IV. EVOLUTION OF GALAXIES AND CLUSTERS

V. BACKGROUND RADIATION

VI. MODELS

VII. THE AGE OF THE GALAXY

VIII. PROSPECTS

POSTER PAPERS

APPENDIX

---

* Invited speaker

PREFACE

Scientists in the late twentieth century are not the first to view galaxy formation as a phenomenon worthy of explanation in terms of the known laws of physics. Already in 1754 Kant regarded the problem as essentially solved. In his *Universal Natural History and Theory of the Heavens* he wrote: "If in the immesurable space in which all the suns of the Milky Way have formed themselves, we assume a point around which, through some cause or other, the first formation of nature out of chaos began, there the largest mass and a body of extraordinary attraction will have arisen which has thereby become capable of compelling all the systems in the process of being formed within an enormous sphere around it, to fall towards itself as their centre, and to build up a system around it on the great scale.... Observation puts this conjecture almost beyond doubt." More than 200 years later, a similar note of confidence was voiced by Zel'dovich at an IAU symposium held in Tallin in 1977: "Extrapolating ... to the next symposium somewhere in the early eighties one can be pretty sure that the question of the formation of galaxies and clusters will be solved in the next few years."

Perhaps few astronomers today would share Kant's near certainty or feel that Zel'dovich's prophecy has been fulfilled. Many, however, will sympathize with the optimistic outlook of these two statements. Unprecedented advances in the past few years have given cause for increased optimism. Traditionally the subject has relied on the limiting sensitivity of available detectors. This has increased impressively in the last decade with the widespread use of CCD cameras, multiobject spectroscopy and, most recently, infrared arrays. Investigations of objects at redshifts $z \lesssim 5$, corresponding to look-back times of up to $\sim 90\%$ of the age of the Universe, have become possible for the first time and have put the subject on a firm observational footing. At the same time, theoretical developments have experienced a substantial surge, powered by new ideas on the physics of the early universe and by the proliferation of computing power which has led to very detailed modelling.

Amongst recent exciting developments is the mounting evidence for substantial activity even at the moderate redshifts, $z \lesssim 3$, which have been most extensively investigated. Optical and infrared observations of radio galaxies, the inferred properties of gas clouds along the line-of-sight to quasars and counts of galaxies and quasars to very faint magnitudes in various passbands all indicate that the Universe at these epochs was quite different from what it is today. Yet, other observations suggest that at least some galaxies at high redshifts are very similar to galaxies observed today. These conflicting lines of evidence —some pointing towards a low redshift of galaxy formation, others pointing in the opposite sense— prompted us to organize a workshop on "The epoch of galaxy formation". When, during the summer of 1986, we began to canvass possible participants, we discovered that groups both at Durham and at the Royal Observatory at Edinburgh had similar ideas. In an unusual instance of Anglo-Scottish cooperation we joined forces and the result was this workshop sponsored by both institutions.

The workshop provided a comprehensive and surprisingly coherent view of the present state of knowledge in the subject. With very few regrettable exceptions, these proceedings contain all the invited and contributed talks. Many more people wished to give contributed talks than could be accomodated in a one week meeting. We attempted to compensate for this by having several sessions in which participants could convey brief ideas or results by means of two viewgraphs. Some of these contributions and the poster papers are collected at the back of this book. The workshop was marked by many lively discussions, some of which we attempted to record in sheets distributed to speakers and their interrogators. The deciphering and transcription of these handwritten sheets was a non-trivial task, but we felt that even a fuzzy flavour of what took place was better than nothing, especially if accompanied by heartfelt apologies from the editors for any imprecisions which we may have introduced.

It became clear from the outset that the concept of galaxy formation can be defined in several ways and that a precise definition is a prerequisite for any discussion of the epoch at which it occured. Equally important is some expectation of what a forming or primeval galaxy should look like. The first contributions address these issues and set the scene by providing a general theoretical background. The discussion then moves on directly to searches for primeval galaxies and to a review of the properties of candidate objects observed at high redshifts. Absorption of quasar light allows a detailed study of the properties of intervening gas clouds at high redshift. Their properties, as well as recent calculations of the thermal history of the intergalactic medium, are reviewed. A subset of these clouds, the damped Ly-α systems seen at $z = 2-3$, contain an amount of mass in HI comparable to the mass in stars in the disk of present day spirals. The first measurement of a low metal abundance in one such system and the possible detection of Ly-α emission in another are reported here. The implications of the observed distribution of quasars and particularly the observed cut-off in their number density at $z \simeq 2.5$ are considered next. The discussion then moves on to galaxies in dense environments at redshifts $z \simeq 0.2 - 1$. Once again, strong evolution both in the stellar populations and in the dynamical properties is the salient theme. Constraints on the epoch of galaxy formation follow from data on distortions and anisotropies in the cosmic background radiation. The implications of these data, including the recently claimed distortion of the spectrum at submillimeter wavelengths are discussed here. Turning to our own galaxy, the difficulties of dating its various subcomponents and some recent results are reviewed.

Theoretical expectations for the epoch and mode of galaxy formation can be formulated to varying degrees of specificity in current cosmogonic models. The most thoroughly investigated is the cold dark matter model which predicts that galaxy formation is a recent and protracted affair which continues until the present day. Alternative models which appeal to explosions, to cosmic strings or to baryon isocurvature fluctuations to seed galaxy formation are also discussed. Predictions for the history of star formation, the growth paths and spatial distribution of galaxies and quasars are presented in various contributions. The final paper provides a glimpse of the challenges to come from some of the exciting observational developments planned for the coming decade.

At the request of several distinguished participants, we held a ballot on the epoch of galaxy formation at the end of the meeting. Although this procedure constitutes a significant departure from established scientific methodology it gives, at least, some indication of the prevailing mood at the time. Two questions were subjected to a vote: (i) What is the redshift by which half the stars that we see today had formed? and (ii) What is the median value of the redshift at which half the mass currently within 5 kpc of the centre of a bright galaxy was in place. The voters were given three possible redshift ranges: $0 - 2.5$, $2.5 - 5$, $5 - \infty$. In this order, the number of votes cast for each option were as follows. Question (i): 28, 54, 4; question(ii): 16, 41, 22. Those readers who peruse the bulk of these proceedings will be able to make an informed judgement on the merits of democracy! No ballot was held to select the preferred theory. This was in keeping with the spirit of Thomas Wright of Durham, a contemporary of Kant's, who whilst speculating on various possibilities for the cosmic centre of gravitation stated: "Which of these is most probable I shall leave undetermined, and must acknowledge at the same time my notions here are so imperfect I hardly dare conjecture."

It is appropriate here to thank various individuals and institutions which contributed to the workshop and to these proceedings. The bulk of the financial support was provided by NATO and complemented by the Physics Department of Durham University, by the Royal Observatory of Edinburgh and by visitor grants from SERC. Thanks are also due to Dr. C. Sinclair, director of the NATO ARW programme for his advice and cooperation. The scientific organising committee consisted of H. Butcher, M. Longair, P.J.E. Peebles and M.J. Rees. I would like to thank them, particularly Martin Rees, for their support and guidance. Several postgraduate students at Durham, especially R.J. Bower, I. Georgantopoulos, B. Moore and N.R. Tanvir played a central role in the smooth running of the meeting and M. Chipchase carried out the most efficient secretarial work while enduring the frayed tempers of the members of the local organising committee: R.S. Ellis, T. Shanks and myself from Durham and A.F. Heavens and J.A. Peacock from ROE. Finally, on behalf of the organising committee, I would like to thank all the participants and contributors to these proceedings who were ultimately responsible for the success of the meeting and for the publication of this book.

Carlos S. Frenk

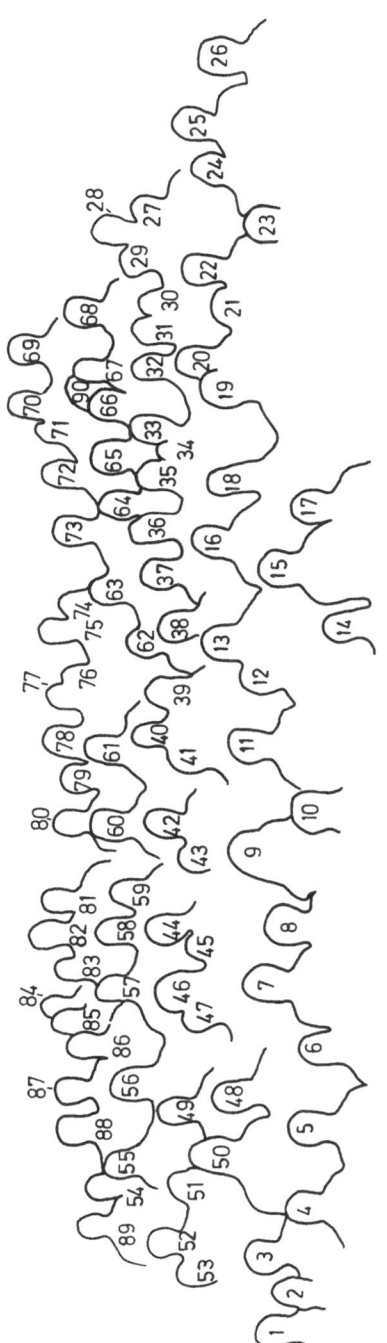

LIST OF PARTICIPANTS

Mr. A. Aragón
Dept. de Astrofísica,
Cuidad Universitaria,
E-28040 Madrid,
SPAIN.

Dr. N. Arimoto
Institut fur Theoretische
Astrophysik der Universitat
Heidelberg,
Im Neuenheimer Feld 561,
D-6900 Heidelberg, F.R.G.

Mr. K.M. Ashman,
School of Maths.,
Queen Mary College,
Mile End Road,
London E1 4NS,
UNITED KINGDOM.

Mr. A. Babul,
Department of Astrophysical
Sciences,
Peyton Hall,
Princeton University,
Princeton, NJ 08544,
U.S.A.

Dr. J.E. Barnes,
Institute for Advanced Study,
Princeton, NJ 08540,
U.S.A.

Dr. J. Barrow,
Astronomy Centre,
Physics Building,
University of Sussex,
Falmer, Brighton,
UNITED KINGDOM.

Prof. J. Bekenstein
Physics Department,
Ben Gurion University,
Beersheve,
ISRAEL.

Dr. J. Bergeron
Institut d'Astrophysique,
98 bis Bld Arago,
75014 Paris,
FRANCE.

Dr. A. Blanchard
Observatoire de Paris-Meudon,
5 Placel Janssen,
92 195 MEUDON Principal Cedex,
FRANCE.

Prof. R. Bond
Canadian Inst. for
Theoretical Astrophysics,
McLennan Labs.,
Toronto, ONT M5S 1A7,
CANADA.

Dr. S. Bonometto
Dip. Fisica,
Via Pascoli,
06100 Perugia,
ITALY.

Dr. R. Brandenberger
Dept. of Physics,
Brown University,
Providence, RI 02912,
U.S.A.

Dr. G. Bruzual
Centro de Investigacion de
Astronomia,
Apartado Postal 264,
Merida 5101-A, VENEZUELA.

Prof. H. Butcher
Kapteyn Lab. PB 800,.
9700 AV Groningen,
THE NETHERLANDS.

Prof. G. Burbidge
Centre for Astrophysics,
Space Sciences,
Univ. of California,
San Diego, La Jolla,
CA 92093, U.S.A.

Dr. R. Carlberg,
Dept. of Astronomy,
University of Toronto,
Toronto, Ontario M5S 1A1,
CANADA.

Dr. B. Carr
School of Math. Sciences,
Queen Mary College,
Mile End Road,
London. E1 4NS
UNITED KINGDOM

Dr. R. Carswell
Institute of Astronomy,
Madingley Road,
Cambridge CB3 0HA,
UNITED KINGDOM.

Dr. P. Coles
Astronomy Centre,
University of Sussex,
Brighton BN1 9QH,
UNITED KINGDOM.

Dr. C.A. Collins
The Royal Observatory,
Blackford Hill,
Edinburgh, EH0 3HJ,
UNITED KINGDOM.

Dr. H. Couchman
CITA,
University of Toronto,
60 St. George Street,
Toronto, Ontario M5S 1A1,
CANADA.

Prof. L. Cowie
Institute of Astronomy,
University of Hawaii,
2680 Woodlawn Drive,
Honolulu, HI 96822,
U.S.A.

Ms. C. Crawford
Institute of Astronomy,
Madingley Road,
Cambridge CB3 0HA,
UNITED KINGDOM.

Mr. J. Dunlop
The Royal Observatory,
Blackford Hill,
Edinburgh EH0 3HJ,
UNITED KINGDOM.

Dr. G. Efstathiou
Institute of Astronomy,
Madingley Road,
Cambridge CB3 OHA,
UNITED KINGDOM

Prof. R.S. Ellis
Department of Physics,
University of Durham,
South Road, Durham DH1 3LE,
UNITED KINGDOM

Dr. A. Evrard,
Institute of Astronomy,
Madingley Road,
Cambridge CB3 0HA,
UNITED KINGDOM.

Dr. S.M. Fall
Space Telescope Sci. Inst.,
3700 San Martin Drive,
Baltimore MD 21218,
U.S.A.

Prof. G. Field
Harvard-Smithsonian Centre
for Astrophysics,
60 Garden Street,
Cambridge MA 02138,
U.S.A.

Dr. K. Freeman
Mt. Stromlo Observatory,
Private Bag Woden P.O.,
Canberra Act. 2602,
AUSTRALIA.

Dr. C.S. Frenk,
Department of Physics,
University of Durham,
South Road,
Durham DH1 3LE,
UNITED KINGDOM

Prof. R. Green
KPNO,
National Optical Astron. Obs.,
950 N. Cherry, P.O. Box 26732,
Tucson AZ 85726, U.S.A.

Dr. R. Griffiths,
Space Telescope Science Inst.,
Homewood Campus,
Baltimore MD 21218,
U.S.A.

Prof. J. Gunn,
Princeton Univ. Obs.,
Peyton Hall,
Princeton, NJ 08540,
U.S.A.

Dr. A. Heavens
The Royal Observatory,
Blackford Hill,
Edinburgh EH0 3HJ,
UNITED KINGDOM.

Dr. L. Hernquist
Institute for Advanced Study,
Princeton, NJ 08540,
U.S.A.

Dr. R. Hunstead
School of Physics,
University of Sydney,
NSW 2006, AUSTRALIA.

Prof. G. Illingworth
Lick Observatory,
Univ. of California,
Santa Cruz, CA 95064,
U.S.A.

Dr. R. Johnstone
Institute of Astronomy,
Madingley Road,
Cambridge CB3 0HA,
UNITED KINGDOM.

Dr. B. Jones,
NORDITA,
Blegdansvej 17,
DK-2100 Copenhagen 0,
DENMARK.

Dr. N. Kaiser
Canadian Inst. for
Theoretical Astrophysics,
University of Toronto,
Toronto, Ontario M5S 1A7,
CANADA.

Mr. N. Katz
Princeton Univ. Obs.,
Peyton Hall,
Princeton, NJ 08544,
U.S.A.

Dr. A.A. Klypin
Institute of Applied Maths.,
Miussakaya Square 4,
Moscow 1215047,
USSR

Dr. D. Koo
Lick Observatory,
University of California,
Santa Cruz, CA 95064,
U.S.A.

Dr. C. Lacey
Centre for Astrophysics,
60 Garden Street,
Cambridge, MA 02138,
U.S.A.

Dr. S. Lilly
Institute for Astronomy,
University of Hawaii,
2680 Woodlawn Drive,
Honolulu, HI 96822,
U.S.A.

Dr. J. Madsen
Institute of Astronomy,
University of Aarhus,
DK 8000 Aarhus,
DENMARK.

Dr. R. Majewski
Yerkes Observatory,
P.O. Box 258,
Williams Bay,
Wisconsin, 53191-0258,
U.S.A.

Dr.E.Martinez-Gonzalez
Departamento de Fisica Moderna,
Universidad de Cantabria,
Avda Los Castros,
39005 Santander, SPAIN

Mr. P.J. McCarthy
601 Cambell Hall,
U.C. Berkeley,
Berkeley, CA 94720,
U.S.A.

Dr. A. Melott
Physics & Astronomy,
University of Kansas,
Lawrence, KS 66045,
U.S.A.

Dr. J. Norris
Mt. Stromlo & Siding Springs
Observatory,
Private Bag,
Woden P.O., A.C.T. 2606,
AUSTRALIA.

Prof. A. Oemler
Dept. of Astronomy,
Yale University,
New Haven, CT 06511,
U.S.A.

Prof. P.J.E. Peebles
Physics Department,
Jadwin Hall,
Princeton University,
P.O. Box 708,
Princeton NJ 08544,
U.S.A.

Dr. A. Pickles
Kapteyn Lab. PB 800,
9700 AV Groningen,
THE NETHERLANDs.

Dr. B. Rocca-Volmerange
Centre National de la
Recherche Scientifique,
Institut d'Astrophysique,
98 Bis, Boulevard Arago,
75014, Paris, FRANCE.

Dr. B. Rudak
Copernicus Astro. Centre,
Ul. Bartycka 18,
00-716, Warsaw,
POLAND.

Dr. R. Sancisi
Kapteyn Lab. POB 800,
9700 AV Groningen,
THE NETHERLANDS.

Prof. W. Sargent
Astronomy Dept., 105-224,
CALTECH,
Pasadena, CA 91125,
U.S.A.

Dr. J.A. Peacock,
The Royal Observatory,
Blackford Hill,
Edinburgh EH0 3HJ,
UNITED KINGDOM.

Dr. M. Pettini
Anglo-Australian Observatory,
Epping Laboratory,
P.O. Box 296,
Epping NSW 2121,
AUSTRALIA.

Prof. M.J. Rees
Institute of Astronomy,
Madingley Road,
Cambridge CB3 0HA,
UNITED KINGDOM.

Dr. J. Rose
Dept. of Phys. Astron.,
University of North Carolina,
Chapel Hill, NC27511,
U.S.A.

Dr. B.S. Ryden
Center for Astrophysics,
60 Garden Street,
Cambridge, MA 02138,
U.S.A.

Dr. J.L. Sanz
Depto Fisica Teoretica,
Univ. de Cantabria,
39005 Santander, SPAIN.

Dr. R. Scaramella
Int. School for Adv. Studies,
Strada Costiera 11,
3404 Trieste, ITALY.

Prof. M. Schmidt
Astronomy Dept., 105-224,
CALTECH,
Pasadena, CA 91125,
U.S.A.

Dr. J. Sellwood
Dept. of Astronomy,
Manchester University,
Manchester, M13 9PL,
UNITED KINGDOM.

Dr. T. Shanks
Department of Physics,
University of Durham,
South Road, Durham DH1-3LE,
UNITED KINGDOM.

Prof. P. Shapiro
Astronomy Dept.,
University of Texas at Austin,
RLM 15.3087,
Austin 78712-1083, U.S.A.

Prof. J. Silk
Department of Astronomy,
U.C. Berkeley,
Berkeley, C.A. 94720,
U.S.A.

Dr. D. Spergel
Institute for Advanced Study,
Princeton, NJ08540,
U.S.A.

Prof. H. Spinrad
Astronomy Department,
University of California,
Berkeley, CA 94720,
U.S.A.

Dr. W. Sutherland
Institute of Astronomy,
Madingley Road,
Cambridge CB3 0HA,
UNITED KINGDOM.

Dr. P. Thomas
CITA,
University of Toronto,
Toronto, Ontario M5S 1A7,
CANADA.

Mr. C. Thompson
Department of Physics,
Princeton University,
Jadwin Hall, P.O. Box 708,
Princeton, NJ 08544, U.S.A.

Dr. M. Tosi
Observatorio Astronomico,
C.P. 596,
I-40100 Bologna,
ITALY

Dr. D. Valls-Gabaud
Institut d'Astrophysique,
98 Bis. Boulevard Arago,
75014 Paris,
FRANCE.

Mr. D. Weinberg
Princeton Univ. Observatory,
Peyton Hall,
Peyton, NJ 08544,
U.S.A.

Prof. S.D.M. White
Steward Observatory,
University of Arizona,
Tucson, AZ 85721,
U.S.A.

Prof. A. Wolfe
Mt. Wilson & Las Campanas Obs.,
813 Santa Barbara St.,
Pasadena, CA 91101,
U.S.A.

Dr. H.K.C. Yee
Department of Physics,
University of Montreal,
P.O. Box 6128, Station A,
Montreal, PQ H3C 3J7,
CANADA.

Dr. B. Yulin,
The Royal Observatory,
Blackford Hill,
Edinburgh EH0 3HJ,
UNITED KINGDOM.

Dr. G. Zamorani
Instituto di Radioastronomia,
Via Irnerio 46,
40126 Bologna,
ITALY.

QUOTATIONS

Things were much the same at $z = 1$ (Peebles)

Now is the epoch of disk formation (Cowie)

If we're incomplete then so is everybody else (Shanks)

I think you've been talking to your lawyer too (White to Peebles)

A little bit blobby and definitely not round (Spinrad)

The models are alright (Klypin)

Beware of mergers in the future (Melott)

I will not give error bars because it would be too awful (Bergeron)

You may be right, however... (Wolfe)

The problem's with the data not the theory (Silk)

The Frenk, Efstathiou and White noise spectrum (Gunn)

Why do people worry about bias? (Kaiser)

I personally don't like black holes (Carr)

There was a forthright discussion in the literature (Norris on Gilmore and Bahcall)

I don't want to rely on violence; I believe in peace (Spergel)

High z is becoming too macho (Ellis)

This is not possible! (Frenk on the final voting on "When did galaxies form?")

Compiled by Richard Ellis

GALAXY FORMATION: HIGH REDSHIFT OR LOW ?

P.J.E. Peebles
Joseph Henry Laboratories
Princeton University
Princeton N.J. 08544
U.S.A.

ABSTRACT. A list of passages in the formation of a galaxy would include the epoch at which the bulk of the mass in the central parts had been put in place; the epoch at which the bulk of the mass in the disc was in place; and the epochs at which most of the seen stars and the heavy elements had formed. I review some lines of evidence by which we might hope to place useful constraints on these epochs.

1. The Epochs of Galaxy Formation

A good deal of the confusion in this subject arises from the fact that people can mean very different things by the epoch of galaxy formation. In particular, we have to distinguish the formation of a galaxy as a physically distinct entity from other important events that may mark its evolution once assembled or the evolution of its components before assembly. Thus a galaxy that formed as a gas-rich system at high redshift and suffered a recent intense burst of star formation might be reckoned to be evolutionarily young and physically old.

Since galaxies are held together by gravity, which depends on mass, a key process in galaxy formation must be the collection of its mass. This need not happen at a definite time. The outer parts could be added after the center was assembled, as in accretion models. The mass in a given part of a galaxy could be added over a wide range of redshifts, in a more or less gentle rain of material, or in a few mergers. To define a characteristic epoch of assembly of the mass let us proceed as follows.

Because the bright central parts of galaxies are the best studied it makes sense to concentrate on the mass within a fixed scale comparable to the optical radius. A reasonable choice for the radius is

$$r_c = 10h^{-1}\text{kpc},\tag{1}$$

where Hubble's constant is

$$H = 100h \text{ km sec}^{-1} \text{ Mpc}^{-1}.$$

It may be relevant also that this choice for r_c is close to the median value of the dark halo core radius (in an isothermal model) in the survey of Athanassoula, Bosma and Papaionnou (1987). In galaxy j observed at $z \sim 0$ the mass within r_c reached half the present value, m_j, at redshift z_j. The median value of z_j for bright galaxies is a characteristic epoch, z_g, of assembly of the mass of the inner parts of galaxies. Other interesting statistics are the width of the distribution

1

C. S. Frenk et al. (eds.), The Epoch of Galaxy Formation, 1–13.
© 1989 by Kluwer Academic Publishers.

of z_j and the correlation of z_j with m_j or with morphological type. And one can of course define similar measures for the outer parts of the galaxy. Most of the discussion in this paper deals with bright galaxies at $r < r_c$, because that is where the observational situation is best, and it is the bright galaxies that contribute most of the light.

The disc of a spiral is a prominent component but according to current ideas it is not the dominant mass even within r_c. It is appropriate therefore to introduce a second redshift, z_d, for the characteristic epoch at which the bulk of the disc mass in a typical large spiral was in place. Since the assembly of the disc could very well be a lengthy process a careful definition would be framed in the manner of z_g.

At $r < r_c$ a considerable fraction of the mass of a galaxy is in stars. Since this is the most readily observed and analysed component it is of practical as well as theoretical interest to have a measure of the epoch of star formation. Let us call z_* the median value of the redshift at which half the mass in seen long lived stars in a bright galaxy had formed.

Finally, the heavy elements in galaxies had to have built up from negligibly small abundances in the primeval material. Let us call z_e the epoch at which there was half of the present mass of heavy elements at $r < r_c$ in bright galaxies.

We may be able to find from theory and observation estimates of or inequalities among these four characteristic redshifts. For example, I gather that most stars at $r < r_c$ are metal rich and most of the metals in a galaxy are in the stars, from which it follows that z_e is greater than z_*. Galaxies undoubtedly are evolving at low redshift, and this could well involve the addition of mass to spheroids by cooling flows or to discs by accretion of gas. This must of course be taken into account in reckoning z_g and z_d. It will be noted, however, that if discs are a secondary phenomenon, as is often argued, then the feeding of discs could make z_d small while leaving z_g large, an example of the difference between evolutionary and physical ages.

Among the four redshifts, z_g is the farthest removed from the observations and so the least likely to be usefully constrained in the forseeable future. That is unfortunate because as I noted above z_g arguably is the fundamental epoch constraining theories of galaxy formation; but it would be wishful thinking to substitute some other redshift just because it can be measured.

In the following sections I discuss three classes of evidence on the characteristic epochs of galaxy formation. By the physical constraints I mean those that follow in a fairly general way if the standard relativistic model for the expanding universe is a useful approximation back to redshifts well in excess of decoupling ($z \gg 1000$). These constraints should be taken seriously, but by themselves they can never yield a believable theory of galaxy formation. The big excitement of this conference is the rapid growth of the observational evidence. My discussion of these astronomical constraints mainly consists of a series of questions that seem timely; I refer to the papers in this volume for the substance. Finally, our ideas are stimulated and even usefully guided by theoretical models of galaxy formation, what I call the cosmogonical constraints. The review in Section 4 is meant to illustrate the wide range of possibilities within the standard models.

2. Physical Constraints

2.1. The Sizes and Densities of Galaxies

In the standard expanding universe cosmology, which I adopt in all this discussion, a mature galaxy has a fixed physical size and the distance between galaxies is increasing with time. This means that galaxies as we know them could only have existed after the time at which the volume they now occupy fills space. The present number density of bright galaxies is $n_o \sim 0.02h^3$ Mpc^{-3}, or mean separation $R \sim 4h^{-1}$ Mpc. The bright parts of a bright spiral can be fitted in a box of width $r \sim 2r_c \sim 20h^{-1}$ kpc. The ratio, R/r, is the maximum redshift at which the inner parts could have existed as stable objects. If protogalaxies collapsed by a factor f_c in radius the size limit would be reduced roughly to

$$z_g < 200/f_c.$$

Almost equivalent to this is the density constraint, that galaxies cannot have existed when the mean mass density of the universe was greater than the density within a galaxy. This is the binding energy (or density) argument Bruce Partridge and I originally used to estimate the epoch of galaxy formation. We can write the mean mass density within the bright parts of a protogalaxy as

$$\text{density} \sim \frac{r_c v_c^2}{G} \frac{3}{4\pi r_c^3} \frac{1}{f_c^3} \sim \frac{9\pi^2}{16} [\Omega \rho_c (1 + z_g)^3].$$

The middle expression is an estimate of the mean density at maximum expansion, before the protogalaxy collapsed by a factor f_c in radius to become a stable object. The factor in square brackets in the right hand expression is the cosmological mass density at redshift z_g, with ρ_c the critical Einstein-de Sitter density at the present epoch and Ω the density parameter. For a spherically symmetric object the numerical prefactor brings this to the mass density within a patch that has just stopped expanding at redshift z_g. With $r_c = 10h^{-1}$ kpc and $v_c \sim 250$ km sec^{-1} this equation gives

$$z_g \sim \frac{30}{\Omega^{1/3} f_c}. \tag{2}$$

If f_c were much greater than unity we would have to allow time for collapse; in the spherical model that would divide z_g by the factor $2^{2/3}$. The essential difference between this and the size constraint is that the latter ignores the material outside the bright parts of galaxies. Equation (2) uses a slightly more sophisticated model for the way a protogalaxy breaks away from the general expansion, but I suspect we should not trust the numerical factor from the spherical model beyond the rough order of magnitude.

It is useful to consider the density argument another way. The baryon density wanted for light element Big Bang nucleosynthesis is $n_o \sim 10^{-6.5}$ cm^{-3} (Yang et al. 1984). If galaxies formed late, starting from a protogalaxy with density comparable to the background at redshift $z = 2$, then to reach the density in our galaxy, $n \sim 1$ cm^{-3}, would require a collapse factor $\sim 10^5$ in density or, in radius,

$$f_c(z_g \sim 2) \sim 50. \tag{3}$$

One gets similar numbers for the collapse factor for dark mass if the galaxy is dominated by dark matter. That is, because galaxies are so dense relative to the cosmic mean, galaxy formation at low redshift would require a impressive collapse factor. At larger z_g, f_c is smaller because the mean density is higher.

Theories of the collapse factor are highly uncertain. On one extreme, the naive estimate for dissipationless collapse is $f_c = 2$, to make the gravitational energy twice the binding energy. On the other extreme, the N-body model simulations of the cold dark matter (CDM) model by White et al. (1987) indicate large collapse factors in halo formation, with compact sub-galactic mass concentrations sinking to make a dense core, perhaps because they are transferring their binding energy to less dense material and driving it out of the protohalo. There are astrophysical limits on such merger scenarios, in particular that disc formation has to postdate any merger phase violent enough to disrupt or sensibly thicken discs. As discussed in § 4.1 the age of our disc may provide a useful constraint here. It may also be possible to reduce the uncertainty by a semiempirical argument, as discussed next.

4

2.2. Continuity

A measure of the mean distribution of mass around a galaxy is provided by the rms line of sight relative velocity of pairs of galaxies as a function of their separation, r:

$$\delta v(r) = \langle (v_1 - v_2)^2 \rangle^{1/2} \sim 300(hr_{Mpc})^{0.1} \text{ km sec}^{-1}.$$

This statistic is fairly reliably estimated at 10 kpc $< hr < 1$ Mpc (Peebles 1984). The value at the small separation end, $hr \sim 10$ kpc, is $\delta v \sim 200$ km sec^{-1}, about what one would expect from the observed motions of matter in galaxies at the same radius. The observation that δv is nearly flat is an extension of the nearly flat rotation curves observed in galaxies and has the same significance: the mass density at distance r from a bright galaxy, averaged over a fair sample, varies as

$$\rho(r) \propto r^{-\gamma}, \quad \gamma \sim 1.8. \tag{4}$$

It is remarkable that this relation with a fixed coefficient describes both the mean density run within a galaxy and the mean concentration of mass in the clustering of other galaxies around it.

The simple interpretation of this scaling relation is that the process by which mass was assembled in galaxies is a scaled version of the process of formation of the pattern of clustering of galaxies. In hierarchical models where formation proceeds from small objects to large the density enhancement due to collapse to virialization is modest on the scale of systems of galaxies because the collapse is nearly dissipationless. To preserve equation (4) we would want a similarly small collapse factor for the formation of galaxies.

If f_c were large, as in equation (3), the enhancement in $\rho(r_c)$ due to collapse would be very large, and in an hierarchical model the scaling behavior of equation (4) would seem to be a remarkable coincidence. Thus if $z_g \sim 2$ a more straightforward interpretation of equation (4) would be that galaxies formed in a non-hierarchical way, as in the pancake or explosion pictures.

Baron and White (1987) emphasize another point, that the mean distribution of luminosity around a galaxy shows a distinct break between galaxies and the clustering on larger scales. That is, in contrast to the mass distribution, galaxies viewed as distributions of light are distinct islands. This effect is in the direction one would expect if the luminous matter collapsed a good deal more within galaxies than it did on on larger scales, which may be connected to the collapse wanted to spin up discs, as discussed next. However, in gravitational theories of galaxy formation I cannot see how the distribution of luminosity could be relevant to the main act, which is the assembly of the mass of the system.

I conclude that, if galaxies formed as part of a clustering hierarchy, then the simple interpretation of equation (4) would be that the collapse factor f_c is less than perhaps 3. In this case equation (2) indicates $z_g > 10$. The other straightforward possibility is that galaxies formed in a 'top-down' scenario, as in the pancake or explosion pictures.

2.3. Angular Momentum

If protogalaxies grew by gravitational instability in a clustering hierarchy then the usual estimate of the angular momentum L produced by tidal torques would be

$$\lambda \sim \frac{LE^{1/2}}{GM^{5/2}} \sim 0.07, \tag{5}$$

where the mass, M, and binding energy, E, have been used to get the dimensionless angular momentum parameter, λ. This value for λ is well below the angular momentum of a rotationally supported disc; the usual presumption is that the disc material was spun up as the result of a considerable collapse factor. The rotation of large ellipticals is comparable to equation (5).

That could mean that the collapse factors for large ellipticals and the spheroid components of spirals are modest, f_c close to unity, consistent with the continuity argument. On the other hand, White et al. (1987) note that in N-body model studies of the gravitational collapse of a system with strong subclustering angular momentum can be efficiently transferred from the denser parts that collect at the center to the less dense outer parts, producing a large collapse factor in a fairly slowly rotating core. If f_c were as large as 50, as in equation (3), then one would want the same thing to happen to a lesser degree in the protodisc material to allow it to get as dense as it is.

It is an important argument for large f_c that White et al. substantiate their interpretation by the results of their numerical N-body model CDM simulations. However, that leaves two interestings puzzles: how do we reconcile such a large collapse factor for protospheroids with the continuity effect discussed in the last section, and why would angular momentum transfer make giant ellipticals rotate so much more slowly than discs?

2.4. The Age of the Elements

The constraint from element ages, which I learned from Jerry Ostriker, is based on the fact that evolution ages for reasonably massive stars are on the order of $10^{7.5}$ years. Since several generations are needed to make the heavy elements a sensible bound on the epoch of formation of the bulk of the elements is

$$t_e = \frac{2}{3\Omega^{1/2} H (1 + z_e)^{3/2}} > 10^8 \text{ yr,}$$

which says

$$z_* < z_e < 20 h^{-2/3} \Omega^{-1/3}. \tag{6}$$

As indicated, the epoch of star formation is later than element formation because the bulk of the seen stars at $r < r_c$ are metal rich.

2.5. Radiation Drag

At redshifts $z > 1000$ the radiation temperature is high enough to ionize hydrogen and radiation drag is strong enough to force plasma clouds with density comparable to the mean to expand with the radiation. At redshifts in the range $1000 > z > 100$ neutral hydrogen can form and decouple from the radiation but any material kept ionized by young stars still suffers an appreciable radiation drag. This might lead to a self-limiting star formation process: neutral regions could tend to break away from the general expansion, but the stars that form in the densest spots would ionize neighbouring material, recouple it to the radiation, and so suppress further star formation. The situation changes at $z \sim 100$ because radiation drag no longer seriously impedes collapse. This means that a gravitationally bound gas cloud hot enough to be ionized is strongly dissipative; at $z \sim 100$ such a cloud has to collapse until star formation or some other astrophysical process restores equlibrium.

This self-limiting process suggests z_* and z_g can be no greater than about 100 (and that if primeval fluctuations in the baryon distribution were large wholesale star formation would commence at $z \sim 100$). A more detailed analysis of the conditions under which the process might actually apply would be welcome.

2.6. Cooling Time

The density of a bound system produced by gravitational instability in the expanding universe is fixed by the global mean value at the time of formation. The mass of the system then fixes the mean internal energy, which translates into a gas temperature if substructure is erased. At redshifts $z < 10$ the behavior of the gas cloud depends on the rate of radiative cooling, as discussed by Rees and Ostriker (1977).

In the scenario pioneered by Faber (1982) clustering develops in a hierarchy of objects of increasing mass at redshifts low enough that cooling by radiation drag may be ignored. At the early low mass end of the hierarchy the material would be cool enough to remain neutral so cooling times would be long and substructure would tend to be erased by mixing as the hierarchy develops. When the matter temperature in objects in the hierarchy reaches $T \sim 10000$ K matter is collisionally ionized, cooling is rapid, and stars presumably form. Given the mass density limit from the assumption $z < 10$ one finds that the mass in a bound system when star formation commences would have to be in the range $\sim 10^{10}$ to 10^{12} $M_\odot$. This argument is developed in more detail by Blumenthal et al. (1984).

The pleasing coincidence of the cooling mass with characteristic galaxy masses requires that cooling by the radiation background may be neglected. That could be taken to argue that z_g and z_* ought to be less than 10.

2.7. Biasing

It is argued, I think rightly, that the simplicity of the Einstein-de Sitter model strongly recommends it. To reconcile this cosmological model with the dynamical estimates of galaxy masses one must assume that galaxies are more strongly clustered than is mass, so the estimates of the local mean mass per galaxy in dense regions are biased low. A standard measure of biasing at density contrast on the order of unity is the bias parameter

$$b = \frac{\delta N/N}{\delta M/M}.$$

The numerator is the rms fluctuation in galaxy counts per unit volume smoothed through a window of radius $10h^{-1}$ Mpc and normalized by the large-scale mean number density. The denominator is the same statistic for the mass. The dynamical estimates are brought into consistency with the Einstein-de Sitter model if $b \sim 3$.

Biasing tends to decrease with time because gravitational instability causes the clustering pattern to grow, gravity drawing together mass and galaxies alike. The excess number of galaxies in a dense region is the sum of the initial number and those added along with mass as the universe expands and density fluctuations grow. When the number added dominates the initial number the initial biasing has been considerably reduced. We can find an approximation to the size of the effect as follows.

In the Einstein-de Sitter model $\delta M/M$ (with a fixed comoving window size) varies with time in proportion to the expansion parameter, $a(t)$, and $\delta N/N$ might be approximated as the sum of a constant term to represent the initial biasing and a term proportional to $a(t)$ to represent the gravitational growth of clustering. In this model the biasing parameter at epoch z_i is

$$b_i \sim 1 + (1 + z_i)(b - 1).$$

An observationally interesting redshift is $z_i \sim 4$, close to the maximum observed for Lyman-α galaxies and quasars. If the present biasing parameter is $b \sim 3$ this gives $b_i \sim 10$. The large initial biasing is reasonable in explosion models, but seems difficult to arrange in astrophysical or autonomous biasing scenarios. My impression is that this is the essential reason why one argues for galaxy formation at low redshift in the biased cold dark matter picture. It is an interesting challenge for this picture to reconcile the the need to preserve appreciable biasing at the present epoch with the indications from Lyman-α galaxies and quasars that galaxy formation was at least underway by $z_i \sim 4$.

2.8. Morphological Segregation

The density argument in Section 2.1 applied to clusters of galaxies shows that these systems could not have existed at redshifts greater than a few. The observation that the abundance ratio of elliptical galaxies to lenticulars is highest in clusters could be taken to indicate that galaxies formed at low redshift, for if they had formed very early it would be difficult to see how the ellipticals knew where they ought to form (eg. Baron and White 1987). It will be noted, however, that the argument applies to disc formation, which could be an event separate from the assembly of the material. Thus we must count this as one of the arguments that z_d is no greater than about three, but to apply it to z_g we will need a better understanding of how galaxies formed.

2.9. Stability of the Galaxy Clustering Pattern

This last constraint is even more vague than the preceeding ones but I think worth mentioning because it leads to a fairly well posed theoretical problem. We have an extreme example of the constraint in the adiabatic massive neutrino model. The coherence length of the mass distribution in this model is set by the smoothing due to free streaming of the neutrinos convolved with the primeval spectrum of density fluctuations. We could adjust the latter so the first generation of pancakes would form a sheet-like arrangement similar to the observed galaxy distribution, which arguably is a good way to account for the tendency of galaxies to lie on two-dimensional surfaces. However, the mass distribution in this model is unstable: mass in the sheets drains into the vertices. Thus to reproduce the observed pattern of galaxy clustering with galaxies produced in the first generation of pancakes one would have to assume these galaxies formed at low redshift.

Reproducing the large-scale galaxy distribution can be problematic in models in which galaxies form at very high redshifts, because the slope of the galaxy two-point correlation function $\xi(r)$ at $\xi \sim 1$ tends to become much too steep when initial conditions are adjusted to make a reasonable slope at larger ξ. This is the indication from N-body model approximations to the scaling solution (for the late time mass autocorrelation function in an Einstein-de Sitter model with pressureless matter and gaussian primeval density fluctuations with power law power spectrum; eg. Efstathiou et al. 1988). The overly steep mass mass autocorrelation function at the transition from linear fluctuations on large scales to non-linear on small reflects the tendency for the mass to be concentrated in tight islands, unlike the fluffy character of the large-scale galaxy distribution.

The theoretical problem I mentioned above is to improve our understanding of the scaling solution. Is it accurately estimated? Is it even unique? Could we make the solution look more promising by going to non-gaussian (fractal) initial density fluctuations? I have spent time on this, without conspicuous success.

There are model-specific ways out of the problem with the shape of the two-point correlation function. In the explosion picture the observed sheets of galaxies could be remnants of the ridges piled up by explosions, and these sheets may be stable enough to resist formation of overly tight islands of galaxies (Ostriker and Cowie 1981). In the primeval baryon isocurvature model the mass autocorrelation function develops a broad plateau at the matter-radiation Jeans length (Peebles 1987). It is conceivable but by no means demonstrated that the plateau counters the tendency of matter to collect in tight islands. And in biased galaxy formation models the scaling solution is of course irrelevant because the galaxy clustering pattern is gravitationally active only in quite dense regions.

2.10. Summary

The central theme of this section is the collapse factor. If protogalaxies were assembled out of material with mean density close to the background at low redshift it would require an impressively large collapse factor, whether a protogalaxy is a nearly homogeneous gas cloud or a collection of dense gas clouds or star clusters. Is this large collapse reasonable? No compelling pattern emerges from the evidence listed above; one can adduce arguments for small collapse factors and $z_g > 10$ or for large collapse factors and $z_g \sim 1$. It could be a suggestive coincidence that the bounds from the arguments in sections 2.1 to 2.5 all saturate at z_g somewhat greater than 10. This is consistent with biasing for special models, notably explosions, or more generally with a low density universe. I would count only the last two constraints as arguments for $z_g \sim 1$, but the last one is very serious. Still more serious, of course, would be useful observational evidence, which is the main subject of this conference. I turn now to a brief summary of observational issues.

3. Astronomical Constraints

3.1. Galaxies at Low Redshift

Have some galaxies formed at redshifts less than unity? Intergalactic HII regions and starburst galaxies are good candidates. If it were shown that these are truly young it would establish an important point: if dwarf galaxies could form at low redshift perhaps giant galaxies could do the same. Of course, the challenge will be to to understand whether the youth is only apparent, a consequence of the more leisurely evolution of less massive galaxies.

The classic problem of determining the relative age of the disc and spheroid of our galaxy is reviewed by Sandage (1988). This could give a very useful constraint on the epochs of galaxy formation. For example, if it were found that the disc of our galaxy is half the age of the spheroid it would show that our disc had to have formed at redshift less than unity, no matter how early the spheroid formed. In fact it appears that there are stars in the thick disc $t_d = 12$ Gyr old. If the age of the disc were no less than this value it would show the disc formed fairly early. For example, taking the age of the universe to be $t_o = 15$ Gyr, and adopting the Einstein-de Sitter model, we would find that our disc started to form at redshift

$$z_d = \left(\frac{t_o}{t_o - t_d} \right)^{2/3} - 1 \sim 2. \tag{7}$$

Late protogalaxy formation by merger and collapse of systems of dense gas clouds or star clusters, as in the CDM picture discussed in § 2.1, presumably would have been violent enough to have disrupted discs, meaning the merger phase would have to have been complete before the disc started forming. We see from the example in equation (7) that this could yield an interesting constraint on z_g; we will be following developments here with interest.

3.2. The Universe of Galaxies at Redshift $z \sim 1$

If we could be taken by a time machine back to the epoch $z \sim 1$ what would we see? One could emphasize the differences: more luminous and younger-looking star populations in more gas rich galaxies merging at a higher rate. I am rather more impressed by the broad similarities. Giant galaxies picked out as radio sources do have infrared magnitudes brighter than would be expected without evolution, but they are only one magnitude or so brighter at $z \sim 1$ than at low redshifts, about what one would expect from passive evolution from significantly higher redshift of formation of the stellar population (Lilly and Longair 1984). A similar remark applies to colours (Eisenhardt and Lebofsky 1987). In some cases, one sees in galaxies at $z \sim 1$ the spectral features one associates with a relatively old stellar population (Hamilton 1985). Were these mature-looking galaxies typical of the progenitors of bright present day galaxies? A useful indicator would be their comoving number density. It may be relevant that the Loh-Spillar (1986) test shows that the comoving number density of all bright galaxies at $z \sim 0.7$ is no more than a factor of about two different from the present density.

Galaxies undoubtedly have been evolving at redshifts less than unity. (How could they avoid it?) However, the simple interpretation of the observations seems to be that the universe of galaxies at $z = 1$ is basically the same as it is now: similar comoving number densities of relaxed systems of stars with ages comparable to the expansion timescale. This would suggest that z_g and z_* are appreciably greater than unity.

3.3. The Universe at Redshifts $z = 2$ to 5

This regime has only recently been opened to observational study. Here is my list of the most interesting controversies that have developed.

If discs of galaxies formed at $z_d > 3$ then one would expect to see a population of gas-rich absorbers in the lines of sight of to quasars at $z > 1.6$. Wolfe (1989) recognized this and looked for and found a set of candidates, the damped Lyman-α absorbers. Are these objects young discs of galaxies? It is consistent with the coincidence between the mean mass density in the Wolfe clouds and the the present mass density in disc material, but perhaps not with the fact that the absorbers are five to seven times more abundant than would be expected if discs at $z \sim 2$ had the same parameters as low redshift discs. These and other observational tests are thoroughly debated in this volume. If it were established that the Wolfe clouds are not discs of galaxies we would have something equally interesting, evidence that $z_d < 1.6$, which might have an interesting bearing on the age of our disc (eq. [7]).

I argued above that we ought to be impressed with the indications that there is nothing all that new at redshifts $z \sim 1$. The observations at $z \sim 2$ to 4 irresistibly draw our attention the other way, to young-looking objects, Spinrad's Lyman-α 'bursters'. Debates on the physical ages of these systems will be followed with great interest: are they young galaxies?

There are also relatively old-looking objects at high redshifts, notably Lilly's galaxy at $z = 3.4$ and those of Miley and Chambers at $z = 3.8$. If the luminosities of such galaxies really are dominated by a star population ~ 1 Gyr old, as is the easiest interpretation of their broadband spectra, it puts the formation of the stars at an impressively high redshift. The constraints on theories will be even more interesting when we have estimates of the abundance of such star systems at $z \sim 3$ to 4.

A standard and reasonable argument has been that if quasars first appeared in substantial numbers at $z \sim 2.5$ then it would naturally suggest that galaxies formed then, $z_g \sim 2.5$. However, as I understand the evidence presented at this meeting by Warren the bright end of the luminosity function is roughly flat at $2 < z < 4$. It would not be perverse to take this as evidence that $z_g > 4$. The other interpretation, discussed by Efstathiou and Rees (1988), would be that the first generations of quasars formed outside galaxies.

The highest redshift quasars show that the intergalactic medium (the Lα forest and the strong depletion of HI between the clouds) is in place at $z \sim 4$. This can be consistent with theories in which $z_g \ll 4$, because massive stars could have formed and ionized the IGM before galaxies formed, as discussed by Couchman and Rees (1986). It may be more challenging to preserve biasing as well as low z_g, however. The stars that formed early could not be expected to know for sure where galaxies were going to form, and so would be in danger of leaving detectable remnants in the voids. For example, Efstathiou and Rees (1988) estimate that in the standard biased cold dark matter model the comoving number density of bound systems with mass $M > 10^{10} M_\odot$ (baryon mass $> 10^9$ $M_\odot$) at redshift $z = 6$ is ~ 0.01 Mpc^{-3} (with Hubble parameter $h = 0.5$), which is ~ 3 times the present number density of bright galaxies. It would be reasonable to expect that such objects formed at $z \sim 6$ would have left detectable remnants since we see lots of galaxies (such as the SMC) with similar masses, but of course they generally are seen in the company of bright galaxies.

3.4. Summary

The dramatic observational advances in the last few years have given us compelling evidence for pronounced evolution of galaxies at $z < 1$ and for the existence of youthful galaxies at $1 < z < 4$. This may mean that galaxy formation is an ongoing process continuing to recent epochs, or it may mean that we have not separated evolution from formation. A powerful test is the search for evidence of old components in objects at high redshift where the time from the Big Bang is considerably compressed. If it were found that the only reasonable interpretation of the continuum seen in objects at $z = 2$ to 4 is the light of a population of stars ~ 1 Gyr old, and if the abundance of such objects were found to be comparable to galaxies, I would count it as strong evidence that $z_* \gg 4$ and $z_g > 4$. Evidence to the contrary taken with all the evidence of youthful morphologies of galaxies at $z \sim 2$ would argue for $z_g \sim 2$.

4. Cosmogonical Constraints

4.1. z_g and Biasing

In the cold dark matter model one assumes that galaxies are biased tracers of mass in order to reconcile dynamics with the Einstein-de Sitter cosmological model. In the standard CDM model or any simple variant, where the initial biasing is modest, it is a generic prediction that galaxies form at low redshifts because, as discussed in Section 2.7, the gravitational growth of mass clustering tends to erase biasing. Late galaxy formation is in qualitative agreement with the unmistakable evidence of significant evolution of galaxies at $z < 1$ and with the highly disturbed appearance of the Lyman α 'formers' at $z \sim 2$. If, consistent with these observations, it were established that z_g is less than perhaps 2, it would have to be counted as a considerable triumph for the theoretical chain that leads from the Einstein-de Sitter model to biasing to late galaxy formation.

Galaxy formation at $z_g < 2$ might not be so easily reconciled with the presence at $z \sim 4$ of the intergalactic medium, very luminous quasars, and at least a few old-looking galaxies. If it turns out that $z_g > 4$ then my impression is that the CDM approach will be in trouble. However there will be no shortage of theories that might accommodate $z_g > 4$ within the Einstein-de Sitter model: Thompson (1989) argues that in the explosion picture a natural epoch for galaxy formation is $z_g \sim 10$. This need not conflict with a high mass density because biasing due to explosions can be arbitrarily large.

It is also possible that the Einstein-de Sitter model is not a useful approximation. A cosmological model with $\Omega \sim 0.2$ (excluding Λ-like terms) has the inestimable advantage that it admits the straightforward interpretation of a considerable body of evidence. The explosion picture seems to be viable in a low density cosmology. Low density would seem to be preferable for the old (pre-1988) massive cosmic string scenario, because it is difficult to see how the formation of a cluster of galaxies by the gravitational attraction of a massive cosmic string loop (Turok 1985) could preferentially accrete galaxies over mass. (In the new cosmic string scenario the mass distribution is much more cut up by passing cosmic strings, which would tend to move the model in the direction of standard CDM.) The baryon isocurvature model (Peebles 1987) may also be worth considering if we abandoned hope for the Einstein-de Sitter model.

4.2. Models for all Seasons

It is striking to see the extent to which this meeting's theoretical considerations of galaxy formation are motivated by the scale-invariant biased cold dark matter model. It is proper that our choice of models should be informed by our intuition; we all can think of cases where a beautiful theory has been redeemed from apparently contrary experimental evidence. However, in the best such success stories the theoretical preferences have had stronger bases than here. Thus it is also reasonable and proper to bear in mind that for any geometrically possible value of z_g we have at least one model with at least some socially redeeming features.

5. Summary

There are serious challenges for models of late galaxy formation, say $z_g < 2$: account for all the activity seen at $z \sim 4$, including the presence of the IGM, very luminous quasars, and old-looking galaxies; and arrange for the enormous collapse factor from the cosmological mean density to that characteristic of a galaxy without leaving any observable seams (eg. eq. [4]). There are challenges also for models of early formation, say $z_g \sim 10$: account for the evidence of youth at low redshifts, from Lα formers to starburst galaxies; and account for the fluffy character of the large-scale galaxy distribution. Out of all this we surely will learn something.

I am grateful to Ruth Daly, Jerry Ostriker and Bharat Ratra for stimulating discussions of this paper. The research was supported in part by the U. S. National Science Foundation.

References

Athanassoula, E., Bosma, A. and Papaionnou, S. 1987, *Astron. Astrophys.*, **179**, 23.

Baron, E. and White, S. D. M. 1987, *Ap. J.*, **322**, 585.

Blumenthal, G. R., Faber, S. M., Primack, J. R. and Rees, M. J. 1984, *Nature*, **311**, 517.

Couchman, H. M. P and Rees, M. J. 1986, *M.N.R.A.S.*, **221**, 53.

Efstathiou, G., Frenk, C. S., White, S. D. M. and Davis, M. 1988, *M.N.R.A.S.*, in press.

Efstathiou, G. and Rees, M. J. 1988, *M.N.R.A.S*, **230**, 5P

Eisenhardt, P. R. M. and Lebofsky, M. J. 1987, *Ap. J.*, **316**, 70.

Faber, S. M. 1982, in *Astrophysical Cosmology*, eds. H. A. Bruck, G. V. Coyne and M. S. Longair (Vatican: Pontifical Academy of Sciences), p. 191.

Hamilton, D. 1985, *Ap. J.*, **297**, 371.

Lilly, S. J. and Longair, M. S. 1984, *M.N.R.A.S.*, **211**, 833.

Loh, E. D. and Spillar, E. J. 1986, *Ap. J.*, **307**, L1.

Ostriker, J. P. and Cowie, L. L. 1981, *Ap. J.*, **243**, L127.

Peebles, P. J. E. 1984, *Science*, **224**, 1385.

Peebles, P. J. E. 1987, *Nature*, **327**, 210.

Rees, M. J. and Ostriker, J. P. 1977, *M.N.R.A.S.*, **179**, 541.

Sandage, A. 1988, in *The Calibration of Stellar Ages*, Van Vleck Observatory.

Thompson, C. 1989, these proceedings.

Turok, N. 1985, *Phys. Rev. Lett.*, **55**, 1801.

White, S. D. M., Davis, M., Efstathiou, G. and Frenk, C. S. 1987, *Nature*, **330**, 451.

Wolfe, A. M. 1989, these proceedings.

Yang, J., Turner, M. S., Steigman, G., Schramm, D. N. and Olive, K. A. 1984, *Ap. J.*, **281**, 493.

DISCUSSION:

WHITE: I think you've been talking to your lawyer too, because it seems to me that the continuity argument is really a scam. When we talk about galaxy formation, most of us probably think about the formation of the galaxies we see, which are made of stars; the continuity comment only applies to the (unseen) mass, however, and fails dismally for the stars. The typical luminosity within 10 kpc of a star is perhaps $L_*/2$. If you use the luminosity density of the universe and the galaxy correlation formation to estimate the scale within which $L_*/2$ is expected as a result of galaxy clustering, you find a radius around 200 kpc. Then the star distribution shows a strong discontinuity on the scale of galaxies which one might naturally interpret as a collapse factor of scale 20 associated with the separation of the visible material from the dark matter during galaxy formation.

PEEBLES: We see here, as in our two prepared talks, a fascinating difference of opinion on what is a primeval galaxy. To my mind, the fundamental criterion is the mass: I would say that a young galaxy has formed when the bulk of the mass within a radius $5h^{-1}$ Mpc is in place. Your definition, while being much more convenient observationally, is I think overly dependent on the way the galaxies evolve after they have formed by my definition.

SHANKS: What is the chance that the continuity in the $V^2(r)$ relation is just a coincidence - how difficult would it be for non scale-free models to give consistency with this result?

PEEBLES: I wish I knew; it certainly is a possibility, but I think not the one we would be inclined to consider first.

FALL: I have the impression that quasar-absorption line studies put some important constraints on the origin of galaxies. In particular, Wolfe and his associates find that at redshifts 1.6 to 3.0, there is as much mass in neutral hydrogen (to within factors of a few) as there is in all luminous material today. The straightforward interpretation of this result is that most of the compression of gas into stars has occurred since a redshift of about 3 or perhaps even 1.6. The mass in galaxies might have been assembled much earlier but if so there was a gap between galaxy and star formation.

PEEBLES: I agree, and would add a comment. An important constraint is the heavy element abundance in the Wolfe clouds. If *high* (which would be contrary to the beautiful results presented by Pettini) it would suggest appreciable star formation and death at $z \geq 3$; if *low*, we would be led back to the old G-dwarf puzzle.

FONG: That there are galaxies at high redshift like nearby galaxies is interesting, but we need a large statistical sample and perhaps the most direct way of looking at this today is through number counts. Indeed, deep counts do require evolution. But, $N(z)$ distributions show a low mean redshift indicating that it is probably the fainter galaxies which are evolving. Thus, as to the epoch of galaxy formation may well depend on the galaxy type and bright galaxies may not be evolving much up to the redshifts to which we can see. So, for these galaxies, results in the near future may be somewhat disappointing. But, the counts are beginning to be interesting!

PEEBLES: I quite agree!

CARLBERG: Doesn't the great increase in the fraction of galaxies with high star formation rates and nuclear activity suggest that redshift one is different from redshift zero and that $z \sim$ 1 may be close to z_*?

PEEBLES: You may well be right but I would urge you to take care to distinguish star formation, which we know is an ongoing process, from the assembly of the mass of the galaxy!

FRENK: The clump you have shown from the N-body simulations is by no means typical. It was in fact the biggest one that formed in that particular simulation. About 1/2 the clumps of galactic size evolve in a relatively quiescent fashion since $z = 1$ and would be good candidates for the sites of spiral disk formation. By contrast we would identify a clump that has undergone a good deal of merging in recent times as the halo of an elliptical. This has the attraction that most of the angular momentum is transferred to the outer halo as substructure is erased. Would you not agree that this process would get around your angular momentum objection?

PEEBLES: It would be very interesting to see maps of these more quiescent halos as they were at redshifts $z = 1$ and 0.5. I would like to see whether there is a reasonably small amount of mass added at small radii and relatively small redshifts, $z \sim 0.5 - 1$. Your point about angular momentum transfer is a good one, but still you have an interesting challenge: to reconcile the chaotic collapse needed to dump angular momentum with the quiescent contraction needed to make a reasonably old thin disc without forming the spheroid at $z > 1$.

OBSERVABLE SIGNATURES OF YOUNG GALAXIES

Simon D.M. White
Steward Observatory, University of Arizona,
Tucson, Az 85721, U.S.A.

ABSTRACT. I review theoretical expectations for the probable appearance of galaxies during their formation phase, placing particular emphasis on the uncertainties in these ideas. Recent models suggest that formation may occur relatively recently, but that young galaxies are less spectacular than previously supposed. They may be analogous to recently discovered high red-shift radio galaxies, and indeed they may have been observed directly in faint galaxy counts. I summarise several other lines of evidence which suggest that galaxy formation may have been a recent process. Finally I give preliminary results from a detailed analytic study of the observable properties of young galaxies in a Cold Dark Matter universe. Predictions are given for faint galaxy counts and redshift distributions, and for the galaxy luminosity function.

1. Introduction

The discovery of galaxies in the process of forming has long been a major goal of extragalactic astronomy. Such objects are expected to be at high redshift, and so very faint. Indeed, detection only seems possible because of the assumption that rapid star formation will produce high UV luminosities. Our ideas of how galaxies form are still very uncertain. Twenty years of active research in this area have alerted us to a wide range of possible formation paths; it is not clear that we would recognise a forming galaxy if we saw it. Even the concept of galaxy formation is open to a variety of definitions. For example, we could identify it with the birth of the first members of the observed stellar population; we could characterise it by the time when half this population had formed; or we could claim a galaxy had formed when its material achieved its present quasi-equilibrium configuration. These definitions can give very different formation epochs; even the ordering can vary from theory to theory. Thus in a hierarchical clustering theory some stars may form very early while the bulk of them form recently. On the other hand, many models assume star formation to occur in a rapid burst during collapse. If ellipticals are merger products, they may have been assembled long after their stars formed, whereas if they condensed from a cooling flow the opposite ranking could apply. The material of spiral disks may have been in place before a large fraction of it was converted into stars.

For the purposes of this talk I'll adopt the definition that a forming galaxy is a system seen during its most rapid period of star formation. With conventional assumptions this is the time when the object has its highest luminosity, and it is likely to correspond to the epoch when most of the stars were made. Notice, however, that my definition does not require the protogalactic material to have been assembled into a single unit. Moreover, with this definition many observed Sd-m and Irr galaxies count as forming systems (see Kennicutt 1983).

15

C. S. Frenk et al. (eds.), The Epoch of Galaxy Formation, 15–30.
© 1989 by Kluwer Academic Publishers.

One obvious way to predict the observational properties of forming galaxies is to look for analogues where similar physical processes are at work. This approach is hampered by the existence of several analogues with very different characteristics. Some of the high redshift objects associated with radio sources may qualify as forming galaxies (see Spinrad, this volume, for a review). They have high inferred star formation rates, a blue continuum, and an irregular morphology. On the other hand, their structure correlates with radio maps and they seem to contain a substantial old population (McCarthy, Lilly, this volume). Their optical properties may therefore be atypical and they may not fulfil my definition. They suggest an optimistic assessment of the visibility of forming galaxies since their very large Ly-α equivalent widths would be relatively easy to detect in more distant or intrinsically fainter systems. This contrasts with extragalactic HII regions, nearby but small objects undergoing their first burst of significant star formation, which have relatively small Ly-α equivalent widths (Hartmann et al. 1984). The largest star formation rates in nearby systems occur in starburst galaxies (e.g. Rieke et al. 1985); however, in these objects almost all the luminosity is processed by dust and emerges in the infrared where it would be undetectable in a high redshift system. Finally, star formation rates sufficient to build a very large galaxy are alleged to occur in cluster cooling flows, and yet the observational signature of this formation is hard to detect (Fabian et al. 1984). Perhaps we are unable to recognise a massive forming galaxy even when it is in our own backyard.

The first detailed models for the appearance of primeval galaxies were those of Partridge and Peebles (1967). Following the ideas of Eggen et al. (1962), these authors assumed that all the stars in a massive galaxy formed in a rapid burst at about the time of maximum expansion of the protogalactic cloud. Rapid formation led to a very bright object; near dissipationless collapse required a small maximum expansion radius and so a high redshift of collapse ($z \sim 20$). The further assumption that stellar light is unaffected by dust then led to the conclusion that primeval galaxies should be diffuse objects with their surface brightness peaking strongly in the near infrared. They would not be observable unless the model substantially overestimated the formation redshift, since very little emission is expected beyond the Lyman limit. An influential set of galaxy formation models by Larson (1974a) suggested that dissipational effects could indeed prolong the time required for galaxy formation. This reduced the predicted luminosity of forming galaxies but brought them to lower redshift so that detectable fluxes were predicted at optical wavelengths. The observational appearance of Larson's models was worked out by Meier (1976). Because stars are formed at the radii where they are currently observed, primeval galaxies are predicted to be quite compact. They are also expected to be quite blue as a result of their low formation redshift.

To get detailed predictions Meier synthesised the spectrum of the Larson models assuming a Salpeter IMF. He found the emission longward of the Lyman limit to be approximately $2.6 \; 10^{27}$ erg/s/Hz, independent of frequency, for a star formation rate of one solar mass per year. This result is very convenient from a theoretical point of view (see, for example, Cowie 1988). Fig. 1 shows how well it holds up when a spectral synthesis is carried out using the much more extensive stellar libraries now available. The predictions are indeed roughly flat over the 1000–3000 Å range for a wide variety of burst ages. However, it is worth noting that the mean level of the spectrum is quite sensitive to the IMF assumed. Thus if the Salpeter IMF is replaced by that of Miller and Scalo (1979) the mean level of emission increases by almost a factor of 3, whereas Scalo's (1986) IMF gives spectra only slightly above those plotted in the figure. The dependence of spectral shape on the IMF is smaller than this but is still significant.

Since the work of Meier, theorists have concentrated on the relation of galaxy formation to the overall growth of structure in the Universe. A variety of alternative pictures have emerged, most of which are strongly conditioned by the growing acceptance of the existence of large amounts of unseen dark matter. Baron and White (1987) surveyed these pictures and tried to assess their implications for the detectability of forming galaxies. There seem to be no predictions common to all theories. While it is likely that galactic disks were both assembled and turned into stars relatively recently (i.e. since $z \sim 2$ or 3), there is no agreement about whether ellipticals and spiral bulges were assembled before or after star formation; characteristic redshifts from 1 to 30 have been suggested for each process. Although roughly equal masses of stars are observed in disks and in spheroidal components (Schechter and Dressler 1987), searches for forming galaxies have concentrated on young spheroidal systems despite the much greater uncertainty in their properties. This is simply because *some* models for spheroid formation predict a population of very luminous high redshift objects, whereas models for disk formation and evolution suggest that they were never very much brighter than they are today (cf. Tinsley 1978, Tinsley and Larson 1979). Null results of searches for primeval galaxies (see the review by Koo 1986), should thus be interpreted as ruling out some very simple models for the dynamics and emission of forming ellipticals, rather than as requiring galaxy formation to occur at high redshift.

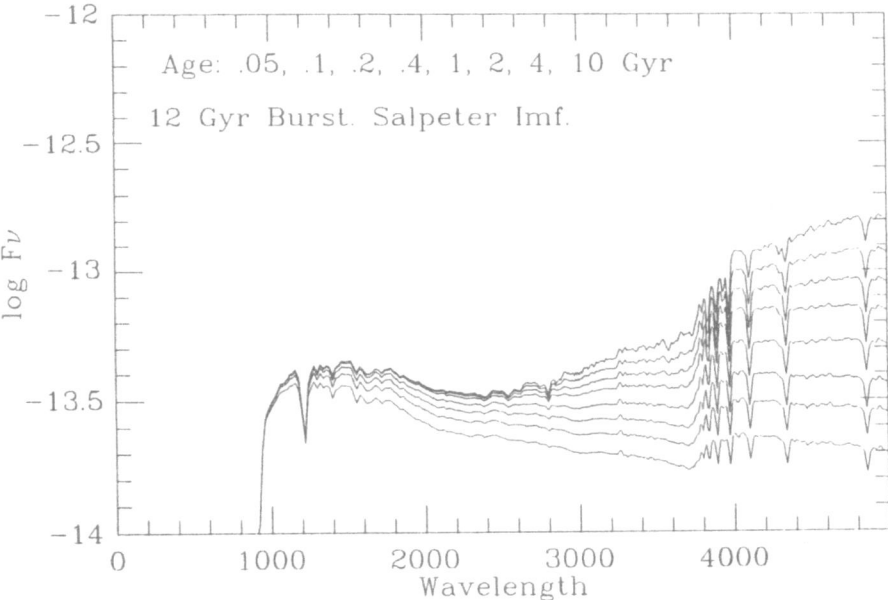

Fig. 1. Synthetic spectra for a region with constant star formation rate at the ages indicated. A Salpeter mass function is assumed with cutoffs at 75 and 0.08 $M_\odot$. These spectra were kindly provided by Gustavo Bruzual and come from the most recent version of his evolutionary synthesis program.

2. Inhomogeneous Primeval Galaxies

A common element which Baron and White (1987) were able to identify in most galaxy formation theories is the idea that the final assembly of a galaxy is a highly inhomogeneous process (an exception would be the condensation of galaxies around loops of cosmic string). In hierarchical models the pregalactic gas, stars and dark matter (WIMPs or Pop III remnants) are present in approximately their cosmic proportions prior to final collapse, whereas early segregation may occur in explosion or pancake models. The final formation of luminous cores within dark halos is a result of dissipative effects which may occur before collapse if star formation is efficient, during collapse if the dissipation and dynamical timescales are comparable, or well after collapse if cooling is slow. These three cases correspond to merging of preexisting stellar systems, to a "standard" but inhomogeneous collapse, and to a quasistatic cooling flow. The fact that half of all known stars are in disks (which were clearly assembled before star formation) argues against an extreme version of the first possibility. Scaling laws for the cooling of gas within halos suggest the third possibility is unlikely to apply to most galaxies (White and Rees 1978). The intermediate case thus seems the most interesting to investigate. In addition, it produces the most easily detectable objects, since they have relatively low redshifts, high star formation rates and irregular morphologies.

A "generic" model of an inhomogeneous protogalactic collapse was computed by Baron and White (1987) and is shown in Fig. 2. They assumed the protogalaxy to start as an expanding ellipsoidal perturbation made of a mixture of gas and dark matter with imposed Poisson density fluctuations. Experimental parameters were chosen so that the gas suffered substantial dissipation and was converted into stars with a timescale comparable to that of the overall collapse. This system settled into a final configuration in which the dark matter and the star distributions had density profiles close to $\rho \propto r^{-3}$ but with half mass radii differing by a factor of 5. In the figure, the model is seen at its nominal collapse time when it is still quite inhomogeneous. This also turns out to be the *end* of its period of most rapid star formation; about 60% of the gas has already been turned into stars. Note that this model is not specific to the Cold Dark Matter model. It is easily rescaled to give plausible predictions for an object at the end of its bright formation phase in, say, the explosion or isocurvature baryon models: the major assumptions are just that the dark matter is dissipationless, pregalactic, and dynamically dominant in the final collapse.

The model as shown in Fig. 2 assumes a system of total mass, $M = 10^{12} M_\odot$, initial gas fraction, $f = 0.1$, and nominal collapse time, t_c, corresponding to $z = 1.8$ in a universe with $\Omega = 1$ and $H_o = 50 \, \mathrm{km/s/Mpc}$. Meier's (1976) model was used to predict the spectrum of the stellar population; this assumes a Salpeter IMF and no dust and so is subject to the uncertainties discussed above. Fig. 2 can easily be scaled to other parameters values provided one retains the asumption that the object is seen at a redshift corresponding to t_c. The angular size of the galaxy scales as

$$\theta \propto M^{1/3} t_c^{2/3} / d_A, \tag{1}$$

where d_A is its angular size distance, while for redshifts in the range $0.5 < z < 3.5$ the B surface brightness scales as

$$S_B \propto f M^{1/3} t_c^{-7/3} (1 + z)^{-3}. \tag{2}$$

At higher redshift the surface brightness begins to fall below this prediction as the Lyman limit moves into the B band. However, the approximate flatness of the spectra in Fig. 1 makes it easy to estimate the surface brightness in a redder band from the prediction of equation (2). These equations show that the most efficient ways to increase the disappointingly low surface brightness predictions are to increase the gas fraction, the Hubble constant, or the number of massive stars in the IMF. Lowering Ω makes the situation worse.

It is clear from Fig. 2 why the failure of "primeval galaxy" searches puts little constraint on the epoch of galaxy formation within most theories. Nevertheless, Baron and White show that such forming systems may contribute significantly to observed faint galaxy counts if the characteristic epoch of galaxy formation is relatively recent (e.g. $z \sim 1$ to 3). Indeed, Tyson (1988) shows explicitly that their predictions can fit his observed counts of faint blue systems quite well. There is some ambiguity here because Tyson would probably count the object of Fig. 2 as a superposition of four fainter systems. However, one can argue on quite general grounds that his observed population contains much of the massive, metal-producing, star formation in

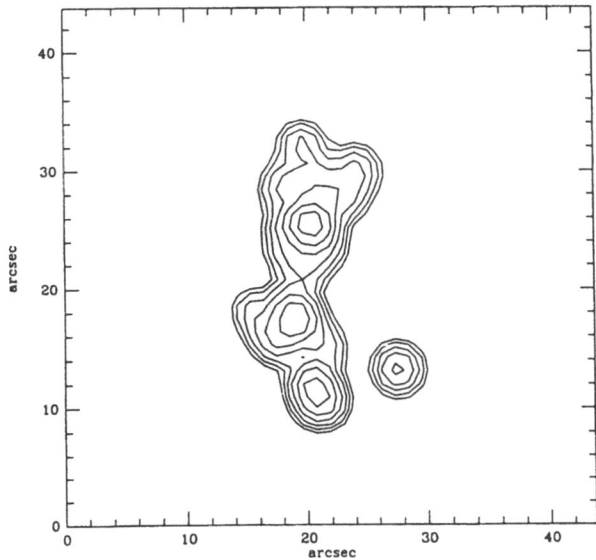

Fig. 2. Surface brightness contours at B for a "generic" model of the collapse of a protogalaxy containing both gas and dark matter. The peak contour is at a surface brightness of 28 mag./arcsec², and contours are at 1 mag. intervals. The total magnitude of this object is $B = 24.5$.

the history of the universe (Cowie 1988 and this volume). It seems that the major epoch of formation of galactic disks, and possibly of all galaxies, may already have been observed. In this view galaxy formation is a more recent phenomenon than has usually been assumed, and it seems worthwhile to review other lines of evidence for and against this idea.

3. Evidence For and Against Recent Galaxy Formation

Within hierarchical clustering theories it is possible to argue that galactic disks must form relatively late in order to acquire their observed angular momenta (see Fall, this volume). Observational support for this comes from the fact that the present star formation rate in our own Galactic disk is not much smaller than its lifetime average value (e.g. Scalo 1986), and that the observed rates in other galaxies are sufficient to make a substantial fraction of their disks in a Hubble time. I illustrate this in Fig. 3 using data for a variety of Sb and later type galaxies from Kennicutt (1983). The diagram shows the ratio of the product of the current star formation rate and the Hubble time to the total luminosity of the galaxy as a function of absolute magnitude ($H_o = 50$ km/s/Mpc is assumed). For many galaxies these "M/L" values are comparable to or larger than typical M/L ratios for disk populations. For such galaxies

20

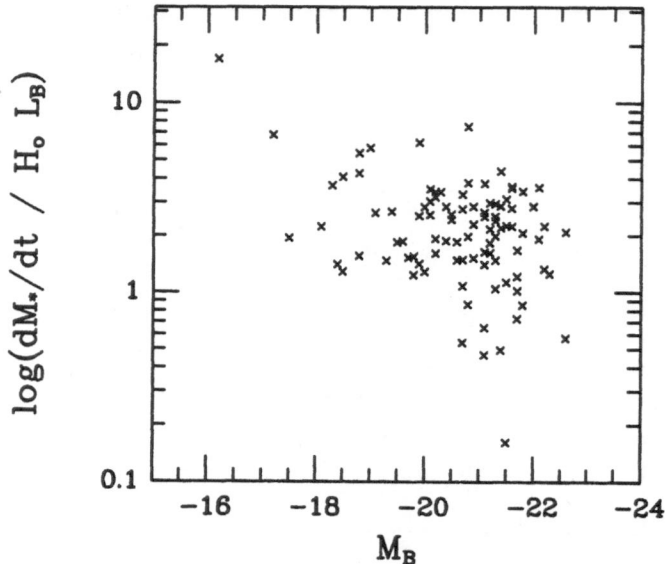

Fig. 3. This plot shows that if star formation continued at its observed rate in spirals for a Hubble time, the ratio of the resulting stellar mass to the present *total* luminosity of the galaxy would be comparable to the currently accepted M/L ratios of disks.

my definition would give a characteristic formation time of order half the current age of the universe.

The damped Ly-α systems seen in absorption in quasar spectra provide a direct measure of the cosmic abundance of neutral hydrogen in high column density objects at a redshift of two (Wolfe, this volume). The result is comparable to the entire mass in stellar disks today, and is much larger than the amount of HI currently seen in galaxies. The most plausible sink for this gas would seem to be the stars in disks. If, as Wolfe argues, the absorption systems are themselves large rotationally supported disks, we must be seeing them before much of their mass has turned into stars, an inference supported by the low apparent metallicity of these systems (Pettini, this volume). Again I infer that the epoch of disk formation (defined by star formation rather than by time of assembly) is more recent than $z = 2$. Since the total mass of stars in the universe is only about twice that in disks, it is conceivable that Wolfe and his collaborators have actually detected the raw material from which most stars in galaxies were made.

A more indirect argument in favour of recent galaxy formation comes from the redshift distribution of quasars. We believe that bright quasars occur in the nuclei of massive galaxies. The apparent decrease in the comoving abundance of quasars beyond $z = 3$ or 4 is then most simply interpreted as reflecting the formation of their parent galaxies (see Green, this volume, for a review of the observational situation). Similarly the decline in quasar numbers at lower redshifts may reflect the fact that the galaxies are using up the available fuel, perhaps by turning it into stars. A recent discussion of this problem in the context of the CDM model is given by Efstathiou and Rees (1988). Of course, other interpretations of the quasar counts are possible, so this argument is far from conclusive.

The strongest argument *against* recent formation of all galaxies comes from observations of high redshift ellipticals ($z \leq 0.8$), some of which have stellar populations which appear already to be old (Hamilton 1985 ; Gunn, this volume). The strength of the 4000 Å break in these objects suggests that the bulk of their stars formed considerably earlier. This argument depends on comparing stellar evolution timescales with cosmological timescales. Thus modest uncertainties in H_o, Ω, and in the lower limits on age derived from spectra, translate into large uncertainties in the redshift of formation. I illustrate this in Fig. 4 which shows the lower limit, z_{lim}, on formation redshift implied at observed redshift, z_{obs}, by a spectrum indicating no significant star formation for a preceding period, Δt. The solid curves are for $\Omega = 1$, the dashed curves for $\Omega = 0$ and the four pairs of curves for $\Delta t / t_o = 0.05$, 0.1 0.2 and 0.4, where t_o is the current age of the universe. Synthesis models suggest that it will not be possible to infer values of Δt greater than about 4 Gyr even with high quality spectra of the 4000 Å break (see Hamilton 1985). Thus current observations out to $z \sim 0.8$ do not restrict the formation epoch to be much beyond $z \sim 2$. Note that the break is not yet observable from the ground beyond $z = 1.5$.

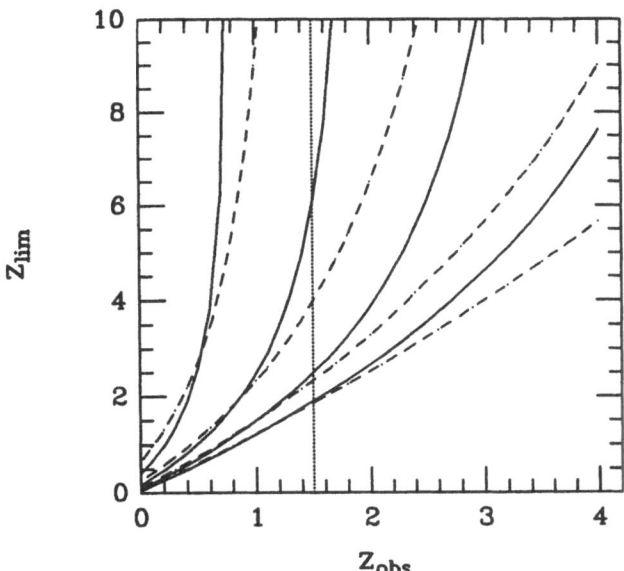

Fig. 4. The lower limits that can be placed on the redshift of formation of an object observed to have redshift, z_{obs}, and to contain no significant young population. See text for details.

The strongest constraint of this kind currently comes from the object at $z = 3.4$ discovered by Lilly (1988, this volume). Lilly does not, of course, have a measurement of the 4000 Å break. However, he infers the presence of a dominant "old" population of mean age at least 1.5 Gyr from the large ratio of 2 micron to optical flux. Comparison with Fig. 4 shows that this interpretation not only implies a high formation redshift, it can also rule out a number of cosmological models. Note that we have little indication of how representative Lilly's object may be, since it was found because of its strong radio emission. Similar but less extreme statistical uncertainties also cloud interpretation of the objects discussed by Hamilton and Gunn because they were selected in the first case by their very red colours, and in the second by their membership in very rich clusters. Nevertheless, it is a striking fact that some ellipticals at high redshift have spectra differing very little from those of their nearby counterparts.

4. Observable Properties of Young Galaxies in a CDM Universe

The linear fluctuation spectrum predicted in a standard Cold Dark Matter universe appears ideally suited for the White and Rees (1978; hereafter WR) galaxy formation scheme to produce galaxies and clusters with scaling properties similar to those observed (Blumenthal et al. 1984). In addition, it does a remarkable job of fitting the detailed statistical and morphological properties of galaxy halos, groups, clusters and superclusters (Davis et al. 1985; White et al. 1987a, 1987b; Frenk et al. 1988). As a result of the program of simulations described in these references, the structure and evolution of the CDM distribution is now understood in considerable detail. This information can be used to extend the techniques of WR and so approach a more detailed and physically realistic treatment of galaxy formation. Such a treatment should lead to estimates of the luminosity function, and of the star formation and merging histories of galaxies, and should check that parameter values which fit large-scale clustering also lead to a realistic galaxy population. It will then be possible to evaluate the common claim that galaxy formation in the CDM model may be too recent to be compatible with observation (e.g. Peebles, this volume). The remainder of this paper is a preliminary report on an attempt to carry through this program. The work has been carried out in collaboration with Carlos Frenk.

The first step is to arrive at a simple description of the structure, abundance, and evolution of the CDM halos within which galaxies must condense. Frenk et al. (1988) show that halo density profiles can be approximated by singular isothermal spheres over the ranges in radius, in mass, and in redshift accessible to their models. They were able to measure nonzero core radii for the largest systems, but this complication is irrelevant for my subsequent analysis. Further Efstathiou et al. (1988) and Efstathiou and Rees (1988) found the abundance of halos by mass to be surprisingly well described by the simple theory of Press and Schechter (1974). These two models can then be combined to give a description of the nonlinear CDM distribution as an evolving set of isothermal halos characterised by their comoving abundance as a function of circular velocity and redshift. The analytic formulae are given by Narayan and White (1988) who verify explicitly that they give reasonable agreement with the large simulations of White et al. (1987b).

In the following I concentrate on models with $H_o = 50\,\text{km/s/Mpc}$, $\Omega_{CDM} = 0.9$ and $\Omega_{bar} = 0.1$. The single remaining parameter is the overall normalisation of the linear CDM spectrum. I use the conventional definition of this amplitude in terms of b, the inverse of the present rms linear fluctuation in a sphere of radius 16 Mpc. The simulation work of our group has all assumed $b = 2.5$, corresponding to a strongly biased galaxy distribution. Observed large-scale streaming motions, if compatible with CDM at all, seem to require a weaker bias, perhaps $b = 1.5$ (Kaiser and Lahav 1988). In Fig. 5 I show the abundance of halos as a function of redshift and circular velocity for these two normalisations. With $b = 2.5$, halos similar to that of our Galaxy peak in abundance at a redshift of 1, and halos corresponding to galaxy groups ($V_c \sim 400\,\text{km/s}$) are peaking today. For the higher amplitude these redshifts move back to $z = 3$ and $z = 1.5$. Comparison of this model with the data of Frenk et al. (1988) shows good agreement at redshift 1 and later, but a noticeable overabundance of high V_c halos at $z = 2.5$. At present it is unclear whether this is a defect of the model or of the simulations.

The next and more difficult problem is the modeling of dissipative processes in the gas within dark halos. Numerical models by Evrard (this volume) show that if cooling is assumed negligible, the distributions of gas and dark matter remain very similar as nonlinear structures

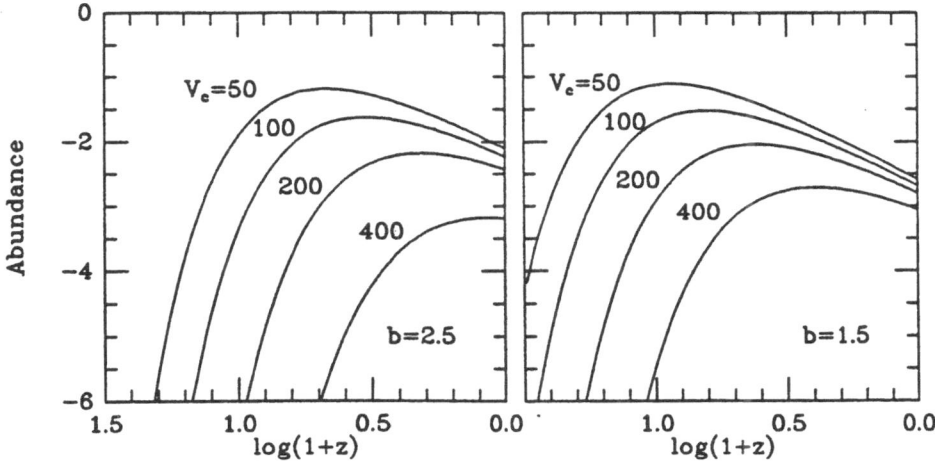

Fig. 5. The abundance of halos (Number/Mpc³/ln(V_c)) as a function of redshift and of circular velocity. The two panels refer to two different normalisations of the CDM fluctuation spectrum.

form. It is thus reasonable to calculate the cooling rate under the assumption that the gas in a halo starts out distributed as $\rho_g = 0.1\rho_{CDM} \propto r^{-2}$, with $T = 36V_c^2$, where T is in degrees Kelvin and V_c in km/s. The cooling time at each radius can then be calculated from the density, temperature and metallicity of the gas, and a cooling radius can be defined as the point at which the cooling time equals the age of the universe. This construction allows a simple definition of the rate at which gas cools off,

$$\dot{M}_{cool} = 4\pi\rho_g r_{cool}^2 dr_{cool}/dt = \Omega_{bar} V_c^2 r_{cool}/2Gt. \tag{3}$$

This rate is quite close to that found in more detailed models for cooling flows (e.g. Fabian *et al.* 1984). Notice that some gas will always be cooling off in every halo. This contrasts with the crude approximation adopted by WR and most subsequent analyses (e.g. Blumenthal *et al.* 1984) that a perturbation collapses to a uniform density system and that no galaxy forms if its cooling time is too long.

The cooling rates given by equation (3) are shown as dashed curves in Fig. 6 for the case $b = 2.5$ and for gas with primordial composition. At any V_c the cooling rates are greater at higher redshift. Unfortunately, over the relevant range of V_c these curves are quite sensitive to metallicity. It is therefore necessary to follow the chemical enrichment of the gas in order to get a realistic model of galaxy formation rates. We have worked out such chemical evolution models, but I will not report on them here because of the additional complexity involved. Rather I will present results assuming no enrichment and merely point out where they are changed substantially by the inclusion of this process. The cooling rates of Fig. 6 are unrealistic for small V_c and high z because they imply that more gas is cooling off than is present in the halo. The mass of baryons within the virialised region of a halo (defined as the region with mean overdensity 200) is simply $0.15\Omega_{bar}V_c^3 t/G$. Hence when cooling is very efficient the maximum rate at which gas can reach the centre is

$$\dot{M}_{inf} = 0.15\Omega_{bar}V_c^3/G, \tag{4}$$

independent of z. This relation is shown as a dotted line in Fig. 6. Clearly, the maximum rate at which gas is made available for star formation in any halo is the smaller of equations (3) and (4).

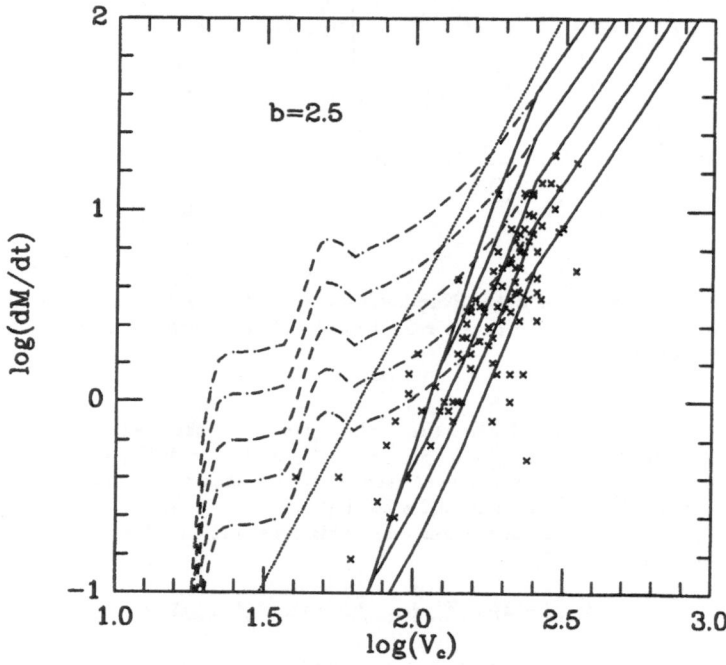

Fig. 6. Cooling, infall, and star formation rates in halos as a function of their circular velocity and of redshift. See text for details. The different dashed and solid curves refer to redshifts 0, 1, 3, 7, and 15 (bottom to top). The crosses are data for spirals from Kennicutt (1983).

Fig. 6 suggests an immediate observational test of the framework so far. If gas is cooling off in the halos of nearby galaxies it must produce significant soft X-ray emission. Now $V_c > 300\,km/s$ implies $T > 0.28$ keV. Thus this emission is too cool to have been detected by the Einstein Observatory for most spirals; however, halos of bright ellipticals should have been quite prominent. While some elliptical halos were indeed detected, their luminosities were substantially less than the values, $L_X \sim 2.5\dot{M}_{cool}V_c^2$, expected for the cooling rates of Fig. 6 (Thomas et al. 1986). We are therefore forced to argue that these objects have lost their gaseous halo, either as a result of interactions with their environment, or because star formation has blown it away. This latter process is related to another difficulty which arises when Figures 5 and 6 are combined.

We can take the gas supply rates of equations (3) and (4) and the halo abundances of Fig. 6 and calculate what happens if all the gas supplied is turned into stars. It turns out that the gas is used up well before the present, producing a cosmic density $\Omega_* = 0.1$ in stars. This is unacceptable; the density contributed by stars in observed galaxies is at least an order of magnitude smaller. This problem of overly rapid star formation was noted by WR. They solved it by invoking feedback processes along the lines of an earlier suggestion by Larson (1974b). Larson noted that the energy input from young stars and supernovae could drive all the gas out of a small protogalaxy before more than a small fraction had been converted into stars. He suggested that this inefficiency might account for the low metallicities of dwarf galaxies. WR showed that it could also cure their gas consumption problem, and Dekel and Silk (1986) gave a detailed reworking and extension of the arguments and applied them to the CDM model. Assume that a fraction ϵ of the energy produced by young stars and supernovae is tranferred to the hot gas in a halo. Then for a standard IMF I estimate the heating rate to be related to the actual star formation rate by

$$dE/dt = \epsilon \dot{M}_* (700\text{km/s})^2. \tag{5}$$

The dissipation rate associated with the cooling gas is $\sim \min(\dot{M}_{cool}, \dot{M}_{inf})V_c^2$. If this exceeds dE/dt I assume that all the gas supplied is turned into stars. Otherwise I assume the mean star formation rate to be such that dE/dt just balances the dissipation rate. This produces the following model for $\dot{M}_*$,

$$\dot{M}_* = \min(\dot{M}_{cool}, \dot{M}_{inf})\min(1, V_c^2/\epsilon(700\text{km/s})^2). \tag{6}$$

The efficiency parameter ϵ hides uncertainties in the IMF in addition to complex details of the hydrodynamic interactions between stellar winds, supernova shells and the hot gaseous halo of a forming galaxy. Dekel and Silk (1986) estimated $\epsilon \approx 0.02$ from a theoretical analysis of the evolution of supernova remnants. On the other hand, the massive X-ray winds driven by some starburst galaxies suggest an efficiency approaching unity (Chevalier and Clegg 1985). I will treat ϵ as a free parameter and choose it so that $\Omega_* = 0.01$ today. This requires $\epsilon = 0.09$ for $b = 2.5$ and $\epsilon = 0.25$ for $b = 1.5$. Solid lines in Fig. 6 show the star formation rate given by equation (6) for this choice of ϵ. The curve for $z = 0$ passes through but somewhat below the data of Kennicutt (1983) which have been translated to this diagram using a standard Tully-Fisher relation. In many spirals the observed star formation rate exceeds the maximum predicted cooling rate. In these relatively quiescent systems the Dekel and Silk estimate of efficiency may apply; heating cannot then halt infall. However, at early times their estimate must be substantially exceeded if overproduction of stars is to be avoided. The cooling effects of heavy elements exacerbate this problem and require a further increase in efficiency.

With these assumptions we can calculate the history of star formation. Fig. 7a shows the total star formation rate as a function of redshift for $b = 2.5$ and $b = 1.5$. For the larger bias star formation peaks at $z = 2.5$ and half of all stars are formed by $z = 1.0$; for the smaller bias the corresponding redshifts are 4.5 and 1.45. For a standard IMF the criteria of Shapiro (this volume) require star formation rates of at least $0.3 M_\odot/\text{yr}/\text{Mpc}^3$ during $6 < z < 3.5$ to photoionise the intergalactic medium and satisfy the Gunn-Peterson constraint. The rates found here fail by up to an order of magnitude. The model apparently requires additional sources of ionising radiation. Fig. 7b plots the mass of current stars as a function of the circular velocity of the halo in which they formed. Most stars are inferred to have formed in systems with V_c lying between 100 and 300 km/s. This agrees well with the corresponding range for the galaxies which contain most observed stars. Note, however, that because of merging, stars may currently be in systems with circular velocities exceeding those of the halos in which they formed. The model thus produces the right number of stars in objects with the right characteristic scale, but

26

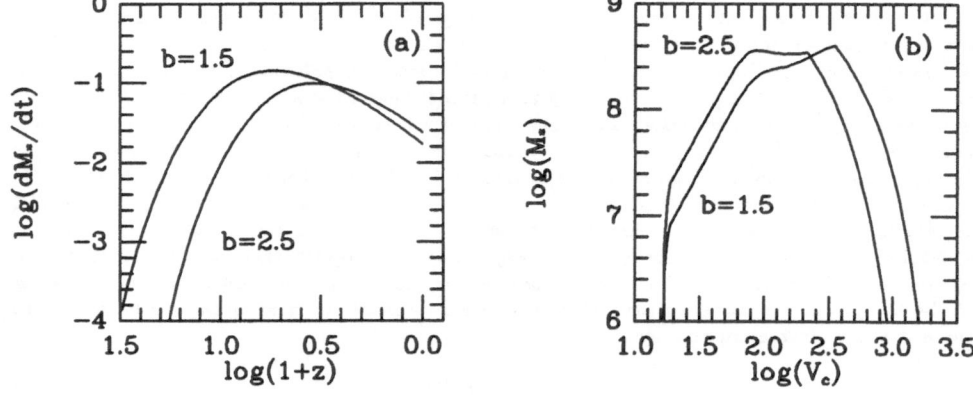

Fig. 7. (a) Star formation rate ($M_\odot$ yr/Mpc^3) as a function of redshift. (b) Mass of stars formed by the present day (in $M_\odot$/Mpc^3/ln(V_c)) as a function of the circular velocity of the halos in which they were formed.

formation is spread over a wide range of redshifts and many stars formed recently. Much of this recent activity (say since $z = 1$) should probably be identified with disk formation.

If as seems likely, the blue objects which dominate Tyson's (1988) faint counts are at moderately high redshift ($z \sim 1$ or earlier), the observed fluxes at B are dominated by young stars. In this case a prediction can be obtained immediately using the cosmological formulae of Baron and White (1987) but replacing their *ad hoc* assumptions about luminosity functions and formation rates by the *a priori* predictions of the CDM theory. In addition, an IMF must be assumed in order to relate the star formation rate to the spectrum. Using a flat approximation to the spectra of Fig. 1, I obtain the results in Fig. 8a. They are rather insensitive to b, and are surprisingly close to the observed counts, represented in the figure by Tyson's power law fit. The small shortfall in the prediction could be attributed to an incorrect IMF, or, especially at the bright end, to the neglect of light from older stars. In Fig. 8b I show the predicted redshift distribution for a series of apparent magnitude intervals. This diagram is for $b = 2.5$; for $b = 1.5$ all the distributions are shifted to somewhat higher redshift. The sharp cutoff beyond redshift 4 is due, of course, to the Lyman break passing through the observed B-band. The median redshifts of these distributions are 0.21, 0.47, 0.85 and 1.13 for the intervals centred on B = 21, 23, 25 and 27 respectively. For comparison, the faint redshift surveys of Ellis (1987) and Koo and Kron (1987) are at typical magnitudes B $\sim$ 21, and have median redshift 0.24. As judged by these plots, the amount of star formation produced by the models agrees well with that inferred from faint galaxy samples.

Once an initial mass function has been chosen, the model can be compared directly with the properties of nearby as well as distant galaxies. The Salpeter mass function of Fig. 1 was assumed to extend from $0.08 M_\odot$ to $75 M_\odot$. For a single population burst Bruzual's models predict a blue mass to light ratio given approximately by $M/L_B = 27.6(t/13\ 10^9 yr)^{1.1}$. This formula, together with the star formation rates of Fig. 7a, leads to a prediction for the current luminosity density of the universe. For $b = 2.5$ I find $7.9\ 10^7 L_\odot$/Mpc^3, corresponding to a mean stellar M/L of 8.6; for $b = 1.5$ the numbers are $6.5\ 10^7 L_\odot$/Mpc^3 and 10.6. The larger M/L in the second case is a result of the greater mean age of stars in the less biased model. For comparison, the recent study of Efstathiou *et al.* (1988) concluded from a variety of data sets

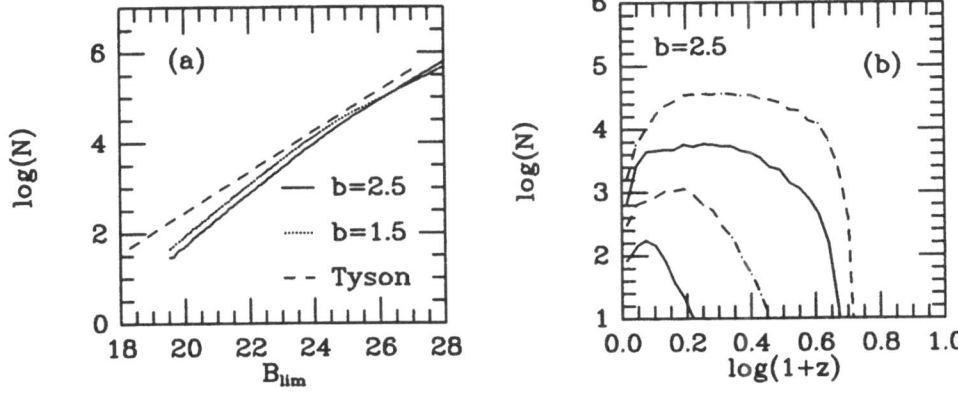

Fig. 8. (a) Predicted integral galaxy counts at B (Number/squ. dgr.) as a function of limiting magnitude. (b) Predicted redshift distributions (in Number/squ. dgr./ln(1+z)) of star-forming regions selected in two magnitude bins centred at apparent magnitude, B=21, 23, 25, and 27 (bottom to top).

that the mean luminosity density of the universe is 9.6 $10^7 L_\odot$/Mpc^3 in B with an uncertainty of almost 40%.

A more detailed comparison would obviously be provided by the current luminosity function of galaxies. Unfortunately this is difficult to calculate within the CDM model, which specifies when and where stars form, but is less clear about where they end up. As discussed by WR, halos of dark matter continually merge as hierarchical clustering proceeds, but galaxies must be more resistant if we are to understand the origin of clusters. Simulation work has confirmed the rapidity of halo merging (e.g. Frenk *et al.* 1988; Efstathiou *et al.* 1988) but has only just begun to address the more difficult problem of galaxy survival. Studies by Carlberg (1988) and Carlberg and Couchman (1988) suggest that galaxies do indeed merge significantly more slowly than their halos. WR were able to calculate a galaxy luminosity function in the absence of galaxy mergers. They assumed that every halo which ever formed lasted for a time about equal to its collapse time, and that the galaxy which condensed within it survives at the present day. Numerical studies have shown that the first assumption, at least, is quite appropriate (Frenk *et al.* 1988). I calculate a similar no-merging luminosity function for the present models as follows. Stars which form at time t in a halo with star formation rate $\dot{M}_*(t)$ are assumed to end up in a galaxy with stellar mass, $\dot{M}_* t$, and luminosity $\dot{M}_* t \langle M/L \rangle$. The value of $\langle M/L \rangle$ is assumed to be that of a stellar population which formed at the earlier of t and $2t_o/3$. By integrating over the history of star formation one obtains the present luminosity density in galaxies of each luminosity, and so, by division, their abundance.

The results of this exercise are shown in Fig. 9. and compared with a standard Schechter function with faint end slope, $\alpha = -1.25$, and the same integrated luminosity density. The agreement is clearly rather poor. The no-merger assumption leads to substantially too many faint galaxies, a problem noted by WR for a wide range of initial conditions. On the other hand the shape of the predicted functions is such that it seems that merging may be able to account for the difference with observation. The excess faint objects are predominantly formed at early times and so are expected to suffer more merging than brighter systems. It is possible to investigate this possibility by constructing models which allow for some merging. However,

28

this requires some understanding of *which* halos merge together as the clustering evolves, and is too complex for discussion in this report. Our preliminary results are encouraging but still far from definitive.

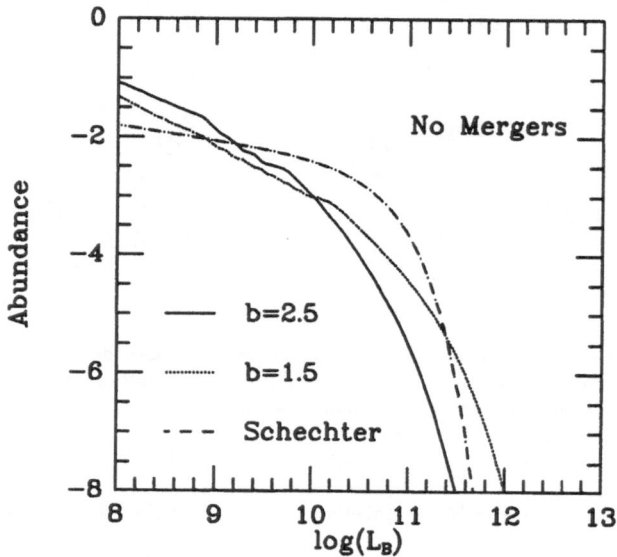

Fig. 9. The luminosity function of galaxies predicted assuming no merging is compared with a Schechter representation of the observed function. Units of luminosity are $L_\odot$, and of abundance are Number/Mpc^3/ln(L) plotted logarithmically.

At this stage of this project it appears that galaxy formation with either value of b may lead to good agreement with the observed properties of faint star-forming systems, and, perhaps, to luminosity functions in agreement with those observed. The inclusion of metallicity effects causes some further difficulties. Although the observed pattern of abundances in galaxies and in the intergalactic gas can be understood relatively easily within the theory, the enhanced cooling requires even more efficient suppression of star formation to avoid overproduction. With the above assumptions this in turn predicts even later formation of the bulk of stars in the universe. We are not yet sure whether parameter sets can be found which lead to good agreement with all the data within this simple theory. However, the results presented in this section suggest that late galaxy formation is not yet a severe problem for CDM, and indeed some of the observational data appear to favour the model over less rapidly evolving alternatives.

I am grateful to the organisers of the Durham workshop for an excellent meeting, to Carlos Frenk for allowing me to describe preliminary results from some of our joint work, and to Gustavo Bruzual for supplying results from his galaxy evolution models. This work was supported in part by NASA Grant NAGW-763.

References

Baron, E. and White, S.D.M. (1987). *Astrophys. J.*, **322**, 585.

Blumenthal, G.R., Faber, S.M., Primack, J.R. and Rees, M.J. (1984). *Nature*, **311**, 527.

Carlberg, R.G. (1988). *Astrophys. J.*, **324**, 664.

Carlberg, R.G. and Couchman, H. (1988). preprint.

Chevalier, R.A. and Clegg, A.W. (1985). *Nature*, **317**, 44.

Cowie, L.L. (1988). In: *The Post Recombination Universe*, p. 1, ed. Kaiser, N. and Laserby, A., Klower.

Davis, M., Efstathiou, G., Frenk, C.S. and White, S.D.M. (1985). *Astrophys. J.*, **292**, 371.

Dekel, A. and Silk, J. (1986). *Astrophys. J.*, **303**, 39.

Efstathiou, G., Ellis, R.S. and Peterson, B.A. (1988). *Mon. Not. R. Astron. Soc.*, **232**, 431.

Efstathiou, G., Frenk, C.S., White, S.D.M. and Davis, M. (1988). *Mon. Not. R. Astron. Soc.*, (in press.)

Efstathiou, G. and Rees, M.J. (1988). *Mon. Not. R. Astron. Soc.*, **230**, 5P.

Eggen, O.J., Lynden-Bell, D. and Sandage, A.R. (1962). *Astrophys. J.*, **136**, 748.

Ellis, R.S. (1987). In: *Observational Cosmology*, p. 367, ed. Hewitt, A., Burbidge, G. and Fang, L.-Z., Reidel, Dordrecht, The Netherlands.

Fabian, A.C., Nulsen, P.E.J. and Canizares, C.R. (1984). *Nature*, **310**, 733.

Frenk, C.S., White, S.D.M., Davis, M. and Efstathiou, C. (1988). *Astrophys. J.*, **327**, 507.

Hamilton, D. (1985). *Astrophys. J.*, **297**, 371.

Hartmann, L.W., Huchra, J.P. and Geller, M.J. (1984). *Astrophys. J.*, **287**, 487.

Kaiser, N. and Lahav, O. (1988). In: *Large-Scale Motions in the Universe*, 1987 Vatican Study Week.

Kennicutt, R.C. (1983). *Astrophys. J.*, **272**, 54.

Koo, D. (1986). In: *Spectral Evolution of Galaxies*, p. 419, ed. Chiosi, C. and Renzini, A., Reidel, Dordrecht, The Netherlands.

Koo, D. and Kron, R.G. (1987). In: *Observational Cosmology*, p. 383, ed. Hewitt, A., Burbidge, G. and Fang, L.-Z., Reidel, Dordrecht, The Netherlands.

Larson, R.B. (1974a). *Mon. Not. R. Astron. Soc.*, **166**, 585.

Larson, R.B. (1974b). *Mon. Not. R. Astron. Soc.*, **169**, 229.

Lilly, S.J. (1988). *Astrophys. J. Lett.*, **333**, 161.

Meier, D. (1976). *Astrophys. J.*, **207**, 343.

Miller, G.E. and Scalo, J.M. (1979). *Astrophys. J. Suppl. Ser.*, **41**, 513.

Narayan, R. and White, S.D.M. (1988). *Mon. Not. R. Astron. Soc.*, **231**, 97P.

Partridge, R.B. and Peebles, P.J.E. (1967). *Astrophys. J.*, **147**, 868.

Press, W.H. and Schechter, P.L. (1974). *Astrophys. J.*, **187**, 425.

Rieke, G., Cutri, R., Black, J., Kailey, W., McAlary, C., Lebofsky, M. and Elston, R. (1985). *Astrophys. J.*, **290**, 116.

Scalo, J.M. (1986). *Fund. Cosm. Phys.*, **11**, 1.

Schechter, P.L. and Dressler, A. (1987). *Astron. J.*, **94**, 563.

Thomas, P.A., Fabian, A.C., Arnaud, K.A., Forman, W. and Jones, C. (1986). *Mon. Not. R. Astron. Soc.*, **222**, 655.

Tinsley, B.M. (1978). *Astrophys. J.*, **222**, 44.

Tinsley, B.M. and Larson, R.B. (1979). *Mon. Not. R. Astron. Soc.*, **186**, 503.

Tyson, J.A. (1988). *Astron. J.*, **96**, 1.

White, S.D.M., Davis, M., Efstathiou, G. and Frenk, C.S. (1987a). *Nature*, **330**, 451.

White, S.D.M., Frenk, C.S., Davis, M. and Efstathiou, G. (1987b). *Astrophys. J.*, **313**, 505.

White, S.D.M. and Rees, M.J. (1978). *Mon. Not. R. Astron. Soc.*, **183**, 341.

DISCUSSION:

SILK: (1) The cooling condition that you mentioned (cooling time less than dynamical time) is a necessary, but not sufficient, condition for star formation. Moreover, disk formation occurs over many dynamical times. How do you reconcile your star formation model with the need for late disk formation? (2) Irregular galaxies are known to form stars sporadically - the LMC being an example. This is also inferred even in the solar neighbourhood for our own Galaxy. Hence one cannot simply infer that disk formation is a late process.

WHITE: My gas supply condition is that the star formation rate cannot exceed the rate at which gas is cooling and flowing in *from the cooling radius* in the halo. This radius is typically ten times the size of observed disks and the timescale for supplying gas is the Hubble time and so is much longer than disk dynamical times. Not all the gas supplied by cooling at large radii may be turned into stars, however, because of the energy input from star formation. In my models, the star formation rate is reduced significantly so that the effective energy input does not exceed the dissipation by cooling. I have looked into models where this supernova/wind regulation results in bursting behaviour. This can substantially modify the shape of the faint end of the luminosity function. However, it seems rather unlikely to me that the global star formation rate in the large spiral galaxies studied by Kennicutt could vary continually on short time scales.

HEAVENS: The rather poor agreement between your luminosity function and the observed luminosity function may arise as a result of your use of the Press-Schechter mass function. John Peacock and I have calculated an improved mass function (see these proceedings) which predicts fewer low-mass objects and more high-mass objects than the Press-Schechter analysis. Using our mass function would improve considerably the agreement between your model and the observation.

WHITE: I agree that the theoretical basis of the Press-Schechter formula is not very convincing. However, we have found that it agrees surprisingly well with direct numerical simulations of evolution from a wide range of initial conditions. Because of this I would be very surprised if the formula were seriously in error in those parts of the mass spectrum which contain a significant fraction of the total mass.

PEEBLES: Have you (or could you not) use your formalism to predict the abundance of dwarf galaxies in the voids? My guess is you will find an uncomfortably large number.

WHITE: It would probably be possible to do this, but I haven't done it yet.

A MULTICOLOR SEARCH FOR FORMING GALAXIES

Lennox L. Cowie
Institute for Astronomy, University of Hawaii
2680 Woodlawn Drive
Honolulu
Hawaii 96822, U.S.A.

ABSTRACT. Some results from a search for forming galaxies are described. The search uses a five color $(U'BVIK)$ deep survey to look for galaxies with the sharp breaks and the nearly flat spectrum (f_ν) regions of wavelength space which should be present in high-redshift young star-forming galaxies. Such a population with a flat spectrum in the BVI bands is found at I magnitudes fainter than 23. Based on the presence of breaks in the U' band most of the members lie in the range $2.8 < z < 3.6$. Model-invariant analysis suggests that this population could easily be responsible for almost all of the early star formation in massive galaxies.

1. Introduction

With the failure of L_α searches to turn up any (dust free) protogalaxy candidates (Koo 1986, Pritchett and Hartwick 1987, Cowie 1988) it has become necessary to turn to other and perhaps more robust search techniques. The obvious alternative diagnostic is the Lyman continuum break which can be searched for with broad-band data. This is the subject of the present talk which is based in large part on collaborative work by myself, S. Lilly, J. Gardner and I. McLean (Cowie et al. 1988, Lilly et al. 1988, Gardner et al. 1988).

Depending on the amount of neutral hydrogen present in a protogalaxy, the 912 Å break should range in depth from a drop of five to essentially infinity (Meier 1976, Bruzual 1983). Thus the first signature to search for in a candidate object is a drop of at least 1.5 magnitudes between two neighboring color bands. Such color drops can also be present in very red objects but these are easily distinquished by inspection of redder bandpasses where the protogalaxy should be nearly flat in f_ν for some substantial wavelength range. Inspection of models such as those of Bruzual (Bruzual and White 1988) suggests that, over a wide range of ages and assumed IMFs, a galaxy undergoing a constant star burst will be flat in f_ν to about 0.1 in the log over a range from about 1000 Å to 3800 Å in the rest wavelength. The reason for this is that the spectrum in this wavelength range is dominated by the ongoing formation of the most massive stars in the burst. We are therefore looking for two diagnostics, a region of flat spectrum and a sudden drop at the short wavelength end.

Even a priori we can be optimistic that such a search technique might succeed. Figure 1 shows the expected AB (equivalent visual) magnitude, defined as $AB = 48.60 - 2.5 \log(f_\nu)$ (cgs units, as given in Oke 1974), for a fairly average-mass galaxy forming as slowly as possible at a given epoch. This graph assumes that a galaxy forms at a constant rate from $t = 0$ to the

31

C. S. Frenk et al. (eds.), The Epoch of Galaxy Formation, 31–38.

Hubble time at a particular z, and as such is almost certainly a gross underestimate of the luminosity of the galaxy which most likely forms over a much shorter timescale. For $q_0 = 0.5$, AB is brighter than 24, corresponding to I brighter than 23.5, even at a z of 7. This is easily detectable even for a quite spatially extended object.

A second reason for optimism is that we can now extend the searches far to the red by means of the newly available IR arrays. Thus we can observe the entire wavelength range from 3200 Å (the short wavelength atmospheric cutoff) to 24000 Å where the thermal background comes in. This corresponds to a range of z for the Lyman continuum break from $z = 2.5$ to around 20 (Table 1).

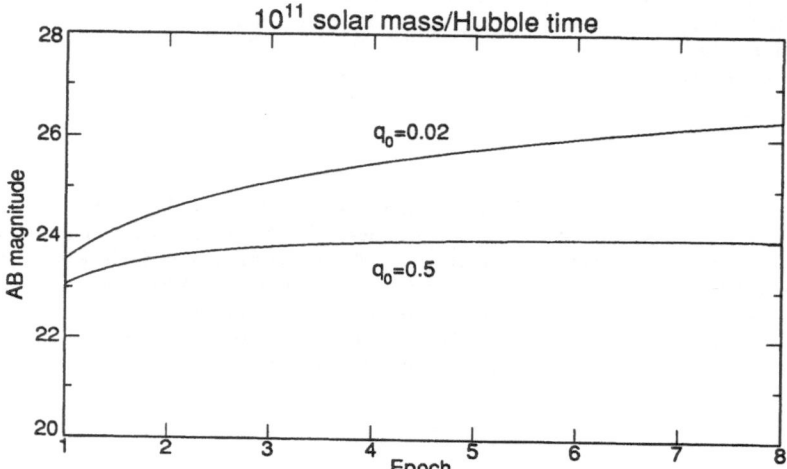

Figure 1 The AB magnitude is shown for the flat spectrum region (912 Å to 4000 Å rest wavelength) of a 10^{11} M$_\odot$ galaxy forming over the full local Hubble time. This is an absolute lower bound to the luminosity of a forming galaxy.

Clearly there are still many ways in which we could fail to find the protogalaxies in such a search. They might be obscured by intrinsic or extrinsic dust or be too spatially extended to detect. They might also be at too high a redshift. One further less serious point is that large amounts of intervening neutral hydrogen would move the break to L$_\alpha$ at 1216 Å owing to the Gunn-Peterson effect for smoothly distributed gas or a very heavy L$_\alpha$ forest. This effect is not significant for $z < 4$ and in any event would not prevent us from identifying the objects. However, it would result in an overestimate of the redshift.

2. Interpretation

In order for a search of this sort to make sense we must also be able to interpret the data in a quantitative and model-independent fashion. Without this we stand a good chance of being fooled and missing or overinterpreting the significance of any given population.

Fortunately such a model-independent interpretation is possible when we are observing the flat spectrum regions of young star-forming galaxies. The light in this wavelength range is dominated by massive stars which also return their synthesised metals. Thus, as outlined by Lilly and Cowie (1987) and Cowie (1988), the average contribution of any such population to the extragalactic background light (EBL) measured in flux per unit frequency at these wavelengths (S) is a direct measure of the total metal density (ρZ) at the current epoch produced in the systems. Numerically,

$$S = 2.1 \times 10^{-25} \left[\frac{\rho Z}{10^{-34} \, \text{g cm}^{-3}} \right] \quad \text{ergs cm}^{-2}\text{s}^{-1}\text{Hz}^{-1}\text{deg}^{-2} \,. \tag{1}$$

Equation 1 is formally independent of the details of cosmology, epoch of formation and timescale of formation, simply as a consequence of energy conservation. It is only very weakly dependent on IMF because for these stars most of the metals are returned and hence are a direct measure of the integrated bolometric luminosity of the population which S directly measures.

So, if we can identify a flat spectrum population and measure S, we can now use Equation 1 to compare it with the current metal density in galaxies and decide if the population is a possible progenitor of current galaxies. The metal density in either spheroids or disks is very roughly $2 \times 10^{-34} h^2 \text{g cm}^{-3}$ (e.g. Cowie 1988) and either population would give $S \sim 4 \times 10^{-25}$ ergs cm^{-2} s^{-1}Hz^{-1}deg^{-2} in any period of rapid star formation. These values are of course substantially uncertain and could also significantly underestimate the total metal production if there were much mass loss in the early galaxies. However, they tell us that a flat-spectrum population must have a contribution to the EBL of this order if it is to be significant.

3. Survey Methodology

Ideally we should choose bandpasses which are uniformly spaced in log wavelength from 3200 Å to 24000 Å. We should then survey them all to a uniform level in f_ν or equivalently in AB magnitude. In practice we have chosen the bandpasses listed in Table 1 which are concentrated to the blue end of the wavelength range. This reflects the still formidable difficulties of working in the IR. In addition while we have been able to obtain quite uniform flux limits from U' to I (to about a factor of 2) we cannot get near this flux level in K because of the high sky background. (All the exposures are purely sky limited and the exposure times in Table 1 are for a 4m class telescope, either CFHT or UKIRT.) However even in K the limit of $AB = 22.8$ is faint enough to be interesting as can be seen by referring back to Figure 1.

Table 1. Adopted Bands

Band	λ_c	$\Delta \lambda$	t_{expo}	Mag.[1]	Mag.[1]	z[2]
U'	3400	275	4.5 hr	$\cdots$	$AB = 26.7$	2.7
B	4470	860	3.5 hr	$B = 27.3$	$AB = 27.1$	3.9
V	5425	900	3.8 hr	$V = 27.2$	$AB = 27.2$	4.9
I	8340	890	6.3 hr	$I = 25.7$	$AB = 26.2$	8.1
K	22000	4800	14 hr	$K = 20.8$	$AB = 22.8$	23

[1] 1σ magnitude in a 3 arc second aperture convolved with the seeing profile.
[2] Lyman continuum break.

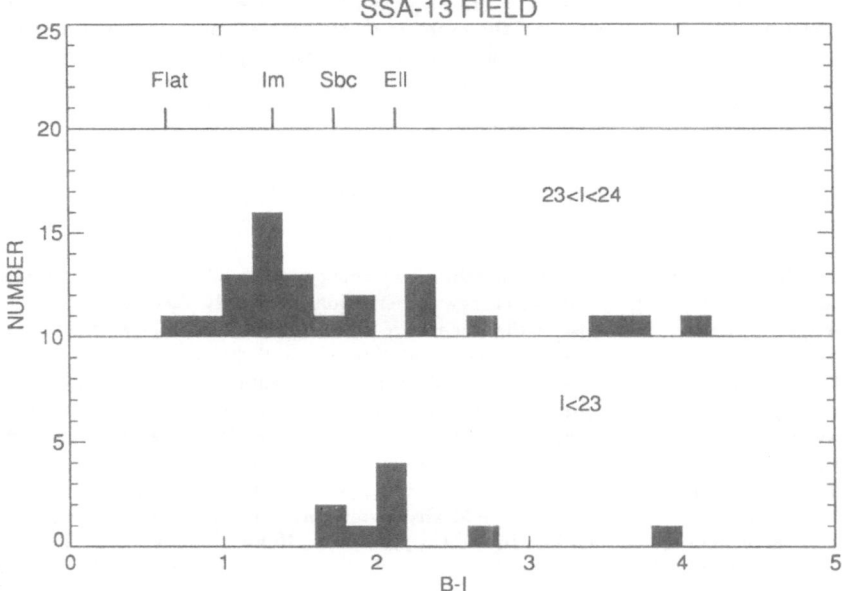

Figure 2 A histogram of the $B - I$ color distribution of sources as a function of I magnitude. Also shown are the roughly expected colors of normal galaxies at zero redshift and the position of a perfectly flat spectrum source.

4. Data

We have almost completed such a survey for three fields covering about 4 square arcminutes. However, at the present time only one field, which we call SSA-13 and which lies near the north galactic pole, is fully complete to the limits of Table 1 and I shall restrict myself to this particular area.

Our analysis of this area has proceeded as follows. We first prepared a sample of objects to $I = 24.5$ in a 3 arcsecond diameter aperture. (Ideally we should choose the sample in the K band but its much poorer limiting magnitude precludes this.) Since this is set at a 3σ level in I (Table 1) the sample is relatively uncontaminated by noise and we can be reasonably confident of the I band magnitudes. The sample consists of 41 objects and at this stage contains a small number of stars as well as galaxies.

We next measured $(U'BVK)$ magnitudes in matched apertures corrected for seeing at the positions of the I band sample. All of the I band objects are detected in V, and all but four in B, giving considerable confidence in the sample. At this stage 8 objects were eliminated as possibly being stars on the basis of having a full width half maximum similar to that of a stellar profile in the B band. (The band where the seeing was best at just under 0.9 arcsec). This may also have eliminated a small number of compact galaxies. It left us with a final sample of 33 objects, 15 of which are detected in U' and 16 in K.

We then searched the sample for flat spectrum sources in the various available colors. Because the sources need not be perfectly flat we required only that a color be bluer than the flat color + 0.4 magnitudes. Objects satisfying this condition in any color were then examined

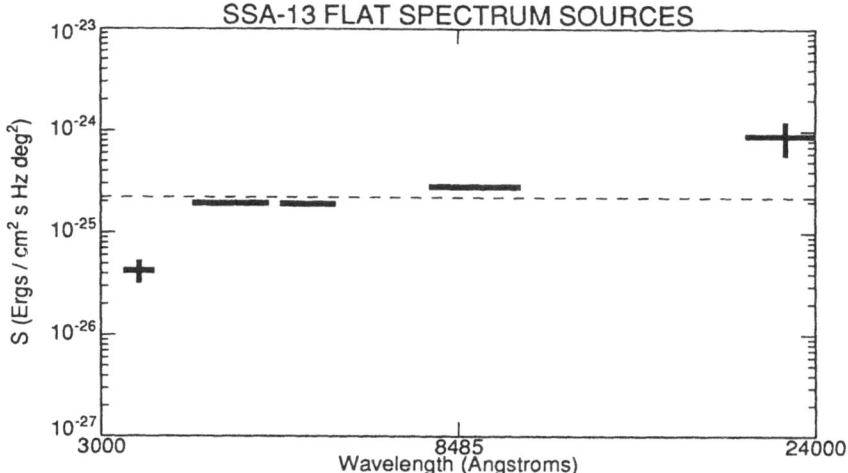

Figure 3 The contribution of the flat BVI sources to the EBL. Note the sharp drop in the U' band suggesting that many of these sources lie near $z = 3$.

for breaks or drops at shorter wavelengths. Only three classes of sources were found by this method, all lying at I magnitudes fainter than 23 :

1. Flat ($U'BVI$) : Only one candidate source was found in this class where we would expect to see lower redshift ($z < 1.8$) rapid star forming galaxies.

2. Flat (BVI), drop by less than 2 in U' : Five sources were found in this category. One possible interpretation of this class is that the U' band falls below redshifted L_α in these sources and is reduced by intrinsic Lyman damping and also by the extrinsic L_α forest. This would place these sources at $1.8 < z < 2.7$. However, some members of this category may lie more legitimately in category 3 and simply lie at the high end of the scatter of the very faint U' sources.

3. Flat (BVI), drop by more than 2 in U' : Ten sources are found in this category. Here the natural interepretation is that the Lyman continuum break has moved through the U' band but not through the B band and hence that these sources lie at $2.7 < z < 3.9$.

There are no sources with flat VI and a substantial drop in B or flat IK and a substantial drop in V, which would correspond to candidates in the ranges $3.9 < z < 4.9$ and $4.9 < z < 8.1$ respectively.

Thus all of the candidate objects have a flat spectrum in BI. For the sake of compactness we shall refer to these objects as the flat spectrum galaxies while noting that by this we mean flat spectrum in at least BVI. The onset and uniqueness of this population can be very clearly seen in Figure 2, where we show a histogram distribution of $B - I$ colors for different ranges of I magnitude. (In this figure we have used a larger data set of all objects for which we have BVI magnitudes. Some of these lie off the K band field and are not included in the main sample.) The flat spectrum galaxies are very distinct from normal galaxies and onset rather abruptly at $I = 23$. They may be identified with the blue galaxies found by Koo (1986) and Tyson (1988) at $B \sim 24$.

From Fig. 2 we can see that the BI color provides a simple discriminant for choosing out flat spectrum sources. If we take $BI < 1.3$ as our criterion for a flat spectrum source we find 11 such sources in our sample of 33 objects. They are of course a substantial subset of the objects in categories 1 through 3 above which were chosen on either flat BV or flat VI. The total contribution to the EBL of the flat spectrum sources chosen on the basis of $B - I$ color is shown in Figure 3.

The first point to note from Figure 3 is that the average contribution, S, of this population to the EBL is $\sim 3 \times 10^{-25}$ ergs cm^{-2} s^{-1}Hz^{-1}deg^{-2}, establishing it as a major population of star-forming galaxies according to the criterion of section 2. It could correspond to the production of almost all of the metals in either the disks or the spheroids.

However, an even more interesting point about Figure 3 is the very sharp drop, by about a factor of 5, of the composite spectral energy distribution (SED) in the U' band. The depth of this drop leaves little doubt that it is caused by the Lyman continuum break lying between the U' and B bands in most of these sources. This places them at $2.8 < z < 3.6$. As we have discussed above there may also be sources in the sample which lie at somewhat lower z and which may indeed contribute most of the weak signal in the U' bandpass that is seen in Figure 3. However, because the signal is so low in U' compared to B we can infer at once from Figure 3 that this lower-z population contains at most 20% of the star formation present at the larger z.

It is much harder to obtain useful information from the red (K band) end of the composite SED. Firstly, while S rises by about a factor of 4, within the 2σ uncertainties this could range anywhere from 1 to 7. However, the more difficult problem arises in interpreting this value because the ratio of red (> 4000 Å) to blue (< 4000 Å) f_ν in a constant star-forming galaxy is very dependent on the age, the time evolution of the burst and the form of the IMF. The reason for this is that in these young galaxies this flux ratio is a measure of the ratio of the number of stars at the main sequence turnoff to the rate of formation of stars at the high mass end of the IMF. The most that can be said from the data is that the SED profiles can be reproduced by a reasonable choice of IMF at an age consistent with the assigned z of 3 (i.e. less than 10^9 years).

Finally we note that at a redshift of $z = 3$ an individual flat spectrum galaxy with an AB mgnitude of 24 is producing stars at a rate of order $100 - 400$ h^{-2} M$_\odot$ per year depending on q_0 (though the exact value is extremely sensitive to assumptions about the IMF). If star formation took place at a constant rate from $z = 3.6$ to $z = 2.8$, such a galaxy would end up with a mass of around 10^{11} M$_\odot$ or more for any reasonable value of h. It is therefore quite plausible to associate these objects with the formation stages of massive galaxies.

5. Discussion

There seems little doubt that the flat spectrum sources described here are galaxies undergoing intense star formation and that many of them lie at around $z = 3$ in redshift. Their contribution to the EBL, which can be seen from Figure 3, is around 3×10^{-25} ergs cm^{-2} s^{-1}Hz^{-1}deg^{-2} and, since this is presumably only the bright-end tail of the luminosity function, there is little doubt that these systems are making nearly as much metals as are contained in either the disks or the spheroids at the present time (cf. Section 2). They are therefore an extremely cosmologically significant population which could easily be the formation stages of present-day massive galaxies.

We know that some radio galaxies and quasar hosts formed at $z > 5$ and hence that the very first galaxy formation occurred at an earlier time than $z = 3$ (Lilly 1988, Warren et al. 1987, McCarthy et al. 1988). (However, the number of such early forming galaxies could be very low indeed). Now, with the detection of the flat spectrum population, which could contain almost all of galaxy formation, and our failure to detect any other candidate objects at higher z, it is very tempting to speculate that the epoch of galaxy formation is strongly peaked at around $z = 3$, but that it stretched over a wide range of z from at least 5 down to around 2. It must be said that it remains perfectly possible that spheroid formation could have occurred entirely at the higher z and that we have simply failed to detect it in the present sample for one of the reasons discussed in the introduction. The flat spectrum population would then correspond to massive star bursts forming much of the stars in the early disks.

It remains to be seen how this final question will be answered as further information is obtained on the flat spectrum population. It will be particularly interesting to see what the distribution of this population looks like in redshift as we gradually obtain spectroscopic redshifts for a substantial sample of these objects.

References

Bruzual, G. A. 1983, *Ap. J.*, **273**, 105.

Bruzual, G. A., and White, S. D. M. 1988, in preparation

Cowie, L. L. 1988, in *The Post-Recombination Universe*, ed. N. Kaiser and A. Lasenby (NATO Advanced Science Institute Series).

Cowie, L. L., Lilly, S. J., Gardner, J., and McLean, I. 1988, *Ap. J. (Letters)*, **332**, L29, (Paper I).

Gardner, J., Lilly, S. J., Cowie, L. L., and McLean, I. 1988, in preparation.

Koo, D. C. 1986, *Ap. J.*, **311**, 651.

Lilly, S. J. and Cowie, L. L. 1987, in *Infrared Astronomy with Arrays*, ed. C. G. Wynn-Williams and E. E. Becklin (Honolulu: U. H. Inst. for Astronomy).

Lilly, S. J., Gardner, J., and Cowie, L. L. 1988, in preparation.

McCarthy, P. J. *et al.* 1988, *Ap. J. (Letters)*, **328**, L29.

Meier, D. 1976, *Ap. J.*, **207**, 343.

Oke, J. B. 1974, *Ap. J. Suppl.*, **27**, 21.

Pritchett, C. J. and Hartwick, F. D. A. 1987, *Ap. J.*, **320**, 464.

Tyson, J. A. 1988, *A. J.*, **96**, 1.

Warren, S. J., *et al.* 1987, *Nature*, **325**, 131.

DISCUSSION:

SHANKS: Since in the $21.^m5$ limited galaxy z survey of Broadhurst, Ellis and Shanks, the blue galaxies have an average redshift of only 0.15 - 0.2, doesn't it seem likely that the Tyson blue galaxies 3 magnitudes fainter are modest redshift z ~0.7 irregulars rather than protospheroids?

COWIE: Yes, this is quite possible. There is of course no way to know anything about the redshift of the blue objects beyond this constraint and the flat spectrum objects could be anywhere from 0 to 2 or 3. However, irrespective of the redshifts the point I have stressed in this talk is that, based on their average sky background contributions, they are making a large fraction of the metals in the universe and constitute a large fraction of the star formation. Given this it may be attractive to speculate, and this is a speculation, that a high z protospheroid population at z ~ 2 or so might be contained in the fainter population. However, it might equally be the case that we are seeing bursts in a more local population.

ELLIS: I'm puzzled. The slope of the blue galaxy counts (dlog N/dm = 0.4 - 0.45) is steeper than that in the I band (0.31) and hence faint galaxies are blue. According to your talk the slope of the K band counts is steeper still at ~ 0.5? Shouldn't you therefore expect to find a population of very red faint objects? Yet you said that this was not the case.

COWIE: The K band counts (not given in the printed text) are very preliminary and are deficient at the bright end because of selection effects in the choice of the fields. However, even if this were not the case it would require very detailed modelling to try to decide whether your point was valid. Simple predictions of the K band number counts based on a passively evolving population of normal galaxies agree extremely well with the K band counts - it is not at all clear to me that one cannot have steeper counts in K than in I over this small magnitude range even without a population of very faint extremely red objects.

WHITE: Have you found any red objects in your fields which are as bright as the two bright z = 0.8 galaxies in the Elston, Rieke and Rieke survey (these appear from spectra and BVRIK photometry to be almost unresolved elliptical galaxies similar in luminosity to the brightest radio galaxies). Isn't it very surprising that they found two such objects in the small area they surveyed.

COWIE: We have surveyed comparable areas at a significantly deeper (about two magnitudes) level and have not identified such objects.

Lyman-Alpha Galaxies in 1988

Hyron Spinrad

Astronomy Department, University of California

Berkeley, CA USA 94720

ABSTRACT. Lyman - α emission dominates the UV spectra of the roughly 18 radio galaxies observed to date with z > 1.6. I discuss three categories of such objects; (a) "normal", large and distant radio galaxies with developed continuum structures, often aligned with the radio source, (b) 2–3 "blobbier", less symmetric objects with weaker continua, but still showing strong Ly α in a large cloud–these are likely to be galaxies in formation such as 3C 326.1, and (c) the few known rare radio–quiet companions to QSO's with fairly strong Ly α. I discuss their structures, possible ionization sources and relevant star–formation–rates (for those objects in which photoionization by hot stars plausibly dominates). I will also comment on the observed or anticipated *extrinsic* spectral phenomena which may make distant (z > 2.5) Ly - α galaxies useful probes of intervening diffuse cosmic matter. Finally I close with a few remarks on the ages of the stars in the "coldest" distant stellar systems; these suggest $z_f \simeq 5$ as a lower bound to the epoch of initial star formation in at least a few large galaxies.

I. Introduction

It would be really valuable if surveys of discrete galaxies in the distant Universe were unbiased toward luminosity, or "special effects", such as nuclear activity or radio emission. Most are not. D. Koo will discuss the available less–biased *field surveys* later in the Conference. Here I concentrate on those special large stellar systems that are located by means of their high radio flux [3C, 4C, and 1 Jy level sources]. The radio bias can lead us to the most luminous galaxies at redshifts far in excess of the field surveys, and almost rivaling those of QSO's. The galaxies I discuss are still the only ones studied in reasonable numbers at z≥1.0, and we are just beginning to sample their diversity!

The look–back times to the most distance galaxies are also impressive; a galaxy at z=3.5 is seen at 0.16 the total age of the expansion for $q_o = + 0.1$, while with $q_o = 0.5$ ($\Omega = 1$) that fractional age is only 0.11. Thus for $H_o = 50$ km s^{-1} Mpc^{-1} (used throughout), in the closed Universe model we are discussing a *maximum time interval* of ~1 Gyr for a galaxy (and its black hole?) to assemble and create a powerful radio source.

Of course the title of my talk implies that the radio galaxies show the Ly α line of H quite strongly in their UV spectra. That will be an important clue to their nature.

II. Classes of Ly α Galaxies

All known radio galaxies with z >1.60 that could show a detectable Ly α above the earth's atmospheric ozone cut–off do show strong Ly α in emission. There are at least 18 known now.

39

C. S. Frenk et al. (eds.), The Epoch of Galaxy Formation, 39–56.

One could ask: are there any intrinsically luminous but apparently faint galaxies *without* Ly α emission? Here we use the current optical statistics on the most–complete radio catalogue, the 3CR sample. Of the 219 identified galaxies, 29 have $z > 1$ and only 4 do not yet have a reliable redshift. There are *no* unidentified sources at $\mid b \mid > 10°$ any longer! Thus these four galaxies are the only *possible* distant 3CR galaxies without Ly α emission. It is likely that obvious alternative explanations are appropriate to these few; e.g., observing them more intensively may show other lines which would lead to a smaller z.

Since the Paris meeting a year ago (Spinrad 1988) the number of published and unpublished Ly α galaxies has almost doubled! 1987 and 1988 have been banner years for record redshifts – startling rewards for some of the faint galaxy observers, who have now pushed to z=3.8. (Chambers and Miley, private communication).

The 3CR and 1 Jy radio galaxies observed by my group, plus 4C 40.36 at z= 2.27 (Chambers, Miley and van Breugel 1988) and Lilly's (1988) amazingly distant galaxy 0902+34 at z=3.391 have fairly good quantitative Ly α data.

The combination of radio mapping, spectroscopy, and continuum line imaging of these galaxies has lead me to classify the Ly α galaxies into 3 categories. The dividing lines between the first two are not yet well–defined.

First, the "regular, normal" Ly α radio galaxies, which include most of the sample because our morphological boundaries are broad. These do resemble concentrated galaxies in their ultraviolet continua; we do *not* require their shape to be round or elliptical in projection (c.f. McCarthy *et al.*1987a). The *second* type is much fainter and more diffuse/irregular in the emitted continuum; however the Ly α cloud in which each object seems to be embedded is both large and turbulent. 3C 326.1 (McCarthy *et al.*1987b) and 3C 294 are the prototypes of this "forming large galaxy" class; our interpretation is largely keyed to the lack of a *condensed* ("mature") galaxy image at the radio core or center, despite the presence of a large (D>100kpc) Ly α cloud. IR imaging data is badly needed to show the older population, if it exists. The emission line is presumably bright due to star–formation (SF) in the forming protogalactic system. The UV continua of these two radio galaxies are blobby, faint, and blue. Perhaps 3C 257 (z = 2.47) belongs to this class as well; line imaging and probably improved continuum imaging will be required (in 1989) to decide.

The *third* category of detected Ly α emission galaxies are the radio–quiet companions of QSO's and radio galaxies, $z \geq 1.8$ (Spinrad 1988). These companions, whose prototype is PKS 1614 + 051 (Djorgovski *et al.* 1985, 1987) are quite rare. As of last year some 45 QSO's and a few galaxies had been examined for companions, and only one or two found. T. Heckman and associates have begun a new search, looking for companions near steep–spectrum radio–loud QSO's at z~2.0. We await their results with a different selection bias. Further Ly α imaging of radio *galaxies* out to z = 3.8 is an obvious follow–up; recall the theoretical CDM models of White *et al.*(1987) strongly imply mass clustering on the 100kpc scale to be very epoch–dependent over the redshift range 5 to 1, or so.

A *fourth* type of Ly α galaxy should have been detected by now, but probably has not been imaged in emission. That is the putative disk object associated with various damped Ly α absorption systems, to be discussed by A. Wolfe later in this conference. These young disk galaxies, if that is what they are, are potentially important, because they may contribute heavily to the baryonic matter density of the Universe (Wolfe 1986).[R. Hunstead announced the spectroscopic detection of a Ly α line within the damped emission feature during the workshop!]

The present limits on the damped system's SFR's and metal–production are fairly severe. Tyson (1988) has suggested that the damped absorbers are a large number of gas–rich dwarfs.

III. Morphological Structures of Ly α Galaxies

The images of the Ly α galaxies are quite diverse in shapes and isophotal size. A few are small and fairly round–looking on the sky (3C 241, 3C 239, 1809 + 407 = 4C 40.36), but most of the distant Ly α galaxies are really quite large, D > 50kpc [for $q_o \simeq 0.2$, a compromise value, 1" $\simeq$ 10kpc for $1.6 \leq z \leq 3.8$]. A few, like 3C 294 and 0902 + 34 are *at least* 100kpc across in their Ly α emission images.

There is also considerably diversity in their shapes; at faint Ly α isophote levels 0902 + 34 is fairly round, while 3C 294 has a "fat-bean" shape The high resolution images of Le Fevre, Hammer and Jones (1988) convincingly show *continuum* sub-components to 3C 238 (z = 1.4), 3C 356 (z = 1.08), and 3C 256 (z = 1.820). I feel the scenario for these segmented galaxies will be a merger of sub-structures that eventually (z ~ 0) becomes a standard giant E system; the authors prefer a lens-amplification interpretation.

3C 194 (Djorgovski *et al.*1988) is a Ly α galaxy with a dramatic bean-like shape (see Figures 1 and 2) in its emitted UV continuum. It looks far different from radio E we know about here and now (z=0)! The galaxies are elongated along their radio axes.

Of major importance in the interpretation of the non-round shapes of the majority of the distant radio galaxies is whether they are merging gravitational sub-units or galaxies formed from radio jets igniting star-formation along a preferred axis that has little to do with gravity or the available mass distribution. There would seem no obvious reason why accumulation of dark or luminous sub-units would occur along the radio axis; probably the procedure is reversed and the radio source somehow channels the axis of new star formation. This must have been so very early, as the oldest detected population is also aligned (Chambers, Miley, & Joyce 1988; Lilly 1988). On the other hand, we know radio galaxies nearby are very round, so that the intervening time interval, from about $\bar{z}$ = 1.5 to $\bar{z}$ = 0.2, (some 8 Gyrs) has been sufficient for the luminous component of the typical large radio galaxy to relax rather completely. One might expect the merging subunits, originally "radio-aligned" to relax and merge on a dynamical time scale. Observations of 3C 368 (Djorgovski *et al.*1987) yield a characteristic crossing-time of ~ 2 x 10^8 yr., quite short compared to either stellar evolutionary changes or our z range of interest. Of course it could be that the "duty-cycling" of strong radio outbursts on a recurring axis determine the true time scale for coherence and smoothing of the baryonic galaxy. Our understanding of these asymmetries could be improved upon observationally with *spatially resolved* spectra of higher quality. In any case, the *implication* is strong that the active nucleus formed early in the accumulation of the galaxy – perhaps well before the bulk of the aligned star-formation proceed. There is some theoretical justification for this time sequence, but it was not formulated with an asymmetry such as that observed in 3C 194 or 3C 368 in mind (c.f. Silk and Szalay 1987 and Baron and White 1987). Perhaps these cores collapsed at z > 5.

Of course it would be very instructive to locate a few *radio-quiet* Ly α galaxies. Their morphologies could be useful counter points to the preceeding discussion. Galaxies with z > 2.5 are reasonable environs to begin such Ly α imaging searches.

The structure of 1 Jy 0902 + 34 is an especially interesting case. We have summed 4 hours of narrow-band Ly α imaging from Lick Observatory (3-m reflector) to form the composite image shown in Figure 3. Note that the Ly α isophotes of Fig 4 show that the high surface brightness inner region is elongated E -W (the radio source angle also: Allington-Smith 1982) but the outer contours are much rounder, with b/a $\simeq$ 0.8. Most impressive is the enormous angular extent of the H gas cloud; the azimuthally-averaged intensity profile for the Ly α cloud can be traced to a full 15" in radius, roughly r = 150 kpc! Recall the line surface brightness is already dimmed by the cosmological $(1 + z)^3$ factor. We show the radial profile in several ways in Figs 5, 6, and 7. To my surprise, the power-law surface brightness fit did not prove viable (Fig. 5), and the data run is definitely curved in log r, log SB space. But a replot in log-linear (Fig. 7) coordinates shows a beautifully exponential decline in Ly α SB with a scale length of 2".3 or some 23 kpc. Can we logically compare this gas profile to real disk galaxies of stars? However the gas "disk" is not the "cold" thin planar distribution we may be accustomed to in spiral galaxies since the Ly α line itself is moderately broad with $\sigma_v \cong 500$ kms^{-1} in its rest-frame.

IV. The Forming Radio Galaxy 3C 294

Since our work on 3C 326.1 has been published (McCarthy *et al.*1987) the second protogalaxy candidate, 3C 294, is next for mention. Observationally 3C 294 is mainly a giant Ly α gas cloud, 100kpc in length, again associated with a large and symmetric radio source. The redshift is z_e = 1.786. 3C 294 has a very extended and non-round Ly α distribution (Fig. 8). It roughly fills the radio source along the latter's inner axis; we have also found a spatially–restricted central zone of higher gas ionization.

As with 326.1, 3C 294 does not have a single central host galaxy of high luminosity. We spectroscopically detected only two weak continuum knots in the near–ultraviolet, each with U $\sim$ 24 mag. This is a primary distinguishing characteristic of the "immature" galaxy in formation. Its main body of stars are still unborn and unconcentrated. The Ly α luminosity is quite high; $J(\alpha) = 5$ x 10^{44} ergs s^{-1}, near the top of the luminosity function, listed in Table 1 . 3C 294 differs quantitatively from 3C 326.1 in that the organized velocity field in the former is exceptionally large. In 3C 294 the Ly α velocity range is 1200 kms^{-1}, while the Ly α line profile has $\sigma = 700$ kms^{-1}!

The radio core region, near the UV knots, has a line spectrum of higher ionization than the remainder of the cloud. It shows NV, CIII], CIV and He II emission. The source of photoionization at the core is uncertain. Hot stars (the knots?) could produce enough ionizing photons there, but whether a buried or anisotropically radiating AGN is needed for the highly ionized species is unclear. If OB stars ionize the large–scale H gas distribution, a total SFR exceeding 300 M$\odot yr^{-1}$ would be required (as per Table 1). Clearly such a copious rate of exhaustion of the interstellar matter in even a large galaxy could last only 10^9 years, or less. Still, 3C 294 has already managed to produce nitrogen and carbon.

V. Not Much New on Ly – α Companions

To my knowledge, no new galaxy or QSO companion objects have been discovered in the last year. Thus my summary from the IAP Conference (Spinrad 1988) suffices for both generalities and statistics. Here I wish to mention an interesting limit Mark Dickinson and I have placed on any Ly α companions to Lilly's 1 Jy galaxy at z = 3.391. No Ly α companion with a flux excess above 2% of the 0902 + 34 Ly α flux was detected; the *limit* on $J(\alpha) = 1.6$ (43). Have any early–life companions been incorporated into the main galaxy body already? Or are they simply "inactive" now? [A SFR like that of the Milky Way would be hard to detect.]

Perhaps a better, but unproven method to select distant galaxy companions to either QSO's or radio galaxies with z $\geq$ 3.5 would be intermediate–band CCD imaging on either side of the Ly α emission line at λ_o1215.7$\mathring{A}$. The amplitude of the extrinsic continuum discontinuity there in QSO's at large z (c.f. Warren *et al.*1987; Schmidt, Schneider, and Gunn 1987) and in 0902 + 34 itself (see also Sect. VII) is large ($\sim$ 1.3–2) and may well also be measurable for putative faint companion objects. Then the Ly α *emission* line in the companion would simply be an extra bonus.

VI. The Ionization Sources for Ly–α Radio Galaxies

Radio galaxies are a subset of AGN in general; it is thought that the thermal gas seen in their optical/UV spectra is ionized by a spectrum with a power–law shape originating from a tiny nuclear object. The evidence for a strong nuclear source is rather compelling for many nearby (z < 1) radio galaxies, as well (c.f. Robinson 1988, Fosbury 1987, McCarthy *et al.*1988). A *hard–spectrum*, perhaps with an anisotropic radiation field may ionize the extranuclear gas patches, too.

But for some of our distant radio galaxies, especially 3C 326.1, 3C 256, and perhaps 0902 + 34 there are hints that a *distributed* stellar ionization source may do the job adequately. This heretical position is worthy of consideration because:

(1) The distributed blue galaxy continua require hot stars, anyway.

(2) The observed deep–UV continua are adequate to produce sufficient Ly–continuum photons.

(3) The galaxy evolutionary models, $z_f \sim 5$, predict stellar Ly – continua fluxes that are consistent with our Ly α observations.

(4) No unusually bright *nucleus* is visible (to us) in the emitted UV of these 3 extended Ly α galaxies, at least.

(5) The radial profile of Ly α (and CIV) is smooth and follows that of the continuum for 3C 256. That means W$_\lambda$ (line) $\simeq$ constant; this would require contrivance in the "central–source" model.

(6) In the case of 3C 256, the ionization parameter U, is roughly constant over the face of this large 120 kpc diameter galaxy.

The best argument against OB star ionization is the strength of the high–ionization lines, NV, CIV, [Ne IV], and He II in our spectra of *most* (but not all) of the radio galaxies. Stars generally produce too soft a radiation field at short λ. Stasinska's (1984) stellar ionization models can ionize carbon the requisite three times, but rare stars with $T_e \geq 55{,}000$ K are needed. More common, later O stars won't produce enough hard ionizing photons. Hotter, metal–poor O stars or even Wolf–Rayet stars could be viable substitutes, perhaps.

The evolutionary models of Bruzual, as computed by Djorgovski (1988) and by Eisenhardt (unpublished) produce too few ionizing photons with $z_f = 5$ to match the Ly α data. This is partly because the models are defaulted to a maximum stellar mass of 25 $M_\odot$. Otherwise similar evolutionary models of Guiderdoni and Rocca–Volmerage (1988), using a maximum star mass of $80 M_\odot$, can do the job, as Fig. 9 shows.

We have also looked into the integrated maximum equivalent width of Ly α from stellar "cluster" sources; a normal observed blue star luminosity function (W. Freedman, 1985) used to integrate the stellar ionizing and non–ionizing UV spectra produces a fairly large ($W_\lambda^o(\alpha) \sim 130\mathring{A}$) Ly α equivalent width. If we use the temporal 30 Doradus (Melnick 1987) luminosity function we find, $W_{\lambda_o} \sim 400\mathring{A}$. Fig. 10 summarizes the Ly α galaxy observed equivalent widths and some stellar/power–law model predictions. While the errors of observation preclude a firm decision, it seems to me that much of the Ly α could easily have come from Ly–continuum photons originating in O stars.

VII. Extrinsic Spectral Features Anticipated for Very Distant Galaxies

Our observations are now taking us to galaxies sufficiently distant that *intervening* (cosmological) clouds may influence the spectral and even the spatial structures of galaxies at $z \geq 3.5$ or so. One extrinsic phenomenon is the integrated effect of the *Ly α forest* on the continuum of 1 Jy 0902 + 34. The local continuum below the emission line is depressed by a factor 0.3 - 0.5. The Ly α forest discontinuity is known well to low–resolution observers of QSO's (Steidel and Sargent 1987) for z ≥ 3. The density of the intervening HI clouds increases rapid with z, so we should anticipate larger Ly α forest discontinuities in the continua of galaxies at z > 3.5, when they become available. It is, I believe, plausible to hope that the "forest edge" will be a discriminant for future *location* of galaxies at z > 4! Obscuration by *intervening dust clouds* (in galaxy disks, presumably) could also be important in the surface brightness studies of radio–selected large galaxies at z > 3. Ostriker and Heisler (1984), Heisler and Ostriker (1988), and Weedman (1987) have suggested that many lines–of–sight to distant QSO's may be obscured. Since our sample of distant galaxies is radio–selected, we can probably test this suggestion when several galaxies at z > 3 are studied thoroughly.

There are many lines of sight to the large Ly α image of 0902 + 34; we note its shape and profile in Ly α to be quite smooth (see again Figs 3 and 4). Let's consider the simple expression in Weedman's analysis; at z = 3.4, with "typical" dusty intervening galaxies of radii R = 5kpc, we find that fraction of the Universe obscured to be f = 0.18. There are about 80 independent "lines of sight" to various portions of the 0902 + 34 Ly α image in Fig. 4. Since we observe no holes on the 1" angular scale this example already has dented the "straw–man" obscuring galaxy model. There is no evidence for dust in the spectrum above $\lambda_o 1216$, either. Eventually complete spectral/imaging data on even more distant (4C) radio galaxies should provide useful quantitative constraints on intervening dust and gas.

VIII. Infrared–Optical–Colors and the Last Epoch of Star–Formation in Some Distant Galaxies

When more photometric 2.2 μ (K–band) imaging directly photometric observations of Ly α galaxies are available, we can directly look at the oldest generation of stars in a very distant stellar system. At the moment the most useful constraints on the epoch of last star–formation come from the almost–passive galaxy (Lilly and Longair 1984) colors for 0902 + 34 (Lilly 1988) and 1Jy 1056 + 39 (Lilly – private communication) at z = 2.17. Both galaxies are red with $(R-K)_o \geq 4.5^{mag}$. This

44

indicates an extant old ($\tau > 10^9$yr.) population at (V/R) emitted wavelengths. Using these data, or even by extrapolating from the reddest radio galaxies at $1 < z \lesssim 1.82$ (3C 65, z = 1.18, R–K = $6^m.0$; 3C 241, z = 1.62, R–K = $5^m.7$) we find $z_f = 5$ barely consistent for the initial star–burst if the longest–time–scale cosmology is adopted. For higher H_o or a larger q_o, $z_f > 5$ must pertain for these systems. This argument was also used by Lilly during the conference. So unless our stellar evolutionary tracks for even middle–age stars are very poor, this red and dead minority of large galaxies suggests a very early initial accumulation and starburst. That event may or may not be followed by future activity. An initial accumulation at $z_f \sim 10$ is quite plausible; we look forward to new observations of galaxies at z>3.4 to test the hypothesis.

I wish to acknowledge observational help and lots of idea exchange with Patrick McCarthy, Mark Dickinson, George Djorgovski, Wil van Breugel, Michael Strauss, Simon Lilly, Ken Chambers, George Miley, and Jim Liebert.

My extragalactic research is supported by the U.S. NSF.

Table 1

Ordered Lyman-Alpha Luminosities and SFR

R.	Galaxy	z	J (α)	Technique	log by C rate (ph.s^{-1})	log SFR* (M$\odot$yr^{-1})
1	3C 256	1.819	16(44)	whole image	55.9	2.9
2	0902+34	3.391	8.7(44)	whole image	55.6	2.6
3	3C 294	1.786	5:(44)	whole image	55.4:	2.4:
4	3C 326.1	1.825	3.6:(44)	whole image	55.3:	2.3:
5	1809+407	2.269	3(44)	whole image	55.2	2.2
6	1141+35	1.781	2.4(44)	spectral	55.1	2.1
7	1056+39	2.171	2.3(44)	image	55.1	2.1
8	3C 241	1.617	2.1(44)	spectral	55.1:	2.1:
9	3C 239	1.781	2.1(44)	spectral	55.1	2.1
10	PKS 1614+51C	3.215	1.3(44)	comp.-image	54.9	1.9
11	3C 454.1	1.840	1.2(44)	spectral	54.9	1.9
12	3C 194	1.779	1.0:(44)	spectral	54.8:	1.8:
13	3C 257	2.474	8:(43)	spectral	54.7	1.7:
14	3C 470	1.652	6.7:(43)	spectral	54.6:	1.6:
15	possible 256 comp.	[1.82]	5:(43)	image?	54.5:	1.5:
16	3C 322	1.681	1.4(43)	spectral	53.9	0.9
17	Putative 0902 comps.	(3.39)	<1.6(43)	lack of images	<54.0	<1.0
18	HI disks**	(2-3)	<3(42)	spectral/im.	<53.2	<0.2

* Assumes a mass-function exponent of -2.4, following Buat et al. in Astr. and Ap., **185**, 35, 1987. On same system, Milky Way Gal., SFR $\simeq$ 0.7.

** From Smith et al., preprint (conversation – mid 1988).

References

Allington–Smith, J.R. 1982, *M.N.R.A.S.* **199**, 611.

Baron, E. and White, S.D.M. 1987, *Ap. J.* **322**, 585.

Chambers, K.C., Miley, G.K. and van Bruegel, W. 1988, *Ap. J.* **327**, L47.

Chambers, K.C., Miley, G.K. and Joyce, R.R. 1988, *Ap. J.* **329**, L75.

Djorgovski, S., Spinrad, H., McCarthy, P. and Strauss, M.A., 1985, *Ap. J.* **299**, L1.

Djorgovski, S., Spinrad, H., McCarthy, P.J., Dickinson, M. van Bruegel, W. and Strom, R. 1988, *A. J.* **96**, 836.

Djorgovski, S., Strauss, M.A., Spinrad, H., Perley, R. and McCarthy, P.J., 1987, *A. J.* **93**, 1318.

Djorgovski, S., Spinrad, H., Pedelty, J., Rudnick, L., and Stockton, A., 1987, *A. J.* **93**, 1307.

Djorgovski, S. 1988 in "Starbursts and Galaxy Evolution", Th. Montmerle (ed.), Paris: Editions Frontieres, (in press).

Freedman, W.L. 1985, *Ap. J.* **299**, 74.

Fosbury, R.A.E., 1987, Talk given at Tenth European Regional SAU Conf., Aug. 1987 (ESO Sci. Preprint 531).

Guiderdoni, B. and Rocca–Volmerage, B. 1988, *Astron. & Astrophys.*, (in press).

Heisler, J. and Ostriker, J.P. 1988, *Ap. J.*, (in press).

Le Fevre, O., Hammer, F., and Jones, J. 1988, *Ap. J.* **331**, L73.

Lilly, S.J., and Longair, M.S. 1984, *M.N.R.A.S.*, **211**, 833.

Lilly, S.J. 1988, *Ap. J.* (in press).

McCarthy, P.J., van Breugel, W., Spinrad, H. and Djorgovski, S., 1987a, *Ap. J.* **321**, L29.

McCarthy, P.J., Spinrad, H., Djorgovski, S., Strauss, M.A., van Bruegel, Liebert, J. 1987b, *Ap. J.* **319**, L39.

McCarthy, P.J., Spinrad, H. and van Breugel 1988, talk presented at IAU Symp. #134 (AGN), Santa Cruz.

Melnick, J. 1987, in "Star–Forming Dwarf Galaxies", D. Kunth, T.X. Thuan, and J. Tran Thanh Van, eds., Editions Froniteres: Paris, p.171.

Robinson, A. 1988, NATO Workshop on Cooling Flows in Galaxies, ed. (in press).

Ostriker, J.P. and Heisler, H. 1984, *Ap. J.*, **278**, 1.

Schmidt, M., Schneider, D., and Gunn, J.E. 1987, *Ap. J.* **316**, L1.

Silk, J. and Szalay, A. 1987, *Ap. J.* **323**, L107.

Spinrad, H. 1988, in IAP Conference "High Redshift and Primeval Galaxies", eds. J. Bergeron, D. Kunth, B. Rocca–Volmerange, and J. Tran Thanh Van, Editions Frontieres, Paris, p. 59.

Stasinska, G., 1984; *Astron. & Ap. Suppl.* **55**, 15.

Steidel, C. and Sargent. W.L.W. 1987, *Ap. J.*, **313**, 171.

Tyson, N.D. 1988, *Ap. J.*, **329**, L57.

Warren, S.J., Hewitt, P.C., Osmer, P.S., and Irwin, M.J., 1987, *Nature*, **330**, 453.

Weedman, D. 1987, in "Star formation in Galaxies", NASA Conf. Public. 2466, p.351.

White, S.D.M., Frenk, C.S., Davis, M. and Efstathiou, G., 1987, *Ap. J.* **313**, 505.

Wolfe, A.M., Turnshek, P.A., Smith, H.E., Cohen, R.D., 1986, *Ap. J.* Suppl. **61**, 249.

Fig. 1 A portion of a red CCD image of the field of the distant, elongated galaxy associated with 3C 194 (z = 1.779, Djorgovski *et al.* 1988). The radio galaxy is the fainter "banana–shaped" image to the left (E) of a foreground galaxy at z = 0.31.

3C 194 (R) z=1.779

KPNO 4-m   31-Dec-1986 UT   800 sec

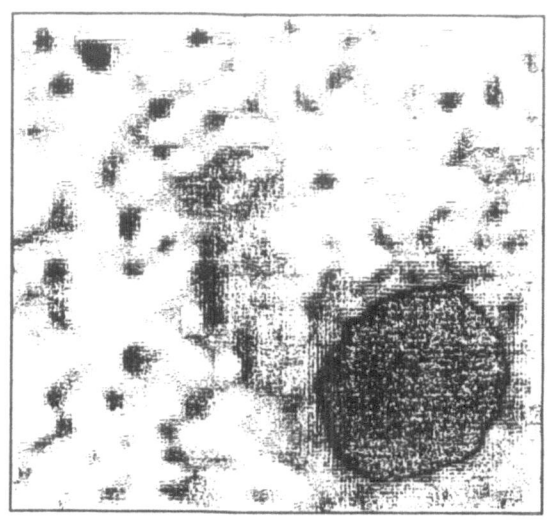

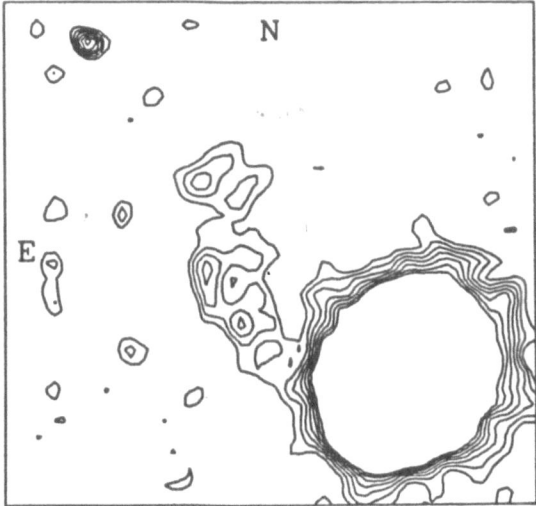

Field size = 19 arcsec

Fig. 2 An isophotal and grey-scale display of the enlarged 3C 194 field (also from Djorgovski); a VLA radio map by van Bruegel and McCarthy shows the radio lobes to just extend beyond the optical extent of the image—in this case, the radio source and starlight contents are roughly aligned (c.f. others in McCarthy *et al.* 1987b).

0902 + 34

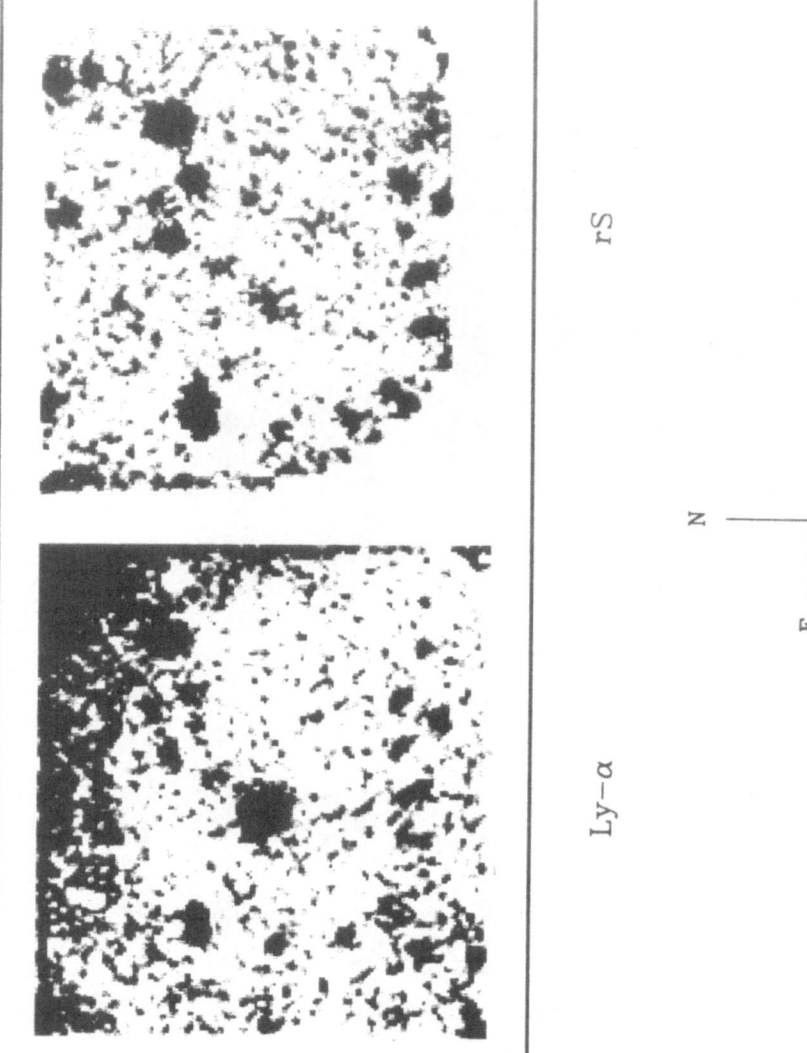

Ly−α

rS

N

E

Fig. 3 Our red (right) and Ly α images (left) of the distant 1 Jy galaxy, 0902+34 (Lilly 1988). Note the large, slightly elongated Ly α gas cloud in the panel. The size of the Ly α frome is ≈96" on a side. The Ly α image is obviously much brighter than the continiuum image.

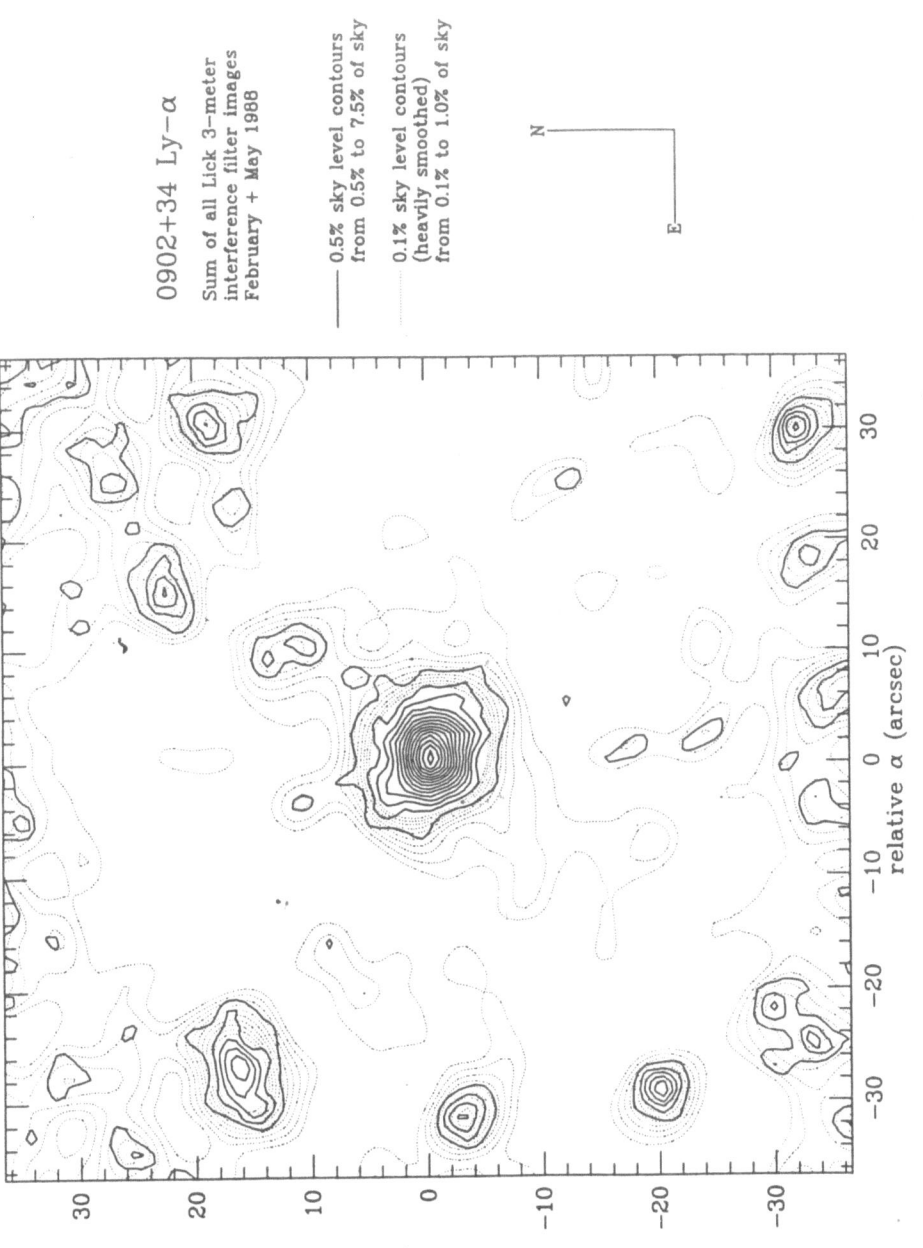

0902+34 Ly−α

Sum of all Lick 3−meter
interference filter images
February + May 1988

—— 0.5% sky level contours
from 0.5% to 7.5% of sky

‧‧‧‧‧ 0.1% sky level contours
(heavily smoothed)
from 0.1% to 1.0% of sky

N

E

Fig. 4 An isophotal plot of the Ly α surface brightness contours of 1 Jy 0902 + 34 (z = 3.391). Note the large angular extent of the Ly α emission cloud!

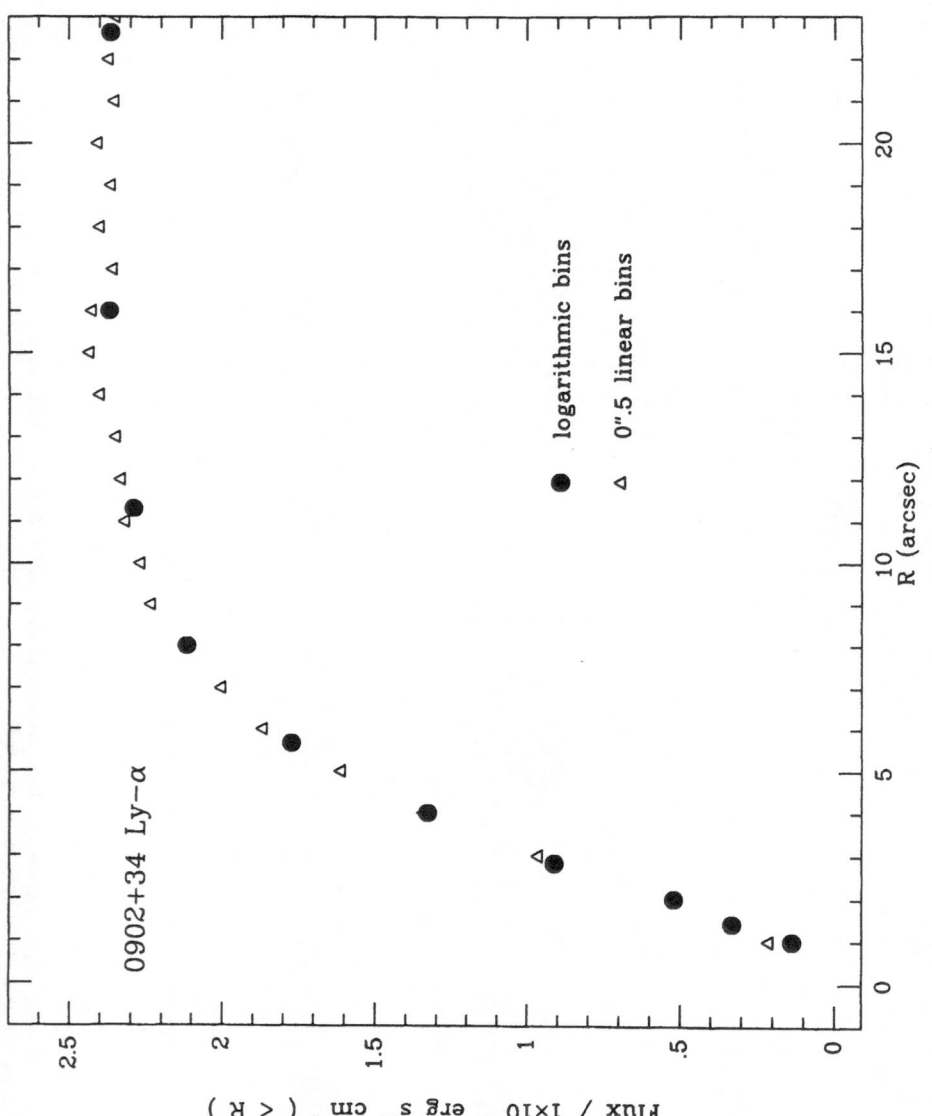

Fig. 5 The linear growth–curve of the 1 Jy 0902 + 34 galaxy Ly α emission (azimuthually averaged).

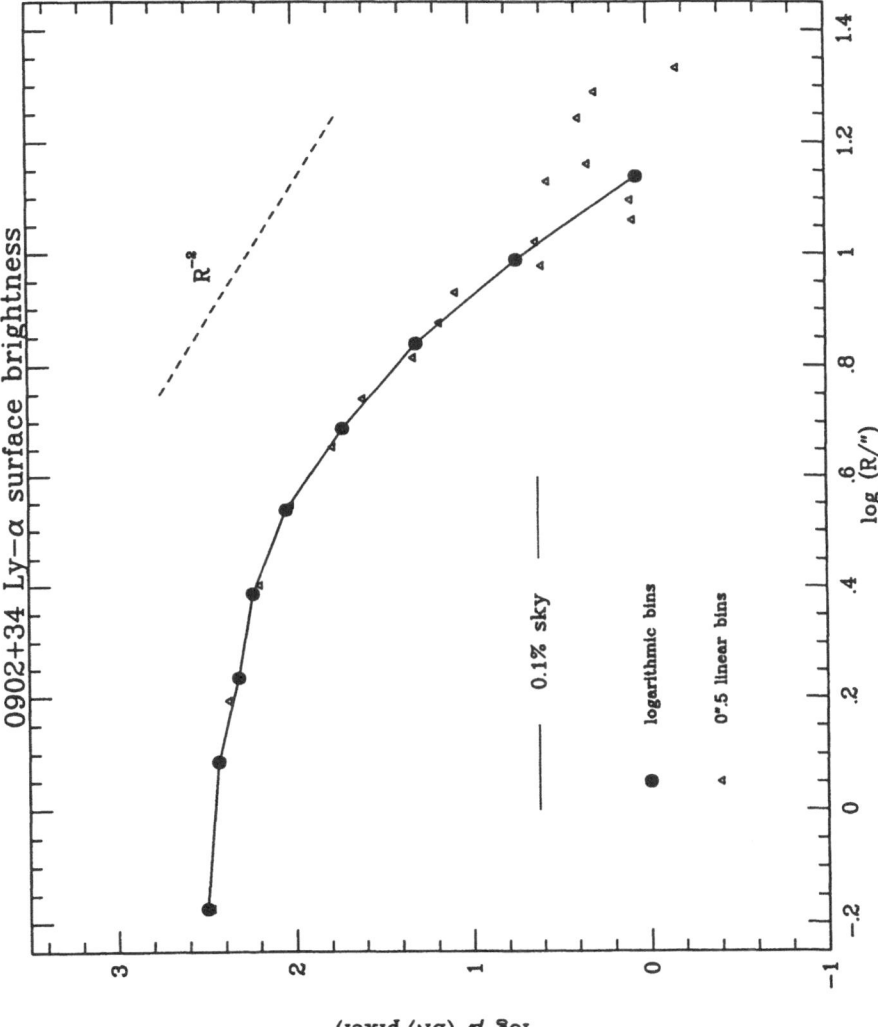

Fig. 6 The Ly α surface brightness distribution in log-log space; note that decline is *not* a power law.

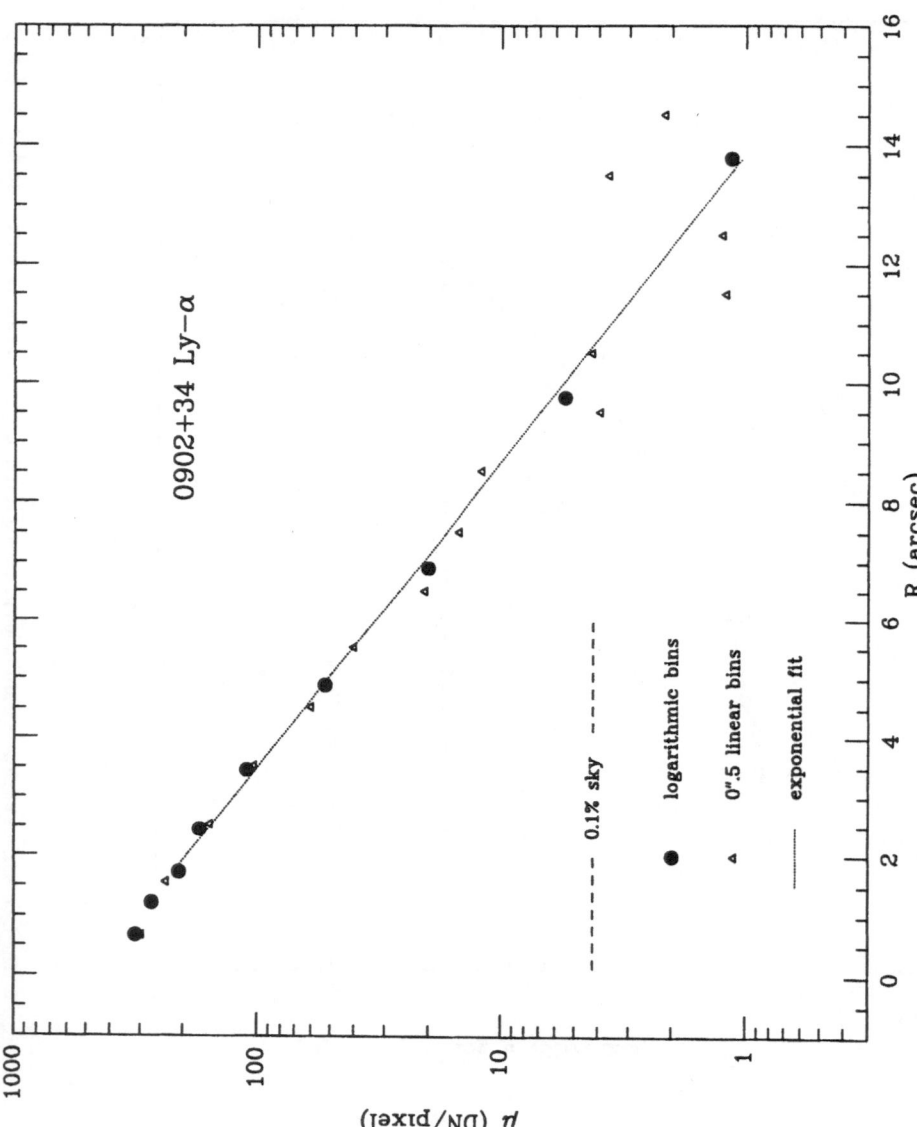

Fig. 7 The Ly α surface brightness distribution for 0902 + 34 in log–linear space. An exponential with a scale length of 2".2 fits quite well.

N

E

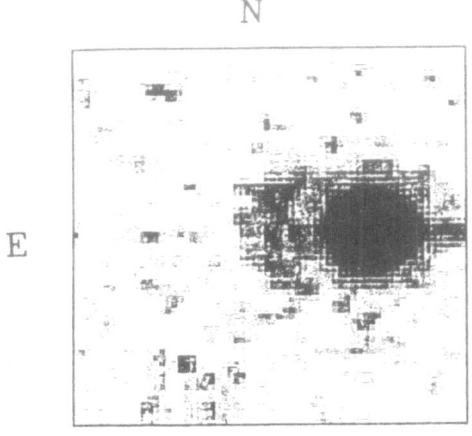

3C 294 Ly−α

Fig. 8 A grey–scale representation of the narrow–band Ly α image of 3C 294. The bright object to the W of the "bean–shaped" galaxy image is a foreground galactic K star.

54

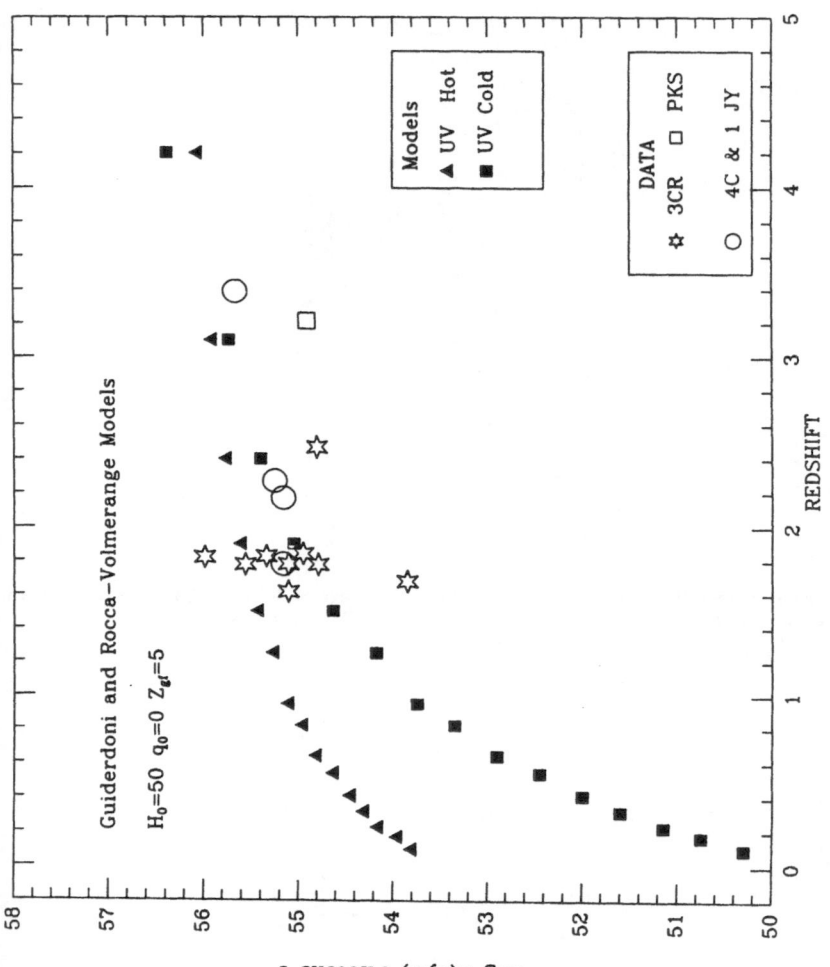

Fig. 9 Model predictions from Guiderdoni and Rocca–Volmerange (1988) of the Ly–continuum photon rate vs. redshift for giant E galaxies which eventually (z = 0) have $M_v = -23.0$. The data points are radio galaxies with Ly α fluxes, plotted here with the assumption that each Ly C photon emitted by an O–star becomes a Ly α line photon we can observe. The agreement is quite good, and suggests stellar photoionization may be important in the extended Ly α emission of radio galaxies.

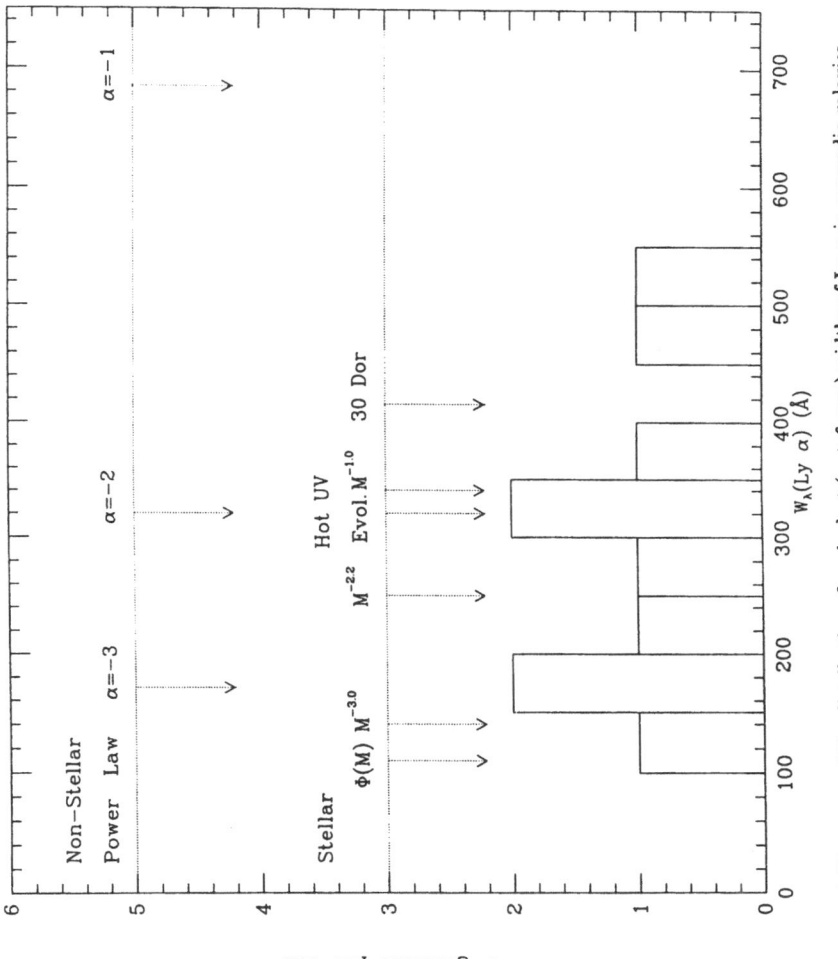

Fig. 10 A histogram of the distribution of equivalent (rest-frame) widths of Ly α in our radio galaxies. The data is noisy, because W_λ depends upon the measurement of very weak continua. The stellar population models with *observed stellar luminosity functions* [Φ(M)] and 30 Dor] or power laws in stellar mass (-exponents), and the AGN-class **straight power law** UV spectral distributions all fit the data moderately well. No clear discriminations between Ly α ionization sources can be made in this manner.

DISCUSSION:

KOO: Can you place any limits on the abundances in the large gas clouds associated with the radio sources?

SPINRAD: Not adequately, but the ionization state for 3C 326.1 may imply a modest ($\sim 1/3$ solar) metal abundance.

ALLINGTON-SMITH: Is there any sign of a difference in Lyα equivalent width between the 3C and 1-Jy galaxies compared with the difference in forbidden line emission between the two samples at lower redshift.

SPINRAD: Not yet in Lyα; both sets (1 Jy data is, of course - still limited) are intermingled, with $W_o \sim 300$ Å .

WHITE: With your improved population models what star formation rates are inferred in these objects?

SPINRAD: From the Ly$\alpha \to$ Ly C photon rates, we find the maximum (3C 256) SFR ~ 700 $M_\odot$ yr^{-1} (normal IMF, total SFR). More typically, SFR $\sim 200\text{-}100$ $M_\odot$ yr^{-1} with 20 or so the observational lower bound.

WHITE: With such high rates in what sense can you say that the galaxy is old - isn't its current star formation rate at least as large as its past average?

SPINRAD: Clearly rates of > 100 cannot be maintained steadily for the $\sim 10^9$ yrs from very large z to z ~ 1.8, as observed! It has to be "bursty"!

REES: You discussed the apparent velocities of the gas in these systems inferred from Lyα, and the fact that C IV gives different velocities. It's worth noting that there will be confusing radiative-transfer effects on the Lyα photons if they undergo multiple scattering. In particular, if they emerge from any systematically-contracting medium, they will be blueshifted. This is likely to be relevant to these young galaxies - the effect doesn't require shocks, or any special kinematics.

SPINRAD: Yes, in fact two "forming" galaxies, 3C 326.1 and 3C 294 both have restricted spatial regions with blue wings on the Lyα emission line. Those may be attributed to either "shocks" (Neufeld & McKee, 1988, *Astrophys. J.* in press), or perhaps your global contraction kinematics. Good point.

SHAPIRO: Have you considered the possibility that shock excitation explains the high ionization lines in your objects rather than photoionization?

SPINRAD: Yes, to some degree we've considered collisional ionization also. Patrick McCarthy found the extant collisional models did not produce nearly a strong-enough He II 1640 Å emission line as needed. Thus we think photo-ionization is the likely ionization mechanism.

COWIE: There appear to be many similarities between 3C 326.1 and the cluster cooling flow systems. These are the similarity of the Lyα equivalent widths, the velocity widths, the presence of central radio sources, and the absence of dust absorption in the Lyα, etc. Given this might it not be more natural to identify 3C 326.1 with a cooling flow cluster system rather than a rapid star forming exotic galaxy, which already has a central radio source.

SPINRAD: Of course it's hard to know; the arguments for formation of an individual galaxy were mainly continuity of $J(\alpha)$ with "mature" Lyα galaxies, and the supposed contracted size at later epoch. Velocities in the present gas cloud could support either final state (cluster or giant galaxy). We thought cooling flows were generally regarded as an evolutionary state appropriate to the current cosmic epoch rather than z > 1.8.

Morphological Evolution of Radio Galaxies

Patrick J. McCarthy and Wil van Breugel

Astronomy Department, University of California

Berkeley, CA USA 94720

ABSTRACT. Powerful radio galaxies are unique at the present time in that they have optically resolvable structures that can be observed over a range of look-back times comparable to the Hubble time. This allows us to examine how their morphologies evolve with cosmic time in their stellar continuum, thermal line-emitting gas, and non-thermal synchrotron emitting plasmas. The continuum shapes of high redshift radio galaxies become more elongated at high redshift and are closely aligned with their radio source axes. The extended emission line regions were larger and more luminous in the past and they too are well aligned with their radio source axes. Finally, the radio sources themselves were more asymmetric in the past both in terms of their structures and in the flux densities, spectral indices and polarization of the diffuse lobes.

Introduction

The 3CR catalog of radio sources is now completely identified and redshifts have been determined for more than 97% of them (see Djorgovski *et al.* 1988 for the latest update). Deep broad band images of a large fraction of the 3CR galaxies are now in the literature (e.g. Baum *et al.* 1988; Djorgovski *et al.* 1988; Le Fevre *et al.* 1988). Emission line images for nearly 100 3CR radio galaxies have recently been obtained (Baum *et al.* 1988; McCarthy *et al.* 1988; Stockton and Lilly 1988). Modern radio data are also becoming available for a large fraction of the high redshift 3CR radio galaxies.

We have used these data bases to examine the changing morphology of the stellar continuum, extended emission lines, and non-thermal radio emission of these galaxies as a function of redshift.

The Emission Lines

Baum *et al.* (1988) have recently completed an emission line imaging survey of nearby radio galaxies. Baum *et al.* find that the extended emission lines in these objects have a median size of 10 kpc and have typical luminosities of $L(H\alpha) \sim 10^{42}$ erg s^{-1}. Morphologically these objects can be split into two classes: small disk-like structures in the inner regions of giant elliptical galaxies, and very extended filaments lying at large distances from the center of the host galaxy. The first class of EELRs have sizes of 1 - 3 kpc and appear to be cool gas that has settled into the central potential of the host galaxy. These regions show no preferred orientation with respect to the radio source axis. An example of a small EELR is shown in figure 1. The second class of EELR occur on scales of up to $\sim$ 100 kpc and often contain a substantial fraction of the total emission line luminosity. While these very extended emission line regions only rarely show a direct correspondence with their radio source structures, they do tend to lie in the quadrants of the sky defined by the radio lobes (Baum and Heckman 1988). Considered as a whole, however, the EELRs in nearby radio galaxies do not have a strong relation to the radio source structures.

Emission line imaging surveys of high redshift 3CR radio galaxies have recently been completed by McCarthy *et al.*(1988) and Stockton and Lilly (1988). These surveys show that

C. S. Frenk et al. (eds.), The Epoch of Galaxy Formation, 57–62.

58

the sizes and luminosities of the EELRs increase systematically with redshift. As Figure 3 shows, for redshifts larger than ~ 0.2 the EELRs the emission line regions nearly always lie within 15° of the radio source axis. In a fair fraction of cases at $z > 0.5$ the emission line regions are spatially co-incident with one of the two radio hot spots.

The luminosities of the EELRs increase strongly with redshift reaching monochromatic luminosities of L([OII]λ3727) = 10^{44} erg s^{-1} at redshifts greater than 1. The total luminosity

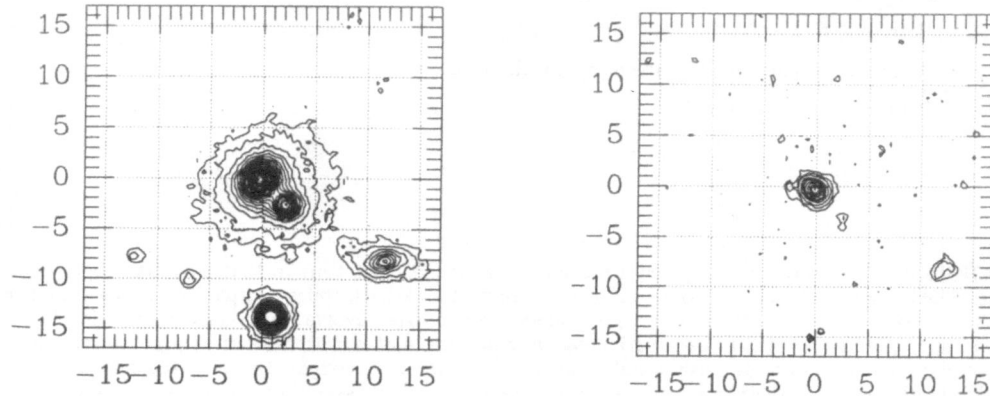

Figure 1. Broad band V and Hα images of the nearby double source 3C 135 (z=0.127). The galaxy lies in a spiral rich cluster and has a number of close projected companions. The Hα is confined to well within the central few kpc of the galaxy, and is slightly extended along the minor axis. This object has continuum and emission line morphologies that are typical of low z powerful radio galaxies. The scales are in arc seconds.

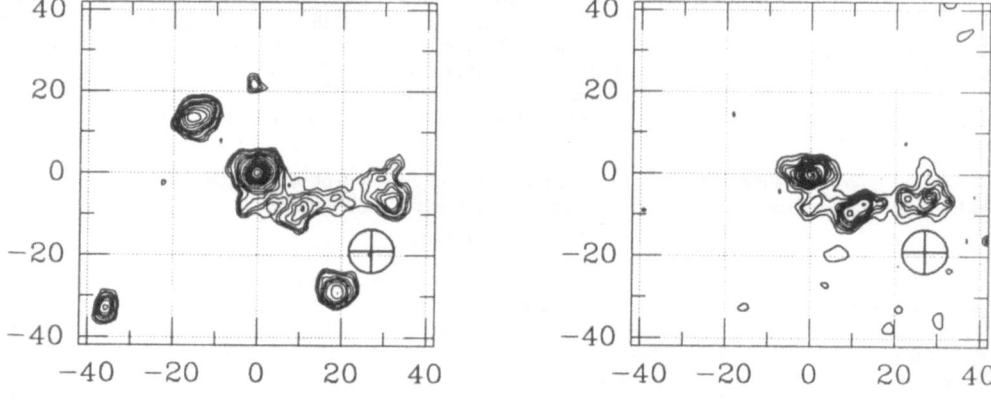

Figure 2. Broad band R and [OII]λ3727 images of 3C 277.2 (z=0.766). This galaxy has a highly extended and distorted morphology in both the emission lines and in the continuum. The emission extends more than 30″ (270kpc, H_0=50, q_0=0.2) from the nucleus. The radio source is highly asymmetric, with the closer of the two hot spots lying adjacent to the emission line gas (marked as a $\oplus$ in the figure). The morphology of this object, both in the lines and continuum is representative of the high redshift ($z > 0.6$) radio galaxies.

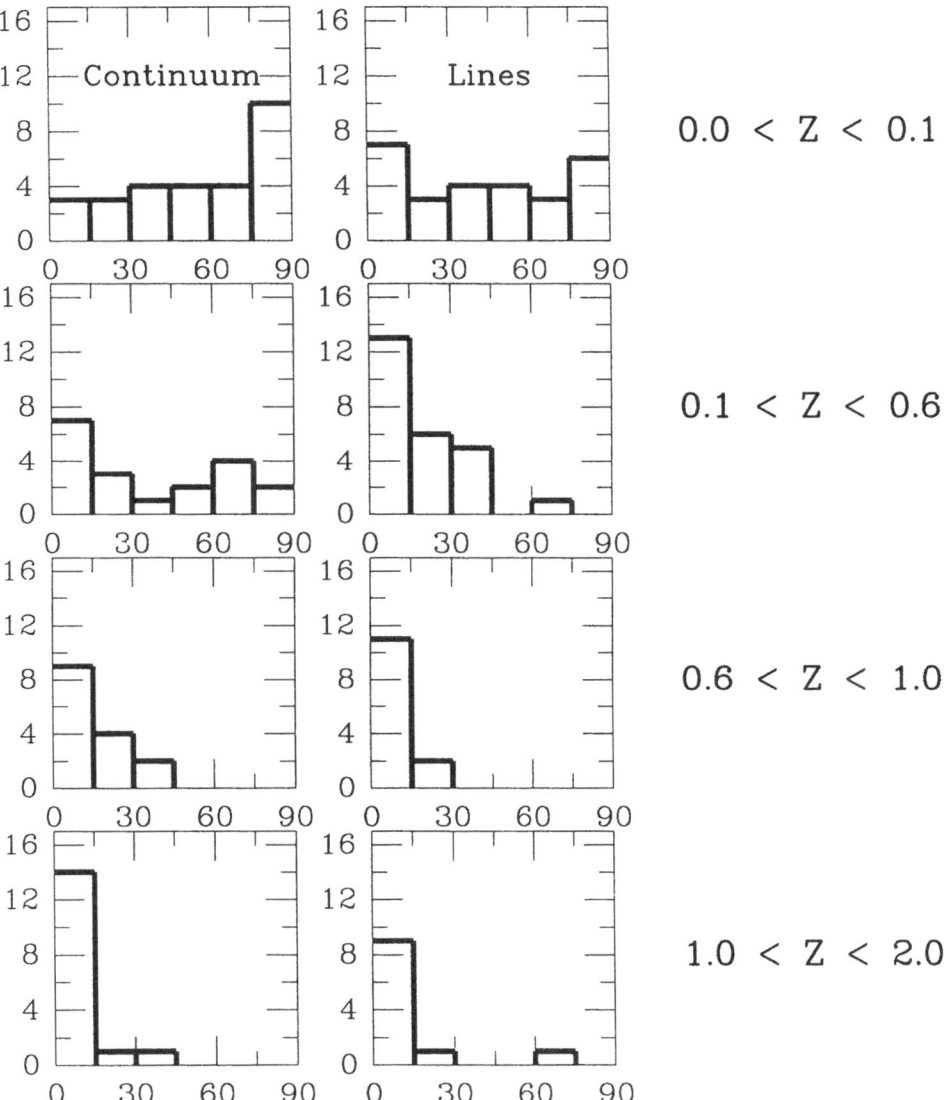

Figure 3. A series of histograms of the alignments of the continua (left) and emission lines (right) with their radio source axes. The number of galaxies per bin is plotted against the position angle difference with respect to the radio source axis. The continua show a slight excess of objects with radio sources oriented along their *minor* axes for $z < 0.1$. For $0.1 < z < 0.6$ the continua show no alignments with the radio sources, but for $0.6 < z < 1.0$ the alignment is quite strong. For $z > 1.0$, virtually all of the galaxies have their continua aligned with their radio source axes. The emission lines show no alignments for $z < 0.1$, but for $z > 0.1$ the emission line regions are well aligned with the radio sources.

of the emission line regions correlates well with luminosity of the extended radio source structures (Baum and Heckman 1988; McCarthy, van Breugal, and Spinrad 1988). The direct correlation between L([OII]) and L(radio) over more than 4 orders of magnitude argues strongly for a causal relation.

The Continuum

The continuum shapes of powerful radio galaxies undergo dramatic evolution for $z > 1$. Locally, powerful radio galaxies are normal giant ellipticals. A substantial fraction of powerful sources with $z < 0.1$, however, show distortions indicative of tidal interactions and mergers (Heckman *et al.* 1987). At larger redshifts, the continuum images become extremely distorted and non-round. High spatial resolution images of $z > 1$ radio galaxies by Le Fevre *et al.* (1988) graphically demonstrate the elongated multi-modal appearance of these objects seen in their *emitted* frame ultraviolet. These elongated continuum structures are aligned with their radio source axes, as shown by McCarthy *et al.*(1987) and Chambers *et al.* (1987). Lilly (1988) shows that those galaxies whose continua are most closely aligned with the radio source axes have the highest starformation rates. This supports the jet-induced star formation hypothesis of McCarthy *et al.* (1987).

Using imaging results from our survey and the work of Baum *et al.* (1988) we have determined the difference between the position angles of the continuum and the radio source for roughly 100 3CR galaxies with redshifts ranging from 0.01 to 2.5. In figure 3 we show a series of histograms of the alignment of the continuum and emission lines for 4 redshift bins. As figure 3 shows, the alignment of the continuum with the radio axis persists until $z \simeq 0.6$. The alignment of the extended emission regions extends to latter times; significant alignment of the EELRs is still seen at $z \sim 0.2$. The longer persistence of the emission line alignment is expected of they are photoionized by a anisotropic continuum source, as proposed by van Breugel and McCarthy (1988).

The Radio Continuum

The sources in this survey are all high luminosity, edge brightened FR II sources. Locally these sources are quite symmetric in both their structures (lobe arm lengths) and in the flux densities of their lobes.

Figure 4.

A series of histograms of the distribution of arm length ratios for 3CR galaxies. The panels cover the redshift ranges:

a) $0.0 < Z < 0.2$,

b) $0.2 < Z < 0.6$,

c) $0.6 < Z < 1.0$,

d) $1.0 < Z < 2.0$.

One source, 3C 299, is not shown in panel b since its arm ratio is > 20. This figure demonstrates the dramatic increase in asymmetry with redshift.

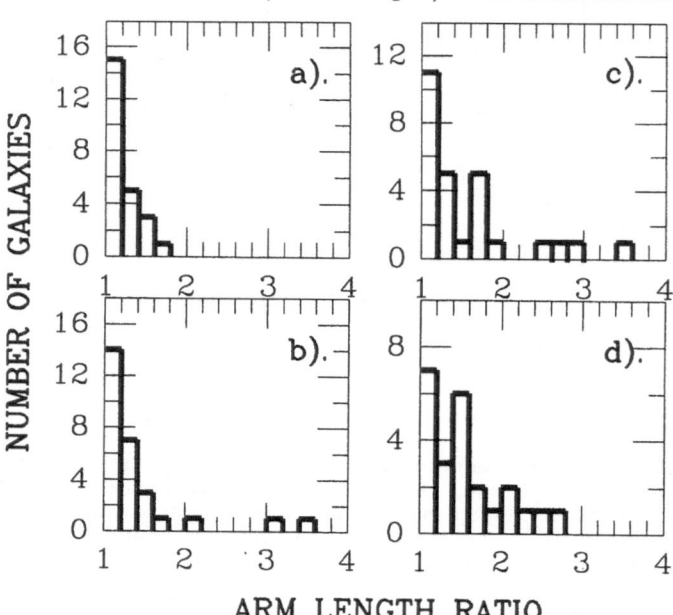

In Figure 5 we show a series of histograms of the arm length ratio, defined as the the distance from the nucleus to the farther of the two lobes divided by that of the closer lobe. As Figure 5 shows, the sources become significantly more asymmetric at large redshifts. Van Breugel and McCarthy (1988) and Pedelty *et al.* (1989) show that the radio sources also become more asymmetric in their lobe flux ratios at large redshifts. Futhermore, Pedelty *et al.* (1989) show that the sources become more asymmetric in their polarization properties with increasing redshift. From our sample of emission line images we find that the asymmetric radio sources more often have extended line emission than symmetric sources. We also find that in *virtually all cases* the brightest regions of line emission lie in the direction of the *closer* of the two radio hot spots. 3C 277.2, shown in Figure 2, is a example of this effect. The near universality of this correlation strongly suggests a causal relation between the radio source asymmetries and the thermal gas. Van Breugel and McCarthy (1988) argue that the arm length asymmetries do not result from orientation and/or light travel time effects, but must be intrinsic to the structure of the source. Deceleration of the hot spots by the thermal gas is the most plausible explanation, but as yet we are not able to rule out non-synchronous ejection from the core.

Conclusions

We have shown that the emission line, stellar continuum, and radio morphologies of high redshift radio galaxies evolve rapidly with cosmic time. These three media were much more closely linked in the past than at the present epoch. The large star forming complexes found along the radio jets, the asymmetries in the radio sources, and the presence of luminous emission line clouds ~ 100 kpc along the radio sources axes are all indicative of a dense, active and rapidly evolving galaxian environment at $z \sim 1 - 2$. These results are more easily accommodated in a high density universe in which galaxy formation, or at least its final stages, occurred fairly late. Lilly (these proceedings) shows that near IR photometry of these objects suggests formation redshifts > 5. Our results show, that while the bulk of the galaxy may have formed at $z \sim 5$, a substantial amount of evolution is still occurring until $z \sim 1$, and this evolution seems to driven by the radio source.

REFERENCES

Baum, S. A. *et al.* 1988, *Ap. J. Suppl.*, *in press.*

Baum, S. A. and Heckman, T. 1988, *Ap. J.*, *in press.*

Chambers, K. C., Miley, G. K., and van Breugel, W. J. M. 1987, *Nature*

Djorgovski, S. *et al.* 1988, *A. J.*, **96**, 836.

Heckman T., *et al.* 1987, *Ap. J.*, **311**, 526.

Le Fevre, O. Hammer, F., and Jones, C. 1988, *Ap. J. Lett.*, **331**, L73.

Lilly, S. J. 1988, *Ap. J.*, *in press.*

McCarthy, P. J., van Breugel, W. J. M., Spinrad, H., and Djorgovski, S. G. 1987, *Ap. J. Letters*, **321**, L29.

McCarthy, P. J., Spinrad H., and van Breugel, W. J. M. 1988. In *IAU Symposium No. 134, Active Galactic Nuclei, D. E. Osterbrock ed.*, Kluwer, *in press.*

Pedelty, J., Rudnick, L., McCarthy, P. J., and Spinrad, H. 1989, *Ap. J.*, *in press.*

Stockton, A. and Lilly, S. J. 1988. In IAU *Symposium No. 134, Active Galactic Nuclei, D. E. Osterbrock ed.*, Kluwer, *in press.*

van Breugel, W. J. M., and McCarthy, P. J. 1988. In IAU *Symposium No. 134, Active Galactic Nuclei, D. E. Osterbrock ed.*, Kluwer, *in press.*

DISCUSSION:

LILLY: A brief comment concerning the optical alignment effects that you mentioned: Within the 3C sample, the optically aligned objects are those with high levels of star-formation activity, as might be expected. The best case of infrared alignment, 3C 368, is the radio galaxy with highest star formation rate, and 40% of the K-band light is expected to come from the aligned young population. I think it is also important to note that rather small aligned components added to a dominant symmetric component will of course produce an overall alignment. For typical 3C galaxies, around 20% of the light at 2 microns may come from the "young" population. Hence some alignment in these systems does not invalidate my conclusion that the 2 micron light is dominated by an old uniform stellar population.

McCARTHY: Your point is well taken, but I'm still quite anxious to see K-band images of the less active 3C galaxies at $z \sim 1$.

HIGH REDSHIFT RADIO GALAXIES AND GALAXY FORMATION

Simon J. Lilly
Institute for Astronomy, University of Hawaii
2680 Woodlawn Drive
Honolulu
Hawaii 96822, U.S.A.

ABSTRACT. On the basis of the observed properties of newly identified faint radio galaxies, including 0902+34 at $z = 3.4$, it is argued that most and possibly all of the powerful radio galaxies seen at $z \leq 3.5$ are massive galaxies ($\sim 10^{12} M_\odot$) that formed well prior to $z \sim 3.5$, at redshifts of at least 5–10.

1. Introduction

The last two years has seen a dramatic improvement in our ability to study stellar systems at redshifts which seemed unimaginably remote only a few years ago. While our knowledge is still patchy in the extreme, we can now begin to constrain our ideas about the formation of structure in the Universe at early times with observational data. While radio galaxies cannot be viewed as representative of the general galaxy population, the relative ease with which very well-defined samples of such objects can be constructed and subsequently studied has meant that a coherent view of these galaxies at high redshift can be obtained. In this paper, I will discuss recent work on the identification of a sample of 1 Jansky empty field sources. These sources remained unidentified ($r \geq 23.5$) after Allington-Smith et al's (1982) Palomar CCD program and constitute about 25% of Allington-Smith's (1982) '1 Jansky' sample, which is selected to have $1 \leq S(408) \leq 2$ Jy. This work has resulted in the discovery of radio galaxies at $z > 3$ and allowed the more complete study of lower redshift systems. On the basis of these data, I believe that we can conclude that the majority of, and possibly all of, the galaxies associated with powerful radio sources are massive systems ($\sim 10^{12} M_\odot$) that formed at $z \geq 5 - 10$.

2. 0902+34 – a massive old stellar system at $z = 3.4$

0902+34 was identified with a faint 24th magnitiude galaxy as part of the systematic study of the 1 Jansky empty fields. Compared with the other new identifications, 0902+34's faint K magnitude and red (J–K) color suggested that it might lie at $z > 2$. The exceptionally blue (V–I) color could be explained by a strong emission line lying within the V-band. The spectrum indeed shows strong Lyman α and weaker [C IV] 1549 at $z = 3.395$ (Lilly 1988). The Lyman α gas is very extended (see e.g. Spinrad's paper at this conference) but has a luminosity similar to that seen in the 3C systems at $z \sim 1.8$.

C. S. Frenk et al. (eds.), The Epoch of Galaxy Formation, 63–69.

Of greater interest is the continuum spectral energy distribution of 0902+34 from rest-frame 1000 A to 5000 A. This is shown in Figure 1 using data from Lilly (1988) and additional B and R photometry. Between 1200 A and 2000 A the continuum is essentially flat. There is a substantial downturn to 1000 A which could be caused by a very thick Lyman α forest. The flat ultraviolet continuum, which is clearly spatially extended, is indicative of a very young stellar population. With a normal IMF, the ultraviolet luminosity translates (Meier 1976) to a star formation rate of order $100 - 300 M_{\odot} yr^{-1}$ ($H_0 = 50 \mathrm{kms^{-1}Mpc^{-1}}, 0 < q_0 < 0.5$). However, the steep rise from 2000 A to 5000 A, indicated by the infrared observations, requires a dominant cool population of stars. This component produces 90% of the light at 5000 A. In terms of conventional population synthesis models (e.g. Bruzual 1983), this can be achieved only by having a dominant population of stars older than 1-2 Gyr. This is shown in Figure 1, where Bruzual's C-model (which has a burst of duration 1 Gyr followed by no further star-formation) has been plotted as a function of age.

Figure 1:

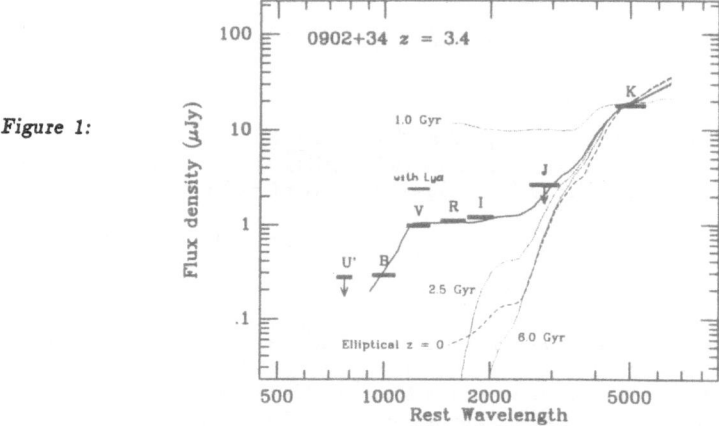

A composite model which fits the observed continuum has been constructed with 3% (by mass) of a 1 Gyr population added to 97% of a 2.5 Gyr population. The mass in the 'young' population could be reduced substantially if the IMF was changed. The mass of the 'old' population must be of order $10^{12} M_{\odot}$. It should be stressed that it is not acceptable, within the framework of these synthesis models, for 0902+34 to have had a uniform star-formation rate over time. Such a model looks similar to the '1.0 Gyr' curve for as long as the star-formation continues. Hence, *however high the star-formation is now, it must have been substantially higher in the past*, and the population of young stars (substantially less than 1 Gyr old) can only represent a very small fraction of the mass of the galaxy. This galaxy is therefore not believed to be a protogalaxy in the process of formation. While it is not possible to produce a unique star-formation history of this galaxy, a conservative statement from the synthesis models is that the mean age of the bulk of the galaxy must be at least 1 Gyr. This may be compared with the Hubble time at $z = 3.4$ which is 4.5 Gyr ($q_0 = 0$) but only 1.4 Gyr ($q_0 = 0.5$) for $H_0 = 50$. Correspondingly, the 'formation redshift' is greater than 4.5 for $q_0 = 0$ and greater than 10 for $q_0 = 0.5$.

Two questions can be raised concerning this conclusion. Firstly, is the K-band light really cool star-light? It has a very steep spectrum (spectral index of around -4) for a 'non-thermal' source and is too hot for thermal dust emission. As regards possible reddening by dust, the flat ($V - I$) and strong Lyman α emission certainly don't support the idea of a very dusty object, although a highly obscured luminous nucleus might be possible. Secondly, can we really trust the population synthesis models? In particular, what happens if the IMF is biassed towards the most massive stars? This may not be as important as might be though since the mass-luminosity relation guarantees that in any IMF flatter than the Salpeter slope, the light is dominated by the most massive stars.

To some degree, both of these questions can be answered most effectively by means of a continuity argument based on the many radio galaxy systems now known at lower redshift, $z \leq 2$. Compared with these objects (see below), 0902+34 has an unexceptional spectral energy distribution and luminosity. In these lower redshift systems, we can be fairly confident that the long wavelength light is indeed dominated by an old stellar population (see section 3.2 below).

3. Complete samples of 3C and 1 Jansky radio galaxies at $1 \leq z \leq 2$

In addition to finding radio galaxies at very high redshift, the multicolor data on the new identifications in the 1 Jansky sample allow for the first time direct comparison of the 3C galaxies with the less radio-luminous 1 Jansky galaxies at $1 \leq z \leq 2$ without the biasses produced by an optical (i.e. rest-frame far-ultraviolet) selection limit. While spectroscopic redshifts for most of the new identifications are not yet available, the VIJK photometry spans a sufficiently long baseline that the redshifts and the amount of star-formation may be usefully estimated. This was done by fitting the two-component model described above for 0902+34 (an 'old' 2.5 Gyr population and a 'young' 1.0 Gyr population in varying proportions) to each of the new identifications. The uncertainties in the fitted redshifts are of course large ($\pm 25\%$) but these estimates are nevertheless useful. The amount of star-formation is parameterized by f_{5000}, the fractional contribution of the 'young' flat-spectrum component to the light at rest-frame 5000 A. Hence 0902+34 (above) has $f_{5000} = 0.1$.

The new identifications bring the Laing et al (1983) 3C and the Allington-Smith (1982) 1 Jansky samples to virtual completeness for $1 < z < 2$ (since the sole remaining 1 Jansky empty field probably has $z > 2$), in a combined sample containing almost 30 galaxies. The following discussion of these two complete samples is based on that in Lilly (1989).

3.1. Star-formation in high z radio galaxies

The values of f_{5000} found in high redshift radio galaxies range from almost zero in quiescent objects such as 3C65 and many of the new 1 Jansky identifications, to about 0.6 in particularly active objects such as 3C368. This wide range is present within both samples, although in the mean the 3C sample is about twice as active as the 1 Jansky sample. Rather surprisingly, the degree of star-formation activity correlates well with the overall radio spectral index measured at 408 MHz, and hardly at all with the radio luminosity. In addition, within the 3C sample alone, (the required data is not available for the 1 Jansky sample as yet) there is also a weaker correlation with the high frequency spectral curvature determined at 5 GHz. These relations are shown in Figure 2.

While the detailed physical explanation for these relations with the radio spectral properties is not yet understood, they further emphasise the close relationship of the star-formation activity to the radio source that has already been indicated by the spatial alignment between the rest-frame ultraviolet emission and the radio source axis seen by McCarthy et al (1987) and Chambers and Miley (1987). In the context of general galaxy formation, there is therefore now clear evidence that the vigorous star-formation seen in many radio galaxies at $z \geq 0.8$ is a phenomenon associated with the interaction of powerful jets with the intergalactic medium, and hence *may not have anything to do with the formation and/or early evolution of 'normal' galaxies*.

It should be noted that, with $f_{5000} = 0.1$, 0902+34 at $z = 3.4$ has a spectral energy distribution that is typical of radio galaxies at $z \geq 1$ and, with a radio spectral index of 0.8, it lies on the correlation shown in the upper panel of Figure 2. Consequently, concerns as to whether the K-band light in 0902+34 is really dominated by an 'old' evolved stellar population can, at some level, be addressed by examining the K-band light in the radio galaxies at $1 \leq z \leq 2$. This is done in the next section.

In passing, it should also be noted that the correlation of star formation activity with low frequency spectral index (above) guarantees that the Miley and Chambers sample of 'ultra-steep-spectrum' radio sources ($\alpha > 1$) will contain preferentially the most active radio galaxies compared with simple flux density selected samples such as those examined here.

Figure 2: (left)

Figure 3: (below)

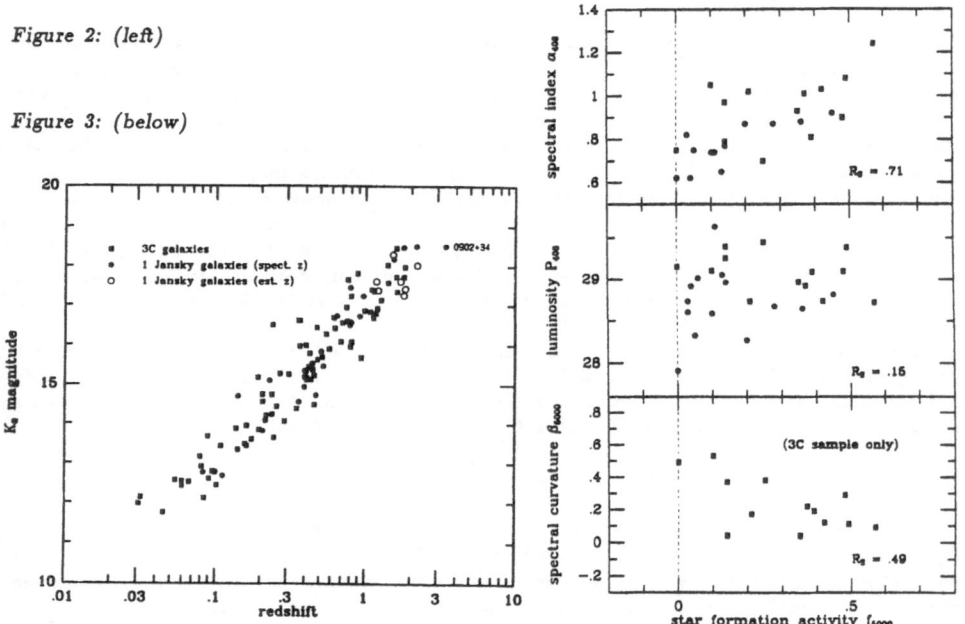

3.2. The infrared K-z Hubble relation

Figure 3 shows the infrared K magnitude – redshift relation for essentially complete samples of 3C and 1 Jansky radio galaxies, omitting only a few 1 Jansky galaxies that do not have a spectroscopic redshift or sufficient photometric data to allow an estimate of the redshift to be made. The data are plotted for a uniform 8 arcsec aperture, but no K-corrections have been applied. The new 'empty field' identifications are represented as open symbols and clearly fit well onto the mean relation despite having the redshift as a completely free parameter in the model spectral energy distribution fits described above. Note in passing that there is very little difference between the 3C and 1 Jansky samples at all redshifts, indicating little evidence for the radio–optical luminosity correlation suggested by Yates et al (1986).

The remarkable thing about Figure 3 is the small dispersion in the Hubble diagram to the highest redshifts sampled. There is no discernable increase in scatter from $z = 0$ to $z = 2$.

The cosmic dispersion in absolute magnitude for radio galaxies at $1 \leq z \leq 2$ may be estimated to be about 0.38 magnitudes for the galaxies with known redshift or 0.41 magnitudes if the estimated redshifts are included. The 3C and 1 Jansky radio galaxies are therefore remarkably uniform at a wavelength corresponding to about 8000 A in the rest-frame, despite a factor of 6 difference in radio luminosity between the two samples and a wide variation in the spectral energy distributions at shorter wavelengths as parameterized by f_{5000} above. This uniformity argues strongly against 'exotic' interpretations of the K-band light, and suggests compellingly that the K-band light is *dominated in the majority of these systems by the photospheric light from a uniform 'old' population of stars*. The small dispersion in absolute magnitude presumably reflects, as at low redshift, some strong mass-selection effect related to the radio activity.

What does the simple two-component model for the spectral energy distribution predict for the composition of the K-band light, and is this consistent with the above conclusion. In the 3C sample, about 20% of the light on average should come from the 'young' flat-spectrum component, with a 1 sigma dispersion of about 10% within the sample. In the 1 Jansky sample, the average contribution is about 10%. These numbers are perfectly consistent with the observed small dispersion in the Hubble diagram and support the conclusion that the K-band light is dominated by the old stars. Note that young stars can account for only a small fraction of the brightening of about 1 mag between redshifts of $0 \rightarrow 1$ that is inferred from the $K - z$ Hubble relation for $q_0 \leq 0.5$ (Lilly and Longair 1984).

The simple two-component stellar population model is also consist with the observed alignment of the infrared image of 3C368 with the radio source axis (Chambers et al 1988) because, in this particularly active galaxy ($f_{5000} = 0.6$), some 40% of the K-band light should be associated with the young, aligned, stellar population. The young stars in more typical radio galaxies will have a much smaller effect and 'weaker' alignments should be found. In testing this prediction, it should however be noted that simple 'alignment' of the infrared and radio axes can be produced by the addition of very small 'aligned' components to an otherwise unaligned object.

The small dispersion in the high redshift Hubble relation can be used to place a limit on the range of galaxian ages within the sample if it is assumed that galaxies get steadily dimmer with time. Using a parameterization of the luminosity evolution of a stellar population that is produced by a normal IMF with $0 < z < 1$ and which is consistent with the radio galaxy infrared Hubble diagram for $0 < z < 1$(Lilly and Longair 1984), the dispersion in the ages of the 'old' component at $z = 1.5$ must be $\Delta \tau / \tau \ll 0.25$. This effectively translates to saying that these systems probably formed at $z > 5$ for $0 \leq q_0 \leq 0.5$. Hence, this constraint on the *relative* ages of the galaxies at $z = 1.5$ is quite consistent with the derived minumum *absolute* age of systems such as 0902+34 at $z = 3.4$.

4. Discussion: The epoch of formation of radio galaxies

It has been argued above that the vigorous star-formation seen in many radio galaxies at $z \geq 1$ is not directly related to 'galaxy formation' and that all the radio galaxies in the two statistically complete samples of 3C and 1 Jansky galaxies have an underlying massive 'old' population of stars that is remarkably uniform from galaxy to galaxy. The uniformity of the absolute magnitudes of radio galaxies at $1 < z < 2$ and, more directly, the redness of 0902+34 between 2500 A and 5000 A in the rest-frame, both suggest that these old components are massive and formed well before $z = 3.5$, most likely at $z \geq 5$ ($q_0 \sim 0$) or even $z \geq 10$ ($q \sim 0.5$). This is the main conclusion from this work.

Are there any exceptions to this conclusion – i.e. are there any radio galaxies that appear to have formed more recently than this? The most interesting candidate is of course the extreme Lyman α object 3C 326.1 at $z \sim 1.8$ (McCarthy et al 1987). This is outside the Laing et al (1983) 3C complete sample and was therefore omitted from the above discussion. At present, an open verdict must be returned on this object. Examination of Figure 1 shows that observations at rest-wavelengths longer than 2500 A are really required to constrain the size of any 'old' population. As far as I am aware, it is quite conceivable that an old population may be present in 3C326.1 at the K = 19 level. A K = 19 component would represent a fainter absolute magnitude than the mean found at $z \sim 1.8$ (see Figure 3), but not worryingly so.

The peak in the AGN luminosity function at $z \sim 2.5$ is evidently not directly associated with galaxy formation at that epoch, since the radio galaxies appear to be already a few billion years old by this time. Thus the peak in comoving density of AGNs is produced by some other phenomenon or is reflecting galaxy formation with a considerable time lag of several billion years. This latter idea ties in with the arguments of Cavaliere and Szalay (1986) and may be physically plausible in terms of the time required to grow a massive black hole.

The relevance of radio galaxies to general galaxy formation is neccessarily limited by their rarity – the density of powerful radio galaxies at $z \sim 2$ is of order 10^{-7} Mpc^{-3}. Furthermore, radio galaxies are clearly amongst the most massive galaxies in the Universe, and it is at least conceivable that there might also be a strong selection in age. Nevertheless the radio galaxies do demonstrate that there are some massive galaxies that are 'well-formed' at $z \sim 3.5$. While the ages mentioned above refer of course to the stars and not to the actual galaxies themselves, the high luminosities and generally ordered appearance of the infrared images suggest that most of the mass of these galaxies has been in place for some time.

For the future, an immediate task is to assess the uniformity of the radio galaxy population at $z = 3 - 4$. With the caveat that they are likely (on the basis of the spectral index correlation) to be particularly active systems, it will be interesting to see whether the new radio galaxies found by Miley and Chambers (private communications) at $z > 3$ are indeed similar in the infrared waveband to 0902+34. At present the uniformity of the K-band light from radio-galaxy to radio-galaxy is the best diagnostic as to what is producing this light — uniformity in the face of widely varying other properties argues strongly in favor of old stellar populations controlled by a mass-selection effect. Unfortunately this test becomes increasingly difficult at $z > 4$ because the 5000 A region of the spectrum will be pushed further into the thermal infrared. Even more challenging will be to determine whether there are a non-radio galaxies that are similarly old at $z \sim 3.5$. Unfortunately these non-radio objects may well not have the vigorous star formation activity in them, and hence may be difficult to detect except in infrared surveys which, at present, are severely limited in areal coverage.

References

Bruzual, G.A.; 1983, Ap.J., 322, 585
Cavaliere, A., and Szalay, A.S.; 1986, Ap.J., 311, 589.
Chambers, K., Miley, G.K. and van Breugel, W.; 1987, Nature, 329, 609
Chambers, K., Miley, G.K., and Joyce, R.R.; 1988, Ap.J.Lett., 329, L79.
Laing, R.A., Riley, J.M., Longair, M.S.; 1983, Mon.Not.R.astr.Soc., 204, 151.
Lilly, S.J., and Longair, M.S.; 1984, Mon.Not.R.astr.Soc., 211, 833.
Lilly, S.J., 1988, Ap.J., in press (Oct 1)
Lilly, S.J., 1989, Ap.J., submitted
McCarthy, P.J., Spinrad, H., Djorgovski, S., Strauss, M.A., van Breugel, W. and Liebert, J.; 1987a, Ap.J.Lett., 319, L9.
McCarthy, P.J., van Breugel, W., Spinrad, H., Djorgovski, S.; 1987b, Ap.J.Lett., 321, L29.
Meier, D.L.; 1976, Ap.J., 207, 343.
Yates, M.G., Miller, L., Peacock, J.A.; 1986, Mon.Not.R.astr.Soc., 221, 311.

DISCUSSION:

REES: Can you be completely sure that the K-luminosity in your highest-z radio galaxy isn't contributed by a short-lived high-mass population dominated by red supergiants? This would then be a special mode of star formation (with unusual IMF) triggered by the radio-activity. This "age" question is so important for cosmogony that it's crucial to exclude all alternative possibilities.

LILLY: This is obviously the crucial question. I can only reiterate the continuity argument based on radio galaxies at $1 < z < 2$. 0902 + 34 has a spectral energy distribution typical of lower redshift systems with $f_{5000} = 0.1$, and lies on the f_{5000} - radio spectral index relation defined by those galaxies. It also has a similar K-band luminosity. In the lower z systems ($1 < z < 2$) the small dispersion of 0.3-0.4 magnitudes in the K-z relation, despite the wide range in radio luminosity and star-formation activity, strongly argues that the K-band light is dominated by a uniform stellar population.

SHANKS: In respect of 0902 + 34, what spatial extent does the angular diameter of 3" correspond to at $z = 3.39$?

LILLY: 3 arcsec corresponds to between 20 and 40 kpc for $0 < q_o < 1$ and $H_o = 50$. The Lyman α emission extends considerably beyond this size.

DEEP SURVEYS OF FIELD AND LOW-FLUX-RADIO GALAXIES

DAVID C. KOO
Lick Observatory, Board of Studies in Astronomy and Astrophysics
University of California, Santa Cruz, CA 95064
U. S. A.

ABSTRACT. We argue that deep galaxy counts are already so high as to favor a low density universe, with perhaps merging as a possible explanation to salvage an $\Omega = 1$ geometry. CCD photometry in the I band suggest many at $B \approx 24$ mag have redshifts $z \gtrsim 1$, while our deep U band photometry imply few have $z \gtrsim 3$. In contrast, spectra and colors of field galaxies to $B \approx 22$ mag indicate increased numbers of galaxies with very active star formation by redshifts of only a few tenths, but with no clear evidence for unusually luminous galaxies. Highlights of our work on the optical identifications, colors, and redshifts of low flux radio sources are also presented to show their relevance to our understanding of the epoch of galaxy formation.

1. Introduction

This presentation will be divided into two sections, one concerned with common field galaxies and the other with galaxies which have been identified from very low radio flux surveys. Since the galaxy formation epoch and process is likely to differ for dissimilar objects, there is danger in assuming that the redshifts and star formation histories of unusual objects are representative of galaxies in general. Traditionally, such peculiar objects as powerful radio galaxies, quasars and N galaxies, central cD cluster galaxies, and perhaps QSO absorption lines, etc., because they were the only objects observable to redshifts z beyond a few tenths, were used as constraints on the history of ordinary galaxies. With the advent of CCDs and multiaperture spectroscopy, however, more common objects are being surveyed to cosmologically interesting limits. The following sections will highlight some of our findings for these less unusual galaxies.

2. Field Galaxy Counts and Colors

For over a decade, a major arena for the study of galaxy evolution and cosmology has been counts and colors of faint galaxies using 4m class telescopes and fine grain emulsion plates (see summaries by Shanks *et al.* 1984 and Ellis 1988). Based only on counts in the blue to depths of $\approx$ 24th mag, any of the following scenarios could be supported to explain why the observed number of galaxies were in excess of predictions by models which included no evolution, i.e. no changes in the

C. S. Frenk et al. (eds.), The Epoch of Galaxy Formation, 71–83.

distribution, luminosity, color, size, dynamics, etc. of galaxies in distant volumes of space from those seen locally today:

- The favored view was that some galaxies were more luminous in the past, observable to greater redshifts on average, and thus appearing in relatively larger numbers as the depth of observations reached cosmologically interesting limits ($B \gtrsim 20$ mag when typical L^* galaxies are beyond $z = 0.1$). Comparisons of such models and data generally favored low density (i.e. open, low omega) cosmologies, but these conclusions were generally accepted as being model dependent.

- Galaxies were simply more numerous in the past with no need to invoke luminosity evolution. Although the required $(1+z)^{1.5}$ increase in comoving volume density is not totally unreasonable (especially if mergers play a substantial role in the life of most galaxies), this idea had much poorer theoretical underpinnings in the late 1970's and early 1980's than e.g. the luminosity evolution models.

- An alternative explanation was that our local observations of the luminosity function of very blue galaxies were so poor, that a low luminosity, low redshift population of such objects could be dominating over more ordinary galaxies in the faint counts. This would be a natural result of such galaxies having small or even negative K corrections due to their blue colors; a steep Euclidian rise in counts towards fainter magnitudes since their redshifts are too small to be dominated by cosmological effects; and the need of observations to reach low surface brightness for detection at all.

- Finally, of course, skeptics could raise arguments and problems based on the likeliness that the counts were affected by superclustering; that the galaxy evolution models were inadequate (always true at some level); that the difference of factors of two among counts of various groups were no larger than the expected differences due to evolution; that errors existed in data analysis or calibrations; etc.

Supplementing the blue photographic counts in a single band with additional bands in the ultraviolet (U), red (R), and near-infrared (I) helped somewhat, largely because intrinsic colors and certain ranges of redshifts could be roughly estimated. For example, models with only simple number increase with lookback time predict far too few blue galaxies; instead, the data suggest an increase in very blue galaxies undergoing active star formation starting at moderate redshifts of only a few tenths (Koo 1986b); the conclusions were otherwise qualitatively similar to, but strengthened, that from counts alone.

The advent of CCDs have had a major impact in the field, not only in allowing genuine spectroscopic redshifts to be acquired for galaxies fainter than $B = 20$ mag, but also in vastly improving the reliability of calibrations and accuracy of photometry to faint limits. Such CCD photometry data of faint field galaxies have led to some interesting results. Using multicolor estimates of redshifts to $z \approx 0.8$, Loh and Spillar (1986), for example, conclude that the universe is closed with little or no luminosity and color evolution. Tyson (1988a) has pushed the CCD to depths of 27th mag and find colors and counts consistent with substantial luminosity evolution in an open universe, thus strengthening the conclusions based on plates alone. Tyson's work also places improved constraints on primeval galaxies, extragalactic optical light, and cosmology in general. Several other deep CCD photometry surveys are underway (see contributions by Cowie and Metcalfe *et al.* in this proceeding).

Our work in this area has largely been an extension of our multicolor surveys, with concentration in improving the depth of our U and I band data and confronting them against our Bruzual galaxy evolution models. In a project that is a collaboration with R. Kron and S. Majewski, U band photometry has been extended from 23rd mag or so using plates to over 25th mag using CCD's. As described by Majewski (1988; this volume) and Cowie (1988), such U data provide powerful constraints on the presence of high redshift galaxies, galaxy formation, and cosmology, since $z > 3$ galaxies would be identified by having a very low U flux as the Lyman continuum break at $912 \mathring{A}$ passes through the band. To $B \approx 24$ mag over a 10 arcmin^2 field, no galaxies were found that were invisible, though several were relatively very faint, on the deep U CCD frames. We thus conclude that few galaxies to $B \approx 24$ have redshifts beyond $z \gtrsim 3$, or that such galaxies have weak or no Lyman continuum breaks.

Are the B $\sim$ 24mag galaxies at high redshift, say $z > 1$? Taking a slightly different approach from that of Cowie (1988; this volume) or Cowie *et al.* (1988), we show how I- band data is also useful in giving us clues on the epoch of galaxy formation. We rely heavily on the galaxy evolution models of Bruzual (1983) and the deep CCD data from Hall and Mackay (1984). Before doing so, let me digress for a moment to clarify the implications of "flat spectra" as defined by L. Cowie or S. White in this conference by direct comparison to Bruzual models of the galaxy spectral energy distributions (SED). Figure 1 shows the spectrum of a galaxy which indeed has a flat SED (i.e. constant in Fν), especially below $3500.\mathring{A}$,

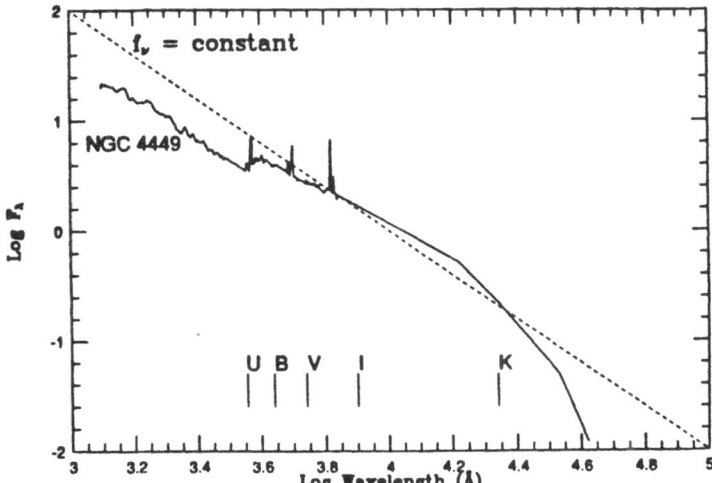

Figure 1. Plot of spectral energy distribution of a nearby Type I Irregular galaxy, NGC4449, versus a "flat" spectrum. Excellent fits to NGC4449, which has B - $V = 0.4$ and U - B = -0.3, can be made using models that undergo constant star formation for up to 2 Gyr (see Figure 2 of Bruzual 1986).

but unlike their claims that such spectra are produced only during very short ($<$ 10^8 years) bursts of star formation, Bruzual achieves excellent fits from the deep ultraviolet to the near infrared K band by using models with star formation that is constant up to 2Gyr for a Saltpeter IMF (see e.g. Fig. 2 of Bruzual 1986). This of course means that even bluer colors than that from "flat" are expected in the deep ultraviolet for younger galaxies or those which have relatively more massive stars. In fact, as emphasized by Meier (1976), three bands such as UBV can be excellent discriminants of high z galaxies undergoing rapid star formation; they would possibly appear with QSO colors (generally regarded as having spectra bluer than flat).

Such very blue colors associated with active star formation (but not necessarily bursts) are expected to be visible for redshifts z $>$ 1, when the restframe ultraviolet enters the optical bands. This can be seen in Figure 2a, which shows the predicted J - I colors versus redshift of a sample of faint galaxies, where J corresponds to standard B, and I is close the the Kron-Cousins I band. In our bandpass system ($UJFN$ - see Koo 1985), flat spectra have U - $J \approx$ -1.22, J - $F \approx$ 0.29, and F - N $\approx$ 0.29, so objects with J - I $<$ 0.58 are bluer than flat. Figure 2a shows that the vast majority of model galaxies that have J - $I \lesssim 1$ (i.e. very blue or close to flat) should have redshifts z $>$ 1. For comparison to the data shown in Figure 2d, Figures 2b and 2c show the different number-color-mag expectations from a no-evolution model versus a mild (high redshift of formation, low Ω) evolution model. These models include photometric errors for the J band (the I band is assumed to have relatively small errors) and incompleteness due to the detection algorithms. If the models are reliable, the presence of substantial numbers of observed galaxies in Figure 2d with J - $I \lesssim 0.6$ strongly support the presence of z $>$ 1 galaxies in our sample for J $>$ 23. Spectroscopy already underway will hopefully be able to reveal some highly redshifted strong emission lines to confirm this conclusion.

3. Is $\Omega = 1$?

One point not already emphasized at this conference is the strong constraint on the cosmological density parameter Ω (assuming $\Lambda = 0$ and $\Omega = 2q_o$) now provided by various deep CCD counts. Such counts serve as a volume test of cosmology, under the assumption that the number of galaxies per comoving volume has largely been conserved. To the extent that luminosity evolution has a relatively small effect on the total number of galaxies after the limiting depth of the survey is fainter than the break of the luminosity function of the most distant galaxies, this test is robust against model assumptions. The key word is limiting depth, for until CCDs broke the 24th mag limit of plates by at least a magnitude, the surveys were still so bright that counts might be expected to continue rising to fainter magnitudes even with a large Ω universe.

What are the expectations and actual observations? Let us start by examining the available volumes. In particular, for one square arcmin, the total available volume to z = 1, z = 3 (U Lyman limit), and z = 4 (B Lyman limit) assuming $\Omega = 1$ (the preferred comology under Cold Dark Matter, of course) is 1200 Mpc^3, 6000 Mpc^3, and 8000 Mpc^3, respectively (Ho = 50 km s^{-1} Mpc^{-1}). This should be contrasted with substantially larger volumes of 2000 Mpc^3, 18600 Mpc^3, and 30800

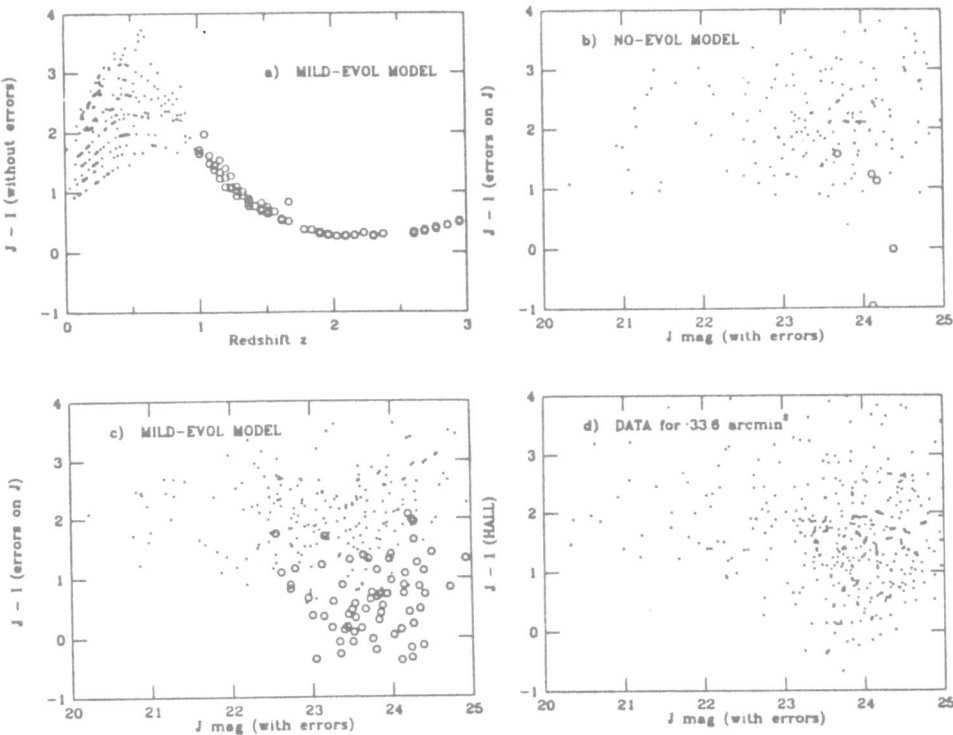

Figure 2. (a) Plot of J - I colors, without any photometric errors, versus redshift for a sample of model galaxies that should be similar to that detected on sky-limited plates taken with a 4-m telescope over a field of $\sim$34 arcmin^2. The models include evolution of the star formation rate with time and are based on those of Bruzual (1983), with the luminosity function parameters adopted from the work of Koo and Szalay (1984). Circles are those with redshifts greater than one. (b) Plot of model J - I colors versus J mag, where J includes photometric errors, the I-band does not, and the whole sample includes detection incompleteness as measured by Koo (1986). The no-evolution model assumes no changes in the rest frame luminosity function of galaxies. (c) Same as for (b) except a mild evolutionary model has been used. (d) Observed J - I colors versus J mag, where the J band was measured from 4-m blue plates and the I-band is from the scanning CCD data of Hall and MacKay (1984). The latter nominally detects objects to $I = 24.4$ at a 4.5 sigma level with standard errors of $\sim$0.2mag at $I = 23$; only objects detected from our plates have been included in the plots.

Mpc^3, respectively, if the universe is open with $\Omega = 0.1$. Assuming an average observable number density of 0.0015 galaxies per Mpc^3 (a conservative estimate, considering that the Anglo Australian Redshift Survey at $B \approx 17$ detected half this, according to Efstathiou *et al.* 1988), about 9 galaxies per arcmin^2 should be observable in the U band and 12 in the B band, if the universe is closed. The raw observed galaxy counts in U and B both peak at 10^5 per mag per deg^2 (see Majewski this volume; Tyson 1988a), corresponding to about 28 per arcmin^2; additional corrections are likely to increase this number.

How can we escape the conclusion that the universe is open? First note that luminosity evolution, the traditional physical effect to explain larger number counts, does not help – e.g., as seen in Fig. 8 of Yoshii and Takahara (1988), the counts in a $q_0 = 0.5$ universe peaks at about 15 galaxies per arcmin^2 regardless of evolution (the differential counts *per mag* are approximately equal to the integral counts at dex about 0.43). Evolution merely pushes the peak to brighter apparent magnitudes; in fact, Yoshii and Takahara claim that a $q_0 = 0.5$ universe is consistent with the data to $B = 24$, but only by proposing that the high observed counts are statistically too high. If they had used the much deeper blue counts of Tyson (1988a) or our U counts, their conclusion would be that $q_0 = 0.5$ (i.e. $\Omega = 1$) models are inconsistent.

Mergers are another possible explanation, of course, but on average, each galaxy today would represent the merging of at least two or more galaxies at redshifts from 1 to 4; as the theoretical models are transformed to observables (see the contribution by White in this conference), we should be in a better position to evaluate if mergers are a viable explanation. Up to now, galaxy evolution models have always presumed number conservation of galaxies, since the strategy was to keep free parameters to a minimum. If more mundane explanations are not found (including the simplest one of a low Ω universe), new (and more complicated) galaxy models which include mergers will need to be developed to explain the observed counts, colors, and redshifts of faint galaxies.

An important source of skepticism regarding the measurement of q_0 from deep galaxy counts is the concern that the numerous faint objects are merely nearby low-luminosity very-blue dwarfs. In other words, the adopted luminosity functions are not accurate at the faint end. Until redshifts are acquired, proof one way or the other would be difficult, but a variety of arguments favor the view that most of the very blue objects at $B > 24$ are indeed at high redshifts, typically at $z \gtrsim 1$. These include colors, observed luminosity functions of local dwarfs, surface brightness and size, and even some redshift data.

Colors: As we had argued before, galaxies with $J - I \lesssim 0.6$ are naturally explained by galaxies undergoing active star formation at redshifts $z \gtrsim 1$. Local dwarfs, even those with extremely blue colors, are not this blue, as can be seen in Figure 1 for NGC4449; this can also be seen in Figure 2a above, in which the bluest low-redshift galaxies have $J - I \gtrsim 0.8$ (corresponds to $B - V \approx 0.1$) and redder such colors up to redshifts of $z \approx 0.5$. Attempts of Tyson and Gullixson (as reported by Tyson 1988b) to measure the colors of local dwarfs indeed yield colors close to $J - I \approx 1$. The trend of bluer colors with fainter magnitudes is also contrary to expectations if we are viewing dwarfs towards greater redshifts.

Luminosity Function: As reported by Sandage *et al.* (1985), the late type dwarfs in the Virgo cluster, which we might imagine are the local counterparts of the presumed dwarfs in the faint counts, actually exhibit a peak and then a drop in their luminosity function beyond $M_V \approx$ -17; in other words, there is no evidence of a continual increase of intrinscially fainter very-blue galaxies (the increase at faint magnitudes is associated with the considerably redder dwarf spheroidals). Searches by Binggeli *et al.* (1985) or Phillipps *et al.* (1987) for low surface brightness dwarfs outside the high density regions in the Virgo or Fornax clusters yield few such objects (it is worth noting the isotropy of the faint blue counts to within 10% observed by Tyson 1988a, which would be unexpected if these were low redshift dwarfs that were as strongly clustered as those in Virgo or Fornax). Moreover, recent evaluations of the luminosity functions by Efstathiou *et al.* (1988) and Phillipps and Shanks (1987) not only show a flatter ($\alpha \approx$ -1) slope for the faint end than the standard value ($\alpha =$ -1.25), but also exhibit error bars that are small to 3 mag fainter than break. Thus any unusual shape to the luminosity function would have to occur fainter than this, corresponding to M_V > -18, or z < 0.1 in a $B = 21$ magnitude limited survey. As discussed in the next section, a large population of such galaxies is not found in faint redshift surveys. Finally, as noted by Tyson (1988a), even the adoption of a Virgo density worth of $M_J =$ -16 galaxies yields a steep (dex = 0.58) count slope from B $\approx$ 22 to 27 (a result that supports the possibility of low luminosity galaxies to dominate faint counts) but such a slope is too steep to match the observed slope of dex = 0.45.

Redshifts: As just argued above, very low luminosity objects of very blue colors should yield a Euclidian slope (i.e. dex = 0.6). If indeed the faintest blue objects are mainly such dwarfs, we can lock their contribution to fractions of such galaxies in completed redshift surveys. For example, if most of the faint galaxies at 25th mag are blue low-luminosity dwarfs, we would expect about 25% to 30% of the galaxies in the Durham redshift survey at $B = 20$ to 21.5 to be such objects; the observed results include 6 out of 170 galaxies with z < 0.1 and none with z < 0.03, which is fully consistent (Broadhurst *et al.* 1988) with extrapolations of the flat faint-end luminosity function as derived by Efstathiou *et al.* (1988). Broadhurst *et al.* (1988) also explored the consequences of adopting much steeper slopes at the faint end of the luminosity function and conclude that results from such modifications are inconsistent with their estimates of the mean redshift. In the models used for my thesis, I actually included a substantial low-luminosity very-blue component (about 3 mag fainter than most galaxies), to force no evolution models to fit the $B \approx 20$ mag counts and colors; such models, however, predict 47 galaxies of our own redshift sample (Koo and Kron 1988a) to have z $\lesssim$ 0.15; actually only 7 were seen and most of these were part of the outskirts of the Coma cluster rather than at random redshifts.

Size and Surface Brightness: At sufficient signal to noise to tell the difference, virtually all objects fainter than $B \approx 22$ remain resolved or extended relative to stars; moreover, this extension continues to fainter limits. Since the sizes of the dwarfs of interest should be $\lesssim 1$ kpc (Binggeli et al 1985, Phillipps *et al.* 1987), most should rapidly shrink to less than the seeing size as they are observed to redshifts of a few hundredths (at the distance of Coma at $z \approx 0.025$, a 25th mag object would have an intrinsic mag of only -11, and a size of 1.4 arcsec per kpc). Since most of the very blue stellar-like *galaxies* found in our QSO survey (Koo and Kron 1988b) to $B \approx 22.5$ mag limits have redshifts of a few tenths rather than a few hundredths (and luminosities of typical galaxies, not dwarfs), this redshift survey constrains the possibility that any such dwarfs are hiding in numerous numbers as compact objects.

4. Field Galaxy Spectra

Although photometric surveys may be very powerful diagnostics of the nature and perhaps distances of faint galaxies, redshifts, perhaps for only a small subsample, are ultimately needed for reliable confirmation. Unfortunately, redshifts are considerably more difficult to measure than broadband colors, unless a large fraction of the flux is being radiated in strong emission lines. To date, several redshift surveys fainter than ~20mag (the depth needed to reach several tenths in redshift and thus a reasonable fraction of the age of the Universe) have been made, including those of the Durham group (see Broadhurst *et al.*1988; poster paper by Colless *et al.* this conference), a group working with the 6m in the USSR (V. Afanasiev, private communication), and ourselves (Koo and Kron 1988a). All three groups are aiming for limits of $B \approx 23$; so far, none of the surveys have achieved high completeness ($\gtrsim 95\%$) to this limit, but the Colless *et al.* survey will presumably reach this goal soon with their new and very impressive low dispersion survey spectrograph system (LDSS).

Here I would like to mention a few of the basic results of at least the Durham and our own surveys. One pleasant surprise is that both groups agree on one important and unexpected result, namely the lack of evidence for the presence of very luminous galaxies at moderately high redshifts of $z = 0.5$ to 1 or greater in the samples. This result is becoming more of a challenge to our standard luminosity evolution models as the redshift surveys reach even fainter limits (Colless *et al.* this conference). We also agree that a substantial fraction of the faint galaxies have [OII] 3727 emission line equivalent widths that are unusually strong when compared to lower redshift samples (Broadhurst *et al.* 1988), but a quantitative estimate of the significance of our confirmation needs to await a more accurate estimate of the areal coverage of our spectroscopic slits. In support of this evidence for relatively more extensive star formation in the recent past, a large fraction of our faint galaxies with redshifts beyond $z = 0.2$ have observed colors corresponding to Sdm or Irr galaxies (see Fig. 2 of Koo and Kron 1988a). One attractive explanation is that many of the blue galaxies are undergoing bursts (Broadhurst *et al.* 1988). Whether this explanation extends to higher redshifts is a crucial problem for observers and theorists alike.

Another finding by both groups mentioned above is that few galaxies of low redshift were actually observed. In our survey, the difference between our original models and the data is probably too large to be explained by statistical fluctuations, even if we grant that substantial large scale structure appear in all three of the fields surveyed. Thus these models will require substantial reduction of low-luminosity blue galaxies. On the other hand, the observed number of red galaxies at redshifts z $\gtrsim$ 0.3 seem to support the evolution models, i.e. the inclusion of the expected luminosity evolution of early type galaxies, but many of the red galaxies are found to lie in only a few high redshift clumps in the same field, and are perhaps strongly affected by statistical fluctuations of the peaks.

5. Low-flux Radio Galaxies

We now turn to galaxies which are active to the extent that they were found from radio surveys, but which may still exhibit optical properties closer to that of normal galaxies than to that from the exotic objects found in the 3CR and 4CR samples. Several such surveys now exist (Windhorst et al. 1985; Fomalont et al. 1984; Condon and Mitchell 1984; Donnelly et al. 1987), but I will concentrate on our fields, since they have the most complete coverage in photometry and redshifts. Our Leiden-Berkeley Deep Survey (LBDS) is characterized by reaching flux limits of ~1mJy at 21cm for an area of ~5.5 deg^2; about 300 radio sources were found in the radio sample. A deeper radio survey reaching well below a mJy was also taken with the VLA in one of these fields (Windhorst et al. 1985), yielding ~100 sources.

These samples are directly relevant to this conference on the epoch of galaxy formation for a variety of reasons:

- The LBDS is about 1000 times fainter than the 3CR sample; thus sources found at redshifts of a few tenths in the 3CR can be detected to equivalent luminosities for sources with redshifts over two. In other words, a direct determination can be made of whether objects of equal radio power at two very different redshifts have similar optical properties, especially the rate of star formation, their size, and relative colors and luminosities.

- This sample may be an efficient and excellent technique to locate distant primeval galaxies, particularly if most radio sources are indeed luminous giant ellipticals. Since such objects are expected to have undergone an intense burst of star formation at early epochs, a sample that reaches the depth of the LBDS has the potential of revealing such a population. In fact, when a sample of blue extended 20th mag objects were found among optical identifications in the early phases of low flux radio surveys, Katgert et al. (1979) proposed that these may be distant ellipticals with active star formation.

- Despite our finding that the blue sources were unlikely counterparts to the early type galaxies found in nearby radio sources, we did discover that the non-blue objects were almost always very luminous and very red. Such an apparently homogenous class of early type galaxies can perhaps serve as tracers of luminosity and color evolution, and thus age.

- As our knowledge of the evolution of optically selected QSO's improve, a key question of whether the radio luminosity function evolves in a similar fashion becomes more important. In other words, what is the relationship between the evolution of different manifestations of active galactic nuclei? As in the case of optical QSOs, reaching faint flux levels is vital towards comparision of samples with equivalent luminosities at different epochs.

- As has been done for a long time, radio surveys can be used to locate very high redshift QSOs. Each redshift record breaker pushes the epoch at which some objects, perhaps unusual, first formed.

- Finally, these low flux radio surveys may provide additional candidates for very extended Lyman alpha emission clouds like those found among much brighter radio sources; reach high enough surface densities to place constraints on clustering at high redshifts; give another high redshift sample useable for tests of biasing theories; and perhaps even yield new populations of objects heretofore undiscovered.

5.1 DATA

To date, we have gathered a variety of data, mostly on the LBDS, which covered nine fields totaling 5.5 deg^2, each previously observed with the 4m prime focus with high quality photographic plates that reach about $B \approx 24$mag . Besides additional radio data (see e.g. Oort *et al.* 1987a, 1987b), most of our recent efforts has been devoted to obtaining spectroscopic redshifts; thus far, nearly 100 redshifts (30% of the entire radio sample) have been measured, largely from Kitt Peak National Observatory and McDonald. To supplement our original *UBVI* plates, we have taken advantage of CCDs to achieve 100 % optical identifications of a subsample of 70 radio sources, using Gunn *gri* bands on the 4Shooter system (four CCD chips) with the 5m at Palomar (Windhorst *et al.* 1987) to reach a limit of $r = 26$.

5.2 HIGHLIGHTS OF RESULTS

- The LBDS and sub-mJy surveys yielded optical counterparts previously unrecognized as being significant in flux limited radio surveys, including some stars, narrow emission line galaxies, spirals, peculiar-interacting systems, and low surface brightness extended objects. The expected populations of early type galaxies, quasars, and broad emission line galaxies were found as well (Kron *et al.*1985; Windhorst *et al.* 1985; Condon and Mitchell 1984; Fomalont *et al.* 1984).

- The 20th mag blue galaxies discussed by Katgert *et al.*(1979) were found to be mainly peculiar galaxies at moderate redshifts of a few tenths rather than high luminosity primeval galaxies at large redshifts (Kron *et al.* 1985, Windhorst *et al.* 1985).

- In contrast to the predictions of Windhorst (1984) that the LBDS would not have radio-resolved sources with redshifts $z \gtrsim 2$ and blue galaxies with $z \gtrsim 1.2$, such objects have subsequently been discovered in our samples (Windhorst 1988). The two most distant blue galaxies have redshifts of 1.34 and 2.39; both have very blue colors suggestive of active star formation.

- Since we have achieved 100% identification in the optical of a subsample of 70 radio sources, and virtually all of the the optical counterparts were galaxies and not point-like QSOs, the proposal of Heisler and Ostriker (1988) for a dust obscured universe is not supported, unless, as they say, *radio sources have an intrinsic redshift cutoff at z less than 3*. In this case, however, the motivation for such a dust theory is weakened, since it was originally proposed to explain the apparent lack of high redshift QSOs.

- Based on the very red colors observed to redshifts beyond $z = 0.8$ for some of the LBDS galaxies, we conclude that their redshift of formation would be very high ($z \gtrsim 5$), especially in a $q_o = 0.5$ universe (Windhorst et al. 1986). This result is fully consistent with that from Hamilton (1985). One unexpected property of the red galaxies is that they are almost always of high luminosity, $M_V \approx$ -22.6 with Ho = 50 km sec^{-1} Mpc^{-1}, even though we could have easily discovered dimmer objects.

- The most exciting development has been the discovery that the faintest optical identifications in the 4Shooter survey were often very low surface brightness objects with diameters often greater than 5" (Windhorst et al. 1987). Such objects may be higher redshift counterparts to the Lyman alpha clouds already found by McCarthy et al.(1987) in the 3CR sample (Spinrad, this volume).

6. Acknowledgements

Virtually all of the work reported above is based upon data and analysis, some of which is unpublished, shared with my collaborators: R. Kron, G. Bruzual, R. Windhorst, H. Spinrad, S. Majewski, J. Munn, and A. Szalay. I would especially like to thank C. MacKay and P. Hall for access to their catalog in SA 57. We are all indebted to the staff of Kitt Peak National Observatory and Palomar Observatory for the many years of support needed to undertake these faint surveys. Finally, I thank the A.A.S. for a travel grant and the Department of Physics at Durham University, United Kingdom, for their hospitality.

7. References

Binggeli, B, Sandage, A. , and Tammann, G. A. 1985, *A. J.*, **90**, 1681.

Broadhurst, T. J., Ellis, R. S., and Shanks, T. 1988, *M.N.R.A.S.*, in press.

Bruzual, A. G. 1983, *Ap. J.*, **273**, 105.

———— 1986, in proceedings of Erice Workshop on *The Spectral Evolution of Galaxies*, ed. C. Chiosi and A. Renzini, p. 263.

Condon, J. J. and Mitchell, K. J. 1984, *A. J.*, **89**, 610.

Cowie, L. L. 1988, *The Post-Recombination Universe*, ed. N. Kaiser and A. Lasenby.

Cowie, L. L., Lilly, S. J., Gardner, J., and McLean, I. S. 1988, *Ap. J. Letters*, submitted.

Donnelly, R. H., Partridge, R. B., and Windhorst, R. A. 1987, *Ap. J.*, **321**, 94.

Efstathiou, G., Ellis, R. S., and Peterson, B. A. 1988, *M.N.R.A.S.*, **232**, 431.

Ellis, R. S. 1988, *High Redshift and Primeval Galaxies*, eds. J. Bergeron, D. Kunth, and B. Rocca-Volmerange

Fomalont, E. B., Kellermann, K. I., Wall, J. V., and Weistrop, D. 1984, *Science*, **225**, 23.

Hall, P. and Mackay, C. D. 1984, *M.N.R.A.S.*, **210**, 979.

Hamilton, D. 1985, *Ap. J.*, **297**, 371.

Heisler, J. and Ostriker, J. 1988, *Ap. J.*, **332**, 543.

Katgert, P., de Ruiter, H. R., van der Laan, H. 1979, *Nature*, **280**, 20.

Koo, D. C. 1985, *A. J.*, **90**, 418.

————1986, *Ap. J.*, **311**, 651.

Koo, D. C. and Kron, R. G. 1988a, *Towards Understanding Galaxies at Large Redshift*, eds. R. G. Kron and A. Renzini, p. 209.

————————1988b, *Ap. J.*, **325**, 92.

Koo, D. C. and Szalay, A. S. 1984, *Ap. J.*, **282**, 390.

Kron, R. G., Koo, D. C., and Windhorst, R. A. 1985, *A. A.*, **146**, 38.

Loh, E. D. and Spillar, E. J. 1986, *Ap. J. Letters*, **307**, L1.

Majewski, S. R. 1988, *Towards Understanding Galaxies at Large Redshift*, eds. R. G. Kron and A. Renzini, p. 203.

McCarthy, P. J., Spinrad, H., Djorgovski, S., Strauss, M. A., van Breugel, W., and Liebert, J. 1987, *Ap. J. Letters*, **319**, L39.

Meier, D. L. 1976, *Ap. J.*, **207**, 303.

Oort, M. J. A., Katgert, P., and Windhorst, R. A. 1987a, *Nature*, **328**, 500.

Oort, M. J. A., Katgert, P., Steeman, F. W. M., and Windhorst, R. A. 1987b, *A. A.*, **179**, p. 41.

Phillipps, S., Disney, M. J., Kibblewhite, E. J., and Cawson, M. G. M. 1987, *M.N.R.A.S.*, **229**, 505.

Phillipps, S. and Shanks, T. 1987, *M.N.R.A.S.*, **227**, 115.

Sandage, A., Binggeli, B., and Tammann, G. A. 1985, *A. J.*, **90**, 1759.

Shanks, T., Stevenson, P. R. F., Fong, R., and MacGillivray, H. T. 1984, *M.N.R.A.S.*, **206**, 767.

Tyson, J. A. 1988a, *Ap. J.*, **96**, 1.

————1988b, *Towards Understanding Galaxies at Large Redshift*, eds. R. G. Kron and A. Renzini, p. 187.

Windhorst, R. A. 1984, *Ph.D. Thesis*, Leiden University.

————————1988, private communication.

Windhorst, R. A., Dressler, A., and Koo, D. C. 1987, in *IAU 124 Symp. 124*, "Observational Cosmology", eds A. Hewitt, G. Burbidge, and L. Z. Fang, p. 383.

Windhorst, R. A., Koo, D. C., and Spinrad, H. 1986, *Galaxy Distances and Deviations from Universal Expansion*, eds. B. F. Madore and R. B. Tully, p. 197.

Windhorst, R. A., Miley, G. K., Owen, F. N., Kron, R. G., Koo, D. C. 1985, *Ap. J.*, **289**, 494.

Yoshii, Y. and Takahara, F. 1988, *Ap. J.*, **326**, 1.

DISCUSSION:

SHANKS: If the local luminosity function turns up at the faint end then surely this could explain the steep number counts even in an $\Omega = 1$ model where galaxy numbers are conserved. The turn up could be hidden locally if it occurred at extremely faint magnitudes where it could be confused by galaxy clustering in fast diminishing volumes.

KOO: You are correct, but this would not be "within the framework of the models", which assume number conservation and a Schechter faint end slope which is quite flat.

McCARTHY: Have you looked at the radio structure of your weak sources? You may be able to use the structural properties to select for high luminosity and hence high-z objects.

KOO: I have not ..... yet.

PEACOCK: How does your $z = 2.39$ object compare with the 3CR galaxies (i.e. is there strong extended star forming activity)?

KOO: Not that we can tell now, but the object is very blue.

GUNN: It seems to me that one cannot say anything about restrictions on numbers from the counts, since all currently fashionable (and observed) luminosity functions are still increasing at the faint end. The Durham group has shown nicely that the counts can be explained well without disturbing the redshift distribution by a small amount of differential evolution at the faint end.

KOO: I qualified the conclusion with the caveat that our assumptions are correct, mainly the conservation of number density and adoption of the standard Schechter luminosity function. "Cannot say anything" is a bit strong, since we do have some limits from the local group on volume densities of tiny systems. Of course, the local group can be argued as being unrepresentative. Also the work of Sandage, Binggeli and Tammann in the Virgo cluster showed that the star forming galaxies, including Im types, had luminosity functions that were more Gaussian than rising at faint luminosities. The dE class may indeed be rising, but they are unlikely to be as blue as those observed among the faint counts. Finally, a recent paper by Phillipps *et al* in MNRAS showed that dwarfs have approximately constant surface brightness of ~ 25 mags arcsec $^{-2}$; thus if the faint objects are such objects, they would have to be essentially stellar in size to be at 26th mag - they are not.

WOLFE: How common are the faint (mJy) blobs compared to the Lyα galaxies found by Spinrad.

KOO: The 3C sample of ~ 170 objects may have a couple such extended Lα objects (i.e. $= 1\%$) whereas our own subsample of 70 radio sources that are completely identified to ~ 26 mag (Windborst, Dressler and Koo 1987) show perhaps a half dozen or 10%.

ROCCA-VOLMERANGE: I agree with the fact that most observational data fitted with evolutionary synthesis models support the idea of high redshifts of formation $z_f > 5$ and low values of q_0; in particular, the recent observations of Dunlop and Longair, 1987, see also our contributions to this conference. However, these analyses are roughly insensitive to H_o.

KOO: If the universe has $q_o = 0.5$ and $H_o = 100$, however, it is difficult to get the reddest galaxies today with the age of the universe only 6.5 Gyr old, even with high metallicity. This problem is only aggravated by the apparent redness of galaxies found at $z \sim 0.8$ by Hamilton (1985) and ourselves (this talk).

CONSTRAINTS ON THE EPOCH OF GALAXY FORMATION FROM DEEP U-BAND COUNTS

S. R. Majewski [1]
Yerkes Observatory
University of Chicago
Williams Bay, Wisconsin 53191-0258
U. S. A.
&
A. T. & T. Bell Laboratories
Murray Hill, New Jersey 07974
U. S. A.

ABSTRACT. Two new galaxy count surveys in the ultraviolet probe the count-magnitude relation to U $\gtrsim$ 25.5. The most straightforward interpretations of the results are consistent with mild evolution and $z_f > 4.5$. To the depth of the surveys, no evidence is seen for the existence of a significant population of $z > 3.5$ galaxies.

1. Introduction

The ultraviolet holds many keys to understanding galaxy evolution since young stellar populations contain massive short-lived stars which emit the most of their energy at these wavelengths. In addition, since the range in rest frame UV color for galaxies is much larger than in the optical, observations that sample short wavelengths offer greater potential sensitivity to galaxy evolution (see, for example, the recent results by Donas *et al.* 1987).

Without observationally confirmed redshifts, deep galaxy count surveys are difficult to interpret. However, the form of the observed count-magnitude relation offers tantalizing clues to the history of galaxy formation and evolution. The slope of the counts as a function of observed wavelength and changes in the slope both depend heavily on the history of star formation, the redshift distribution of the galaxies, and the epoch of galaxy formation. In these respects, deep optical surveys longward of 4000 Å to date have yielded two intriguing results: 1) an increasing slope with decreasing wavelength, and 2) no apparent features in the counts except possibly in the blue where Tyson's (1988) deep counts seem to show a bend at B ~ 25. The former reflects the fact that faint field galaxies are bluer in color and it is of interest to test whether this trend continues into the ultraviolet, which is the most sensitive passband to galaxy evolution. The bend in the B counts may be due to the presence of significant numbers of high redshift (z >3) galaxies with dark Lyman continua redshifted into the B passband. If so, the same should occur for the U

---

1 Visiting Astronomer, Kitt Peak National Observatory, National Optical Astronomy Observatories, operated by the association of Universities for Research in Astronomy, Inc., under contract with the National Science Foundation.

C. S. Frenk et al. (eds.), The Epoch of Galaxy Formation, 85–88.

passband, and at brighter magnitudes since lower redshift galaxies would contribute to the effect.

2. Deep U-Band Surveys

Unfortunately ground-based observations in the U-band have been limited because of the difficulty in working with the narrow bandpass (~600 Å, defined on the short end by the atmospheric cutoff at 3300 Å) and relatively poor ultraviolet detector quantum efficiencies. The deepest U-band survey to date (Table 1) has been the photographic work of Koo (1981) which yielded a steep log count-magnitude slope (dlogA/dm = 0.68± 0.02) for $20 \le U \le 22$, significantly greater than that for the B-band. Koo showed that faint galaxies in general are not only bluer but more ultraviolet in color. In addition, he found that the color-magnitude effect was greatest in U-B.

To study the U-band count and color-magnitude relations to still fainter magnitudes, two new surveys have been undertaken in collaboration with D. Koo and R. Kron (Table 1). The first survey (Majewski, Koo, and Kron 1988; hereafter MKK) supplements the data of Koo (1981) with the addition of several more Mayall 4-m photographic plates, digitally coadded to increase depth. The second survey (Koo, Kron, and Majewski 1989; hereafter KKM) consists of deep CCD imaging to U $\gtrsim$ 25.5 in two smaller areas (~10 arcmin^2 each) with the Mayall 4-m.

survey	image medium	limit (peak counts)	fields	area (arcmin^2)
Koo (1981)	Mayall 4-m photographic plates	U ~ 22.8	SA 57 (NGP) SA 68	1021 769
Majewski, Koo, and Kron (1988)	Five digitally coadded Mayall 4-m photographic plates	U ~ 24.7	SA 57 (NGP)	80
Koo, Kron, and Majewski (1989)	RCA-CCD + Mayall 4-m	U ~ 25.7	SA 57 (NGP) Hercules	10 10

Table 1. U-BAND SURVEYS

The counts from these surveys are shown in Figure 1. Several qualitative features of the data are worth noting. First, the data are consistent. This is encouraging given that the Koo (1981) data employed different photometric algorithms than the other surveys (which were reduced using the FOCAS package at AT&T) and the KKM survey used a different detector. (The offset in logA between the surveys in the magnitudes of overlap is consistent with expectation from different sample areas.)

The new data confirm to fainter magnitudes the steep ultraviolet galaxy-count slope found by Koo for $20 \le U \le 22$. A line of slope dlogA/dm = 0.6, which corresponds to a homogeneous, isotropic, nonexpanding Euclidean universe, has been included in Figure 1 as a guide. The U-band counts rise more steeply than this out to U ~ 24. Since typical values of the blue count-magnitude slope hover around 0.50 ± 0.05 to at least B = 25 (cf. Tyson 1988), it is clear that the U-B bluing trend seen by Koo to his survey limit continues for several more magnitudes.

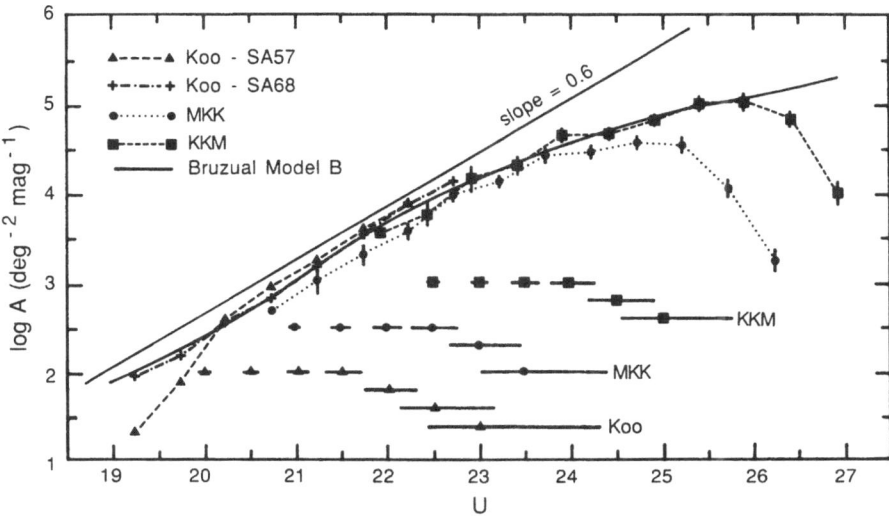

Figure 1. Count-magnitude diagram for deep U-band surveys, and Bruzual's (1981) model B. Both the MKK and KKM data are for the SA57 field. The one-sigma photometric errors as a function of magnitude for the various surveys are given at the bottom of the diagram; the larger error bars have been offset for clarity.

An additional qualitative feature in the count-magnitude data is a possible bend at $U \sim 24.5$, seen in both the MKK and KKM data. Since the effects of incompleteness and Malmquist bias have not been included in the data presented here, it may be premature to speculate on the reality of this feature. On the other hand, it is unlikely that incompleteness would be significant in the KKM survey so far from its apparent limit.

3. Simple Tests

The data as presented lend themselves to a variety of simple tests which constrain models of galaxy formation and evolution:

3.1 MILD EVOLUTION, HIGH z_f

Figure 1 includes Bruzual's (1981) model B ($H_0 = 50$ km s^{-1} Mpc^{-1} $q_0 = 0$, $t_g = 16$ Gyr, exponentially-declining star formation rate). The similarity to the data, including the apparent bend at $U \sim 24.5$ is quite good. Model B corresponds to a formation epoch of $z_f = 4.7$. Bruzual's model A (same as B, but with $t_g = 9$ Gyr), which makes a very similar U-count prediction, corresponds to $z_f \gg 10$. Within the limits of the Bruzual models, the ultraviolet data are consistent with both mild evolution and a relatively high epoch of galaxy formation.

3.2 LOW z_f

The lack of a bend in the U-band counts for U < 24 is an important constraint on cosmological models favoring low redshifts of galaxy formation. If one assumes that galaxies obey a Schechter luminosity function at all redshifts, and one assumes $U^* \sim B^* \sim -21.5$ ($H_0 = 50$ km s^{-1} Mpc^{-1}), which is reasonable for flat spectra typical of late spirals (U-B ~ 0), then galaxies forming at z = 1 in a $q_0 = 0.5$ universe would have a distance modulus of 44.0 and would require a bend as bright as U = 22.5. One might expect some evolution of U^* with redshift. A dimming of U^* with age drives the required break to still brighter magnitudes. (The situation is more complicated for a U^* evolution in the opposite direction.) Clearly the U-band data are *inconsistent* with at least one or more of these assumptions: Schechter function (having a constant or brightening U^* with z), $z_f \sim 1$, or $\Omega = 1$.

3.3 LACK OF z > 3.5 PROTOSPHEROIDS AND THE VALUE OF q_0

Assuming galaxies have sharp Lyman limits, one could search for high redshift (z > 3.5) galaxies by looking for objects with normal optical colors but extremely red U-B colors due to the shifting of the 912 Å limit past the 4000 Å edge of the U waveband (Cowie 1988, Majewski 1988). Cowie (1988) has suggested that at least half of the faint optically blue galaxies in a $q_0 = 0.5$ universe must be z > 3.5 protospheroids in order to explain the fact that the blue counts seem to require significant numbers of z > 0.8 galaxies. Moreover, he determines that the depletion of gas due to star formation, as deduced from the damped Lyman α systems, is too small throughout the observed range $1.7 \leq z \leq 3.2$ to account for the blue excess. We have obtained deep blue data in our SA57 U-band survey area and we find no object to B = 24.5 which is *not also* present on our U images (which are limited to U $\sim$ 26). Thus our deep U-band imaging to date has not revealed the presence of any z > 3.5 population. If a population of z > 3.5 galaxies does exist, they must be fainter than our current survey limits. The easiest resolution of this discrepency would be a lower value of q_0, since this would dim galaxies by increasing luminosity distances and lengthening the mapping of timescales to redshift.

I would like to thank Tony Tyson and AT&T Bell Labs for support during this work. I am greatly indebted to Dave Koo and Rich Kron for assistance with this report. This research was partially supported by NSF grant AST-8705517.

References

Bruzual, G. 1981, PhD. Thesis.
Cowie, L. L. 1988, in *The Post-Recombination Universe* , eds. N. Kaiser and A. Lasenby, in press.
Donas, J., Deharveng, J. M., Laget, M., Milliard, B., and Huguenin, D., 1987, Astron. Astrophys., **180**, 12.
Koo, D. C. 1981, PhD. Thesis.
Koo, D. C., Kron, R. G., and Majewski, S. R. 1989, in preparation.
Majewski, S. R. 1988, in *Towards Understanding Galaxies at High Redshift*, eds. R. G. Kron and A. Renzini, Kluwer Academic Publishers, p. 203.
Majewski, S. R., Koo, D. C., and Kron, R. G. 1988, in preparation .
Tyson, J. A. 1988, A. J., **96**, 1.

LYMAN LINE ABSORPTION SYSTEMS

R. F. Carswell
Institute of Astronomy
Madingley Road
Cambridge CB3 0HA
England

ABSTRACT.
    High resolution quasar spectra can provide a wealth of detailed information on the Lyα absorbing systems seen at high redshifts. The distribution of the numbers of these systems with redshift, Doppler width, and HI column density, and any relationship between them, derived from such data is discussed. The inferred properties of the clouds derived from such data depend on size estimates, which rely on a single key observation of a gravitationally lensed quasar. The data available provide some constraints for various cloud models, but do not yet allow any to be clearly ruled out.

1. Introduction

    The spectra of high redshift quasars are usually rich in absorption lines, most of which are due to HI absorption in systems where lines from heavier elements are not seen. The natural explanation for the absence of the heavy element lines is that the corresponding elements are significantly underabundant, and so significant star formation has not occurred. Thus one of the hopes that one has in studying these systems is that we are observing the remnants of the process of galaxy formation. Perhaps these are the signatures of either those gas clouds which were too small to form galaxies or perhaps the dwarf galaxy end of the distribution function. From these one then hopes to be able to contribute to the overall picture of galaxy formation. However, despite the significant improvement in the observational material over the past few years, and active theoretical interest, we can still not differentiate between possible models for their nature. Here, I shall summarize our present understanding of the observational material, and the inferences we can draw from it using theoretical considerations.

2. Measurable Quantities

    A brief examination of the Lyα forest region of any high redshift quasar reveals the reason the term 'forest' has come into general use. The line density is so high that the spectrum is almost completely filled with them at wavelengths shortward of the Lyα emission line. This highlights the first problem which has to be addressed in studying these systems – where is the continuum level which we must use to determine any of the absorption line parameters? At the highest redshifts this is not a trivial question, and the best you can do is to somehow

C. S. Frenk et al. (eds.), The Epoch of Galaxy Formation, 89–99.

connect regions where the noise level above the continuum corresponds to that expected from the photon counts obtained, after allowing for instrumental effects. This point is best illustrated by a section of the spectrum of the redshift $z = 4.11$ quasar $0000 - 26$ (Webb *et al.*, 1988) (Fig 1.). At lower redshifts the problem is less severe, but it is worth bearing in mind for the subsequent discussion that continuum uncertainties can affect the parameters measured for the weaker (and shallower) lines.

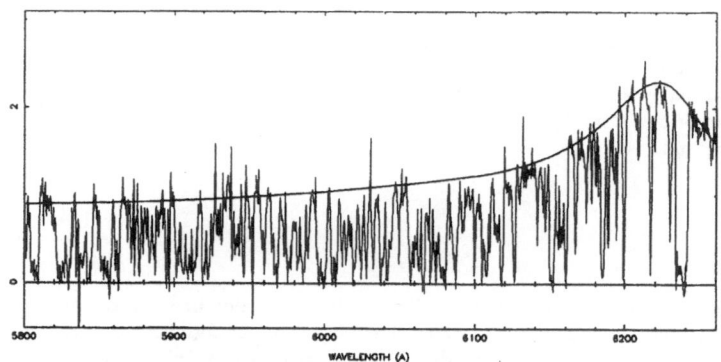

Figure 1: *Part of the spectrum of Q0000-26 obtained by Webb et al. (1988) using CASPEC at the ESO 3.6-m telescope, with an estimate of the continuum.*

Having established the continuum, we may then measure the wavelengths, equivalent widths and, where the spectral resolution is sufficiently high, profiles for the absorption lines in the Lyα forest. Some of these will be due to heavy elements in a relatively small number of red-shift systems, and those identified as such are obviously excluded from any analysis aimed at establishing the properties of the Lyα systems. The Lyα lines from the heavy element systems may come from a separate population (Sargent *et al.*, 1980, but see Tytler, 1987), so these too are excluded from further consideration unless otherwise indicated.

If the spectral resolution is $R \gtrsim 10000$ then there is little line profile information so the best one can hope to do is to determine the redshifts and equivalent widths of the Lyα and higher order Lyman lines, and determine their distribution functions. As shown first by Peterson (1978), the numbers of Lyα lines is a strong function of the redshift. Recent studies indicate that the redshift evolution for the numbers of lines per unit redshift, $\frac{d\Psi}{dz}$, for Lyα lines with rest equivalent widths > 0.32 Å can be approximated by

$$\frac{d\Psi}{dz} \propto (1 + z)^\gamma, \tag{1}$$

where $\gamma = 2.30 \pm 0.42$ for redshifts $1.5 \lesssim z \lesssim 3.8$ (Hunstead *et al.*, 1988). This should be contrasted with the form expected if there is no evolution:

$$\frac{d\Psi}{dz} \propto \frac{(1 + z)}{(1 + 2q_0 z)^{\frac{1}{2}}}. \tag{2}$$

Such estimates require spectra of a large number of quasars covering a range of redshifts, and so the data have been obtained at only intermediate dispersion. It is possible under these circumstances that undetected line blending will affect the results, but it is not clear if blending would make the apparent line density at high redshifts higher or lower. Weak lines below the equivalent width limit will blend to form features above the limit, while strong lines above the limit may merge to be counted as single features instead of a few lines. Simulations suggest that, luckily, the two effects seem to roughly cancel for the equivalent width limits used in the studies to date (Parnell and Carswell, 1988; see also Liu and Jones, 1988), but other limiting values may bias the γ estimator (Webb, in preparation).

The same data also serve to provide estimates for the Lyα equivalent width distribution. This has been studied by Sargent *et al.* (1980) and more extensively by Murdoch *et al* (1986), who find that the data is adequately described by an exponential distribution with an excess at low equivalent widths. It is not clear how blending affects the results in this case, but Murdoch *et al.* indicate that any corrections are not large.

More detailed information is coming from higher resolution studies ($R \gtrsim 10000$) where the Lyman line profiles are resolved. Obtaining the data requires a large commitment of time on large telescopes, so it is coming rather slowly. Determining the Doppler parameters and HI column densities by fitting Voigt profiles convolved with the instrumental profile to the data, which often involves fitting multiple components to complex blended features, is also very time consuming and so publication is even slower. Some results are available from work by Carswell *et al.* (1984, 1987) and Atwood *et al.* (1985), and unpublished material by Webb (1986, 1987). The earlier published data involved searching parameter space for minimum χ^2 solutions for each component of any blend, while the two later ones come from unconstrained minimization of the value of χ^2 for the whole blend. Both techniques use all the available Lyman lines in the fit, and both provide error estimates for the parameters derived. The minimum χ^2 is used as a guide to goodness of fit, and the global distribution of this quantity serves as a check that we have not fitted more components than are required by the data, but of course that does not necessarily mean that we have found all components to the lines.

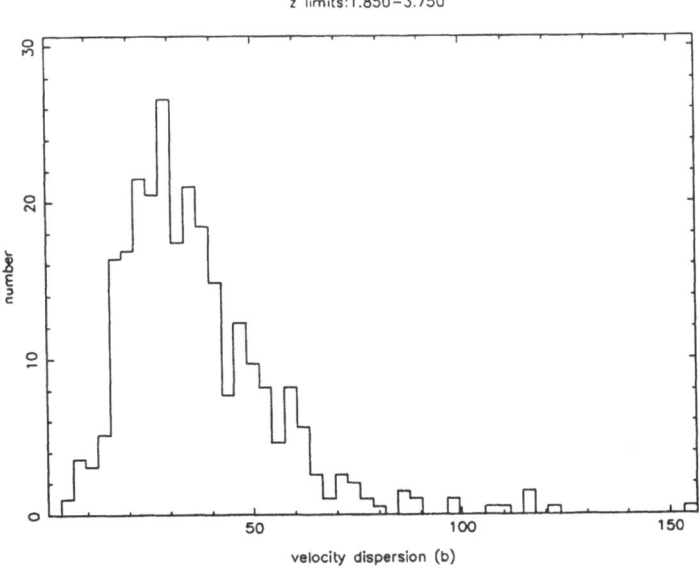

Figure 2: *The distribution of Doppler parameter b (km s^{-1}) for 487 Lyα lines selected so that the error estimate $\sigma_b < b$.*

The Doppler parameter b ($= \sqrt{2}\sigma$) shows a broad peak centered at about 30 - 35 km s^{-1} (fig 2), and the mean value shows no strong redshift dependence. The values range from close to zero to about 100 km s^{-1}, but it is difficult to say how much of the spread is intrinsic and how much is due to measurement error. We have done some trials with simulated data, and it appears that there is at least some intrinsic spread in the Doppler parameter corresponding to a first moment about the mean of about 7 - 10 km s^{-1}. Higher signal - to - noise and higher resolution spectra are needed to examine this point properly.

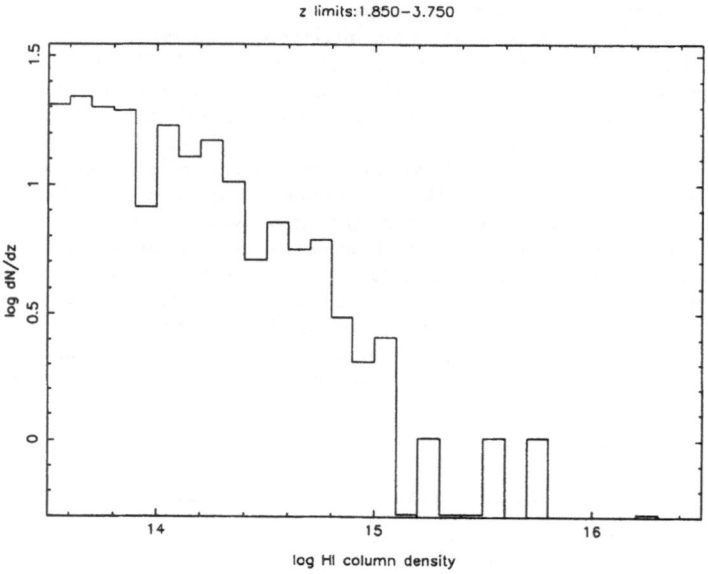

Figure 3: *The distribution in HI column density (cm^{-2}) for 355 Lyα systems for which the error estimate $\sigma_{logN} < 1.0$.*

The HI column density distribution (fig. 3) is well approximated by a power law with the number of systems N given by

$$dN \propto N^{-\beta} dN, \tag{3}$$

where N is the HI column density and $\beta \sim 1.75$. This applies roughly over the range $13 \lesssim \log N \lesssim 15$, for N in cm^{-2}; for higher values too few systems are found, and for lower values they are below the detection limit for the available data. Of course, higher HI column density systems are known (*e.g* Wolfe *et al.*, 1986), but where there have been detailed investigations of these heavy element lines are present and so they have been excluded from consideration.

The slope of the power law is not strongly dependent on the redshift, but there has been a suggestion that at high redshifts the distribution flattens somewhat at low column densities (Carswell *et al.*, 1987). Any departure from a power law is of crucial importance in testing models of the Lyα cloud evolution, since it is only by following such departures as a function of redshift that we can provide useful constraints. However, the suggested change of slope is not very significant, and depends on some uncertain corrections to allow for the loss of weak lines

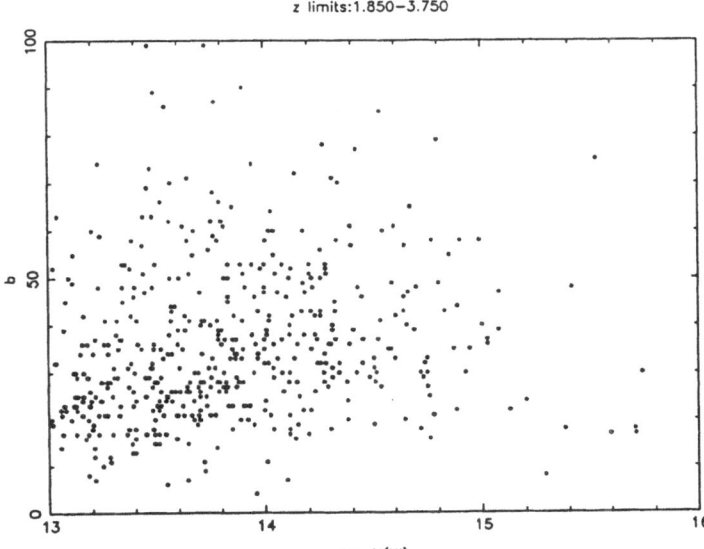

Figure 4: *The Doppler parameter (b km s⁻¹) vs HI column density distribution for a subsample of the Lyα systems for which the errors $\sigma_b < b$ and $\sigma_{\log N} < 1.5$.*

in stronger ones, so further data are required for this point also.

There may be a weak correlation between Doppler parameter and HI column density. The data we have is shown in figure 4, where we have chosen to restrict the sample so that the error estimates $\sigma_b < b$ and $\sigma_{\log N} < 1.5$ so that any possible correlation is not lost in the noise. This may be compared and contrasted with a similar figure given by Hunstead *et al.* (1987), who show a stronger correlation. There are two possible causes for a spurious correlation. Noisy features will tend to be resolved into too few components under any scheme where the the S/N limits the analysis, as it must unless one has a model description for the real behaviour. These will have higher Doppler parameters, and higher total column densities, than the individual components, so can lead to an apparent excess of high b, high N(HI) systems. Similarly one would find an excess number of low b, low N(HI) systems if there is a tendency to try to fit too many components to features through underestimating the noise level.

There has also been recent interest in the degree of clustering of the Lyα systems, since work by Sargent *et al.* (1980) showed little evidence for clustering on scales of 300 km s⁻¹ or more as would be expected if they were associated with galaxies in clusters. An analysis of the higher resolution data by Webb (1987) reveals that there is significant clustering on scales of about 150 km s⁻¹ at lower redshifts. Figure 5 shows the two - point correlation functions for two datasets, one for redshifts $1.9 \lesssim z \lesssim 2.7$ with resolution ~ 20 km s⁻¹ (Webb, 1986), and a higher redshift sample from Cerro Tololo with resolution ~ 30 km s⁻¹ (Atwood *et al.*, 1985; Carswell *et al.*, 1987). The difference between the two samples is interesting, in that it suggests that clustering in the Lyα systems increases as the redshift decreases. Since the two data subsets were obtained at differing resolutions, it is worth considering whether or not this could be an instrumental effect, but this is unlikely since in both cases the line profiles at the (same) average value of $b \sim 35$ km s⁻¹ are adequately sampled. Further evidence that the Lyα

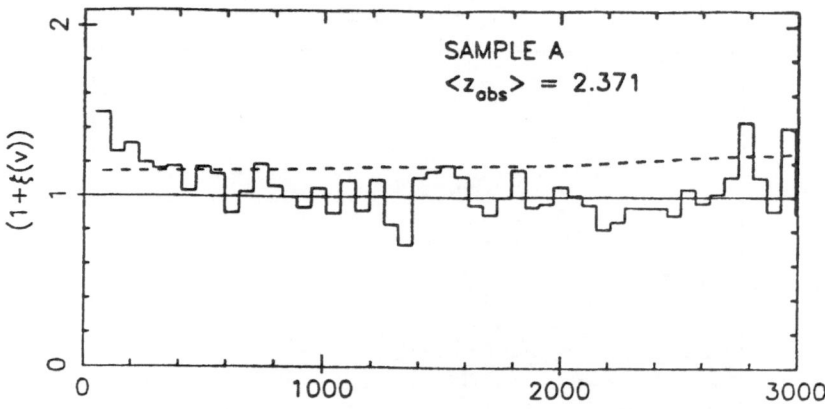

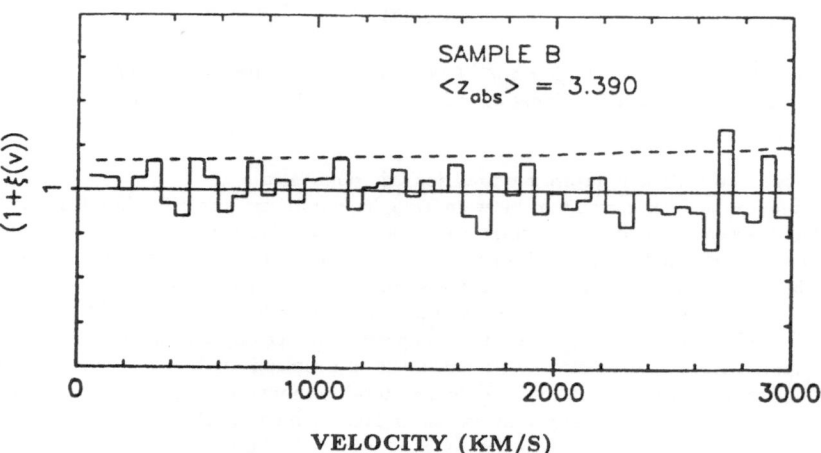

Figure 5: *The two - point correlation functions for Lyman line systems at intermediate and high redshifts (from Webb, 1986).*

systems are clustered on scales of ~ 100 km s^{-1} has come from studies over the Lyman lines in close pairs of quasars (Crotts, 1988). From all these studies it is clear that the Lyman line systems are less strongly clustered than clusters galaxies today. However, depending on how galaxy clustering evolves, it is possible that the level is comparable to galaxy - galaxy clustering at redshifts $z \sim 2$ (Shaver, 1988).

While the description 'Lyman systems' carries with it the suggestion that these are different from the systems in which heavy element lines are found, this is not firmly established. Tytler

(1987) found that the HI column density distribution for systems which show heavy elements follows the same power law which has been found for the Lyman line systems, and so suggested that both types of system are from a single population. However, subsequent work by Bechtold (1987) shows a deficit of systems with $16 \lesssim \log N(\mathrm{HI}) \lesssim 18$ relative to the power law, so the true situation is not clear. It is worth noting that if Tytler's description is correct, then it would suggest that the Lyα clouds are mainly neutral, since otherwise there would be a change in character at $\log N \sim 17.5$ where the clouds become optically thick to any ionizing radiation.

A further suggestion that the Lyα systems may not be very different from the heavy element systems comes from the work of Meyer and York (1987), who found a number of weak heavy element lines corresponding to what had previously been regarded as Lyα systems. Again there is some uncertainty as to how to interpret this result. It may still be that most of the systems do not show heavy element lines, and they have detected lines from the lower column density end of the heavy element distribution. From stacking several Lyman systems in an attempt to detect weak heavy element lines, Williger et al. (1988) infer that at most about 20% of these systems can have heavy element lines as strong as those found by Meyer and York.

3. Inferred Properties

If we are to determine the conditions in the absorbing clouds we need to know more than the HI column density, Doppler parameter and redshift. The most critical unknown quantity is the typical matter density in the clouds, and here we have to rely on a single key observation by Foltz et al. (1984). They obtained spectra of two of the images of the gravitationally lensed quasar 2345+007, and found that the two lines of sight contained a number of common Lyα absorption lines with rest equivalent widths above their limiting value of ~ 1Å. From this they inferred that the lower limit to the size of the corresponding clouds is somewhere in the range 5 - 25 kpc, and since some of the lines are not seen in both spectra this value could be a rough size estimate. If we assume that these are each single systems with Doppler parameters of about 30 - 35 km s^{-1}, then the corresponding HI column densities are $\sim 10^{15}$ cm^{-2}. Thus, if the clouds are roughly isotropic and are this size, then the HI density is $\sim 10^{-8}$ - 10^{-9} cm^{-3}.

If we wish to obtain the density from this rough figure, then we need to estimate the neutral hydrogen fraction. If the clouds are ionized by an ultraviolet background, as seems the most likely option, then knowledge of this background flux, and the HI density above, allows an estimate of the total density. The ionizing background due to QSOs and young galaxies has been computed in detail by Bechtold et al. (1987), and applying this yields number densities of order 10^{-3} cm^{-3} and neutral hydrogen fractions of order 10^{-5} or so, with uncertainties of about an order of magnitude, and total masses of order $10^{7} M_{\odot}$.

Given the density in the clouds, and the ionizing flux, then the question of heavy element abundances can be addressed quantitatively. The absence of higher ionization heavy element lines can then place quite strong limits to the abundances, with the most severe so far in a system towards S5 0014+81, where the heavy element to hydrogen ratio is $\lesssim 10^{-3.5}$ solar (Chaffee et al., 1986).

However, these results depend quite strongly on the assumption that the clouds are roughly spherical. If they are highly flattened then the density may be considerably higher, and the neutral hydrogen fraction and heavy element abundances higher. Under these circumstances the heavy element abundances may be $10^{-1.5}$ solar, though then the clouds would have to be about 1pc thick. The distribution of numbers of clouds with respect to HI column density then depends on the cloud orientation with respect to the quasar to observer sightline, but this does not give rise to any conflict with the distributions inferred from the observations (Barcons and Fabian, 1987).

Recently there has emerged a possible way of determining the ionizing background flux relatively directly, by noting how the number of Lyman systems changes as the redshift approaches that of the background quasar against which they are seen. For clouds close to the quasar, the ionizing flux will higher and so the HI fraction less than in those ionized predominantly by the integrated background, so the HI column density will be lower and there will be fewer clouds per redshift interval down to some equivalent width limit. The number density of lines, on average, is then related to the ratio of the quasar flux to the background flux at any point. The total quasar ionizing flux may be estimated be extrapolating from the value at longer wavelengths, and the distance of the clouds from the quasar from the redshifts, and so the flux at a cloud due to the quasar evaluated. Then from the above argument, the background flux derived.

The work so far has been primarily to establish whether or not such a 'proximity effect' is observed, and the data are as yet too sparse to use to estimate the background with any precision. The most detailed study has been by Bajtlik et al. (1988), who find that the background flux is roughly constant, with a value $\log J_\nu = -21.0 \pm 0.5$ (ergs cm^{-2} s^{-1} Hz^{-1} sr^{-1} at the Lyman limit) over the range $1.7 < z < 3.8$. This is rather higher than estimates from an integrated quasar background at redshifts $z > 3$, so it suggests there are other sources of ionizing photons at that stage.

There are, as often, some complications which have to be considered. The main one concerns the true distance between the absorbing cloud and the quasar, as determined from the difference in redshift. The problem here is to assign the correct redshift to the quasar, since the high ionization emission lines (which are the ones that are used) have systematically lower redshifts by an average of about 1000 km s^{-1} than the low ionization lines (Wilkes, 1987; Espey et al., 1988), with a range in values from 0 - 4000 km s^{-1}. It is not clear which of these redshifts corresponds to the systemic redshift of the quasar, though future studies where forbidden line redshifts may be compared with high ionization line redshifts in bright quasars may provide the answer. In the meantime, since the background and the quasar flux are approximately equal for a distance corresponding to a velocity difference of 2 - 3000 km s^{-1}, this introduces a large uncertainty in the results. Since the true redshift may be higher than that determined by measuring the Lyα and CIV1549 emission lines, The J_ν estimate above should be treated as a lower limit. For a higher ionizing flux the total hydrogen number density within the clouds is higher, and the mass estimate also rises. This is because the neutral hydrogen density is fixed by determining the HI column density and the size, and is proportional to the (electron) density and inversely proportional to the ionizing flux.

4. Models

The properties of the Lyα clouds are still uncertain, and so provide few clues as to their nature. There is a wide range of possibilities. They could be primordial intergalactic clouds (Sargent et al., 1980), dwarf galaxies (Fransson and Epstein, 1982), the results of shocks generated by collisions of protogalaxies (Hogan, 1987), or clouds confined by cold dark matter (Rees, 1986, 1988), to list a few.

The suggestion that they are pressure - confined intergalactic primordial clouds gained considerable support from the work by Sargent et al. (1980), who found that the Lyα systems are not clustered as strongly as galaxies. This, and the absence of heavy element lines, strongly suggests a location outside galaxy clusters. Since self-gravity is unlikely to be strong enough to confine the gas, pressure - confinement by a hot intergalactic medium is an attractive possibility. If the confining medium is homogeneous, then this leads to a prediction that the density and temperature in all clouds at a given redshift is the same, and that they should evolve with redshift in a way which can be computed (e.g. Carswell et al., 1984, 1987, Baron et al., 1988).

Such a model now has to face the clustering on small velocity scales (Webb, 1987) and the range of Doppler parameters found. A possible difficulty is that the temperatures predicted are less than 30,000 K, while the Doppler parameters, if interpreted as thermal motions, indicate temperatures of order 70,000 K. The obvious answer to this is that bulk motions are responsible for the difference, but in this case the clouds would tend to disrupt on timescales $\sim 10^8$ years unless they are confined in some other way. This timescale corresponds to an evolution in cloud numbers $\sim (1 + z)^\gamma$ with $\gamma \gtrsim 10$, so is hard to reconcile with the observations.

If the clouds are gravitationally bound, then (Ostriker, 1988) self - gravity cannot be responsible since the collapse times are too short. However, stars or cold dark matter could provide the necessary gravitational potential. Rees (1988) has shown that if the clouds are gravitationally bound spheres then different lines of sight through the sphere lead to power law neutral hydrogen column density number distributions, with $f(N) \propto N^{-1.5}$ to N^{-2} for a realistic range of parameters. Thus the observed distribution is explained, and the range of $\sim 10^4$ in HI column density for the Lyα systems does not require a large range of cloud masses.

References

Atwood, B., Baldwin, J.A., and Carswell, R.F., 1985. *Ap. J.*, **292**, 58.

Bajtlik, S., Duncan, R.C., and Ostriker, J.P., 1988. *Ap. J.*, **327**, 570.

Barcons, X., and Fabian, A.C., 1987. *M.N.R.A.S.*, **224**, 675.

Baron, E., Carswell, R., Hogan, C. and Weymann, R., 1988. Preprint.

Bechtold, J., 1987. in *High Redshift and Primeval Galaxies*, ed. J. Bergeron, D. Kunth, B. Rocca - Volmerange and J. Tran Thanh Van, (Editions Frontieres: Gif sur Yvette), p397.

Bechtold, J., Weymann, R.J., Lin, Z, and Malkan, M., 1987. *Ap. J.*, **315**, 180.

Carswell, R.F., Morton, D.C., Smith, M.G., Stockton, A.N., Turnshek, D.A., and Weymann, R.J., 1984. *Ap. J.*, **278**, 486.

Carswell, R.F., Webb, J.K., Baldwin, J.A., and Atwood, B., 1987. *Ap. J.*, **319**, 709.

Chaffee, F.H., Foltz, C.B., Bechtold, J., and Weymann, R.J., 1986. *Ap. J.*, **301**, 116.

Crotts, A., 1988. Preprint.

Espey, B.R., Carswell, R.F., Bailey, J.A., Smith, M.G., and Ward, M.J., 1988. Preprint.

Fransson., C., and Epstein, R., 1982. *M.N.R.A.S.*, **198**, 1127.

Hogan, C.B., 1987. *Ap. J.*, **316**, L59.

Hunstead, R.W., Pettini, M., Blades, J.C., and Murdoch, H.S., 1987. in *Proceedings of IAU Symposium 124, Observational Cosmology*, ed. A. Hewitt, G. Burbidge and L.Z. Fang (Dordrecht: Reidel), p799.

Hunstead, R.W., Murdoch, H.S., Pettini, M., and Blades, J.C., 1988. *Ap. J.*, **329**, 527.

Liu, X.D., and Jones, B.J.T., 1988. *M.N.R.A.S.*, **230**, 481.

Meyer, D.M., and York, D.G., 1987. *Ap. J.*, **315**, L5.

Murdoch, H.S., Hunstead, R.W., Pettini, M., and Blades, J.C., 1986. *Ap. J.*, **309**, 19.

Ostriker, J.P., 1988. *in "QSO Absorption Lines - Probing the Universe"*, ed. J.C. Blades, D. Turnshek and C.A. Norman, (Cambridge University Press), p319.

Parnell, H.C., and Carswell, R.F., 1988. *M.N.R.A.S.*, **230**, 491.

Rees, M.J., 1986. *M.N.R.A.S.*, **218**, 25P.

Rees, M.J., 1988. in *"QSO Absorption Lines - Probing the Universe"*, ed. J.C. Blades, D. Turnshek and C.A. Norman, (Cambridge University Press), p107.

Peterson, B.A., 1978. IAU Symp. 79, "The Large Scale Structure of the Universe", ed. M.S. Longair and J. Einasto (Dordrecht: Reidel), p389.

Sargent, W.L.W., Young, P.J., Boksenberg, A., and Tytler, D., 1980. *Ap.J. (Suppl.)*, **42**, 41.

Shaver, P.A., 1988. in *IAU Symposium 130, Evolution of Large Scale Structures in the Universe.*

Tytler, D., 1987. *Ap. J.*, **321**, 49.

Webb, J.K., 1986. Ph.D. thesis, Cambridge University.

Webb, J.K., 1987. in *Proceedings of IAU Symposium 124, Observational Cosmology*, ed. A. Hewitt, G. Burbidge and L.Z. Fang (Dordrecht: Reidel), p803.

Webb, J.K., Parnell, H.C., Carswell, R.F., McMahon, R.G., Irwin, M.J., Hazard, C., Ferlet, R. and Vidal-Madjar, A., 1988. *ESO Messenger*, **51**, 15.

Wilkes, B.J., 1987. in *Emission Lines in Active Galactic Nuclei*, ed. P.M. Gondhalekar, (Rutherford Appleton Laboratory, RAL-87-109), p79.

Williger, G.M., Carswell, R.F., Webb, J.K., Boksenberg, A., and Smith, M.G., 1988. Preprint.

Wolfe, A.M., Turnshek, D.A., Smith, H.E., and Cohen, R.D., 1986. *Ap.J. Suppl.*, **61**, 249.

DISCUSSION:

FIELD: You mentioned blending. Can you rule out the hypothesis that the large b values are entirely due to blending rather than to broadening of individual lines?

CARSWELL: No.

SILK: What is the present situation with regard to the deuterium abundance in Lyα line systems?

CARSWELL: There's really no information - the HI column densities for those found so far are too low for DI to be detected for any plausible D/H ratio. Heavy element systems offer better prospects, but the interpretation is less clear because in these nuclear processing has occurred.

PETTINI: While we are discussing deuterium, I would like to suggest that the *damped Lyα galaxies*, rather than the intergalactic clouds giving rise to the Lyα forest, may offer the best chance to determine the primordial abundance of deuterium. Firstly, the column density of H I is sufficiently high to make the detection of the weaker D I Lyα feasible (and indeed measurements of deuterium in the Galactic ISM refer to sight–lines where the H I Lyα is damped). Secondly, if the damped systems are generally metal–poor and at an early stage of galactic evolution, their deuterium abundance, compared to the solar neighbourhood value, would give measure of the degree of astration of deuterium from which in turn it would be possible to extrapolate to the primordial abundance. Ironically, in the case of the $z_{abs} = 2.3091$ system in PHL 957, which I discussed on Monday, the H I column density is *too high* to allow the isotope shifts to be resolved, even for very low values of the velocity dispersion of the gas (b = 1 - 2 km s^{-1}).

SHAPIRO: Within the context of the pressure-confined, intergalactic cloud model for the Lyα forest, local variations in the confining pressure can also affect the observed HI column density distribution. A higher pressure in the vicinity of a quasar, for example, might explain the proximity effect without requiring that the quasar ionizing radiation flux dominate over the background flux throughout the redshift range which shows the absorption line density decrease along the line of sight to the quasar. Similarly, apparent clustering of the absorption lines could result from pressure inhomogeneities in the confining IGM.

CARSWELL: I agree.

SANCISI: Is there any evidence of voids in the distribution of Lyα absorption-systems?

CARSWELL: The most recent work suggests that there is nothing significant.

FALL: Could you say what the prospects are for determining or setting limits on the abundance of heavy elements in the Lyα forest clouds?

CARSWELL: For low column densities the abundance determination depends critically on ionization corrections, particularly to hydrogen. Since we can only make informed guesses at the ionizing flux and density these are very uncertain.

As Sargent points out, optically thick systems allow reasonable abundance estimates because you don't have to make such large ionization corrections, but it is not clear yet if any of these are Lyα forest as opposed to disk systems.

JONES: Why do you use two-point correlation functions as a measure of clustering: it is known to be a poor discriminator of structure when the clustering is weak. The line interval distribution is likely to be a more sensitive indicator.

CARSWELL: We use correlations functions because of tradition! One viewgraph shown displayed line interval distribution to be consistent with Poisson.

DAMPED LYMAN ALPHA ABSORBERS: THE PROGENITORS OF GALACTIC DISKS

A. M. Wolfe
University of Pittsburgh

ABSTRACT. Damped Lyα absorption lines arise in cold HI layers that contain most of the known baryonic matter in the Universe at large redshifts. A review of their properties suggests that the individual objects are the progenitors of galactic disks. The absence of Lyα emission from these objects indicates that the rate of heavy element production by nucleosynthesis in UMS stars was suprisingly low during the early phases of disk chemical evolution. A survey for new damped Lyα systems confirms our previous result that they occur more frequently than galactic disks down to $\log N(HI)=21$.

1. Introduction

QSO absorption systems with strong damped Lyα lines comprise a unique population of objects with large redshifts. They have attracted attention because (i) they contain most of the known baryonic matter in the Universe in the redshift interval $z=[2,3]$, (ii) the mass per unit comoving volume of the baryons is comparable to that contained in the stellar disks of known spiral galaxies, and (iii) physical conditions in individual objects bring to mind the HI disks of spiral galaxies. For these reasons the damped Lyα systems are prime candidates for the progenitors of stellar disks in galaxies (cf. Wolfe 1988).

The aim of this paper is to first review the essential properties of the damped Lyα systems, and then to give a progress report concerning current investigations.

2. Damped Lyα Review

The first systematic study of the damped Lyα systems began with a low-resolution survey for absorption features in the Lyα forest that were strong enough to be candidates for damped Lyα lines (Wolfe et al. 1986). Analysis of 68 QSO spectra produced a list of 47 candidates. Follow-up spectroscopy at intermediate resolution confirmed 18 of these as damped Lyα systems (Turnshek et al. 1989; Wolfe et al. 1989).

The gas content of the 18 absorption systems has the following properties: First, the redshifts of the detected systems are in the interval $z=[1.8,2.8]$. The redshift distribution is consistent with constant comoving density and cross section; i.e., there is no positive evidence for evolution. Second, the HI content of the gas is characterized by a mean column density $\log\langle N(HI)\rangle =21$. By comparison the H_2 content is exceedingly low. Studies of two absorbers in the sample show that $f \leq 10^{-5}$ and 10^{-4} respectively where $f=2n(H_2)/n(HI)$ (Black et al. 1986;

C. S. Frenk et al. (eds.), The Epoch of Galaxy Formation, 101–105.

Lanzetta *et al.* 1988). By contrast f $\geq 10^{-1}$ in the disk of the Galaxy for lines of sight encountering comparable column densities. Third, heavy elements are present in all 18 systems. Low ions such as CII, SiII, and FeII are always detected, whilst high ions such as CIV and SiIV are less common. The equivalent widths of the high-ion transitions are generally lower than the low ions. Fourth, a comparison between the optical continua of QSOS located behind the damped systems with a control sample reveals no evidence for reddening. As a result the gas-to-dust ratio of the damped systems, $(D/G)_{damp} \leq 0.5 \times (D/G)_{Galaxy}$ (Fall and Pei 1988). Fifth, curve of growth studies of the metal lines show that the effective velocity dispersion of the gas, σ_{metals}, ranges between 10 and 100 kms^{-1}. This σ_{metals} distribution is indistinguishable from the one inferred for the MgII absorbers (Lanzetta *et al.* 1987). However, studies of the HI kinematics of the 7 damped systems with 21 cm absorption show that, with one exception, $\sigma_{HI} \leq 17$ kms^{-1}. The difference in kinematics suggests a two-component model in which the bulk of the gas resides in a quiescent component, while a smaller fraction is contained in a turbulent component. The quiescent component creates strong damped Lyα lines and saturated low-ion lines that are weak. The turbulent component gives rise to saturated lines of low-ion and high-ion species that are strong. In order to determine metal abundances, it is necessary to sort out the the quiescent component with high-resolution spectroscopy. Sixth, a recent VLBI study of the 21cm absorption feature arising in the damped Lyα system toward PKS 0458-02, reveals that the quiescent HI extends across the 2 arcsecond extent of the background radio source. The corresponding linear scale subtended at z=2.04 is 8h^{-1} kpc, where h=H$_0$/100 kms^{-1}Mpc^{-1}. The kinematics of the gas further indicate that it is confined to a disk-like structure with radius of curvature R $\geq$ 40h^{-1} kpc (Briggs *et al.* 1988).

There is increasing evidence that the gas which creates damped Lyα lines is a magnetoionic medium with $|\mathbf{B}|$ fields exceeding 1 μG (cf. Wolfe 1988). The evidence stems from the incidence of Faraday rotation in a large sample of radio-selected QSOs (Welter, Perry, and Kronberg 1984). While Faraday rotation is detected in less than 20 % of the sample, it is found in all 5 QSOs located behind absorbers with 21 cm absorption. A sixth object was added to the sample with Turnshek's (1988) identification of a z=1.3 damped Lyα system in Altner and Heap's (1988) IUE specturm of Q0957+561, a lensed qso with strong Faraday rotation.

The global properties of the damped Lyα systems can be inferred from their incidence along the line of sight. The number of systems per unit redshift interval is dN$_{damp}$/dz=0.29±0.08 for logN(HI) $\geq$ 20.3. By contrast the prediction for spiral galaxies is dN$_{spiral}$/dz=0.05±0.03 at the same average redshift $\langle z \rangle$=2.4. Combining this statistic with log$\langle$N(HI)$\rangle$ =21 and the average look-back time we find that the current density paramater Ω_{damp}=(1.5-2.5)$\times 10^{-3}$/h. By comparison the mass density of luminous matter (stars) in stellar disks is $\Omega_{lum} \approx 2 \times 10^{-3}$/h.

3. Lyα Emission from the damped systems

Detection of metals such as Mg, Si, O, and Fe suggests that nucleosynthesis in the cores of massive UMS stars occurred prior to the absorption epoch. The energy released as a byproduct of nucleosynthesis is emitted by the stars as ionizing radiation which is subsequently converted into Lyα in the opaque HI gas. The Lyα photons random walk until they escape from the HI layer, provided that the dust content is low; i.e., $(D/G)_{damp} \leq 0.03 \times (D/G)_{Galaxy}$. The photons also random walk in frequency space and emerge in a symmetric double-hump feature centered on (1+z)$\times$1216 Å. The humps are characterized by peaks which are displaced from either side of line center by $\Delta\lambda_*$, and FWHM given by (2/3)$\times \Delta\lambda_*$ where

$$\Delta\lambda_* = (1 + z) \times 1.63 \times [(\sigma_{HI}/10 km s^{-1})(N(HI)/10^{21} cm^{-2})]^{1/3} \AA \qquad (1)$$

(Urbaniak and Wolfe 1981). The integrated mean intensity of the emergent radiation, $J(0)$, is obtained by adopting the Eddington approximation and then solving the transfer equation. We find that

$$J(0) = [\sqrt{3}/(8\pi)](m_H \Delta E)(N_\perp |dZ/dt|), \tag{2}$$

where $N_\perp$ is the HI column density perpindicular to the plane of the layer, ΔE is the nuclear energy released per gram, and Z and X are the metal and H abundance (by mass). As a result limits on the observed Lyα surface brightness, $I_{obs} = J(0) \times (1+z)^{-4}$, place limits on the rate at which massive stars enrich the gas with heavy elements. Note that the limits are independent of cosmological paramters (a similar point regarding continuum radiation was made by Cowie 1988).

My colleagues (Turnshek, Lanzetta, Gunn, and Oke) and I are currently searching for Lyα emission from selected damped systems. The search is performed with an interference filter tuned to the wavelength centroid of the damped Lyα trough. The 25 Å width of the filters is comparable to the ≈ 15 Å extent of the features predicted by eq. (1). The filters are placed at the prime focci of the KPNO 4m and Palomar 5m reflectors. CCD image frames are acquired both with the narrow Lyα filter and with standard broad band filters. So far we have not found any statistically significant emission features. However, Hunstead and Pettini (these proceedings) report the detection, with slit spectroscopy, of a narrow (FWHM ≤ 1.5 Å) feature in Q0836+113, an object observed by us. Since their detection is tentative, the remaining discussion will instead focus on the z=2.04 absorber in PKS 0458-02.

In order to set upper limits on $|dZ/dt|$ for the HI layer toward PKS 0458-02 it is necessary to estimate the spatial extent of the expected Lyα emission, and then anaylze the narrow-band CCD frame with an optimally matched search window. The angular diameters are expected to range from the ≈ 1 arscec seeing diameter to the ≈ 15 arcsec predicted for rarely occuring progenitors of L. galaxies. The latter is computed by assuming that the comoving density of the damped Lyα systems equals that of spiral galaxies and then computing the HI cross section from dN_{damp}/dz. Accordingly we analyzed the narrow-band CCD frame with square windows of 1, 2, 4, and 8 arcseconds in length. After removing obvious point sources we used the frequency distribution of sky brightnesses to estimate the following 4-σ upper limits: $I_{obs} \leq 1.4 \times 10^{-17}$, 9.3×10^{-18}, 6.8×10^{-18}, and 5.2×10^{-18} ergs(sec cm^2 arcsec^2)$^{-1}$ respectively. Equation (2) was then used to set corresponding limits on $|dZ/dt|$. Fig. (1) illustrates the limits along with predictions inferred from Twarog's (1980) best-fit model for the age-metallicity relationship of the Galaxy. The look-back time was calculated with a $q_0=0.5$ cosmology.

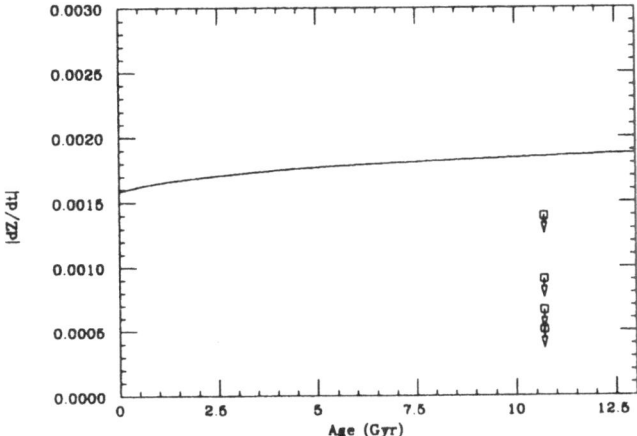

FIGURE 1. Comparison between upper limits on $|dZ/dt|$ and inferred AMR of Galaxy.

Fig. (1) shows that the 4-σ upper limits on |dZ/dt| are incompatible with results inferred for the Galaxy. If these upper limits are typical throughout the redshift interval of the damped sample, and if damped Lyα systems do evolve into galactic disks, then in order to explain the current metallicites of galactic disks |dZ/dt| must have exceeded the rates predicted by Twarog's model for a time interval exceeding 2 Gyr.

4. New Survey for damped Lyα systems

A new survey for damped Lyα systems is underway. In order to identify candidates for damped Lyα lines, the Palomar 5m telescope was equipped with the double spectrograph to acquire spectra for a total of 99 QSOs with average emission redshift $\langle z_{em}\rangle \geq 3$. The spectral resolution $\Delta\lambda = 4$ Å in the blue and 6 Å in the red. Sargent, Steidel, and Boksenberg (1988; SSB) obtained 59 of the spectra in their search for high-redshift Lyman-limit absorption systems. The high quality of the data enabled SSB to *confirm* 8 candidates as damped Lyα systems with logN(HI)$\geq$20.3 along a redshift path, z_{path}=68. Six of the 8 systems have logN(HI)$\geq$ 21. However, in a follow up study of the original sample (Wolfe *et al.* 1986) of damped Lyα candidates Turnshek *et al.* (1988) and Wolfe *et al.* (1989) used intermediate-resolution spectra ($\Delta\lambda \leq$ 2Å) to confirm a complete sample of 16 damped systems with log(N(HI)$\geq$ 20.3 along z_{path}=55. In this case 5 of the 16 systems have LogN(HI)$\geq$21. This shows that the 2 surveys are both complete and consistent down to logN(HI)$\geq$21, and that the SSB survey is incomplete down to logN(HI)$\geq$20.3. In any case at logN(HI)$\geq$21, dN$_{damp}$/dz=0.089$\pm$0.027, while dN$_{spiral}$/dz$\leq$0.024 along the combined z_{path}=123. Thus the preliminary results of the survey support the conclusion of Wolfe *et al.* (1986) that damped Lyα systems occur more frequently than the intervening HI disks of spiral galaxies. SSB have generously made their spectra available to us, so that we can make a complete search for damped Lyα candidates down to logN(HI)$\geq$20.3

The remaining 40 QSOs in the new survey were observed with the same data acquisition system used by SSB, and with the 100 inch telescope at Las Campanas. This sample is heavily populated with high-z radio-bright QSOs. The principal goal of this study, which is still in progress, is to find new 21 cm absorption systems, and to increase z_{path} from 123 to $\sim$ 200. Hopefully this will quadruple the known sample of damped Lyα systems, and allow us to carry out meaningful statistical tests on their z distribution etc.

I wish to thank Kem Cook, Ken Lanzetta, and Dave Turnshek for valuable discussions. This research was supported by NSF grant AST 87150-70.

References

Altner, B. and Heap, S.R. 1988, *10 years of the IUE Sattelite*, NASA preprint.
Black, J.H., Chaffee, F.H., and Foltz, C.B. 1987, *Ap.J.*, **312**, 50.
Briggs, F.H., Wolfe, A.M., Liszt, H. ,Davis, M.M., and Turner, K. 1988, *Ap.J.*, submitted.
Cowie, L. 1988, preprint.
Fall, S.M. and Pei, Y.C. 1988, *Ap.J.*, submitted.
Lanzetta, K.M., Turnshek, D.A., and Wolfe, A.M. 1987, *Ap.J.*, **322**, 739.
Sargent, W.L.W., Steidel, C.C., and Boksenberg, A. 1988, *Ap.J.*, accepted for publication.
Turnshek, D.A., Wolfe, A.M., Lanzetta, K.M., Briggs, F.H., Cohen, R.D., Foltz, C.B., Smith, H.E. and Wilkes, B.J. 1988, *Ap.J.*, submitted.
Twarog, B.A. 1980, *Ap.J.*, **242**, 242.
Urbaniak, J.J. and Wolfe, A.M. Z1981, *Ap.J.*, **244**, 406.
Welter, G.L., Perry, J.J. and Kronberg, P.P. 1984, *Ap.J.*, **279**, 19.

Wolfe, A.M. 1988 in *QSO Absorption Lines: Probing the Universe* , eds. C. Blades, D. Turnshek, C. Norman (Cambridge), in press.

Wolfe, A.M., Turnshek, D.A., Smith, H.E. and Cohen, R.D. 1986 *Ap. J. Suppl.*, **61**, 249.

Wolfe, A.M., Turnshek, D.A., and Lanzetta, K.M. 1989 *Ap.J.*, to be submitted.

DISCUSSION:

COWIE: The Hunstead–Pettini observation of emission would seem to raise at least potential questions about the interpretation of the damped Lyα systems as protodisks. Firstly, the location of the emission so close to the quasar would suggest on probabilistic grounds that this cloud is much smaller than needed to have a disk mass. Secondly even a relatively low metagalactic UV flux (10^5 photons cm^{-2}) would produce the emission seen by Hunstead and Pettini simply by ionizing the surface of the cloud. At these levels the emission would then map the surface of the system which then has a size $< 1 \times 2$ arcsec or so. Again, this is a good bit smaller than expected. Maybe an interesting speculation is that at least some of the damped Lyα systems are just the high mass end of the Lα forest clouds where the clouds are capable of self shielding themselves against the extragalactic UV flux.

WOLFE: The probability of finding the Lyα emission patch close to the quasar is not all that small. That is, the average size of a fitted protodisk is about 2" $\times$ 4" compared with the 1" $\times$ 2" pixel size used by Hunstead and Pettini. Thus the probability of finding 1 is 0.25. However, if 3 or more are found close to the quasar, I would agree with you.

Your idea of ionizing the surface of the absorber with the UV background is an interesting one. However, since we don't know the magnitude of the background flux very well and because one can think of various ways in which Lyα can be destroyed, I don't think this argument can be used to rule out an absorber with 4 times the area corresponding to a Pettini–Hunstead pixel.

I think that there are several lines of evidence that argue against the last idea. The most important one stems from the lack of detectable metals in the Lyα forest clouds. If the Lyα forest clouds have a total column density equal to the HI column density of the damped Lyα systems i.e. $< N > = 10^{21}$ cm^{-2}, then the detection of Lyα forest clouds with N(HI) $\leq 10^{13}$ cm^{-2} implies a neutral fraction of $X_H \sim 10^{-8}$. Because the upper limits on N(C IV), N(Si IV), etc. are likewise $\sim 10^{13}$ cm^{-2}, the upper limits on the metal abundance (by number) is Z $< 10^{-8}$. However, many authors have detected minimum column densities of $\sim 10^{15}$ cm^{-2} for CII in damped systems with N(HI) $\sim 10^{21}$ cm^{-2}. As a result $Z_{damp} > 10^{-6}$ which contradicts the upper limit imposed by the Lyα forest clouds.

A CHEMICALLY YOUNG GALAXY AT z = 2.3

MAX PETTINI
Anglo-Australian Observatory

ALEC BOKSENBERG
Royal Greenwich Observatory

RICHARD W. HUNSTEAD
School of Physics, University of Sydney

ABSTRACT. We present the first results from a programme to measure the metal content of QSO absorption systems with damped Lyα lines, thought to be the progenitors of present-day disk galaxies. Observations of the $z_{abs} = 2.3091$ system in PHL 957 with the Hale and Anglo-Australian telescopes have shown that it resembles closely a galaxy in its early stages of evolution by being metal-poor ($Z = -1.3$) and relatively dust-free. The star formation rate, deduced from the upper limit to Lyα *emission*, is less than 3–30 M$_{\odot}$ yr^{-1}. These properties are typical of the H II galaxies we see today.

1. Introduction: the Damped Lyman α Systems

The great potential of QSO absorption lines for studying the evolution of matter at earlier epochs was realised soon after their discovery. However, progress in this area has been relatively slow, partly because it was first necessary for astronomers to learn to recognise different classes of absorbers and the matter they trace. A major step forward in this direction has been provided recently by the survey by Wolfe *et al.* (1986) of "damped Lyman α" systems at redshifts $z_{abs} = 2 - 3$. These are systems where the column density of neutral hydrogen is sufficiently high, log N(H I) ≥ 20.3 (cm^{-2}), for the Lyα line to develop prominent damping wings and, with a rest equivalent width $W_0 > 10$ Å, be easily identified even in low resolution QSO spectra. By analogy with the local interstellar medium, where clouds with log N(H I) > 20 are essentially confined to the disk, Wolfe *et al.* identified the damped Lyα systems with disk galaxies at high redshifts. Furthermore, the frequency with which damped systems occur in QSO spectra indicates that: (a) they are ~5 times more numerous than expected from the cross-sections in H I of present-day disk galaxies and (b) the total mass they trace, $\Omega \simeq 2 \times 10^{-3}$ h$_0^{-1}$, is comparable to that of luminous matter in galaxies at the present epoch (Wolfe 1986). Based on this evidence, Wolfe and collaborators put forward the suggestion that the damped Lyα systems

C. S. Frenk et al. (eds.), The Epoch of Galaxy Formation, 107–113.

arise in the progenitors of present-day disk galaxies, still undergoing gravitational collapse.

It is important to remember, however, that a damped Lyα line is not in itself indicative of the *morphology* of the host galaxy. Sight-lines through gas-rich irregulars, for example, also commonly intersect regions with log N(H I)= $20 - 22$, and indeed there have been recent suggestions (Tyson 1988) that dwarf galaxies dominate the Wolfe *et al* sample. It is perhaps more appropriate to think of the damped Lyα systems as arising in the inner, mainly neutral regions of high redshift galaxies — of whatever morphology — and therefore forming a distinct population from, for example, the CIV absorption systems, which are generally attributed to more extended, ionized galactic haloes. Although somewhat obvious, it it worth stressing that, unlike the high-redshift radio galaxies discussed elsewhere at this workshop, these are unexceptional galaxies which are accessible to observation only because they fortuitously lie along the sight-line to a background QSO.

2. Abundance Determinations

A key property of the damped Lyα systems is their chemical composition; a measure of how far metal enrichment through stellar nucleosynthesis has progressed in these high-z galaxies would be a major clue to their evolutionary status. Although the damped nature of the Lyα line ensures that in most cases the column density of H I is well-determined, the measurement of metal abundances is beset by two major difficulties, well known from analogous work on the local interstellar medium (e.g. Jenkins 1987). Firstly, most elements are depleted from the gas phase onto interstellar grains, some by large and varying amounts. Secondly, for log N(H I) > 20, the commonly observed absorption lines of the most abundant astrophysical elements are heavily saturated, to the point that the equivalent widths are no longer useful indicators of column densities.

These difficulties can be largely circumvented by searching for the Zn II $\lambda\lambda 2025, 2062$ doublet associated with the damped Lyα lines. This species is a particularly good indicator of the metallicity of the gas for the following reasons:

1. In H I regions Zn is predominantly singly ionized, thus avoiding the need to account for unobserved ion stages.

2. Zn has a near-solar abundance in the interstellar medium. A recently completed survey with IUE by Van Steenberg and Shull (1988) over 188 sight-lines has shown that interstellar Zn has a mean depletion of only $[\mathrm{Zn/H}]_{ISM} \equiv (\log(\mathrm{Zn/H})_{ISM} - \log(\mathrm{Zn/H})_{\odot})= -0.23$.

3. The low solar abundance of Zn, $(\mathrm{Zn/H})_{\odot} = 3.8 \pm 0.7 \times 10^{-9}$ (Aller 1987), and the doublet nature of the transition generally result in a small and quantifiable degree of saturation of the absorption lines.

4. Zn is thought to be produced by the same nucleosynthetic process responsible for the iron peak elements (Cameron 1981) and its abundance tracks that of Fe in metal-poor stars.

5. For redshifts $z \simeq 2 - 3$ the Zn II doublet is redshifted into an easily observable part of the optical spectrum, where the efficiency of CCD detectors is high and absorption by the earth's atmosphere does not generally present serious problems. Furthermore, in all known damped Lyα systems the Zn II lines fall redward of Lyα emission and are therefore free of contamination by Lyα forest lines.

As an extra bonus this spectral interval also includes the Cr II triplet $\lambda\lambda\lambda 2055, 2061, 2065$. All of the above considerations apply to this species as well, apart from point (2), Cr being among the most heavily depleted elements in interstellar gas, with typically $[Cr/H]_{ISM} \simeq -2$ (e.g. Morton 1975). Thus, less than 200 Å of the red spectrum of QSOs with damped Lyα systems gives us two important parameters relevant to the evolutionary status of these high redshift galaxies. While [Zn/H] is a direct measure of the degree of metal enrichment, [Zn/Cr] depends on the level of depletion of refractory elements and therefore provides an indication — albeit a qualitative one — of the dust content of these systems.

3. Observations

To this end we have begun a survey of this spectral region using the 5 m Hale telescope at Palomar Observatory and the 3.9 m Anglo-Australian telescope at Siding Spring, Australia. Here we present the first results from this programme — also the subject of a forthcoming paper in the Astrophysical Journal — relating to the $z_{abs} = 2.3091$ system in the bright (V = 16.57) QSO PHL 957 at $z_{em} = 2.681$. This system has been studied extensively by Black et al. (1987) who measured log N(H I) = 21.40, near the upper end of the column density distribution of the Wolfe et al (1986) sample. This is in fact the *total* column density of neutral hydrogen in all forms in the system, since Black et al. found no evidence of H_2 absorption $(2N(H_2)/N(H) \leq 4 \times 10^{-6})$.

At Palomar we recorded the red spectrum of PHL 957 with Oke's double spectrograph and the red TI CCD chip at a resolving power $\lambda/\Delta\lambda = 3600$ (FWHM = 84 km s^{-1}); with a total exposure time of 12,000 s we achieved a S/N = 90 and a 3σ detection limit $W_0 = 20$ mÅ in the rest frame at $z_{abs} = 2.3091$.

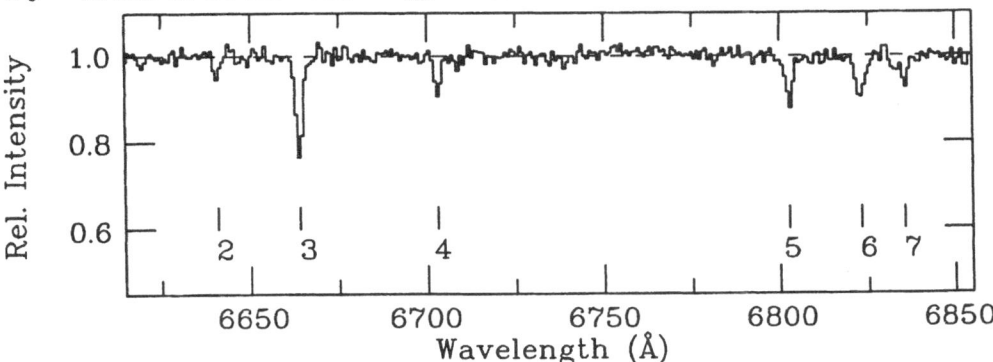

Figure 1: Portion of the normalised Palomar spectrum of PHL 957, encompassing the Zn II and Cr II absorption lines in the $z_{abs} = 2.3091$ system. Note the expanded vertical scale.

Line 4 is the stronger member of the Zn II doublet, $\lambda2025.512$, at $z_{abs} = 2.3091$, while lines 5 and 7 are Cr II $\lambda2055.59$ and $\lambda2065.46$ in the same system. Line 6 is a partially resolved blend of the weaker member of the Zn II pair, $\lambda2062.016$, and Cr II $\lambda2061.54$, while features 1 (not shown), 2 and 3 are FeII lines in a different absorption system, at $z_{abs} = 1.7975$ (all redshifts are vacuum heliocentric).

The Zn II and Cr II lines are weak, with $W_0 = 51$ and 80 mÅ for the strongest member of the doublet and triplet respectively; the ratios of equivalent widths within each multiplet (after deblending line 6) indicates that the lines fall on the linear part of the curve of growth and the corresponding column densities are well determined: $\log N(Zn^+) = 12.65$ and $\log N(Cr^+) = 13.35$ respectively, with an uncertainty of approximately 10 %. These in turn imply abundances in the $z_{abs} = 2.3091$ system [Zn/H] $= -1.3\pm0.1$ and [Cr/H] $= -1.8\pm0.15$ relative to solar. In Figure 2 we compare these values with analogous measurements for interstellar clouds near the Sun (ζ Oph; data from Morton 1975, and Snow, Weiler and Oegerle 1979) and in the Large Magellanic Cloud (R136a; de Boer, Fitzpatrick and Savage 1985); all three sight-lines sample similar column densities of neutral gas.

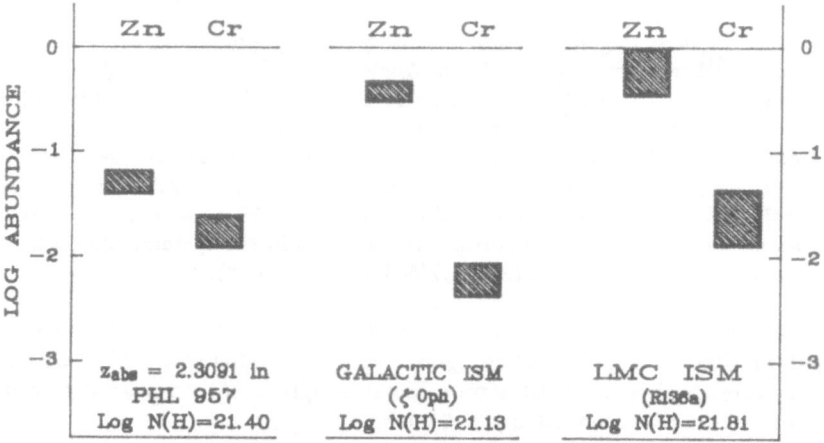

Figure 2: The abundances of Zn and Cr in the $z_{abs} = 2.3091$ system are compared with the values appropriate to interstellar clouds in the Galaxy and the LMC.

4. Results

The major results are as follows:

Firstly, the galaxy at $z_{abs} = 2.3091$ is genuinely metal-poor, since the abundance of Zn is $\sim$ 20 times lower than in the Sun, and $\sim$ 10 times lower than in the ISM of the Galaxy and the LMC. Clearly chemical evolution has not yet progressed very far in this system.

Secondly, the depletion of refractory elements onto interstellar grains is severely reduced in the $z_{abs} = 2.3091$ system relative to diffuse clouds in the Galaxy and the LMC, since we find Cr to be less abundant than Zn only by a factor of ~ 3, rather than by 1-2 orders of magnitude as is the case in nearby interstellar clouds. If other refractory elements also show similarly reduced depletions, this would suggest that in the high redshift system the dust-to-gas ratio is significantly lower than in the Galaxy and the LMC. Since dust grains are thought to be produced primarily in the atmospheres of red supergiants, the lack of dust may be taken as another indication — together with the low metal content — that the cycling of interstellar gas through stars has not proceeded as far. Furthermore, it is likely that the absence of molecular hydrogen in the $z_{abs} = 2.3091$ system reported by Black *et al.* (1987) is related to the low dust-to-gas ratio, since interstellar grains act as catalysts for the production of H_2.

In a related programme, described in Dick Hunstead's talk at this meeting, we have been searching for Lyα *emission* associated with the damped systems, and have been successful in at least one case. For PHL 957, AAT observations show excess flux at the bottom of the damped Lyα line at $z_{abs} = 2.3091$ at the 3σ significance level. However, the emission is broad and of low contrast, and we prefer to reserve judgement on its reality until better data are obtained. Our current upper limit to the integrated flux is 9×10^{-17} erg cm^{-2} s^{-1} over an area of sky 1.88 arcsec^2; this corresponds to an average surface brightness $\leq 5 \times 10^{-17}$ erg cm^{-2} s^{-1} arcsec^{-2} (> 24 mag arcsec^{-2}), ~ 2 times lower than previous upper limits for this system (Foltz *et al.* 1986). For $H_0 = 50$ km s^{-1} Mpc^{-1} and $q_0 = 0.5$ this in turn implies a Lyα luminosity $\leq 3.5 \times 10^{42}$ erg s^{-1}, from which we deduce an upper limit to the star formation rate in the galaxy at $z_{abs} = 2.3091$ of only 3–30 $M_\odot$ yr^{-1}, depending on the extent to which Lyα emission is suppressed by dust and H I (Hartmann *et al.* 1984).

5. Conclusions

The properties determined for the $z_{abs} = 2.3091$ system in PHL 957, namely: (a) high column density of neutral hydrogen (log N(H I) $= 21.40$); (b) low metallicity ([Zn/H] $= -1.3$); (c) low dust-to-gas ratio and (d) Lyα luminosity ($\leq 3 \times 10^{42}$ erg s^{-1}) are typical of present-day H II galaxies (see for example the recent review by Terlevich, 1987). We can take this similarity as an indication that either the damped Lyα systems occur predominantly in dwarf galaxies, as suggested by Tyson (1988) or, within the context of the Wolfe *et al.* interpretation of the damped systems, that the progenitors of spiral galaxies resemble closely the H II galaxies we see today. In the latter case, we note that a metallicity $Z = -1.3$ at an epoch when the universe was ~ 0.17 of its present age ($q_0 = 0.5$) is broadly consistent with the age–metallicity relation for the solar neighbourhood (e.g. Nissen *et al.* 1985; Tosi 1988). Clearly, it is not possible to arrive at firm conclusions as to the origin of the damped Lyα systems on the basis of a single sight-line. Nevertheless, the results presented here have shown that the absorption system at $z_{abs} = 2.3091$ in PHL 957 resembles closely a galaxy in its early stages of evolution, and have demonstrated the potential of chemical abundance studies. It now remains for the wider survey of the damped Lyα systems to provide, through statistical results, the clues they hold to the epoch of galaxy formation.

References

Aller, L.H. 1987, in *Spectroscopy of Astrophysical Plasmas*, ed. A. Dalgarno, and D. Layzer, (Cambridge University Press), p. 89.

Black, J.H., Chaffee, F.H., and Foltz, C.B. 1987, *Ap. J.*, **317**, 442.

Cameron, A.G.W., 1981, in *Essays in Nuclear Astrophysics*, ed. C.A. Barnes, D.D. Clayton and D.N. Schramm, (Cambridge University Press).

de Boer, K.S., Fitzpatrick, E.L., and Savage, B.D. 1985, *M.N.R.A.S*, **217**, 115.

Foltz, C.B., Chaffee, F.H., and Weymann, R.J. 1986, *A.J.*, **92**, 247.

Hartmann, L.W., Huchra, J.P., and Geller, M.J. 1984, *Ap.J.*, **287**, 487.

Jenkins, E.B. 1987, in *Interstellar Processes*, ed. D.J. Hollenbach and H.A. Thronson, (Dordrecht:Reidel), p. 533.

Morton, D.C. 1975, *Ap.J.*, **197**, 85.

Nissen, P.E., Edvardsson, B., and Gustafsson, B. 1985, in *ESO Workshop on Production and Distribution of C, N, O Elements*, ESO Conf. and Workshop Proc. No. 21, p. 131.

Snow, T.P., Weiler, E.J., and Oegerle, W.R. 1979, *Ap.J.*, **234**, 506.

Terlevich, R. 1987, in *High Redshift and Primeval Galaxies*, ed. J. Bergeron, D. Kunth, B. Rocca-Volmerange and J. Tran Thanh Van, (Ed. Frontières-Paris), p. 281.

Tosi, M. 1988, *Astr. Ap.*, **197**, 33.

Tyson, N.D. 1988, *Ap. J. (Letters)*, **329**, L57.

Van Steenberg, M.E., and Shull, J.M. 1988, *Ap.J.*, **330**, 942.

Wolfe, A.M. 1986, *Phil. Trans. Roy. Soc. Lond. A*, **320**, 433.

Wolfe, A.M., Turnshek, D.A., Smith, H.E., and Cohen, R.D. 1986, *Ap.J.Suppl.*, **61**, 249.

DISCUSSION

Cowie: Given the narrowness of H I in radio measurements of the damped Lyα systems, the Zn lines from the bulk of the gas must be highly saturated and the abundances of this gas may be much higher.

Pettini: On the contrary, even for b values as low as a few km s^{-1}, the Zn II lines are optically thin; this is confirmed by their doublet ratio, even allowing for the partial blending of the $\lambda 2062$ line. If the velocity dispersion within the $z_{abs} = 2.3091$ system is similar to that shown in 21 cm absorption by the $z_{abs} = 2.0395$ system in PKS 0458–020 (Wolfe *et al.* 1985, *Ap. J. (Letters)*, **294**, L67) — to which I think you are referring — the Zn II lines would be totally unsaturated. I consider it very unlikely that line saturation may have led us to underestimate significantly the column density of Zn$^+$ and hence the metallicity of the gas.

Sellwood: I was surprised to see that your measurements of LMC vs the galactic source seem to indicate that the LMC has, if anything, a higher metallicity than the Galaxy.

Pettini: The abundances of Zn in the interstellar medium of the Galaxy and the LMC are indeed the same to within a factor of ~ 2. This is in line with recent work on the metal content of the LMC, based on abundance analyses of F supergiants (Russell *et al.* 1988, *Mount Stromlo and Siding Spring Observatories preprint*, and references therein), which indicates a mean $[\text{Fe/H}]_{LMC} = -0.3 \pm 0.2$.

LYMAN α EMISSION FROM A POSSIBLE PRIMEVAL GALAXY AT $z = 2.5$

RICHARD W. HUNSTEAD
School of Physics, University of Sydney

MAX PETTINI
Anglo-Australian Observatory

ABSTRACT. We present the first detection of Lyα emission from a galaxy recognised by its absorption signature. The narrow velocity width of the emission line and its apparently small spatial extent argue in favour of the galaxy being a gas-rich dwarf. The inferred star-formation rate is $\sim 1\,M_\odot\ \mathrm{yr}^{-1}$.

1. Introduction

Absorption spectra of QSOs provide crucial information about the distribution of diffuse matter in the universe over cosmologically long timescales, matter which cannot be observed by any other means. In addition, if the spectra have sufficiently high resolution and S/N ratio, it is possible to probe directly the physical environment of the absorbers and look for trends with redshift. It is worth emphasising that most of the sight-line to a high-z QSO samples passively intervening material which is expected to be typical of *normal* galaxies and intergalactic gas at that redshift. Primeval galaxies, for instance, must leave some characteristic absorption fingerprint on a QSO spectrum; it is likely that our failure to recognise this signal simply reflects our general ignorance of the structures responsible for the narrow absorption lines.

We have recently begun a spectroscopic program directed at an important class of QSO absorption systems, the so-called "damped Lyman α systems" (Wolfe *et al.* 1986). These systems, with $N(\mathrm{H\ I}) \geq 2 \times 10^{20}\ \mathrm{cm}^{-2}$, have been interpreted as arising in the H I disks of intervening galaxies which may still be at an early stage in their evolution and therefore prime candidates for primeval galaxies. Two complementary approaches are being used to examine the evolutionary status of these putative disk galaxies:

1. To determine the degree of chemical enrichment of the galaxy's ISM via stellar nucleosynthesis. Many of the usual difficulties associated with obtaining reliable abundance estimates can be overcome by using the Zn II $\lambda\lambda$2025,2062 doublet (Pettini *et al.*, this volume).

2. To place direct limits on the star-formation rate by searching in the base of the

C. S. Frenk et al. (eds.), The Epoch of Galaxy Formation, 115–120.

damped Lyα absorption line for Lyα emission from H II regions in the galaxy or protogalaxy responsible for the absorption.

The second approach has been pursued spectroscopically by Foltz *et al.* (1986) and, through narrow-band imaging at the wavelength of the damped Lyα, by Smith *et al.* (1987). The damped Lyα lines in these two studies fall in the range $z_{abs} \sim 2 - 3$; in some theoretical models for galaxy evolution (e.g. Meier 1976, Shull and Silk 1979, Cox 1985) this epoch is characterised by rapid star formation and, therefore, high Lyα luminosity. In neither study was there a positive detection of Lyα emission. Smith *et al.* conclude that the lack of detectable emission from nine damped systems suggests that disk galaxies in this redshift range are not strong Lyα sources and that the inferred star formation rates are probably $\leq 1 \, M_\odot \, yr^{-1}$.

2. Observations

Our original plan was to reobserve a selection of QSOs with damped Lyα systems at substantially higher spectral resolution than the finding survey of Wolfe *et al.* (1986) or the follow-up survey by Smith, Cohen and Bradley (1986) with the principal aim of obtaining more reliable estimates of $N(H\,I)$. The first object to be observed was Q0836+113 (Hazard *et al.* 1986) which has $z_{em} = 2.67$ and $z_{abs} = 2.465$; although reported as B = 18.8, our estimate from the Palomar Sky Survey, together with our spectrophotometry, suggests that B = 19.5 ± 0.3.

Long-slit observations of Q0836+113 were made in 1987 April at the 3.9 m Anglo-Australian Telescope using the RGO spectrograph and IPCS detector. With a slit width of 1.2 arcsec (which matched the seeing), the spectral resolution was 1.5 Å FWHM; spatial increments along the slit were 2.3 arcsec and the QSO was guided into one increment. The wavelength coverage ~ 1000 Å included Lyβ ($z_{abs} = 2.465$) towards the blue edge. Total exposure time was 14000 s spread over two nights; data from each night were combined using weights based on S/N ratios. The slit was oriented at the parallactic angle which ranged from p.a. $140 - 160°$.

The region of the damped Lyα line at $z_{abs} = 2.465$ is shown in Fig. 1 together with our fitted profile for $N(H\,I) = 4.2 \times 10^{20} \, cm^{-2}$. A narrow emission line is marked in the base of the damped Lyα. We have carefully assessed the reality of this weak feature in relation to the noise in the base of Lyα and the noise in the adjacent slit increments; a conservative estimate of its significance is 4.0σ. The excess signal can not be explained by velocity structure in Lyα absorption (Lyβ is single with zero central intensity) or by poor sky subtraction (sky is smooth in this spectral region). We conclude, therefore, that we are seeing Lyα *emission* from the intervening galaxy responsible for the damped absorption system.

Does the Lyα emission extend along the slit? To test this, we formed a sky-subtracted image from the spatial data in the region of the damped Lyα line. This image is shown in grey-scale form in Fig. 2. It is not surprising to find the QSO light confined to one spatial increment but the fact that the Lyα emission does not extend outside the same increment means that the emission region is within 1 arcsec of the QSO sight-line. At $z_{abs} = 2.465$, assuming $q_o = 0.5$, $H_o = 50 \, km\,s^{-1}\,Mpc^{-1}$, 1 arcsec corresponds to 7.8 kpc. The measured Lyα emission therefore comes from a region no larger than 1.2×2.3 arcsec or 9×18 kpc.

The main properties of the Lyα emission are listed below. Due to the uncertainties of narrow-slit spectrophotometry, the flux-related quantities may be in error by up to 30%.

z_{em} (vac-helio) : 2.4654 ± 0.0001
Δv (em − abs) : 25 ± 30 km s^{-1}
Lyα flux : 2.9×10^{-17} erg cm^{-2} s^{-1}
Lyα luminosity : 1.2×10^{42} erg s^{-1}
Lyα linewidth : ≤ 50 km s^{-1} FWHM

The average Lyα surface brightness is 1.0×10^{-17} erg cm^{-2} s^{-1} arcsec^{-2}; this is a factor ~ 8 below the upper limits set by Foltz *et al.* (1986) but comparable to the non-detection levels quoted by Smith *et al.* (1987).

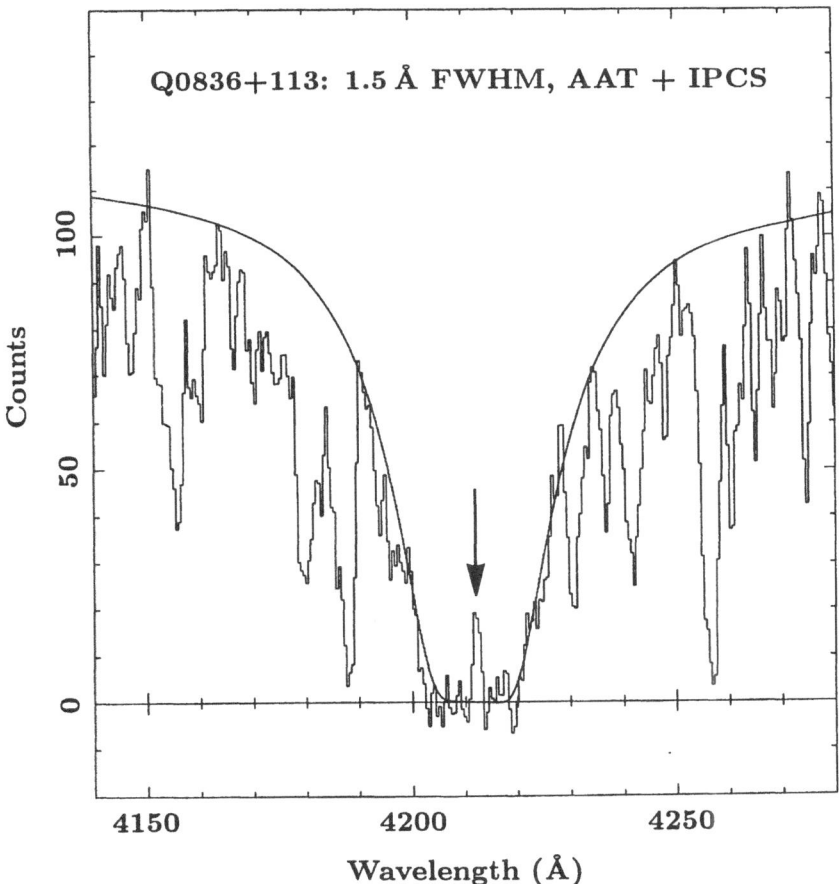

Figure 1: Spectrum of Q0836+113 in the region of the damped Lyα line at $z_{abs} = 2.465$ with its fitted profile; wavelengths are observed air values. The excess emission in the base of the line is arrowed.

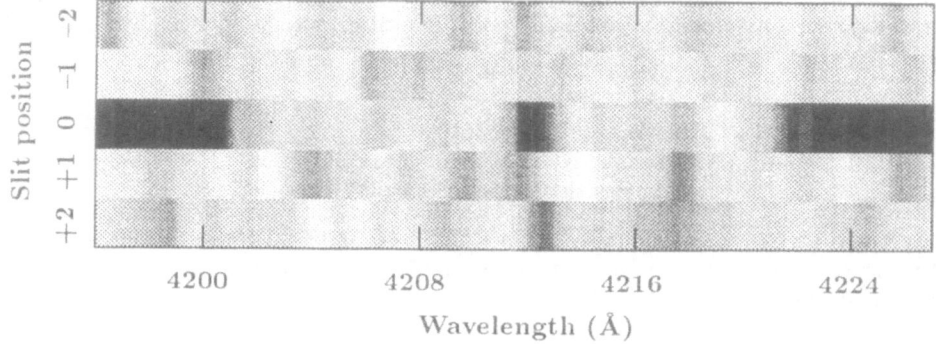

Figure 2: Grey-scale representation of the spectral data in the vicinity of damped Lyα for the slit increment containing the QSO plus the two adjoining increments on either side. The slit coverage shown is equivalent to 90 kpc at $z = 2.465$ ($q_o = 0.5$, $H_o = 50$).

3. Discussion

The most notable features of the Lyα emission from the $z = 2.465$ galaxy are its narrow velocity width, its apparently small spatial extent and the close correspondence between z_{em} and z_{abs}. These can be understood if we are dealing with a dwarf galaxy similar to the nearby H II galaxies (Terlevich 1987) rather than a large disk system, as proposed by Wolfe *et al.* (1986); indeed, Tyson (1988) has recently suggested that damped Lyα systems may occur predominantly in dwarf galaxies. Lyα emission has only been detected in a few H II galaxies and fluxes tend to fall well below the predictions of recombination theory; this is attributed in part to dust scattering of Lyα photons in H I regions, a view supported by a trend for Lyα/Hβ to decrease with increasing metallicity (Hartmann *et al.* 1988). On the other hand, no evidence for dust has been found in the damped Lyα systems (Pettini *et al.*, this volume; Fall and Pei 1988); if we assume, therefore, that we are measuring the *total* Lyα luminosity of the galaxy at $z = 2.465$, the inferred star-formation rate (based on the IMF adopted by Kennicutt, 1983) is $\sim 1 M_\odot \, \mathrm{yr}^{-1}$. This value is typical of present-day H II galaxies and low-luminosity spirals.

H II galaxies are usually underabundant in heavy elements with an optical and UV continuum dominated by O and B stars. There is no discernible continuum under the emission line in Fig. 1. Based on the noise level, we estimate a rest-frame equivalent width, $W_o \geq 10$ Å, which is consistent with IUE data for H II galaxies (Hartmann *et al.* 1988). The heavy-element column densities in the $z_{abs} = 2.465$ system are difficult to determine at this resolution due to confusion with Lyα forest lines. With the usual caveats, profile-fits to Si II, N I, O I and Fe II indicate an abundance ~ -1 dex relative to solar, consistent with that seen in H II galaxies (Terlevich 1987) and in the $z_{abs} = 2.3091$ damped system in PHL 957 (Pettini *et al.*, this volume).

It is interesting to compare the $z = 2.465$ galaxy with the substantial body of data for H II galaxies at Hβ (Terlevich 1987). If we assume no dust and case B recombination, the 2.465 galaxy has $\log L(H\beta) = 40.1$ (for $H_o = 90$, as used by Terlevich), close to the boundary between giant H II regions and H II galaxies. For the upper-limit linewidth of 50 km s^{-1} (log σ = 1.33), the z = 2.465 galaxy falls very close to the well-defined L : σ trend line (Terlevich, Fig. 6). While this evidence is suggestive, a higher resolution spectrum is needed to settle the issue.

The accumulated evidence, therefore, points towards an interpretation of *this* damped Lyα system at z = 2.465 as arising in a dwarf galaxy very similar to the H II galaxies we see today. Is it a primeval galaxy? All we can say at present is that H II galaxies are known to be young systems and, for that reason, they are often suggested as the prototypes for primeval galaxies. It remains to be seen whether future observations of the z = 2.465 system at higher spatial and spectral resolution support the H II galaxy interpretation and also whether the objects underlying other damped Lyα systems show similar emission properties.

References

Cox, D.P. 1985, *Ap. J.*, **288**, 465.

Fall, S.M. and Pei, Y. 1988, *Ap. J.*, in press.

Foltz, C.B., Chaffee, F.H. and Weymann, R.J. 1986, *A.J.*, **92**, 247.

Hartmann, L.W., Huchra, J.P., Geller, M.J., O'Brien, P. and Wilson, R. 1988, *Ap. J.*, **326**, 101.

Hazard, C., Morton, D., McMahon, R., Sargent, W.L.W. and Terlevich, R. 1986, *M.N.R.A.S.*, **223**, 87.

Kennicutt, R.C. 1983, *Ap. J.*, **272**, 54.

Meier, D.L. 1976, *Ap. J.*, **207**, 343.

Pettini, M., Boksenberg, A. and Hunstead, R.W. 1988, this volume.

Shull, J.M. and Silk, J. 1979, *Ap. J.*, **234**, 427.

Smith, H.E., Cohen, R.D. and Bradley, S.E. 1986, *Ap. J.*, **310**, 583.

Smith, H.E., Cohen, R.D. and Burns, J.E. 1987, poster paper in *QSO Absorption Lines: Probing the Universe*, eds. J.C. Blades, C. Norman & D. Turnshek, p. 148 (Space Telescope Science Institute).

Terlevich, R. 1987, in *High Redshift and Primeval Galaxies*, ed. J. Bergeron, D. Kunth, B. Rocca-Volmerange & J. Tran Than Van (Ed. Frontières:Paris), p. 281.

Tyson, N.D. 1988, *Ap. J. (Letters)*, **329**, L57.

Wolfe, A.M., Turnshek, D.A., Smith, H.E. and Cohen, R.D. 1986, *Ap. J. Suppl.*, **61**, 249.

DISCUSSION:

YEE: You claimed that the star formation rate inferred from the Lyα flux is similar to that found for low redshift spirals or HII galaxies. If that is the case, why would you refer to this object at z = 2.5 as a "primeval galaxy". Wouldn't this be just an ordinary galaxy at high z?

HUNSTEAD: I suppose it's a matter of semantics. A couple of years ago, David Koo (in *Spectral Evolution of Galaxies*, ed. Chiosi & Renzini, p 419, 1986) defined primeval galaxies as "high-redshift galaxies undergoing their initial stage of star formation". Here we have a putative disk galaxy at high z which is forming stars, albeit at a rate somewhat less than theoretical predictions. If it has a genuinely low metal abundance of 1/10 solar or less (like the z = 2.3 damped system in PHL 957) then it is surely a good candidate for a primeval galaxy.

REES: The statistics of the "damped Lyman α" absorption systems suggests that the covering factor over the sky of high column density clouds with z in the range 2 – 3.5 is quite substantial. How does the sensitivity of your limit compare with the limits set by narrow-band searches for emission–line objects in "blank fields"?

HUNSTEAD: Len Cowie's "blank field" limits are somewhat brighter than our detection level. Furthermore, his searches were directed towards objects of higher redshift and larger angular size. If the emission regions generally have small angular size (< 2 arcsec) and small velocity dispersion (< 100 km sec^{-1}), I suspect that previous searches would not have expected to see them.

WOLFE: Three comments: (1) The occurrence of a narrow Lyα emission line centered at the absorption redshift, z_{abs}, is surprising. If the Lyα photons resonantly scatter through the N(H I) = 4. 10^{20} cm^{-2} which you derived, the emergent Lyα would consist of 2 broad humps (FWHM ~ 500 km sec^{-1}) displaced symmetrically by ~ 1000 km Sec^{-1} from z_{abs}. The implication is that you have detected a "naked" HII region. (2) I doubt whether the abundances in 0836 are really low. The b = 30 km sec^{-1} you derived from the metal lines reflect turbulent conditions in low column gas rather than the quiescent gas responsible for damped Lyα . I think your low metallicities are actually lower limits to the disk abundances. (3) To answer Martin Rees' comment, the damped Lyα statistics imply that a few of these emission features should be present in typical Palomar 160 × 160 arcsec narrow band (25 Å) CCD frames. So far, the interference filter technique has produced no detections to a 4 σ limit of ~ 5 10^{-18} ergs cm^{-2} sec^{-1} arcsec^{-2}.

HUNSTEAD: (1) I would simply make the point that there is no necessary correspondence between N(HI) along the sightline to the QSO and N(HI) in front of the HII region producing the Lyα emission. (2) Of course, the b-value of the dominant HI component could be much less than 30 km sec^{-1}. Clearly we need higher spectral resolution to address the question of abundances with any confidence ... but remember this is a faint object.

THE QUASAR REDSHIFT CUT-OFF

Richard F. Green
Kitt Peak National Observatory
National Optical Astronomy Observatories
P.O. Box 26732
Tucson, Arizona 85726

ABSTRACT. Evidence for a cut-off in the comoving density of quasars at z>2.5 is examined. The basis for comparison is the luminosity function of quasars at lower redshifts, which is becoming increasingly well determined. The change of the LF with cosmic time can be parameterized by luminosity evolution or luminosity-dependent density evolution. New complete surveys for high-redshift quasars by slitless spectroscopy and color selection indicate the presence of a strong peak in the integrated density of luminous quasars near z=2. The question is whether the deficit of high-redshift quasars reflects the true distribution, consistent with the results from radio source counts, of whether it is apparent because of increasing obscuration by intervening dust. The damped Lyman α systems alone cannot produce the observed decrease. A simple exercise shows that optically thick systems with column densities down to log N (H I) = 19.7 and Galactic dust/neutral gas ratios can reproduce the observations for a population of high-redshift quasars as numerous as that at z=2, provided the dust has a very weak 2200 Å feature. Physical consequences of various scenarios are discussed.

I. INTRODUCTION

From the time of their discovery, quasars were recognized as valuable probes of conditions in the early Universe. Because of their unique energy distributions and spectral signatures, they are relatively easy to distinguish from other stellar objects. The advent of the capability to digitize and analyze large areas of photographic plate has led to the construction of statistically complete samples of quasars. These have been used to determine the evolution of the Luminosity Function (LF) and total space density of Active Galactic Nuclei (AGN's) with redshift.

If the epoch of quasar formation can be identified, it will obviously have a direct relationship to the epoch of galaxy formation. Establishment of the physical connection requires a more secure grasp

121

C. S. Frenk et al. (eds.), The Epoch of Galaxy Formation, 121–133.
© 1989 by Kluwer Academic Publishers.

on the nature of the central energy source. In the meanwhile, directly observed population trends have intrinsic value in setting limits on how late relatively massive galaxy structures coalesced. A redshift cut-off implies a framework for evaluating density expectations, and will depend on assuming a cosmological model. The approach here is to adopt a world model, then evaluate the samples to derive the LF at lower redshifts. The results of the new surveys for the highest redshift quasars are then compared to evaluate the strength of the evidence for an observed redshift cut-off. Such a cut-off has been modeled by intervening obscuration or as a direct consequence of physical source evolution with cosmic time.

II. QUASAR SURVEYS

The solid observational basis for the derivation of the quasar LF is the result of a tremendous amount of careful work, leading to a consistent picture for luminous objects at redshifts less than 2-2.5. Three principal techniques have been used to distinguish quasars from galactic stars: ultraviolet excess, slitless spectroscopy, and multi-color separation. Each is powerful in sharply limiting contaminant populations, while each may be subject to its own bias. Ultraviolet excess excludes most higher redshift objects once the Ly α emission has shifted into the B band; colors must be measured to sufficient accuracy to include the redward excursion of the color locus around $z=0.5$, when the 3000 Å bump is centered in B. The UVX technique would also exclude a population with very red emitted energy distributions; neither radio nor slitless spectroscopic surveys show evidence for a subsantial population of this nature. Slitless spectroscopy must have high enough spectral resolution and contrast to detect the weaker emission lines from low-redshift objects, and compensate for the weakening of the ultraviolet lines with higher luminosity and increasing absorption. The flat redshift distribution of the CFHT survey shows that the technique can be quite successful. Mutli-color selection assumes that quasars will be separated from the stellar locus for all spectral forms; cross-checking is required to learn what fraction of quasars fall directly into the stellar locus. The following table lists the surveys from which complete statistical samples can currently be derived (several more are in progress (Foltz and Osmer 1988)).

III. RESULTS FOR LOW REDSHIFT

The faint survey limits of Boyle et al. (1987) and Koo and Kron (1988) showed that the very steep increase in surface density with fainter magnitude does not continue indefinitely. There is a pronounced flattening of that slope beyond B=19.5, such that $\frac{d \log N}{dB} = 0.86$ changes to $\frac{d \log N}{dB} = 0.32$. This change in slope must reflect a feature in the Luminosity Function itself.

Quasar Surveys

Name	Number	B lim	Area (sq.deg.)	Reference
		UVX (z < 2.2)		
BQS	92	16.16	10,714	Schmidt & Green 1983
AB	16	18.0	36	Braccesi et al. 1970
US-SA 29	12	18.5	7.7	Usher et al. 1983
BF	28	19.65	1.72	Marshall et al. 1984
MBQS	32	17.65	87.3	Mitchell et al. 1984
Durham	170	20.9	4.2	Boyle et al. 1987
Durham II	420	20.9	11.3	Boyle et al. 1988
Marano et al	23	20.9 (J)	0.69	Marano et al. 1988
		Slitless Spectra		
4-m Grism	14	19.5	5.1	Hoag & Smith 1977
4-m Grism	5	19.5	2.6	Sramek & Weedman 1978
CFHT	71	19.8	4.2	Crampton et al. 1987
Schmidt prism	7	18 (R)	60	Hazard et al. 1986
5-m transit	44	lines	14.3	Schmidt et al. 1988
APM	192	18.5	102	Foltz et al. 1987
		Multi-Color		
Koo-Kron	11	21.1	0.29	Koo & Kron 1988
	(30	22.6)		
Warren et al	65	20.0	30	Warren et al. 1988
	(24 with z>3)			

The surveys listed above fill in the Quasar Hubble Diagram
sufficiently that parametric modeling is no longer required to fill in
large gaps. The luminosity distribution is given by the distribution
of objects in apparent magnitude for a slice in redshift, provided
that the difference in areal coverage and limiting magnitude is taken
into account for the different surveys. To get the space density
contribution of objects within the redshift slice, the union of
available volumes from all surveys is computed; $1/V_a$ is summed for the
ensemble (Avni and Bahcall 1980). The results do depend on the choice
of redshift limits for the slices, but not on the absolute magnitudes
chosen for display purposes. For this review, the calculation was
performed for the set of surveys listed in the table (with the
exception of the material from the APM and Marano et al., for which
more information was required to produce complete subsamples). The
quasar magnitudes were corrected for the effect of a galactic cosecant
extinction law; that correction made negligible difference. Objects
with one-line redshifts were generally not included, so the detailed
shape of the LF may vary from published analyses of similar

material. The derived LF as a function of cosmic epoch is presented in Figure 1.

QUASAR LUMINOSITY FUNCTION

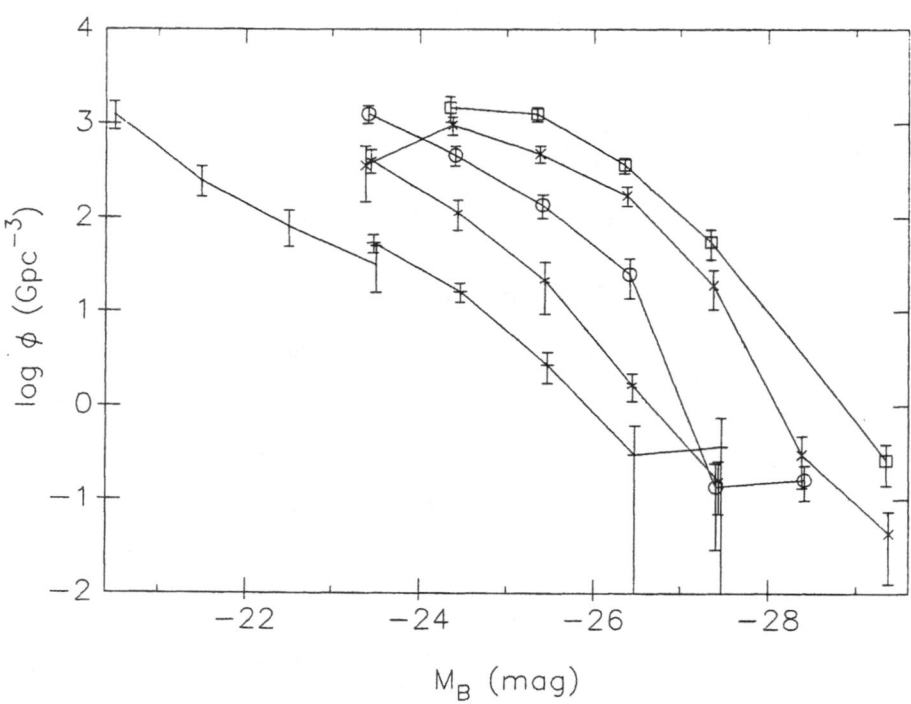

Figure 1. Quasar Luminosity Function for different cosmic epochs. The lowest luminosity points are the Seyfert galaxy LF from Cheng et al. (1985). The + symbols are for 0.0 <z< 0.3 ; * -0.3 <z< 0.7 ; o - 0.7 <z< 1.2 ; x-1.2 <z< 1.7 ; open squares -1.7 <z< 2.2. The density is given in units of Gpc^{-3} Mag^{-1}; volumes and absolute magnitudes were calculated based on H_0 = 50 km/s/Mpc, and Ω = 1.

Many workers in this field have been struck by the possibility that these LF's may be described by a family of self-similar curves varying with cosmic time. Marshall (1987) paramaterized the form as a sum of two power-laws, normalized at a characteristic luminosity. Boyle et al. (1987) fit the two sections with

$L^{-3.7}$ (luminous end) and $L^{-1.4}$ (low-luminosity end).

There is also a clear shift of the function with redshift, and by implication, cosmic time. The change of the ensemble with cosmic time can be modeled parametrically; this technique provides predictions for higher redshift or lower luminosity, and timescales within the given model context from which physical properties may be inferred. At least four models have been proposed to fit the LF as a function of redshift.

The first is included as a reminder, that the assumption of the cosmological model may be wrong. If the luminosity function of quasars were assumed to be a constant with cosmic time, then the changing size of volume elements in a Friedmann Universe is inconsistent with that assumption and the data. The volume elements and luminosity distances in chronometric cosmology do provide a relatively uniform distribution of quasars with cosmic time (Segal and Nicoll 1986). The remainder of the present discussion proceeds on the assumption of conventional cosmological models.

One construct to use in evaluating the evolution of the LF is density evolution, which corresponds to a vertical shift of the LF with increasing look-back time. The analogy is to Galactic structure problems, in which the density of a tracer population of known luminosity is followed to determine scale lengths. Schmidt and Green (1983) used the form

$$\rho = \rho_0 \, e^{k(Mo - M)\tau} \qquad \text{where } \tau \text{ is the look-back time.}$$

Since the rate of evolution depends on the luminosity of the source, this is called luminosity-dependent density evolution. Pure density evolution is ruled out. The original parameterization of their HH5 model did not provide a good fit to the new information from the faint-limit surveys, having too strong an evolution for high luminosities, and too weak for low-luminosity quasars and high-luminosity Seyferts. A better approximation is now

$$\rho(M_B, z) = \rho(M_B, 0) e^{0.78 \, (-17.13 - M_B)\tau} \qquad H_0 = 50, \ \Omega = 1.$$

This density increase applies only for z<2.5. Crampton et al. found that they could fit their sample from the CFHT survey with luminosity-dependent density evolution laws that were a modification of the HH5 model. They replaced k with

$$k' = \frac{K \text{ model}}{e^{2(z-z_c)} + 1} \qquad \text{where } z_c \sim 2.9$$

The redshift cut-off parameter yields a maximum comoving density at z ~1.7, after which it declines.

An alternative view is that of luminosity evolution, which has its analog in studies of the galaxy luminosity function. In this picture, there is a constant integral number of sources, which fade as an ensemble with cosmic time. This process corresponds to a horizontal shift of the LF in the figure. Such models have been investigated by Mathez (1978), Braccesi et al. (1980), Cheney and Rowan-Robinson (1981), Koo (1983), Weedman (1986), and Marshall (1987). Boyle et al. found a best fit

$$L (z) \propto L_0 (1+z)^{3.2 \pm 0.1}$$

The fit within the quasar luminosity range seems to be very good; the agreement in detail in comparing the higher redshift low-L quasars with the zero-redshift Seyfert galaxy LF is not perfect with respect to the location of M*. Both the faint, high redshift sample and the local objects require further observational work to reduce systematic uncertainties.

A different approach was taken by Koo and Kron (1988) to model the region around M*, determined from their own faint object survey. It is based on the extrapolation that the redshift distribution is independent of magnitude for 19.5 <J< 22.5. The counts are fit by a power-law, which defines the power-law faint end of the LF. Higher luminosity objects are assumed to fade out of the distribution, with the more luminous objects disappearing first. The addition of data from other surveys then defines the high-luminosity roll-off as a function of redshift. The impact of this approach will be discussed further in relation to the highest redshift quasars in the next section.

IV. EVIDENCE FOR A HIGH-REDSHIFT CUT-OFF

There has been preliminary evidence for several years that the density of high-redshift quasars did not follow from an extrapolation of the density or luminosity increase observed at lower redshifts. Such evidence is found in density estimates by Osmer (1982), Weedman (1985), and Veron (1986). The composite LF constructed here from the CFHT, 4-m grism, and Koo-Kron surveys also points in that direction.

The three systematic surveys concentrating on high-redshift objects now allow quantitative estimates of the observed effect. They are by Warren, Hewett and Osmer (1988) using multi-color selection; Schmidt, Schneider, and Gunn (1988) with a CCD transit grism survey; and Hazard, McMahon and Sargent (1986) with Schmidt telescope objective prism plates. These techniques are governed by complex observational selection effects, for the estimation of which we must in first instance rely on the calculations of the investigators.

Preliminary values of space densities are available for the 5-m transit survey (Schmidt, Schneider and Gunn 1988). These are based on flux limits for the Lyα emission; a limit in Lyα luminosity > 10^{45} ergs/s corresponds approximately to M_B <-26. Warren, Hewett, and Osmer (1988) also plot preliminary space densities of their multi-color selected objects. If the integral LF to M_B = -26 is plotted as a function of cosmic time, the result in Figure 2 is obtained. A narrow, very strong peak occurs around z=2, with a width in fractional look-back time of less than 0.1. Although the analyses of Schmidt et al. and Warren et al. still show disagreements at the factor of two level, they both delineate a strong trend of lower density than the peak values.

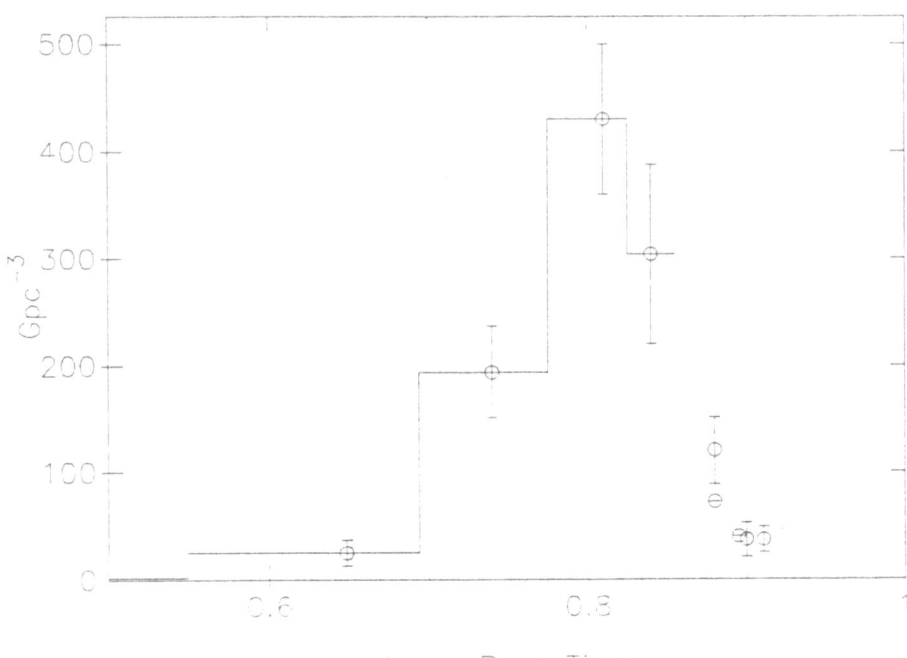

Figure 2. Integrated density of luminous quasars as a function of
fractional look-back time. Solid bars connect values derived from
lower redshift LF presented in Figure 1. Circles without error bars
are values from Schmidt et al.; circle at τ = 0.90 is from Hazard et
al.; the other two circles with error bars come from Warren et al.

A discrepancy appears to exist between these surveys and that of
Hazard et al., in that they seem to yield approximately similar volume
densities, but the Hazard objects are reported to be two magnitudes
more luminous. Extrapolating along the very steep high-luminosity
tail, the total would be much higher to M_B = -26. The discrepancy may
be reduced in two ways. If the apparent magnitudes estimated by
Hazard et al. are too bright, the numbers come closer to agreement.
The initial determination of Warren et al. indicates that the LF in
this absolute magnitude range may be much flatter than it is at lower
redshifts. This result raises many physical questions, but further
reduces the difference in integrated values, compared to a steeper
extrapolation. The much higher space density estimates of Koo and
Kron (1988) come from assuming that the constant normalization power-

law for lower-luminosity objects still applies at z>3 and smoothly connects to the data of Hazard et al. at $M_B \sim -28$.

V. DOES THE APPARENT REDSHIFT CUT-OFF REFLECT AN INTRINSIC CUT-OFF?

The direct interpretation of a redshift distribution is that it mirrors the true distribution of objects with cosmic time. Such a result would be consistent with the estimates for radio quasars by Peacock (1985), which also seem to show a peak at z~2 followed by a decline in comoving density. Radio and X-ray flux would be unaffected by intervening dust; the production of this apparent trend by optical obscuration would require a significant shift to much fainter magnitudes of the optical counterparts for faint radio quasars.

Could the observational techniques be introducing substantial incompleteness at high redshifts? Slitless spectroscopy may miss objects with weak emission line intensities and color selection based on indices with significant measuring uncertainties may lose objects in the stellar color locus. On the other hand, both techniques would have to be 90% incomplete to produce a space density of luminous quasars equal to that at z=2. Such a low level of completeness is unlikely.

A relatively late formation epoch for quasars eases constraints in most cosmological scenarios for the timescales to develop deep gravitational potentials. Efstathiou and Rees (1988) investigated quasar formation in the Cold Dark Matter scenario; this model predicts very few quasars at z>5, but a peak at z=2 allows some cosmic time between the coalescence of massive host galaxies and the initiation of quasar activity.

An alternative explanation of the observed redshift distribution is that it is strongly shaped by the effects of intervening obscuration as a function of increasing pathlength. The presence of many absorption systems in the spectra of high-redshift quasars, including several metal-line systems, suggests that associated dust would play a role in changing the observed energy distributions of the quasars.

The problem has been explored from a theoretical viewpoint by Heisler and Ostriker (1988). They examine the effect of intervening galactic disks with an exponential distribution of optical depth with radius and a distribution of tilts. The blue optical depth at the center of the disk is assumed to be ~.5. Such a population of absorbing galaxies leads to two underestimates for optical flux-limited quasar samples: that of the true quasar density at high redshifts, amounting to a factor of 10 at z=3, and that of the average absorption along the line of sight.

An empirical approach can also be taken, by examining the observed number of absorbers along the line of sight, and estimating the amount of associated dust. Wolfe et al. (these proceedings) report on the damped Ly α systems, which seem to contain a significant fraction of the baryonic matter at high redshifts. For a total redshift interval of 55, they discovered 16 systems with log N (H I) > 20.3 at ⟨z⟩ = 2.4. A uniform distribution of objects in Ω = 1 cosmology scales as $\sqrt{(1+z)}$ per unit redshift, implying that there are 1.1 systems per line of sight out to z=4. This value is insufficient to account for any major absorption effects. A more careful analysis by Fall and Pei (1988) finds that the damped Ly α systems cannot create reddening sufficient to obscure a substantial fraction of the high-redshift quasar population.

The question can be turned around to ask what kind and frequency of absorbers are required to hide a space density of high-redshift quasars equal to that at z=2? To estimate this requirement, the assumption is made that the true luminosity functions at z=3 and z=4 are identical to that observed at z=2. The condition to be met is that the integrated comoving density of quasars to an apparent M_B = -26 mag decreases by a factor of 3 between z = 2 and z = 3, and by another factor of 3 from z=3 to z=4. Systems with A_B ~0.44 mag cause an apparent drop in the integrated number of high luminosity objects (because of the steep slope of the LF) by almost a factor of 3, while introducing a negligible change in the power-law slope of the observed ultraviolet spectral energy distribution. This relatively modest total extinction produces a large effect because of the wavelength dependence of the extinction and the fact that the observed B magnitudes come from an emitted wavelength range around 1300 Å. For a Galactic gas/dust ratio, such a system would have log N (H I) = 19.7. To produce the apparent decreases in integrated density, it is sufficient to have a uniform volume density of these absorbers with 2 per line of sight out to z=2, 3 out to z=3, and 4.4 out to z=4.

What is the observed frequency of systems with this column density? A power-law distribution of column densities fit to the high column density end of the observed systems is significantly flatter than the power-law found for low column density absorbers (Bechtold 1987). Using this relation, it is found that systems with log N (H I) = 19.7 are ~4 times more frequent than systems with log N (H I) = 20.3. Thus, the observed absorbers with column densities 5 times lower than the damped systems of Wolfe et al. could distort the integral density of high-luminosity quasars to an apparent fixed luminosity limit as a function of redshift. This construct requires only a normal Galactic dust/neutral gas ratio. The major caveat is that the dust cannot produce a nominal 2200 Å feature; its strength must be significantly diminished, since such features are not typically detected in connection with the high column density absorbers. It is also valid to ask whether a normal dust/gas ratio is expected in a physical regime in which the molecular/neutral gas ratio is very different from that observed locally (Black et al. 1987). Spectral energy distributions over a wide range of observed wavelengths will be required to test for the presence of more pronounced reddening in high redshift objects relative to their lower redshift counterparts.

VI. INTERPRETATIONS OF A CHANGING LF

The ultimate goal of statistical studies is to relate ensemble behavior to the evolutionary histories of individual objects. Derivation of the true quasar LF at high redshifts, when combined with the statistical samples of high-redshift galaxies, will offer a preliminary picture of the development of the central gravitational potential and its delay from the epoch of initial star formation and coalescence. Inferences drawn from the decreasing activity between z=2 and the current epoch give us general insight into the fraction of sites that hosted nuclear activity.

Cavaliere et al. (1988) give a clear picture of the physical consequences of different interpretive assumptions of the changing LF. The straightforward mathematical formalism for the evolution of the LF at lower redshifts is luminosity evolution. The simplest view is that the total number of quasars is statistically conserved and amounts to a few percent of the brightest galaxies. The population dims by a factor of 35 on a natural timescale of about 3 Gyr, independent of source luminosity. That value requires a currently unknown physical regulatory mechanism, since it is intermediate between a Hubble time and the timescale for a Black Hole energy source. The consequence is that today there will be a small number of sources, about 6×10^3 Gpc^{-3}, with very large Black Holes and a very large integrated energy output. An average local luminosity greater than 1/40 of the Eddington luminosity could rule out this simple picture. David et al. (1987) proposed a physical model of stellar mass loss following the collapse of a dense star cluster into a Black Hole as a candidate to fit this scenario.

An alternative physical view is that of density evolution. The extreme example given by Cavaliere et al. is that of quasars as single events per host, with many generations of different hosts. This picture requires a spread of central potential formation or latency times. The density of hosts would then be 7×10^5 Gpc^{-3} for z>1.5, with local and weak emitters contributing substantially more by number. This scenario would predict luminosities near Eddington and high Black Hole masses for inactive objects.

An intermediate case is that of recurrent activity triggered by interactions. In that case, the rate of interactions is expected to decrease with cosmic time; one model is given by Roos (1985) based on scattering stars into loss-cone orbits. In his model, the combination of luminosity and density evolution reproduces the observed shape of the LF with time. Cavaliere et al. work out that the density of host galaxies is $\sim 3 \times 10^6$ Gpc^{-3}, which includes all galaxies to L(B) ~ 0.1 L*. The prediction for this scenario is that L/L_{Edd} for AGN's will increase toward 1 as redshift increases.

It may therefore be possible to gather observational evidence to distinguish these physical models. Evidence for interactions and availability of new quasar sites as a function of epoch comes from imaging surveys of quasars and AGN's (e.g., Yee and Green 1987, Yee 1987, Stockton and MacKenty 1987). The existence of currently inactive Black Holes in local galaxies has been proposed (Tonry 1987, Kormendy 1988, Dressler and Richstone 1988). The census and mass distribution of such objects is critical to validate the gravitational model of energy production and to estimate the number of potential AGN hosts. Physical estimates of the central mass and accretion rate of quasars as a function of epoch also discriminate between evolutionary models within the Black Hole model context (Sun and Malkan 1988, Czerny and Elvis 1987). Finally, the determination of observed spectral energy distributions over a broad emitted wavelength range for high-redshift quasars can begin to answer the question of whether the observed deficit reflects directly the epoch of quasar formation or whether it represents the gradual obscuration of that era from our view.

REFERENCES

Avni, Y. and Bahcall, J.N. 1980, Ap. J., 235, 694.
Bechtold, J. 1987, MWLCO Preprint #3058.
Black, J.H., Chaffee, F.H. Jr. and Foltz, C.B. 1987, Ap. J., 317, 442.
Boyle, B.J., Fong, R., Shanks, T. and Peterson, B.A. 1987, M.N.R.A.S., 227, 717.
Boyle, B.J., Shanks, T., and Peterson, B.A. 1988, in 'Proceedings of a Workshop on Optical Surveys for Quasars,' Eds. Osmer, Porter, Green, and Foltz, A.S.P. Conf. Series. Vol. 2, p. 1.
Braccesi, A., Formiggini, L., and Gandolfi, E. 1970, Astr. Ap., 5, 264.
Braccesi, A., Zitelli, V., Bonoli, F. and Formiggini, L. 1980, Astr. Ap., 85, 80.
Cavaliere, A., Giallongo, E., Padovani, P. and Vagnetti, F. 1988 in 'Proceedings of a Workshop on Optical Surveys for Quasars', eds. Osmer, Porter, Green, and Foltz, P.A.S.P. Conf. Series, Vol. 2, p. 335.
Cheney, J.E. and Rowan-Robinson, M. 1981, M.N.R.A.S., 195, 497.
Cheng, F.-Z., Danese, L., De Zotti, G. and Franceschini, A. 1985, M.N.R.A.S., 212, 857.
Crampton, D., Cowley, A.P., and Hartwick, F.D.A. 1987, Ap. J., 314, 129.
Czerny, B. and Elvis, M. 1987, Ap. J., 321, 305.
David, L.P., Durisen, R.H. and Cohn, H.N. 1987, Ap. J., 313, 556.
Dressler, A. and Richstone, D.O. 1988, Ap. J., 324, 701.
Efstathiou, G. and Rees, M.J. 1988, M.N.R.A.S., 230, 5P.
Fall, S.M. and Pei, Y.-C. 1988, Ap. J., in press.
Foltz, C.B., Chaffee, F.H. Jr., Hewett, P.C., MacAlpine, G.M., Turnshek, D.A., Weymann, R.J., and Anderson, S.F. 1987, A.J., 94, 1423.

Foltz, C.B. and Osmer P. 1988, in 'Proceedings of a Workshop on Optical Surveys for Quasars,' eds. Osmer, Porter, Green, and Foltz, A.S.P. Conf. Series, Vol. 2, p. 361.

Hazard, C., McMahon, R.G., and Sargent, W.L.W. 1986, Nature, 322, 38.

Heisler, J. and Ostriker, J.P. 1988, Ap.J., 332, 543.

Hoag, A.A. and Smith, M.G. 1977, Ap. J., 217, 362.

Koo, D.C. 1983 in Proc. 24th Liege Ap. Colloq., 'Quasars and Gravitational Lenses,' ed. Swings (Liege, Institut d'Astrophysique), p. 240.

Koo, D.C. and Kron, R.G. 1988, Ap. J., 325, 92.

Kormendy, J. 1988, Ap. J., 325, 128.

Marano, B., Zamorani, G., and Zitelli, V. 1988, M.N.R.A.S., in press.

Marshall, H.L. 1987, A.J., 94, 628.

Marshall, H.L., Avni, Y., Braccesi, A., Huchra, J.P., Tananbaum, H., Zamorani, G. and Zitelli, V. 1984, Ap. J., 283, 50.

Mathez, G. 1978, Astr. Ap., 68, 17.

Mitchell, K.S., Warnock, A. III, and Usher, P.D. 1984, Ap. J.(Letters), 287, L3.

Osmer, P.S. 1982, Ap. J., 253, 28.

Peacock, J.A. 1985, M.N.R.A.S., 217, 601.

Roos, N. 1985, Ap. J., 294, 486.

Schmidt, M. and Green, R.F. 1983, Ap. J., 269, 352.

Schmidt, M., Schneider, D.P. and Gunn, J.E. 1988, in 'Proceedings of a Workshop on Optical Surveys for Quasars', eds. Osmer, Porter, Green, and Foltz, A.S.P. Conf. Series, Vol. 2, p. 87.

Segal, I.E. and Nicoll, J.F. 1986, Ap. J., 300, 224.

Sramek, R.A. and Weedman, D.W. 1978, Ap. J., 221, 468.

Stockton, A. and MacKenty, J.W. 1987, Ap. J., 316, 584.

Sun, W.-H. and Malkan, M.A. 1988 in 'Supermassive Black Holes', ed. Kafatos, Reidel, p. 273.

Tonry, J.L. 1987, Ap. J., 322, 632.

Usher, P.D., Green, R.F., Huang, K.L. and Warnock, A. III 1983, in 'Quasars and Gravitational Lenses,' Proc of the 24th Liege International Astr. Colloq., Universite de Liege.

Veron, P. 1986, Astr. Ap., 170, 37.

Warren, S.J., Hewett, P.C. and Osmer, P.S. 1988, in 'Proceedings of a Workshop on Optical Surveys for Quasars,' eds. Osmer, Porter, Green, and Foltz, P.A.S.P. Conf. Series, Vol. 2, p. 96.

Weedman, D. 1985, Ap. J. Suppl., 57, 523.

Weedman, D. 1986, in Trieste Symp. on 'Structure and Evolution of Active Galactic Nuclei', ed. Giuricin, Reidel.

Yee, H.K.C. 1987, A. J., 94, 6.

Yee, H.K.C. and Green, R.F. 1987, Ap. J., 319, 28.

DISCUSSION:

SILK: If the picket fence model of Heisler and Ostriker is correct, then one would expect considerable patchiness in the redshift distribution over a wide range of angular scales. Can you comment on any evidence for this?

GREEN: This effect should lead to non-Poisson fluctuations for a given redshift. The Durham group has looked for this in the context of clustering. Perhaps Tom Shanks could comment on their results?

SHANKS: Although our QSO number counts over the sky are reasonably Poisson, we do detect strong QSO clustering on scales of $10\,h^{-1}$ Mpc. To check these results agains the Heisler–Ostriker model I'd need to know, for example, the expected angular scales of the patchiness. On the other hand we have already detected dust absorption of line-of-sight QSOs by low redshift $(z < 0.3)$ galaxy groups in 2-D cross-correlation analyses of the galaxy and QSO catalogues.

SILK: The expected angular scale would be several degrees at low $(z \sim 2)$ redshift, reducing to a scale of arcminutes at high $(z \geq 3)$ redshift.

FALL: Yichuan Pei and I have made a detailed search for dust in the damped Lyα systems in the Wolfe *et al* survey. Our 95% confidence limits, $(D/G)_{DLy\alpha} < 0.5\ (D/G)_{Gal}$, when combined with the column densities and redshifts of the damped Lyα systems, imply that the mean optical depth to a redshift $z \sim 3$ is fairly low. As a consequence, the observed luminosity function of quasars must be very close to the true luminosity function.

SHAPIRO: I believe that the dust obscuration model for producing an apparent quasar redshift cut-off has a problem with the X-ray background. Since the intervening obscuring objects are transparent to X-rays the X-ray emission of the dust-obscured high redshift quasars is not obscured. Apparently, Heisler and Ostriker find that these quasars would contribute a substantial fraction of the observed intensity of the X-ray background. Unless high-redshift quasars have a significantly different X-ray spectrum, however, from that observed for lower redshift quasars, this would result in a spectrum for the X-ray background in conflict with the observations. Could you comment on this?

GREEN: I avoided using the X-ray source counts as a diagnostic for dust. Depending on the method used to relate X-ray to optical source counts, there is considerable dispersion (more than a factor of 3) in the predictions of the fraction of the X-ray background contributed by AGN's. A modest amount of evolution in the X-ray properties of the observed objects can lead either to a contribution exceeding 100% or to a very low value, leaving considerable room for an unseen component.

REES: Would you care to comment on whether the observed luminosity function is steep enough for gravitational lensing to modify and confuse your conclusions?

GREEN: This was a topic of lively discussion at the Tucson Quasar Workshop last January, which did not lead to a clear resolution. $L^{-3.7}$ is somewhat too shallow for lensing to be a major contributor. As a benchmark, there are six quasars with redshifts greater than 1.5 in the Palomar Bright Quasar survey, only one of which (the "triple") has been shown to be lensed.

PEACOCK: I don't understand how you can look at the evolution of the total numbers of QSOs above some threshold relative to M^*, as we don't know how M^* evolves at $z > 2$. Since at $z < 2$ $L^* \propto (1 + z)^{3.5}$, any extrapolation would be dangerous.

GREEN: The test for the continuation of the same evolutionary form in L^* beyond $z = 2$ is exactly to integrate to some threshold relative to $M^*(z)$ and test for the constancy of the volume density. Integrating to a fixed value of M_B compares the value at $z = 2$ to that at higher redshifts in the case of no further evolution in L^*. Even then, there seems to be a deficiency.

FIFTY-THREE MULTICOLOUR SELECTED QUASARS WITH REDSHIFTS GREATER THAN THREE

S. J. WARREN, P. C. HEWETT & P. S. OSMER

Institute of Astronomy,
Madingley Road,
Cambridge CB3 OHA,
UK

Kitt Peak National Observatory,
National Optical Astronomy Observatories,
Tucson, Arizona 85726,
USA

ABSTRACT. We present the absolute magnitude v redshift diagram for the 53 quasars of redshift $z \geq 3.0$ found in two UK Schmidt Telescope (UKST) fields (effective area 50 square degrees) by a UB_JVRI multicolour technique. We describe the criteria adopted for selecting candidates for spectroscopic observation and explain how the well-defined selection criteria will enable the incompleteness of the sample to be quantified precisely.

1. Introduction

We are engaged in a wide-field survey to investigate the nature of the evolution of the quasar luminosity function (QLF) over the redshift interval $2.2 < z < 4.5$, for quasars brighter than an apparent magnitude $m_R = 20.0$. The survey is sensitive to quasars of intermediate luminosity, $M_B \leq -25.5$ at $z = 3$ ($H_o = 50, q_o = 0.5$), and complements the deeper surveys undertaken by Koo and Kron and others. The essentials of the technique have been described by Warren, Hewett and Osmer (1988), but to summarise the main points: scans by the Automated Plate Measuring facility of pairs of UKST direct plates produce magnitudes or limits to magnitudes in five passbands, U, B_J, V, R and I. A multicolour photometric catalogue is formed, of stellar images detected in the R passband (but possibly absent from other passbands), and brighter than the survey limit. Candidate quasars are selected by identifying objects with unusual colours: previously this was achieved by picking off outliers in four-dimensional (4-D) multicolour space, but we now work in 5-D magnitude space. Candidates are ranked by their 5-D distance to the 10th nearest neighbour (which is effectively the distance to the edge of the densely populated locus of common stars), and the number of candidates is determined by an arbitrary cutoff distance. This cutoff then defines one of the selection criteria used in establishing the sample incompleteness, as explained below.

C. S. Frenk et al. (eds.), The Epoch of Galaxy Formation, 135–139.

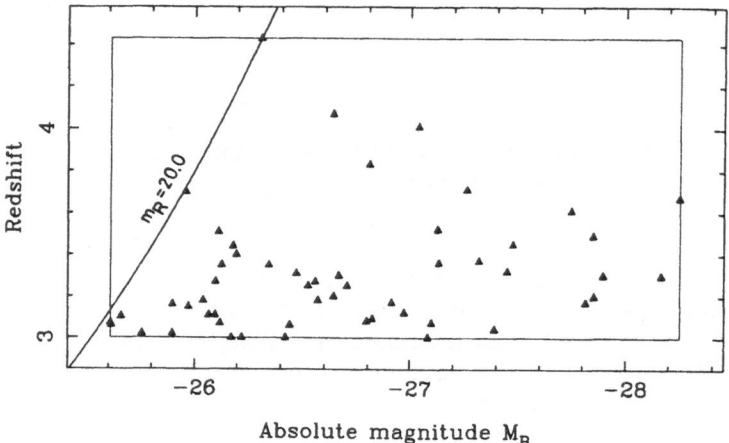

Figure 1: The Hubble diagram for the 53 multicolour selected quasars with redshifts $z \geq 3.0$ found in two UKST fields, the SGP and F401.

2. Observations

Candidate lists have been compiled for two fields, the South Galactic Pole (SGP, 0h 53m, $-28°$ 03') and UKST Field 401 (F401, 20h 48m, $-35°$ 00'). Follow-up spectroscopic observations have been undertaken through 1986 to 1988 at the Anglo-Australian Telescope and the Cerro Tololo Inter-American Observatory 4-metre, and have proceeded in two phases. In the first, we have sparse-sampled the parameter space, selecting candidates from the top of the ranking list, with a very broad colour range, in order to assess whether previous searches for high-redshift quasars were in fact seriously incomplete, due for instance to some unusual spectral evolution of the quasar population. This approach has the important advantage that preconceived ideas concerning the likely location of high-redshift quasars in the multicolour space were not a factor in the selection. Practical limitations, related to the availability of telescope time for follow-up spectroscopy, suggest a modification to this extremely general approach! Thus, once a sufficiently large sample of high-redshift quasars had been identified, we used these to define the multicolour domain occupied by quasars, and selected all objects from our candidate lists whose colours were similar to those of the known quasars. Using this approach it was possible to gradually extend the search radius around the known quasars, whilst monitoring the identification success rate. The candidate selection was repeated as each new quasar was identified. The hyperspheres in multicolour space around each quasar largely overlap, and define a large domain in the parameter space where we have obtained spectra of all candidates to the adopted threshold distance.

In the SGP and F401 our sample contains respectively 35 and 18 quasars of redshift $z \geq 3.0$. The absolute magnitude v redshift diagram for these

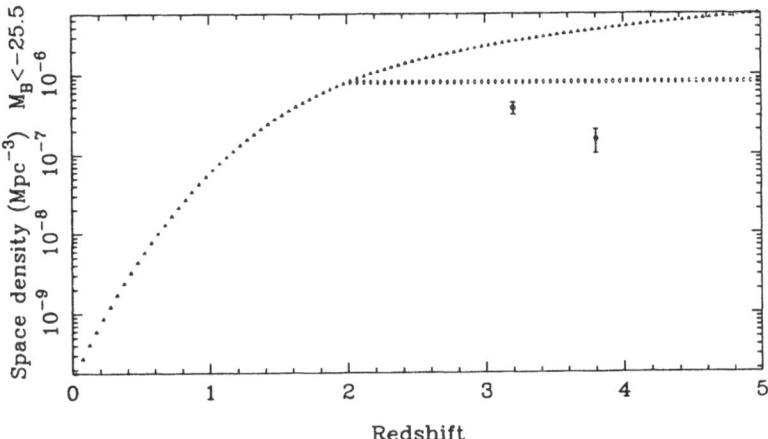

Figure 2: The variation of the space density of quasars brighter than $M_B = -25.5$ as a function of redshift, showing two alternative evolutionary models (see text), and the data for the SGP field computed with a preliminary assessment of the sample incompleteness.

53 objects is shown in Fig. 1, for adopted cosmological parameters $H_0 = 50, q_0 = 0.5$, and spectral index $\alpha = -0.5$. There are two likely reasons for the apparent discrepancy in surface density between the two fields. For the data of Fig. 1 we have made no correction for interstellar absorption for F401, which may account for a part of the effect, although based on the maps of Burstein and Heiles (1982) this is unlikely to amount to more than $A_R = 0.2$. At the revised F401 sample limit, $m_R = 19.8$, the number of quasars of redshift $z \geq 3.0$ in the SGP reduces to 30. A more likely explanation is due to the fact that F401 lies over the galactic bulge and contains about four times as many stars as the SGP, so the volume of the 5-D magnitude space occupied by stars (*i.e.* in which quasars are effectively undetectable) is considerably larger than for the SGP. Note that this source of incompleteness will be accounted for in deriving the selection functions for each field.

In Fig. 2 we present an estimate of the integral space density of quasars brighter than $M_B = -25.5$, for two redshift intervals centred on $z = 3.2$ and $z = 3.8$, based on our SGP data alone. The upper dotted line shows the evolutionary model of Boyle, Shanks and Peterson (1988), derived from their analysis of the QLF for redshifts $z < 2.2$, but here extrapolated to higher redshifts. The lower line represents an alternative picture, where the evolution ceases abruptly at a redshift $z = 2$, with both the shape and normalisation of the luminosity function held constant thereafter. The two data points are derived from a preliminary analysis of the sample incompleteness. Also, because our sample does not reach to $M_B = -25.5$ over the complete redshift range of Fig. 1, we have followed a procedure similar to the analysis

presented by Warren, Hewett and Osmer (1988): assuming simple density
evolution over the redshift range of interest with no change in the shape
of the QLF. The respective density and luminosity functions that describe
this behaviour have been computed following the method of Chołoniewski
(1986).

3. Quantification of incompleteness

It now becomes straightforward, in principle, to compare the sample of
Fig. 1 against the predictions for a particular QLF evolution model, and to
determine the range of models consistent with the data, for the well-defined
selection criteria and a precise knowledge of the photometric errors enable us
to compute the probability of detection of any type of quasar. For a quasar of
a particular intrinsic spectral-energy distribution, continuum absolute magnitude, and redshift, we compute the apparent magnitudes for the different
passband filter/emulsion combinations. We then vary the magnitudes in a
statistical manner and enter the artificial quasars into the original data-sets.
The candidate selection routines are then run for each realisation. The frequency with which an artificial quasar is detected (i.e. would have been
selected for spectroscopic observation, as defined by the foregoing criteria)
and the distribution of observed apparent magnitudes define the 'selection
function' for a specified type of quasar of a given continuum absolute magnitude and redshift. The set of arrays for a range of absolute magnitudes,
redshifts and quasar types defines the selection function for the whole survey. We can use this selection function to predict the observed numbers of
quasars as a function of apparent magnitude and redshift for any assumed
QLF and cosmological model. A suitable statistical test (e.g. 2-D K-S)
can be used to decide on the quality of the fit between model and data. It
should be noted that this procedure possesses the decided advantage that
Malmquist bias and k-corrections for the presence of emission lines in the R
passband are accounted for automatically. A detailed study of the sample
incompleteness along these lines is currently underway.

References

Burstein, D. & Heiles, C., 1982. *Astr. J.* **87**, 1165.
Chołoniewski, J., 1986. *Mon. Not. R. astr. Soc.* **223**, 1.
Warren, S. J., Hewett, P. C. & Osmer, P. S., 1988. In: *Optical Surveys for
    Quasars*, p.96, eds. Osmer, P. S., Porter, A. C., Green, R. F. & Foltz, C.
    B., *Publs astr. Soc. Pacif.* Conference Series No. 2.

DISCUSSION:

PEACOCK: From your $z - M_B$ diagram, there appears to be some evidence for a lack of high–z, bright–M_B objects suggesting that the decline to high z may be *more* rapid at the highest luminosities. Have you performed any test for correlation on the $z - M_B$ plane? In any case, does the density search discriminate against bright objects where low numbers mean that the stellar locus is less well-defined?

WARREN: I doubt that the effect is significant, but this has not yet been checked. Bright redshift four quarars do exist however: Q0000-26 (z = 4.11) has $M_B \sim -28$. As regards the density search, bright quasars will always be easier to find that faint quasars of the same colours.

GREEN: How do you define your limiting magnitude operationally? Is it done from object counts as a function of magnitude? What is the signal to noise ratio (net flux to sky + object fluctuations) of an object at the limiting magnitude?

WARREN: The sample limiting magnitude, $m_R = 20.0$, was chosen as this corresponds to the point at which the stellar locus starts to swell dramatically. The quoted magnitude comes from calibration against a standard sequence i.e. there will be a Malmquist bias in the measured quasar surface density. The survey limit is about 1 magnitude above the nominal limit of the R plates. At $m_R = 20.0$ the standard error is $\sigma = 0.08$.

THE LUMINOSITY FUNCTION AND CLUSTERING OF QSOS

T. Shanks, R. Fong
Physics Dept., Univ. of Durham, South Road, Durham, DH1 3LE
B.J. Boyle
Anglo-Australian Observatory, PO Box 296, Epping, NSW, Australia
B.A. Peterson
Mount Stromlo & Siding Spring Observatories, Woden, ACT 2606, Australia

ABSTRACT. We describe results from our survey of 423 $B \leq 21^m$, $z \leq 2.2$ QSOs, made using the FOCAP fibre optic coupler at the Anglo-Australian Telescope (AAT). The QSO luminosity function is found to show a turnover at faint magnitudes and the form of its evolution is consistent with pure luminosity evolution (PLE) models; this result is either an accident or implies that the comoving space density of QSOs has been constant since z=2.2. We have also found evidence for strong QSO clustering on scales smaller than $10h^{-1}$Mpc. At larger scales, the overall QSO correlation function shows no evidence for clustering in the range $(10 \leq r \leq 1000h^{-1}$Mpc). However, if attention is restricted to the low redshift QSOs, some evidence for trough-peak structures in the range $10 \leq r \leq 300h^{-1}$Mpc may be seen.

1. Introduction

The QSO survey of Boyle et al 1988 is based on COSMOS machine measurements of deep U and B UK Schmidt (UKST) photographs. These give B magnitudes and U-B colours for $\approx$20000 $B \leq 21^m$ stars per UKST field. The ultraviolet excess (UVX) criterion (U-B$\leq$-0.^{m}35) is then used to define 1600 QSO candidates per UKST field. The crucial advantage of UVX colour selection is that it allows complete QSO samples with $0 \leq z \leq 2.2$ to be selected; the vast majority of QSOs that have been detected by objective prism or radio methods have U-B$\leq$-0.^{m}35 (eg Schmidt & Green, 1983, Veron, 1983). These UVX QSO candidates were then observed spectroscopically using the AAT fibre optic coupler, FOCAP (Gray, 1986) which allows 50-60 QSO candidates to be observed in a 40 arcmin diameter field in a 3hr exposure.

The combination of the COSMOS/UKST QSO candidates and FOCAP proved very powerful. In 12 clear AAT nights we observed redshifts for 423 $B \leq 21^m$ QSOs in 34 fibre fields, spread over 8 UKST 25deg^2 fields. We identified 9 broad absorption-line QSOs as well as 61 narrow line compact galaxies ($z \leq 0.4$), 43 white dwarfs and 810 Galactic stars. Only 3% of our candidates were not positively identified and we obtained unambiguous redshifts for 86% of our QSOs. Currently this survey is the largest 'complete' QSO catalogue. At $B \leq 20.^m 9$ we find a QSO sky density of 37$\pm$3 deg^{-2}.

As Boyle et al 1988 show, the derived QSO number-redshift relation is smooth with no sign of preferred peaks and flat in the range $0.6 \leq z \leq 2.2$. The cut-off seen beyond z=2.2 is caused by a selection effect; once the Lyman alpha 'forest' is redshifted into the U band QSOs show much less UV excess. The QSO n(m) relation shows a turnover at B$\approx$19.^{m}5. At brighter magnitudes

C. S. Frenk et al. (eds.), The Epoch of Galaxy Formation, 141–145.

the d(logN)/dB slope is 0.86±0.04; at fainter magnitudes the slope is 0.31±0.05 (Boyle et al 1987,1988). Thus we have confirmed the result of Koo & Kron(1982) who previously detected this turnover from an analysis of QSO counts on KPNO 4-metre photographs.

2. The QSO Luminosity Function

With 397 QSOs in the complete, $z \leq 2.2$ sample, it is possible to divide the data into discrete redshift bins and directly inspect the QSO luminosity function and its evolution with redshift.

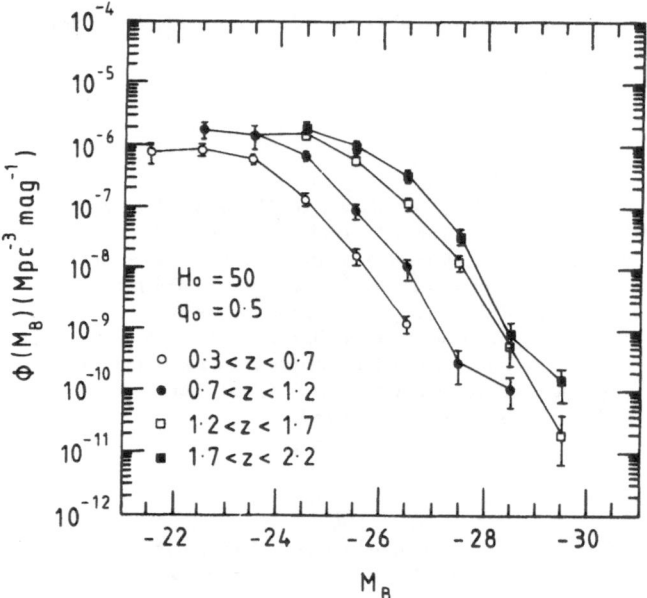

Fig. 1. The QSO luminosity function as a function of redshift from our survey, including data from other surveys at brighter limits (see Boyle et al, 1988).

In Figure 1 we show the results, where we have also used surveys complete to brighter limits (references given by Boyle et al 1988) to provide the data at the bright absolute magnitudes. It can be seen that a turnover is detected in each of our QSO luminosity functions. Secondly it is obvious that the luminosity function evolution takes the form of 'pure luminosity evolution' with the luminosity function seeming to move uniformly toward fainter magnitudes at lower redshifts. Using maximum likelihood methods to fit an evolving 2 power-law model we find that the knee of the luminosity function, L*, obeys

$$L^*(z) = L^*(0)(1 + z)^{3.15}$$

with $M^*(B) = -22.^m4$ ($H_o = 50 \mathrm{kms}^{-1}\mathrm{Mpc}^{-1}$, $q_o = 0.5$) at $z=0$, and $d(\log\phi)/d(\log L) = -3.8 \pm 0.15$ at bright absolute magnitudes , flattening to -1.4 ± 0.2 at fainter magnitudes.

Elsewhere (Boyle et al 1988) we show that when our PLE model fit is extrapolated to $z=0$, it approximately matches the Seyfert galaxy luminosity function (eg Cheng et al 1985).

At higher redshifts, however, there is increasing evidence that this model predicts too many QSOs at $z \geq 2.5$ (eg Schmidt et al 1987, Warren et al this volume). Taken together, there is an attractively simple interpretation of the apparent evolutionary behaviour of the QSO luminosity function, as previously discussed by Koo (1983) and Marshall (1985). The PLE form of the evolution allows for the possibility that QSO numbers may be conserved between $z=2.2$ and $z=0$ with no births or deaths of QSOs. Thus QSOs may have formed at an epoch $z \geq 2.2$ and then dimmed uniformly with time to become the Seyfert galaxies we see around us at the present day. Under this interpretation, the implication would be that QSOs are much longer lived than, for example, required by a galaxy interaction mechanism for QSO fuelling.

However, the above argument may be naive because it implies very large black hole masses in the centre of some Seyferts at the present day. Assuming $L^* \propto (1+z)^{3.15}$ and an efficiency of 10% for the conversion of mass into radiation it is easy to show that an M^* Seyfert at the present day contains a black hole of mass $\approx 5 \times 10^{10} M \odot$. This assumes a bolometric luminosity of $1.5 \times 10^{46} ergs^{-1}$ for an M^* QSO at the present day; the implied mass would drop by a factor of ≈ 7 if the QSO luminosity only evolves in the optical and UV parts of its spectrum. At the NGC4151 luminosity ($M_B \approx -19^m$) the implied black hole mass is $\approx 10^9 M \odot$. Although high, these masses have not yet been ruled out on the basis of emission line or X ray variability. The alternative interpretation is that the apparent PLE form for the evolution is simply accidental, representing a coincidence between the birth and death rates of much shorter lived QSOs. In this case the luminosity function results can still be used to place constraints on short-lived evolutionary models after the fashion of Cavaliere et al (1985). But in this circumstance a price has to be paid; without the underlying basis of the $(1+z)^{3.15}$ evolution and a constant QSO space density, it becomes more difficult to justify a simple extrapolation of the QSO luminosity function evolution to predict QSO densities at higher redshifts. This makes less straightforward the identification of any apparent high redshift QSO 'cutoff' with an epoch of QSO or galaxy formation.

3. QSO Clustering.

We have analysed the QSO survey using correlation techniques to place constraints on QSO clustering (Shanks et al 1987, 1988). Despite the survey's large size there are relatively small numbers of QSO pairs with comoving separation ($r \leq 10h^{-1}$Mpc). Nevertheless we have still the most significant detection of clustering in a complete QSO survey. For the $q_o=0.5$ model we observe 25 pairs of QSOs with $r \leq 10h^{-1}$Mpc whereas we would have expected 11 if the QSOs were distributed at random. Since more than 90% of the QSO pairs are independent, Poisson statistics apply and this represents a 4.2σ rejection of the random null hypothesis. Although the form of the QSO correlation function at small scales is not inconsistent with the -1.8 power-law slope of the galaxy correlation function, the amplitude is 2.6 ± 0.5 times higher than would be expected if the QSOs were randomly sampling a 'stable' galaxy clustering distribution. This tends to fit the observation of Yee & Green (1987) who found in CCD imaging experiments that low redshift, radio-quiet QSOs are found in 2.6 times richer than average galaxy environments.

At separations larger than $10h^{-1}$Mpc, the QSOs are unmatched as probes of the large scale structure of the Universe. The QSO correlation function ($q_o = 0.5$) is shown in Fig. 2. Errors here are estimated from field-to-field variations. The correlation function shows no deviation from the homogeneous $\xi = 0$ result which is significant at more than the 2σ level. As can be seen the level of clustering at $r \geq 10h^{-1}$Mpc is significantly less than that shown by Abell clusters (Bahcall & Soneira 1983). Thus the initial impression is that the results give little indication of any large scale inhomogeneity. However there is one important caveat here. If the Universe is Einstein-de Sitter then gravitational instability causes small features in ξ grow quickly with redshift as $1/(1+z)^2$. In Figs. 3a and b we show the correlation function divided

144

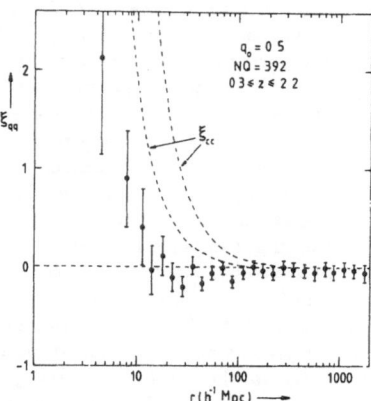

Fig. 2 The QSO 2-point correlation function for $0.3 \leq z \leq 2.2$.

into high and low redshift bins. In the high redshift bin the correlation function at large scales is very flat and close to zero. However, in the low redshift bin the correlation function gives a different impression. Although there is no tendency to develop a tail of positive correlation like ξ_{cc}, there does seem to be increasing evidence for a trough-peak structure developing in the range 10-300h^{-1}Mpc. Although some points appear significantly different from zero (eg the point at r=85h^{-1}Mpc) it may be too early to conclude that there is some indication of large scale structure in the QSO data at low redshift. But determining the reality of these large scale features and investigating whether they evolve with redshift provides a solid motivation to enlarge our QSO survey; we are currently considering technical improvements that may allow an order of magnitude increase in our QSO survey size.

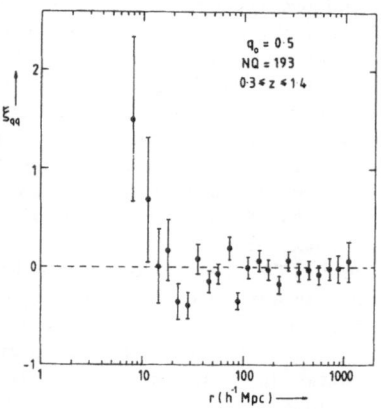
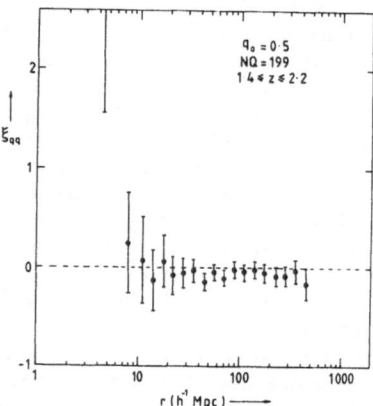

Fig. 3a. As Fig. 2 for the range $0.3 \leq z \leq 1.4$. **Fig. 3b.** As Fig. 2 for the range $1.4 \leq z \leq 2.2$.

4. Conclusions

Our redshift survey of 423 faint QSOs has produced important constraints on the QSO luminosity function and its evolution. A turnover in the QSO luminosity function has been detected at all redshifts in the range (0.3≤z≤2.2). The evolution of the luminosity function takes the form of pure luminosity evolution, allowing the possibility that for z≤2.2 QSO numbers have been conserved. However, this leads to a very large implied mass for the black hole at the nucleus of a present day Seyfert and the pure luminosity evolution form may simply be accidental. Using correlation analysis, we have detected strong QSO clustering at $r \leq 10h^{-1}$Mpc in our survey. At larger separations there is no definite evidence for large scale inhomogeneity, but if attention is restricted to the lower z QSOs tentative evidence for trough-peak structures in ξ in the range $10 \leq r \leq 300h^{-1}$Mpc may be seen. We have noted that the evolution of the large scale correlation function with redshift, could provide a powerful cosmological probe.

References

Bahcall,N.A. & Soneira, R.M. 1983, Astrophys. J., 270, 20.
Boyle, B.J., Fong, R., Shanks, T. & Peterson, B.A. 1987 Mon. Not. R. astr. Soc., 227, 717.
Boyle, B.J., Fong, R.,Shanks, T. & Peterson, B.A, 1988 in preparation.
Boyle, B.J.,Shanks, T. & Peterson, B.A. 1988, Mon. Not. R. astr. Soc., in press.
Cavaliere, A., Giallongo, E. & Vagnetti, F. 1985, Astrophys. J. 296, 402.
Cheng, F.Z., Danese, J., De Zotti, G. & Francheschini, A. 1985, Mon. Not. R. astr. Soc., 212, 857.
Gray, P.M. 1986, Proc. SPIE, 627, 96.
Koo, D.C., & Kron, R.G. 1982, Astron. & Astrophys., 105, 107.
Koo, D.C. 1983, In 'Quasars and Gravitational Lenses', ed. J.P. Swings, pp. 240-244, Liege: Universite de Liege.
Marshall, H.L. 1985, Astrophys. J., 299, 109.
Schmidt, M. & Green, R.F. 1983, Astrophys. J., 269, 352.
Schmidt, M., Schneider, D.P. & Gunn, J.E. 1987, Astrophys. J., 316, L1.
Shanks, T., Fong, R., Boyle, B.J. & Peterson, B.A. 1987, Mon. Not. R. astr. Soc. 227, 739.
Shanks, T., Boyle, B.J. & Peterson, B.A. 1988 Publ. Astr. Soc. Pacif. in press.
Veron, P. 1983 In 'Quasars and Gravitational Lenses', ed. J.P. Swings, pp. 210-235, Liege: Universite de Liege.
Yee, H.K.C. and Green, R.F. 1987 Astrophys. J., 319, 28.

DISCUSSION:

PEACOCK: If you want to, you can make a PLE model fit the radio quasar and radio galaxy data. However, we know directly that these objects have lifetimes $\ll H_o^{-1}$. By analogy, therefore, I don't feel the long-lived interpretation of optical PLE should be taken very seriously.

SHANKS: I take the warning although I suspect that the suggestion of PLE in the radio data is not as explicit as in the optical.

HIGH REDSHIFT QUASARS IN THE COLD DARK MATTER COSMOLOGY

G. Efstathiou
Department of Astrophysics,
Oxford, England.

ABSTRACT. We argue that the observed space density of luminous quasars between $z = 2$ and $z = 4$ is compatible with the cold dark matter (CDM) model if quasars are short lived and radiate at about the Eddington limit. However, according to the CDM model the abundance of high luminosity quasars must decline exponentially at higher redshifts. Even if all proto-galaxies form quasars, and about 1% of the baryons within a protogalaxy collapse into a compact object, we would expect a steep fall in the density of quasars with $L > 10^{47} \mathrm{erg\,s^{-1}}$ at redshifts $z \gtrsim 5$. The existence of a cut-off in the quasar numbers at high redshift could therefore supply an important test of the CDM theory.

1. INTRODUCTION

Quasars offer important clues to conditions in the Universe at redshifts $z \sim 2 - 4$. Before any quasars can turn on, some galaxies must have evolved at least to the stage of harbouring enough concentrated mass-energy to provide the observed luminosity. The question then arises of whether the CDM cosmogony, with its late epoch of galaxy formation, is compatible with the discovery of quasars at $z > 4$.

The comoving density of quasars is well established to be much higher at $z \approx 2$ than at the present epoch. For a cosmological model with $q_0 = 0.5$ the number density of quasars at $z = 2$ brighter than $L = 10^{47} \mathrm{h_{50}^{-2}}$ erg s^{-1} is approximately

$$N_Q \approx 1.2 \times 10^{-7} \mathrm{h_{50}^{-3}} \quad \text{per comoving Mpc}^3 \tag{1}$$

(Schmidt and Green 1983; Weedman 1985; Boyle et.al. 1987; Koo and Kron 1988). In deriving equation (1), we have assumed that bolometric luminosities are given approximately by $3(\nu F(\nu))_\mathrm{B}$ (e.g. Phinney 1983).

The prodigous rise in the comoving number density of luminous quasars seen at low redshifts does not continue beyond $z \gtrsim 2.5$ (e.g. Schmidt, Schneider and Gunn 1986), but the quantitative details of the change in evolutionary behaviour (e.g. the luminosity dependence) are controversial. There seems to be no incontrovertible evidence for a decline in the density of luminous quasars below the number given in (1) at redshifts $\lesssim 3.5$ (Hazard and McMahon 1985; Koo and Kron 1988). As discussed at this meeting, there is no absolute redshift cut-off at $z \leq 4$ since some intrinsically very powerful quasars are seen at higher redshifts (Warren et.al. 1987).

C. S. Frenk et al. (eds.), The Epoch of Galaxy Formation, 147–152.

The central mass involved in a quasar depends on its luminosity, its lifetime t_Q, and the efficiency ϵ defined as the fraction of the central rest mass energy that is converted into radiation. The central mass associated with each $10^{47}h_{50}^{-2}$ erg s^{-1} quasar may therefore be written as

$$M_Q \approx 2 \times 10^9 h_{50}^{-2} t_{Q_8} \epsilon_{0.1}^{-1} \, M_\odot, \tag{2}$$

where t_{Q_8} denotes the lifetime in units of 10^8 years and $\epsilon_{0.1} = \epsilon/0.1$. At present there is no way of firmly estimating t_Q (though see Kovner and Rees, 1988), thus we do not know whether there were one, or several, generations of high redshift quasars.

The density of quasars implied by (1) is of course very low compared to the present density of bright galaxies. However, according to the CDM model, galaxies could have formed only at progressively rarer upward fluctuations in the initial density field as we extrapolate back in time. At high redshifts we must expect a precipitous drop in the number of galaxies capable of harbouring a central object as massive as required by (2). Our goal then is to calculate the critical redshift z_{crit} above which the abundance of luminous quasars would be expected to fall steeply below the value (1) and to check whether this is compatible with observational constraints.

2. THE HOST GALAXIES OF HIGH-Z QUASARS

The value of z_{crit} depends on how big a galaxy has to be in order to develop a central mass concentration of the requisite size (2). The ratio of the quasar's mass to the total galactic mass involves three factors:
(i) *The fraction* f_b *of the matter in baryonic form*, which is ≈ 0.1 if the $\Omega = 1$ CDM model is to be compatible with primordial nucleosynthesis (Yang et.al. 1984).
(ii) *The fraction* f_{ret} *of the baryons originally associated with the dark halo which are retained within the galaxy rather than being expelled via a supernova-driven wind.* This is likely to depend steeply on the galaxy's characteristic circular velocity v_c (Dekel and Silk 1986); it may well be $\lesssim 0.1$ for $v_c \lesssim 100$ km s^{-1}.
(iii) *The fraction* f_{hole} *of the baryons retained in the galaxy which participate in the runaway gravitational collapse processes manifested in the quasar phenomena.* This depends on the route whereby the central mass accumulates and evolves (*cf.* Begelman and Rees 1978), and on how efficiently gas in the outlying parts of the galaxy can lose angular momentum and sink towards the centre - two perennial uncertainties bedevilling all attempts to model active galactic nuclei quantitatively.

If for instance, one were to suppose $F = f_b f_{ret} f_{hole} = 0.01$ - likely to be a highly optimistic estimate (from the point of view of efficient formation of a central mass concentration) - then equation (2) implies a total halo masse of $\sim 10^{12} \epsilon_{0.1}^{-1} h_{50}^{-3} M_\odot$ if $t_Q = 5 \times 10^8 h_{50}^{-1}$ years, the elapsed time between $z = 4$ and $z = 3$. More pessimistic, and probably more realistic, suppositions about the efficacy of black hole formation would imply correspondingly larger galactic masses.

3. ABUNDANCE OF HIGH-Z QUASARS

To compute the number density of rare objects we use the Press-Schechter (1974) theory. The details of this calculation are given by Efstathiou and Rees (1988) together with a comparison of the results with N-body simulations. A summary of these results is shown in figure (1a) where we have assumed a normalization of the CDM spectrum identical to that adopted in the N-body studies of White et.al. (1987) and

Frenk et.al. (1988). A change in the normalization from σ_o to σ_o' would cause the curves in figure (1a) to shift to redshifts $(1 + z') = (1 + z)\sigma_o'/\sigma_o$. The abundances of rare objects are therefore extremely sensitive to the normalization of the CDM spectrum.

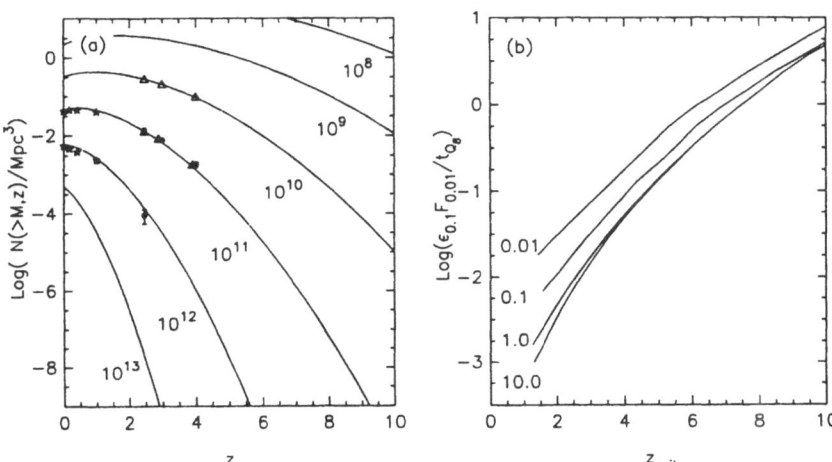

Figure 1. (a) shows the comoving number density of groups with mass $> M$ (indicated in $M_\odot$) plotted against redshift for the adiabatic, scale-invariant, CDM model with $h_{50} = 1$. The solid lines have been computed according to the Press-Schechter (1974) theory. The symbols show results from a series of N-body simulations. (b) shows the redshift z_{crit} at which the comoving number density of luminous $(L > 10^{47} \text{ erg s}^{-1})$ quasars must fall below (1) in the CDM model. Results are illustrated for several values of the lifetime t_Q which is indicated in units of 10^8 years.

If we now assume that every halo of mass M forms a quasar with lifetime t_Q, the expected number density of quasars is given approximately by

$$N_Q(> L_{47}, z) \approx 1 \times 10^{-3}(1 + z)^{5/2} t_{Q8}(L_{47}t_{Q8}\epsilon_{0.1}^{-1}F_{0.01}^{-1})^{-0.866}$$
$$\exp(-0.21(L_{47}t_{Q8}\epsilon_{0.1}^{-1}F_{0.01}^{-1})^{0.266}(1 + z)^2) \text{ Mpc}^{-3}, \qquad (3)$$

where $L_{47} = L/10^{47}\text{erg s}^{-1}$, $F_{0.01} = F/0.01$. Equation (3) is based on a power-law approximation to the density fluctuation spectrum in the CDM model with $h_{50} = 1$ and is accurate for halos in the mass range $10^{12} \lesssim (M/M_\odot) \lesssim 10^{10}$ provided that $(t_Q/N_Q)(\partial N_Q/\partial t) \ll 1$.

In figure (1b), we show z_{crit} plotted against $(\epsilon_{0.1}F_{0.01})/t_{Q8}$ for several values of t_{Q8} (Efstathiou and Rees, 1988). This figure shows that to avoid a catastrophic decline in the number density of luminous quasars at $z \lesssim 4$ requires

$$(\epsilon_{0.1}F_{0.01})/t_{Q8} \gtrsim 0.1. \qquad (4)$$

This condition can be fulfilled, for reasonable values of the lifetime t_Q, but not by a wide margin. As discussed in section 2, the value of $F_{0.01}$ depends on complex and uncertain astrophysics; however we consider any value exceeding 1 (corresponding to more than 10% the galactic baryons collapsing into the central object) to be highly implausible, and that 0.1 may be a more realistic upper limit. The radiative efficiency factor $\epsilon_{0.1}$ cannot substantially exceed 1 for any gravitational collapse or accretion process, nor even for more exotic processes such as electromagnetic extraction of energy from spinning black holes (Blandford and Znajek, 1977; Phinney, 1983).

The quantity $(\epsilon_{0.1} F_{0.01})$ is therefore unlikely to be much larger than 0.1. Inequality (4) then implies that very high-z quasars could more readily exist if t_Q were short. However, unless the quasars radiate with super-Eddington luminosities, their lifetimes must be longer than $t_E = \epsilon t_S = \epsilon \sigma_T c/(4\pi G m_p) \approx 4 \times 10^7 \epsilon_{0.1}$ years (Salpeter, 1964). We thus expect that $(\epsilon_{0.1} F_{0.01})/t_{Q_s} \lesssim 2 F_{0.01}$. Comparing this with figure (1b) we conclude that an exponential decline in the number densities of luminous quasars should be seen at $z \gtrsim 8$ for the highly optimistic value $F_{0.01} = 1$, or at $z \gtrsim 5$ if $F_{0.01} = 0.1$.

The preferred hypothesis is therefore that quasars have lifetimes $t_Q \approx t_E \lesssim 4 \times 10^7$ years, and radiate at around the Eddington limit. Their masses would then be $\sim 10^9 M_\odot$, and they would reside in galaxies with masses $10^{11} - 10^{12} M_\odot$ (for $F_{0.01} \approx 1 - 0.1$) and circular velocities $v_c \lesssim 200(1 + z)^{1/2} \mathrm{km\,s}^{-1}$. This hypothesis has the following further implications:

(i) In the above discussion we have assumed that a quasar can switch on as soon as a dark halo reaches virial equilibrium. Clearly to avoid low values of z_{crit} there cannot be a long latency period between the virialization of a galaxy and the onset of activity in its nucleus (cf. Cavaliere and Szalay, 1986).

(ii) At high redshifts $z \gtrsim 4$, we might expect to see strong differential evolution in the quasar numbers. This is because high luminosity quasars must be associated with massive proto-galaxies whereas low luminosity quasars could form in less massive and hence more abundant systems.

(iii) In the CDM model, luminous quasars at high redshifts would be associated with rare high peaks in the density field. They should therefore be correlated ab initio (see e.g. Bardeen et.al. 1986) with a clustering amplitude comparable to that of local bright galaxies.

(iv) At lower redshifts, $z \lesssim 2.5$, the comoving density of luminous quasars declines rapidly as the universe expands (e.g. Schmidt and Green 1983). In the CDM model, where galaxy formation continues until epochs $z \lesssim 2$, this decline must reflect a change in the probability that a bright quasar is associated with a newly formed galaxy. The physical mechanisms underlying such a change are poorly understood, though some authors have speculated that it may be related to the frequency of galaxy interactions (see e.g. Lin, Pringle and Rees, 1988; Sanders et.al. 1988).

4. CONCLUSIONS

A high space density of luminous quasars at $z \gtrsim 5$, could prove to be an embarrassment for the CDM theory. This would require one or more of the following: (a) that an extremely high fraction of the baryons within a protogalaxy can collapse and form a compact object, (b) exotic models in which quasars radiate at super-Eddington luminosities, (c) a reappraisal of the normalization of the CDM spectrum,(d) a nonstandard initial fluctuation spectrum.

I thank Martin Rees with whom this work was done and the Institute of Astronomy, Cambridge, where I have enjoyed many happy years.

REFERENCES

Bardeen, J.M., Bond, J.R., Kaiser, N. and Szalay, A.S., 1986, *Astrophys. J.*, **304**, 15.

Begelman, M.C. and Rees, M.J., 1978, *Mon. Not. R. astr. Soc.*, **185**, 847.

Blandford, R.D. and Znajek R.L., 1977, *Mon. Not. R. astr. Soc.*, **179**, 433.

Boyle, B.J., Fong, R., Shanks, T. and Peterson, B.A., 1987, *Mon. Not. R. astr. Soc.*, **227**, 717.

Cavaliere, A. and Szalay, A.S., 1986, *Astrophys. J.*, **311**, 589.

Dekel, A. and Silk, J., 1986, *Astrophys. J.*, **303**, 39.

Efstathiou, G. and Rees, M.J., 1988, *Mon. Not. R. astr. Soc.*, **230**, 5p.

Frenk, C.S., White, S.D.M., Davis, M. and Efstathiou, G., 1988, *Astrophys. J.*, **327**, 507.

Hazard,C, and McMahon, R., 1985, *Nature*, **314**, 238.

Koo, D.C. and Kron, R.G., 1988, *Astrophys. J.*, **325**, 92.

Kovner, I. and Rees, M.J., 1988, preprint.

Lin, D.N.C., Pringle, J.E. and Rees, M.J., 1988, *Astrophys. J.*, **328**, 103.

Phinney, E.S., 1983, Ph.D. thesis, Cambridge University.

Press, W.H. and Schechter, P.L., 1974, *Astrophys. J.*, **187**, 425.

Salpeter, E.E., 1964, *Astrophys. J.*, **140**, 796.

Sanders, D.B., Soifer, B.T., Elias, J.H., Madore, B.F., Matthews. K., Neigebauer, G. and Scoville, N.Z., 1988, *Astrophys. J.*, **325**, 74.

Schmidt, M. and Green, R.F., 1983, *Astrophys. J.*, **269**, 352.

Schmidt, M., Schneider, D.P. and Gunn, J.E., 1986, *Astrophys. J.*, **310**, 518.

Warren, S.J., Hewett, P.C., Irwin, M.J., McMahon, R.G., Bridgeland, M.T., Bunclark, P.S. and Kibblewhite, E.J., 1987, *Nature*, **325**, 131.

Weedman, D.W., 1985, *Astrophys. J. Suppl*, **57**, 523.

White, S.D.M., Frenk, C.S., Davis, M. and Efstathiou, G., 1987b, *Astrophys. J.*, **313**, 505.

Yang, J., Turner, M.S., Steigman, G., Schramm, D.N. and Olive, K.A. 1984, *Astrophys. J.* **281**, 493.

DISCUSSION:

MARTINEZ-GONZALEZ: Do you have any reason to explain why the Press-Schechter multiplicity function fits so well the one found in the simulations?

EFSTATHIOU: It isn't that surprising that the Press-Schechter function does so well - but we have recently analysed some large Cray simulations of CDM universes which show that the Press-Schechter formalism deviate by about 30 – 40% from the numerical results.

PEEBLES: Your CDM model yields two remarkable predictions. The comoving abundance of the most luminous quasars would be expected to drop sharply between $z = 2$ and $z = 4$, which seems to conflict with the results presented by Warren. And the abundance of Lilly-type objects would be expected to be exceedingly small at $z > 3.5$, which should be testable. I expect you would predict that there are very few Lilly-type objects.

EFSTATHIOU: I have focused on the abundance of ultra-luminous quasars ($M_J < -26.8$) for which the comoving density should fall rapidly at redshifts $z \gtrsim 5$. I don't think that this conflicts with Warren's data, but I agree that it would not be surprising to see strong luminosity dependent evolution in the QSO luminosity function if the CDM model is correct - at high redshifts, the most luminous quasars should decline in density much faster than less luminous quasars. The comoving abundance of Lilly-type objects is even less than the abundance of the ultra-luminous quasars that I discussed ($< 10^{-8}$ Mpc^{-3}). It is crucial that we obtain estimates of the mass-to-light ratios of these objects so that we can assess the amount of gas and dark-matter involved. If the objects are typically as massive as Simon argues ($\sim 10^{12}$ M$_\odot$ of stellar material for $0902 + 34$ at $z = 3.395$), then we can see from my diagram (Figure 1a) that there might be a problem with the CDM model, especially if the bulk of the stars formed at a still higher redshift. However, can we really rule out massive star formation (and correspondingly lower inferred masses) in such galaxies? The statistics and nature of these rare objects is definitely worth detailed investigation.

The Intergalactic Medium and the Epoch of Galaxy Formation

Paul R. Shapiro* and Mark L. Giroux
Department of Astronomy
The University of Texas at Austin
Austin, Texas 78712 USA

ABSTRACT. The hypothesis that photoionization of the IGM by high redshift radiation sources explains the absence of a detectable Gunn-Peterson effect in the spectra of all known quasars is investigated. This hypothesis has strong implications for the epoch and nature of galaxy formation as it relates to the origin of the required ionizing radiation background. New calculations are described of the ionization and thermal history of the IGM which determine the minimum required ionizing photon emissivity of the universe at $z \gtrsim 4$. The inability of the observed high redshift quasars to produce this emissivity is discussed, as are the consequences for primordial galaxy luminosity and metal production if early-type stars provide the emissivity instead. Other alternatives, such as radiation from protogalactic shocks and the radiative decay of unstable 'inos, are also discussed.

1. Introduction

The observed spectra of high redshift quasars show no evidence of the H Ly α absorption trough (Gunn-Peterson effect) expected from H atoms in a smoothly distributed intergalactic medium along the line of sight to the quasars. As we shall see, this *transparency* of the IGM places strong constraints on the epoch of galaxy formation as it relates to the generation of enough radiation and energy to fully ionize the IGM.

The optical depth τ_{GP} through such an intergalactic medium between us and a quasar at redshift z_{GP} observed just to the blue side of the quasar's Ly α emission line is given by

$$\tau_{GP} = 3.31 \times 10^4 (1-\chi)_{GP} \left(\frac{\Omega_b h}{0.1}\right)(1+z_{GP})^2[1-2q_0+2q_0(1+z_{GP})]^{-1/2} \qquad (1.1)$$

where $(1-\chi)_{GP}$ is the neutral fraction of the H atoms in the IGM at $z = z_{GP}$, and Ω_b is the density of the IGM in units of the critical density. The observed upper limits on τ_{GP}, therefore, imply an upper limit to the product $(1-\chi)_{GP}(\Omega_b h)$. For $z_{GP} = 3.5$, an upper limit of $\tau_{GP} = 0.1$ requires that $(1-\chi)_{GP}(\Omega_b h/0.1)$ be less than 3.2×10^{-7} (1.5×10^{-7}) for $q_0 = 1/2$ $(q_0 = 0)$. In fact, there is apparently no detectable τ_{GP} at roughly this level even out to $z_{GP} = 4.43$, the highest redshift yet observed for a quasar. Steidel and Sargent have attempted to tighten this this constraint, finding $\tau_{GP} \lesssim 0.05$, but at somewhat lower redshift ($z_{GP} \simeq 2.8$).[1] As we shall see, the higher the value of z_{GP}, the more significant are the requirements for ionizing the IGM. We

*Alfred P. Sloan Research Fellow

C. S. Frenk et al. (eds.), The Epoch of Galaxy Formation, 153–161.
© 1989 by Kluwer Academic Publishers.

shall compromise here by taking $z_{GP} = 3.5$ and $\tau_{GP} = 0.1$ in the discussion which follows.

The density of the IGM at $z \gtrsim 3.5$ is highly uncertain. Most current theories of galaxy formation require $0.03 \lesssim \Omega_b h^2 \lesssim 0.1$ for the IGM at high redshift. Furthermore, an interpretation of the "Ly α forest" clouds as pressure-confined by the ambient IGM requires $\Omega_b h^2 \sim 0.03$ for the confining medium.[2] The "Ly α forest" clouds themselves have been estimated to give a density $\Omega_c h^2 \sim 2 \times 10^{-2}$. [2] An estimate of the mean density of the IGM based upon the ratio of X-ray-emitting gas mass to galaxy stellar luminosity inside galaxy clusters and the mean luminosity density of galaxies yields $\Omega_b = 0.22$.[3]

In short, therefore, the Gunn-Peterson (GP) upper limit implies that the IGM was highly ionized by $z \sim 4$. We focus here on the explanation of this high degree of ionization of the IGM in terms of *photoionization* by high redshift sources. In what follows, we discuss the radiation flux levels required to satisfy the GP constraint and the implications of the origin of such flux levels.

2. Requirements for Photoionizing the IGM

We have previously described the requirements for ionizing the IGM by the overlap of the intergalactic H II regions formed around discrete sources of ionizing radiation, such as quasars.[4-7] Here we focus on the stronger requirement that, after the H II regions overlap, the neutral fraction in the IGM is below the GP limit. After the H II regions have overlapped and the IGM is fully ionized, the approximation of photoionization equilibrium is valid. This permits us to write the minimum required mean intensity $J(z_{GP})$ of ionizing background radiation (in *photons* $cm^{-2}s^{-1}Hz^{-1}ster^{-1}$) as

$$4\pi J(z_{GP})/n_H^0 > 2.2 \times 10^{10} \Sigma^{-1} R(\tau_{GP}^{max}/0.1)^{-1}(\Omega_b h/0.1) A(q_0, z_{GP}) \tag{2.1}$$

where

$$A(q_0, z_{GP}) = (1 + z_{GP})^5 [1 - 2q_0 + 2q_0(1 + z_{GP})]^{-1/2}, \tag{2.2}$$

$R = \alpha/\alpha(10^4 K)$, α is the Case A H recombination rate, $\Sigma = (\sigma_{th} J)^{-1} \int_{\nu_{th}}^{\infty} d\nu \sigma_\nu J_\nu$, and $\sigma_\nu = \sigma_{th}(\nu/\nu_{th})^{-3}$ is the H photoionization cross section. The emissivity required to generate such a mean intensity can be expressed in terms of a dimensionless parameter ζ defined according to

$$\zeta = \frac{2n_x^0 N_{ph}(\geq 13.6 eV)}{3 H_0 n_H^0} \tag{2.3}$$

where n_x^0 is the number of sources per cm^3 (present comoving volume) assuming $n_x(z)/n_x^0 = (1 + z)^3$, $N_{ph}(\geq 13.6 eV)$ is the luminosity per source in ionizing photons, and $n_H^0 = 8 \times 10^{-6} \Omega_b h^2$ cm^{-3} is the mean baryon density of the IGM at present (assuming $n_H(z)/n_H^0 = (1 + z)^3$). The value of ζ required to produce the mean intensity of inequality (2.1), ζ_{GP}, assuming constant luminosity sources of constant number per comoving volume and ν^{-1} power-law energy spectra, $z_{GP} = 3.5$, and $T_{IGM} = 2.5 \times 10^4 K$ is plotted in Figure 1 for a fully ionized pure H IGM in photoionization equilibrium.[8] We assume $q_0 = 1/2$ hereafter. We find $10^{2.1} \lesssim \zeta_{GP} \lesssim 10^3$ for $0.025 \lesssim \Omega_b h^2 \lesssim 0.1$ for $4 < z_{turn-on} < \infty$, assuming no cut-off in the source density out to $z_{turn-on}$.

2. 1. The Observed Quasar Contribution

For $z \sim 3$, Koo and Kron[9] report $\zeta_{obs} \sim 2(\Omega_b h^2/0.1)^{-1}$ from the results of a survey of high redshift quasars to faint magnitudes. In order to take account of possible uncertainties in this estimate, we take $\zeta_{obs} \lesssim 10(\Omega_b h^2/0.1)^{-1}$ as a conservative upper limit. For $z > 3$, this ζ_{obs} may significantly overestimate the actual value in view of the growing evidence that the

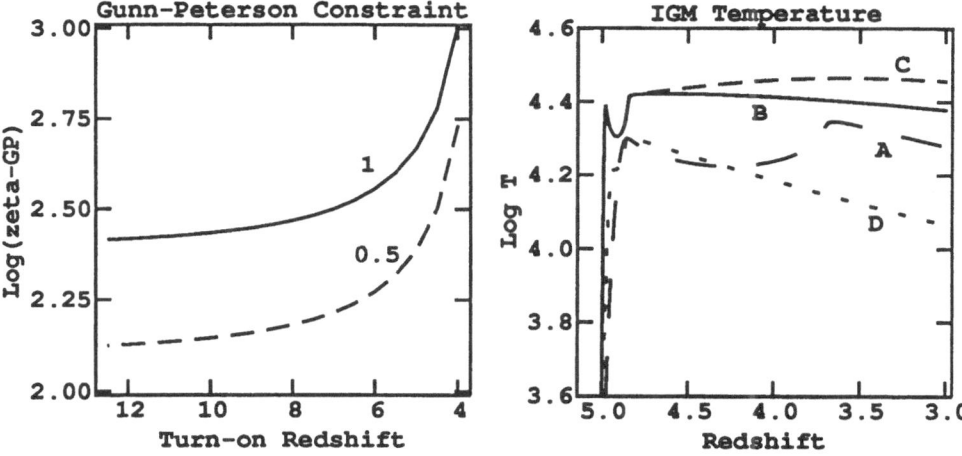

Figure 1. (left) ζ_{GP} versus $z_{turn-on}$, for $\tau_{GP} = 0.1$ at $z_{GP} = 3.5$, calculated analytically for pure H IGM with $T = 2.5 \times 10^4$ K, $\Omega_b = 0.1$, $q_0 = 1/2$, and $h = 1$ and 0.5, as labelled.

Figure 2. (right) Temperature of IGM versus z for $z_{turn-on} = 5$, $\Omega_b = 0.1$, $q_0 = 1/2$, for Case A ($\zeta = 40$, $h = 0.5$, quasar-like), Case B ($\zeta = 450$, $h = 1$, quasar-like, no cloud opacity), Case C (Case B but *with* cloud opacity), Case D ($\zeta = 450$, $h = 1$, stellar). "Quasar-like" sources have ν^{-1} energy spectra. "Stellar" sources have Planck spectra with $T = 47,000$ K.

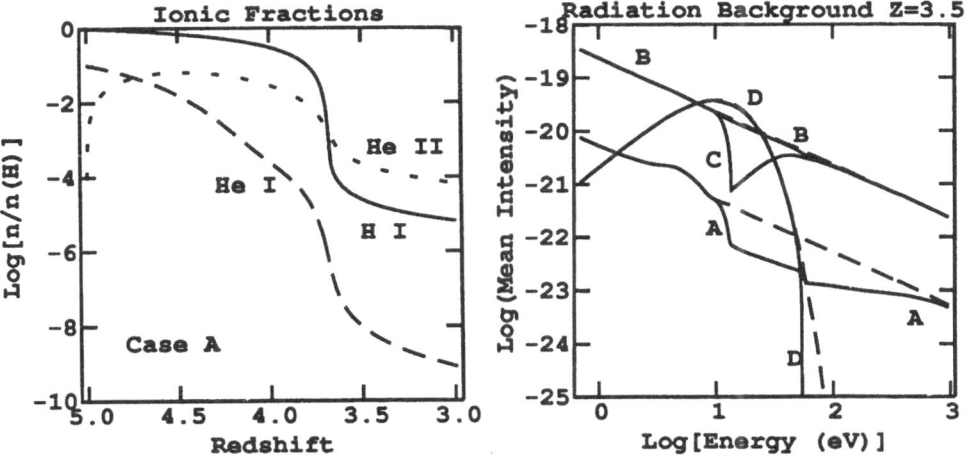

Figure 3. (left) Ionized fractions n_i/n_H versus z for Case A, for $n_i = n(HI), n(HeI), n(HeII)$ as labelled. All cases assume $n_{He}/n_H = 0.1$.

Figure 4. (right) J_ν ($erg\ cm^{-2}\ s^{-1}\ Hz^{-1}\ ster^{-1}$) versus photon energy (eV) at $z = 3.5$ for the cases defined in Fig. 2, as labelled (solid curves). Dashed lines show what J_ν would have been if optical depth attenuation were ignored.

quasar number per comoving volume declines at higher redshift.[9-12] Apparently, the observed quasars are not sufficient to explain the GP upper limit *even if there is no redshift cut-off for the quasars*. Either the present observations are missing the true number density of quasars (see, for instance, Heisler and Ostriker[13]) or else something else must ionize the IGM at these redshifts.

3. Thermal and Ionization History of the IGM

3. 1. Quasar Ionization

In order to demonstrate how serious the discrepancy is between the current estimates of the observed quasar contribution and the requirements for ionizing the IGM, we have performed a new, detailed calculation of the ionization and heating of a uniform, cosmologically expanding IGM of H and He, including the effects of radiative transfer, diffuse emission by the gas, radiative and Compton cooling, and nonequilibrium chemistry.[14] Our results for the case with $\Omega_b h^2 = 0.025$, $h = 0.5$, and $\zeta = 40$ (the estimated upper limit for the observed quasars at $z \sim 3$), a turn-on redshift of $z = 5$, and a ν^{-1} energy spectrum between 0.7 eV and 12.4 keV per quasar, are shown in Figs. $2 - 6$, referred to as "Case A". The gas in this case is not fully ionized until $z \approx 3.7$. Prior to this, the continuum optical depth through the IGM to ionizing photons is substantial, and the H Ly α absorption trough in the spectrum of any quasar observed at $z > 3.7$ would be completely black for wavelengths $4.7 \leq \lambda_{obs}/1215$ Å$\leq 1 + z_{QSO}$. Even after this, however, at $z = z_{GP} = 3.5$, τ_{GP} for H atoms exceeds 4, while that for He II ions exceeds 8.

In order to satisfy the H GP constraint of $\tau_{GP} = 0.1$ at $z_{GP} = 3.5$, in fact, a value of $\zeta \approx 250(500)$ is required if $z_{turn-on} = 5$, $\Omega_b = 0.1$, and $h = 0.5(1)$. In Figures 2 and $4-7$ we show the results for the case (Case B) with $\zeta = 450$, $\Omega_b = 0.1$, and $h = 1$, which marginally satisfies the GP constraint at $z_{GP} = 3.5$. We note that, at $z = 4.43$, where a quasar has recently been observed, $\tau_{GP} = 0.6$ for H Ly α is predicted in this case, a value large enough to have been detectable. The absence of a detectable τ_{GP} at this redshift, therefore, suggests that the value $\zeta \sim 500$ necessary to satisfy the GP constraint at $z = 3.5$ is a conservative lower limit. This value of ζ_{GP} is not very sensitive to the value of $z_{turn-on}$ if $z_{turn-on} \gtrsim 5$, but if $z_{turn-on} < 5$, ζ_{GP} is even higher that this.

3. 2. The Effect of Quasar Ly α Absorption-Line Cloud Opacity

The minimum value of ζ required to satisfy the GP limit may increase substantially if we take into account the H continuum opacity of the observed Ly α absorption line clouds. Observations of the Ly α forest have been fit with a distribution $f(N_{HI}, z)$ of cloud number N_c of column density N_{HI} per cloud given by

$$f(N_{HI}, z) \equiv \frac{\partial^2 N_c}{\partial z \partial N_{HI}} = A(1 + z)^\gamma N_{HI}^{-\beta} \tag{3.1}$$

for $1.5 < z < 4$ and $10^{14} < N_{HI} < 10^{17} cm^{-2}$, with $\gamma = 2.3 \pm 0.4$ and $\beta = 1.7 \pm 0.2$ and, if we take $\gamma = 2.3$ and $\beta = 1.7$, $A \approx 1.8 \times 10^{10}$. [15,16] An extension of the range of equation (3.1) to $N_{HI} \gtrsim 10^{21} cm^{-2}$ conservatively estimates the additional absorption due to the highest column density Ly α absorption clouds, the damped Ly α clouds.[15,16] These are the clouds, it turns out, which dominate the total opacity, although the correct value of γ to use for these large column density absorbers is uncertain. Our calculations of the previous case with $\zeta = 450$ described in §3.1, which marginally satisfied the GP constraint at $z_{GP} = 3.5$ when no cloud opacity was included, were generalized to include the cloud opacity in equation (3.1) for

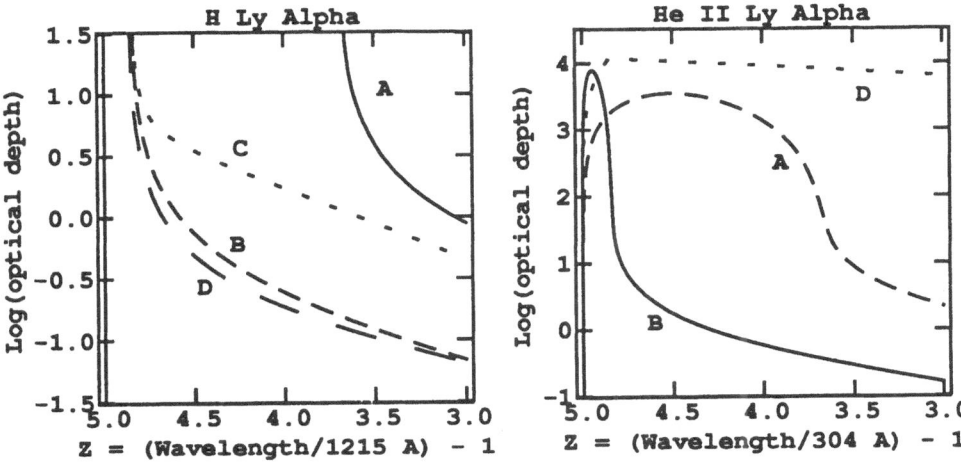

Figure 5. (left) τ_{GP} versus z for H Ly α for cases defined in Fig. 2, as labelled. Same curves show wavelength dependent Ly α optical depth τ_λ [which modifies observed quasar flux by factor $\exp(-\tau_\lambda)$] for a quasar seen today at z_{QSO} if $(\lambda_{obs}/1215\text{Å}) = 1 + z$ and only $z \leq z_{QSO}$ is used.

Figure 6. (right) Same as Fig. 5, except for He II Ly α at 304 Å, instead.

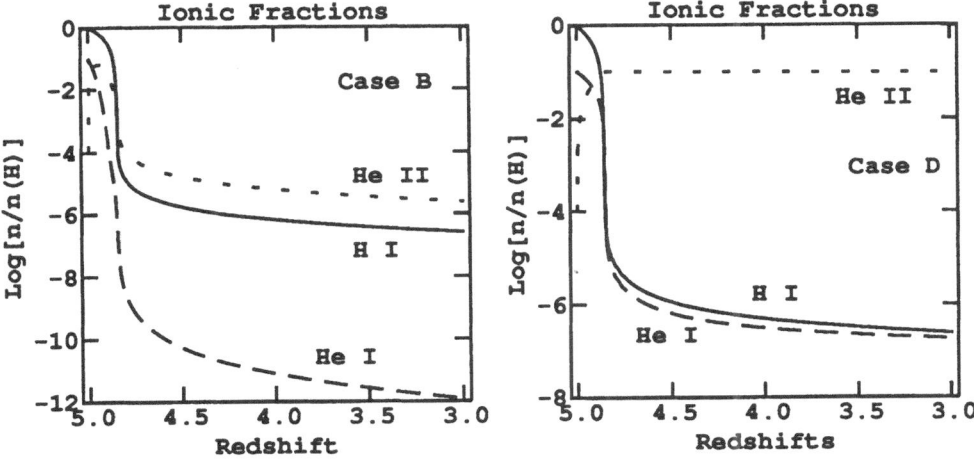

Figure 7. (left) Same as Fig. 3, except for Case B.

Figure 8. (right) Same as Fig. 3, except for Case D.

$10^{14} \leq N_{HI} \leq 10^{21} cm^{-2}$.[14] The results, shown in Figs. 2, $4-6$ (Case C), indicate that τ_{GP} at 3.5 is ~ 0.9. In order to satisfy the GP limit with this cloud opacity taken into account, therefore, we require $\zeta_{GP} \gtrsim 4 \times 10^3$! We note that this value of ζ_{GP} is an underestimate in that we have ignored the He continuum opacity of the clouds.

3. 3. Stellar Sources of the Ionizing Background

We have also calculated the thermal and ionization history of the IGM in the presence of a *stellar* UV background, such as might emerge from primeval galaxies at high redshift. As an illustration, we have taken sources with a Planck spectrum characterized by a temperature of 4.7×10^4 K, corresponding roughly to O5 stars. We find that, for $\Omega_b h = 0.1$ and $z_{turn-on} = 5$ as before, $\zeta \sim 500$ is required to satisfy the GP limit at $z_{GP} = 3.5$, yielding a background mean intensity J_ν at 13.6 eV at that redshift of about $2 \times 10^{-20} erg\ cm^{-2}\ s^{-1}\ Hz^{-1}\ ster^{-1}$ [see Figs. 2, $4-6$, and 8 (Case D)]. We note that previous attempts to calculate the ionizing background contributed by star formation in young elliptical galaxies assuming a mass-to-light ratio $M/L = 4$, a Salpeter mass function, and $z_{formation} = 5$ found a maximum value of this J_ν at $z \sim 3.8$ of only $\sim 10^{-21.3}$ in these units, well below the required level.[17]

What are the implications of so large a required ζ in stellar UV emissivity at high redshift? If these stars reside inside primeval galaxies, then the average ionizing photon luminosity per galaxy from the epoch of galaxy formation at $z \gtrsim 5$ until z_{GP} is given by

$$N_{ph,gal}(\geq 13.6eV) > 10^{55} \left(\frac{\Omega_b}{0.1}\right) \zeta_3 \left(\frac{n_{gal}^0 h^{-3}}{10^{-2}Mpc^{-3}}\right)^{-1} photons/sec/galaxy \qquad (3.2)$$

where $\zeta_3 \equiv \zeta/10^3$, $n_{gal} = n_{gal}^0 (1+z)^3$ is the number density of galaxy sources at z, and 10^{-2} Mpc^{-3} corresponds roughly to the present density of L_* galaxies. This implies a minimum luminosity per L_* galaxy of $L_{gal} > (13.6eV)N_{ph,gal}$, or,

$$L_{gal} > 2.5 \times 10^{44} \left(\frac{\Omega_b}{0.1}\right) \zeta_3 \left(\frac{n_{gal}^0 h^{-3}}{10^{-2}Mpc^{-3}}\right)^{-1} erg/sec. \qquad (3.3)$$

Regardless of whether the required stellar emissivity occurs in already-assembled L_* galaxies or in more numerous, smaller systems each of lower luminosity, the space-averaged emissivity itself implies the generation of a large metallicity, as follows. The ionizing photon luminosity per solar mass of stars for a Salpeter mass function in the range O4 - O9 (which dominates the total ionizing photon emissivity) is 3×10^{47}.[18] The space-averaged mass density of such stars required to generate a given ζ then implies an "instantaneous contribution" to the total mass density of the universe given by $\Omega_* > 1.5 \times 10^{-6} \zeta_3 (\Omega_b h/0.1)$. In order to calculate the total amount of mass which must have cycled through such stars in order to maintain the required ζ throughout the time from the source turn-on epoch until $z \sim 3$, we multiply Ω_* by the ratio of this time interval $t_{duration}$, which is roughly the age of the universe at $z \sim 3$, or $t_{age} = 8.2 \times 10^8 h^{-1} yrs(q_0 = 1/2)$, to $t_{lifetime}$, the lifetime of the dominant contributor O stars ($\sim 40 M_\odot$), which is about 4×10^6 years.[19] This yields

$$(\Omega_*)_{tot} = \left(\frac{t_{duration}}{t_{lifetime}}\right) \Omega_* \gtrsim 3 \times 10^{-4} \zeta_3 \left(\frac{\Omega_b}{0.1}\right) \qquad (3.4)$$

For every solar mass M_* of stars which contribute to this total, a fraction $\varepsilon_M M_*$ is turned into metals. For the O stars which dominate the ionizing flux, $\varepsilon_M \sim 30\%$ (e. g. $\varepsilon_M M_* =$

$0.4 M_* - 4.2, M_* \approx 40$).[20] Together with equation (3.4) this implies that the metal production associated with the minimum required stellar UV emissivity corresponds to

$$\Omega_{metals} = \varepsilon_M (\Omega_*)_{tot} \gtrsim 9 \times 10^{-5} \zeta_3 (\Omega_b / 0.1). \qquad (3.5)$$

Since the mean mass density of the luminous matter in galaxies corresponds to $\Omega_{gal,lum} \approx 10^{-2}$, equation (3.5) implies $\Omega_{metals} / \Omega_{gal,lum} \gtrsim 9 \times 10^{-3} \zeta_3 (\Omega_b / 0.1)$. But the mass fraction of these metals in a solar elemental abundance mixture is about 10^{-2}. Hence, if $\zeta_3 > 1/2$ is required to ionize the IGM, then this implies that the equivalent of more than half the Population I metallicity was generated before $z = 3$!

4. Alternative Sources of Ionizing Radiation

4. 1. Protogalactic Shock Radiation

Suppose every baryon seen inside galaxies today had to have cooled radiatively from roughly the virial temperature of the protogalaxy in order to be available for star formation. What ζ does this radiation imply? The generic form for this cooling is that behind a radiative shock of velocity of order the protogalaxy virial velocity. Let us define ϕ as the number of ionizing photons radiated per shock-heated baryon. In that case,

$$\zeta_{shocks} \approx \frac{2 n_{baryon}^{gal} \phi}{3 H_0 t_{duration} n_H^{IGM}}, \qquad (4.1)$$

where $n_{baryon}^{gal} / n_H^{IGM} = \Omega_{gal,lum} / \Omega_b$ and $t_{duration} \sim t_{age}(z_{GP}) = (2/3) H_0^{-1} (1 + z_{GP})^{-3/2}$. Hence, for $z_{GP} = 3.5$, $\zeta_{shocks} \approx 9.5 (\Omega_{gal,lum} / \Omega_b) \phi$, and if $\Omega_{gal,lum} / \Omega_b \approx 0.1$, $\zeta_{shocks} \sim \phi$. Our detailed radiative shock calculations for primordial composition gas indicate that, for $v_{shock} \sim v_{gravitation} \sim 300 \ km \ s^{-1}$, $\phi \approx 5$.[21] Since the required ζ exceeds this by $\sim 10^2$, we conclude that protogalactic shock radiation fails (as long as $\Omega_{IGM} > \Omega_{gal,lum}$), unless each baryon is shocked many times.

4. 2. Exotic Alternatives: Decaying 'Inos

It has been suggested that unstable cosmic 'inos which decay to ionizing photons might contribute to the ionization of the IGM.[22-24] Melott finds, for example, that an emissivity of decaying neutrinos $n_\nu^0 N_{ph} (\geq 13.6 eV) = 2 \times 10^{-22} photons \ cm^{-3} s^{-1}$ is possible.[22,23] This implies $\zeta \approx 50 (\Omega_b h^3 / 0.1)^{-1}$, which is still too small to satisfy the GP constraint. However, if $h \approx 1/2$ and $\Omega_b \lesssim 0.03$, it is difficult to rule this possibility out on the basis of the GP requirements alone. A similar conclusion applies to the photino suggestion by Sciama.[24]

5. Summary and Conclusions

We have shown that the ionizing flux levels required to photoionize the IGM to the degree implied by the GP limit at $z = 3.5$ exceed the contribution estimated from the observations of quasars at $z > 3$ This result is not very sensitive to the redshift $z > 5$ at which the ionizing sources are assumed to have begun emitting. A more recent mean epoch of source turn-on makes the discrepancy worse. The known presence of large column density Ly α absorption clouds at high redshift increases this quantitative discrepancy even further, by a substantial factor.

We have considered the possibility that the ionizing radiation was produced by massive O stars in primordial galaxies instead. The star formation rate required in this case, starting at $z \gtrsim 5$, is high enough to imply a very large bolometric luminosity per L_* galaxy if these are the basic star forming entities. Regardless of the size of the mass concentrations which are the typical sites of this star formation, the required space-averaged, time-integrated star formation rate implies that a significant fraction of the metal abundance of Pop I stars would have been produced before $z \sim 3$.

We have also considered the radiation emitted during the radiative cooling of the protogalactic shocks which dissipate gravitationally-induced gas motions inside protogalaxies. We show that this radiation falls far short of producing the required ionizing emissivity unless either the atoms which eventually become stars inside galaxies are shocked many times by $z \gtrsim 4$ or else the mean density of such shocked baryons inside galaxies at $z \gtrsim 4$ exceeds that of the IGM at that epoch.

In considering a range of alternative radiation sources, we have shown that recent suggestions that decaying neutrinos or photinos photoionize the IGM are marginally consistent with the GP requirements, but only for low values of the IGM density and of H_0.

This work benefitted from the support of Robert A. Welch Foundation Grant F-1115, Texas Advanced Research Program Grant 4132, and NASA Training Grant NGT-50316.

References

1. Steidel, C. C. and Sargent, W. L. W. 1987, *Ap. J.* **313**, 171.
2. Ostriker, J. P. and Ikeuchi, S. 1983, *Ap. J.(Letters)*, **268**, L63.
3. Ikeuchi, S. and Ostriker, J. P. 1986, *Ap. J.*, **301**, 522.
4. Shapiro, P. R. 1986a, in *Galaxy Distances and Deviations from Universal Expansion*, eds. B. F. Madore and R. B. Tully (Dordrecht: Reidel), p. 203.
5. Shapiro, P. R. 1986b, *P. A. S. P.*, **98**, 1014.
6. Shapiro, P. R. and Giroux, M. L. 1987, *Ap. J. (Letters)*, **321**, L107.
7. Shapiro, P. R., Giroux, M. L., and Kang, H. 1987, in *High Redshift and Primeval Galaxies*, eds. J. Bergeron, D. Kunth, B. Rocca-Volmerange, J. Tran Thanh Van (Paris: Editions Frontiéres), p.501.
8. Shapiro, P. R., Giroux, M. L., and Ostriker, J. P. 1988, in preparation.
9. Koo, D. C. and Kron, R. G. 1988, *Ap J.*, **325**, 92.
10. Koo, D. C. 1986, in *Structure and Evolution of Active Galactic Nuclei*, eds. G. Giuricin, F. Mardirossian, M. Mezzeti, and M. Ramella (Dordrecht: Reidel), p. 317.
11. Osmer, P. S. 1982, *Ap. J.*, **253**, 28.
12. Schmidt, M., Schneider, D. P., and Gunn, J. E. 1986, *Ap. J*, **310**, 518.
13. Heisler, J. and Ostriker, J. P. 1988, *Ap. J.*, **332**, 543.
14. Shapiro, P. R. and Giroux, M. L. 1988, in preparation.
15. Duncan, R. and Ostriker, J. P. 1988, Princeton Observatory Preprint POP-261.
16. Bechtold, J. 1987 in *High Redshift and Primeval Galaxies*, eds. J. Bergeron, D. Kunth, B. Rocca-Volmerange, J. Tran Thanh Van (Paris: Editions Frontiéres), p.403.
17. Bechtold, J., Weymann, R. J., Lin, Z., and Malkan, M. A., 1987, *Ap. J.* **315**, 180.
18. Mezger, P. G., Smith, L. F., and Churchwell, E., 1974, *Astr. Ap.*, **32**, 269.
19. Güsten, R. and Mezger, P. G. 1983, *Vistas in Astronomy*, **26**, 159.
20. Weaver, T. A. and Woosley, S. E. 1986, *Ann. Rev. Astron. Ap.*, **24**, 205.
21. Kang, H. and Shapiro, P. R. 1988, submitted to *Ap. J.*
22. Melott, A. L., McKay, D. W., and Ralston, J. P. 1988, *Ap. J.(Letters)*, **324**, L43.
23. Ralston, J. P., McKay, D. W., Melott, A. L., 1988, *Phys. Lett. B*, **202**, 40.
24. Sciama, D. 1988, *M. N. R. A. S.*, **230**, 13P.

DISCUSSION:

FIELD: Can the problem you described be resolved by adopting Ω_{IGM} about 10 times smaller? After all, some baryons are needed to make galaxies and damped Lyα clouds, so Ω_{IGM} could be substantially less than Ω_B.

SHAPIRO: Adopting a value of Ω_{IGM} 10 times smaller (i.e. $\Omega_{IGM} \sim 10^{-2}$) will help but will not resolve all of the problems. For $H_o = 100$, $\Omega_{IGM} = 10^{-2}$ requires $\zeta_{GP} \sim 100$ if the sources turned on at $z = 5$, even if we ignore the opacity of the Lyα clouds. Koo and Kron estimate that quasars at $z \sim 3$ contribute in this case only $\zeta_{QSO} = 20$. As we have heard at this meeting, in fact the Koo and Kron estimate may be too high, and there is a growing body of evidence that the quasar density drops off significantly as redshift increases past 3. Hence, the quasars are likely to contribute a ζ much less than 20 at $z \sim 5$. Adopting $H_o = 50$ (and $\Omega_{IGM} = 10^{-2}$), which further lowers ζ_{GP} to ~ 50 and raises ζ_{QSO} at $z = 3$ by a factor of 4, will also not resolve the discrepancy in the face of such a redshift cut-off.

The Lyα clouds themselves provide a lower limit to the required ζ even if $\Omega_{IGM} = 0$ between the clouds, as follows. Let us assume that the Lyα forest is caused by clouds photoionized by background radiation which are either pressure or gravity-confined. It is necessary then for the HII regions created by turning the ionizing radiation on in the midst of the initially neutral Lyα clouds to overlap by $z = 4$. These HII regions are comprised of a clumpy gas in which the clumps are the Lyα clouds. Since, as Ostriker and Ikeuchi have estimated, Ω h$^2 \simeq 0.03$ for the clouds, the ζ required to produce this overlap exceeds the observed ζ_{QSO} even if $\Omega_{IGM} = 0$ outside the clouds, it can be shown.

The other aspect of the problem I have pointed out involves evaluating the chances that other sources of the required radiation background can be found. The ζ required for these other sources, such as early star formation or radiation from protogalactic shocks, does go down in direct proportion to Ω_{IGM} so lowering Ω_{IGM} to 10^{-2} does help these other sources solve the problem. Such a low Ω_{IGM} however, is already lower than the value assumed by most currently popular models of galaxy formation, such as cold and hot dark matter, so such a constraint on Ω_{IGM} would be very significant.

A COOLING FLOW AROUND THE QUASAR 3C196

C.S. Crawford
Institute of Astronomy Madingley Road Cambridge CB3 OHA U.K.

ABSTRACT. Observations made with the Faint Object Spectrograph on the 4.2m William Herschel Telescope are presented. Extended line emission around the quasar 3C196, at a redshift of 0.87, was detected and its spectrum measured. The intensity ratio of oxygen emission lines indicates a high pressure for the gas, which is assumed to be ionized by the quasar luminosity. The properties of a hot confining medium are then consistent with a cooling flow around the quasar.

It is well-known from optical observations that low-redshift QSO are embedded in faint nebulosity, which often shows a detailed structure. Imaging studies in continuum bands (Hutchings, Crampton & Campbell 1984; Malkan 1984) show the morphology of this 'fuzz' resembles a host galaxy, usually elliptical- or spiral-shaped in the case of radio-loud or -quiet respectively. Narrow-band imaging (Smith et al. 1986; Stockton & MacKenty 1987) shows patchy structures emitting in forbidden oxygen lines, often stretching many tens of kpc from the QSO itself. Such emission-line gas is a vital diagnostic of the environment directly surrounding a QSO, as its ionization state is dependent on the ionizing flux and the gas density. Other environmental clues to the nature of QSO are that radio-loud quasars are preferentially found in clusters, albeit poor ones (Yee & Green 1984, 1987). To determine the ionization state of the external emission-line regions, good signal-to-noise spectroscopic observations are needed, as it is often difficult to remove the contamination of the nebulosity emission by the bright light spilling over from the nucleus.

The indications of spectroscopy of a previous sample of intermediate redshift objects (Crawford, Fabian & Johnstone, 1988) are that in order to explain the observed extended emission line ratios in gas around radio-loud QSO, a high gas density is required; this is not necessary for the radio-quiet objects of the sample. We followed this work up by observing a second sample of 35 radio-loud (preferably steep-spectrum) QSO over a wide range of redshift and luminosity, taken from the X-ray quasars published by Worrall et al. (1987). The aim is to search for extended line emission in these QSO, and if found, to use the deduced ionizing spectrum of the QSO and the projected distance of the nebulosity from the nucleus to give an estimate of the density, and hence pressure of the emitting gas.

We report here on our observations of the quasar associated with the steep-spectrum extended source, 3C196 (0809+433), which at a redshift of 0.871, was the most obvious case of extended emission at a higher redshift than has previously been studied. Kristian (1973) first noticed from photographic plates a faint patch of nebulosity extending 1.5–2 arcsec south-east

163

C. S. Frenk et al. (eds.), The Epoch of Galaxy Formation, 163–166.

from the quasar. We took a 5000 second exposure across the nucleus in 1 arcsec seeing, at a position angle of 163°.

Figure 1 shows the fully-reduced spectra both on and off the QSO nucleus, showing that the equivalent width of the [OII]λ3727 doublet and the [OIII]λ5007 line both increase further away from the nucleus; this unambiguously indicates that the line is more spatially extended than the continuum. The data are of very high signal-to-noise, with thousands of counts in the raw off-nucleus data. We investigate these forbidden oxygen lines by taking a spatial intensity strip over the wavelength range of each emission line. We assess the contribution from the continuum to the intensity in that strip for each cross-section from neighbouring wavelength bands of continuum. This continuum contribution is then subtracted, so we are left with the spatial profile of the oxygen line. We assess the spatial profile of the continuum from neighbouring continuum bands of spectrum and assume the spill-over of nuclear light into the off-nuclear spectra to follow this same distribution. The continuum profile is scaled to match the line profile at the nuclear cross-section, and subtracted. The residual distribution represents any line emission extended over the continuum profile (Figure 2); both oxygen lines show extended emission. The result for [OIII] is noisier because it is very near the edge of the chip where the instrumental response is dying, and it lies exactly on a prominent sky line.

We can form the ratio of the lines at the cross-section where the lines have the highest equivalent width, at a radius of 32 kpc ($H_0 = 50 \, \mathrm{km\,s^{-1}\,Mpc^{-1}}$). We estimate the photo-ionization from the nucleus with a power-law spectrum obtained by extrapolation between the measured ultraviolet (2100Å) and X-ray (2keV) fluxes of the QSO, and assume the gas lies at the radial distance of the projected separation of where we have measured the line ratio. We run Ferland's CLOUDY (Ferland 1987) at constant pressure for varying values of the ionization parameter until we obtain a fit to the line ratio – we can then infer the density and hence pressure of the gas that is emitting these lines. For 3C196, we infer a density of $25^{+10}_{-8} \, \mathrm{cm^{-3}}$ for the gas at a radius of 32 kpc.

Such high density clouds would disperse very rapidly, (on a timescale 10^6 yr) if not confined in some way; otherwise, this must represent a very small percentage of a large mass of cold dense gas being continually burnt away by the ionizing flux (Fabian et al. 1987). The simplest interpretation for the continued existence of the emitting clouds is that they are thermally confined by intracluster gas. Such intracluster gas does not just occur in rich clusters of galaxies, but in poor groups (Schwartz, Schwarz & Tucker 1980; Canizares, Stewart & Fabian 1983), so this is consistent with the findings that radio-loud QSO are preferentially found in poor clusters of galaxies (Yee & Green 1984, 1987).

Unless the surrounding cluster is particularly rich (i.e. $T > 10^8$ K), the radiative cooling time of the gas in the core of the cluster will be less than the Hubble time, and a cooling flow will have formed in the intracluster gas (Fabian, Nulsen & Canizares 1984). The extensive filaments and clouds of warm gas seen around the central galaxy in nearby cooling flow clusters are indeed similar to the narrow-band images of the extended emission-line regions around intermediate redshift QSO. The key diagnostic of a cooling flow, as deduced from the X-ray observations, is a high surrounding gas pressure, $nT > 10^5 \, \mathrm{cm^{-3}}$ K, i.e. the bremsstrahlung cooling time is less than the Hubble time if $T < 5 \times 10^7$ K (Fabian et al. 1986). The gas density we derive for the extended gas around 3C196 is consistent with being in confinement in a cooling flow. If we compare the deduced gas pressure with that in nearby cooling flows, it suggests that it is compatible with that seen in nearby poor groups, such as MKW3s, which has a cooling flow of $\sim 100 \, \mathrm{M_\odot \, yr^{-1}}$ (Canizares, Stewart & Fabian 1983).

There is further evidence for a confining medium from the nature of the extended radio structure – 3C196 has a bent jet on a scale of 2 arcsec (Brown, Broderick & Mitchell 1986). Usually such a sharp bend in a radio jet is explained by the collision of the jet with a massive gas cloud, and the interaction of the two could lead to further ionization in the form of shocks. (However, the p.a. of our observation means we do not sample the bends of the radio jet, so we are not sampling any of this potentially shock-ionized gas). 3C196 is thus consistent with the hypothesis that clusters, intracluster gas and cooling flows can occur around radio-loud quasars at higher redshifts (Fabian *et al.* 1986). A preliminary analysis of the rest of the sample indicates that at least a third show such high pressure extended emission as described for 3C196, so this quasar is certainly not an exceptional case.

The implications of this study is that cooling flows exist at least to redshifts of 0.9, and hence are a long-lived phenomenon, important for the evolution and continued formation of the central cluster galaxies. They could also be the source of gas both fueling the QSO nucleus and confining the steep-spectrum radio source – the demise of present-day radio-loud quasars are perhaps thus due to the evolution of cooling flows.

References

Brown, B.L., Broderick, J.J. & Mitchell K.J., 1986. *Astrophys. J.*, **306**, 107.

Canizares, C.R., Stewart, G.C. & Fabian A.C., 1983. *Astrophys. J.* , **272**, 449.

Crawford, C.S., Fabian, A.C. & Johnstone, R.M., 1988. *Mon. Not. R. astr. Soc., in press.*

Fabian, A.C., Arnaud, K.A., Nulsen, P.E.J. & Mushotzky, R.F., 1986. *Astrophys. J.*, **305**, 9.

Fabian, A.C., Crawford, C.S., Johnstone, R.M. & Thomas, P.A., 1987. *Mon. Not. R. astr. Soc.*, **228**, 963.

Fabian, A.C., Nulsen, P.E.J. & Canizares, C.R., 1984. *Nature*, **310**, 733.

Ferland, G.J., 1987. *Ohio State University, Astronomy Department, Internal Report, no.87-001.*

Hutchings, J.B., Crampton, D. & Campbell, B., 1984. *Astrophys. J.*, **280**, 41.

Kristian, J., 1973. *Astrophys. J. Lett.*, **179**, L61.

Malkan, M.A., 1984. *Astrophys. J.*, **287**, 555.

Schwartz, D.A., Schwarz, J. & Tucker, W., 1980. *Astrophys. J. Lett.*, **238**, L59.

Smith, E.P., Heckman, T.M., Bothun, G.D., Romanishin, W. & Balick, B., 1986. *Astrophys. J.*, **306**, 64.

Stockton, A. & MacKenty, J.W., 1987. *Astrophys. J.*, **316**, 584.

Worrall, D.M., Giommi, P., Tananbaum, H. & Zamorani, G., 1987. *Astrophys. J.*, **313**, 596.

Yee, H.K.C. & Green, R.F., 1984. *Astrophys. J.*, **280**, 79.

Yee, H.K.C. & Green, R.F., 1987. *Astrophys. J.*, **319**, 28.

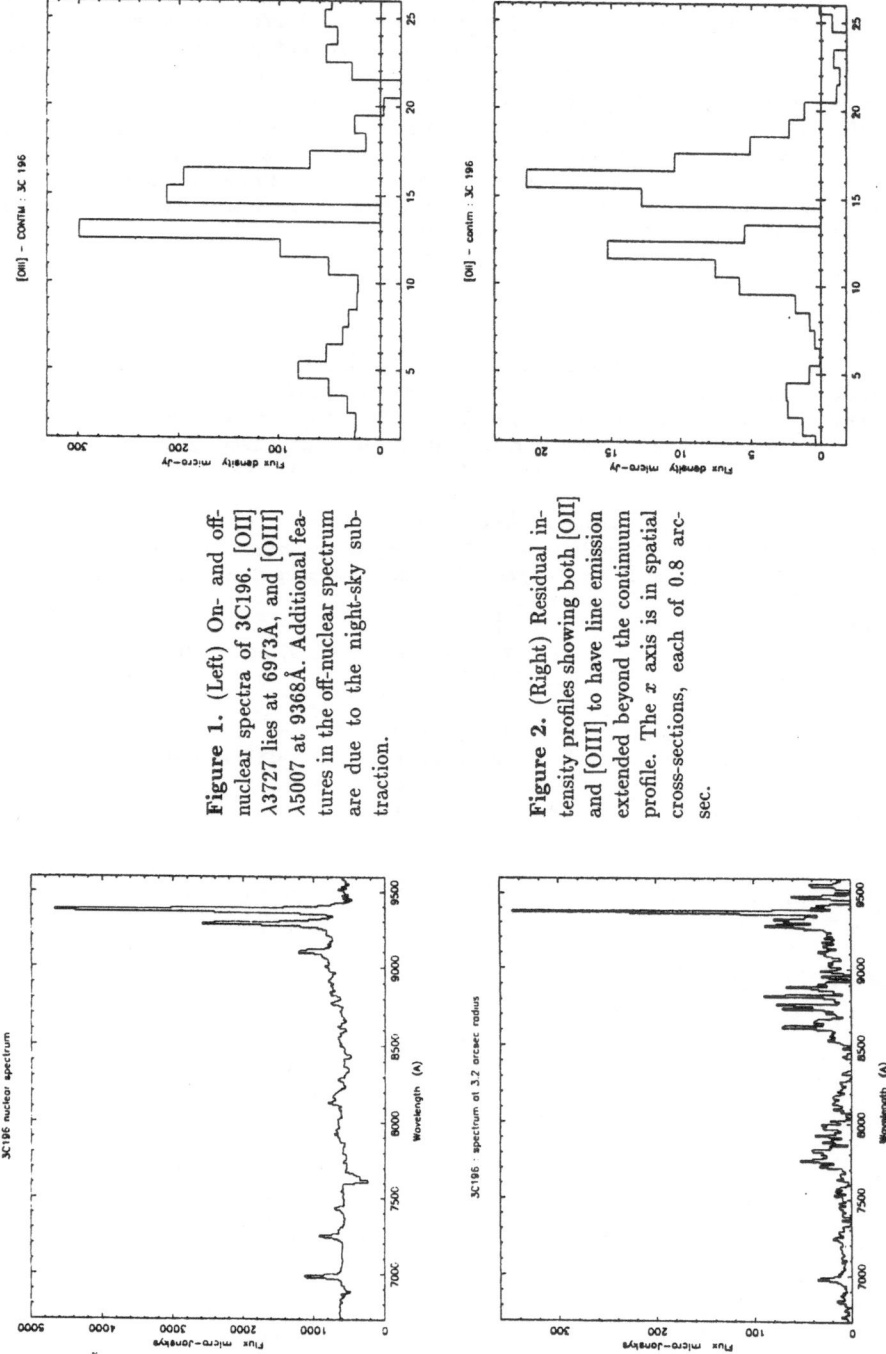

Figure 1. (Left) On- and off-nuclear spectra of 3C196. [OII] λ3727 lies at 6973Å, and [OIII] λ5007 at 9368Å. Additional features in the off-nuclear spectrum are due to the night-sky subtraction.

Figure 2. (Right) Residual intensity profiles showing both [OII] and [OIII] to have line emission extended beyond the continuum profile. The x axis is in spatial cross-sections, each of 0.8 arcsec.

GALAXY EVOLUTION IN HIGH DENSITY ENVIRONMENTS

James E. Gunn
Princeton University Observatory
Princeton, NJ 08544
USA

ABSTRACT. There exists by now almost unequivocal evidence that there has been strong evolution of the population of galaxies in clusters over the relatively short span of the last third of the age of the universe ($z \approx 0.5$), and what data there is suggests strongly that even much more marked effects occur between then and $z = 1$. We discuss the mechanism for this strong evolution, and argue that large galaxies were very much more gas-rich even a relatively short time ago than now. There is hope that some galaxies in clusters have evolved sufficiently "passively" that their formation can be dated, but the relevance of the derived ages to the epoch of galaxy formation in general is clouded by statistical considerations.

1. SURVEYS AND CLUSTER CATALOGS

One of the most difficult issues involved in investigating the evolution of cluster galaxies is the homogeneity of the samples. It has been known for a long time, and has been quantified by the fundamental work of Dressler (1980) on the subject, that the populations in clusters are different from the field population in the sense that clusters have a much higher proportion of early-type galaxies. Thus the invesitigation of such matters as the Butcher-Oemler (1978) effect, the apparent increase in *photometrically* later-type galaxies in clusters at moderate redshifts, must address the issue of whether the aggregates one is looking at at a redshift of 0.5, say, are the same as the Abell clusters one investigates today. This is further complicated by the inevitable dynamical evolution of the cluster population itself; since clusters typically have dynamical times of the order of the Hubble time, it follows that they are dynamically young objects, and can be expected in many cases to have had quite different properties when the universe was half its present age.

The present situation regarding catalogs of clusters is not very reassuring. The Abell catalog, certainly the most homogeneous catalog of relatively nearby rich clusters, is known to have serious statistical shortcomings, though that knowledge has not sufficiently daunted a large number of investigators to staunch the flow of far-reaching conclusions based on analyses of that sample. A catalog of somewhat more distant clusters is under construction by Efstathiou and colleagues at Cambridge by objective criteria from digital scans of southern hemisphere survey plates, but is not yet available. Shectman (1985) has produced a catalog based on surface density enhancements in the counts of Shane and Wirtanen (1967), but there is, of course, no magnitude information in the counts, and the counts themselves were done by hand. Hoessel, Oke, and the author (1986) have produced a catalog of distant clusters, in the redshift range 0.25 to about 1.0, from a series of photographic and image-tube surveys. The photographic bandpasses at each stage were chosen in order that the rest-frame wavelength at the center of the band would be roughly constant for the bulk of the clusters found at that stage, and that

C. S. Frenk et al. (eds.), The Epoch of Galaxy Formation, 167–178.

was, on the whole, successful. The clusters were chosen by visual inspection of the plates, as were the Abell clusters, but the contrast against the background drops precipitously for clusters of a given richness at high redshifts, and it is clear that our completeness drops as well, as is suggested by the counts, which are illustrated in Figure 1. It is clear that at redshifts beyond about 0.5 the catalog is seriously incomplete (unless, of course, the universe contains very few rich clusters at higher redshifts, which is a distinct possibility, but with the quality of the current data incompleteness seems much more likely to be the culprit.) The clusters we have found at larger redshifts are, therefore, certainly the very high-richness tail of the distribution and one must be cognizant of that fact when comparing their populations with clusters nearer by; fortunately, that fact for the most part strengthens our conclusions regarding evolution, since richer clusters in general have earlier-type populations.

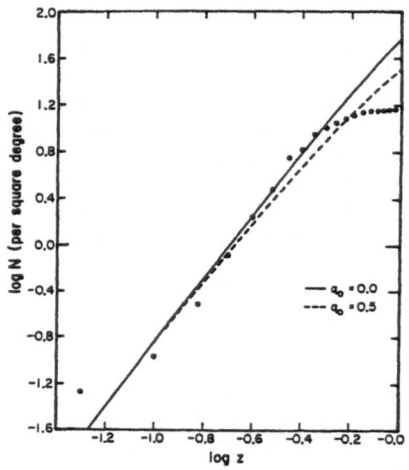

Figure 1. The cumulative areal counts from the GOII cluster survey. Severe incompleteness sets in at about $z = 0.5$. From Gunn $et. al.$ 1986.

If one wants to do the job correctly it is imperative that the clusters be found from machine-generated catalogs with algorithms which can be tested on realistic simulations. Efstathiou's group are attempting this for the Southern Schmidt data, and we (Postman, Oke, Hoessel, Schneider, and the author) are constructing such a catalog from survey data obtained with the Four-Shooter CCD camera (Gunn et al 1987) in slow drift mode. The survey will when complete cover six square degrees in six one-degree square regions distributed over the sky to about V=24.5 and one square degree in six 1 degree by 8 arcminute strips to about V=25.8. Data are obtained in bands which are essentially Johnson V and I (the Space Telescope WF/PC F555W and F785LP bands) simultaneously, using the fact that the two leading detectors of the Four-shooter can be equipped with one kind of filter and the two trailing ones with another. A galaxy catalog will be extracted from these data, and clusters will be found using a matched detector algorithm in both space and magnitude. The data-taking is almost complete, and the cataloging software for object finding and image classification is finished; the processing is done for one field. We have not yet begun seriously to test the cluster-finding algorithm, but expect to have done by the beginning of the year. Figure 2 shows two I-band magnitude slices (galaxies

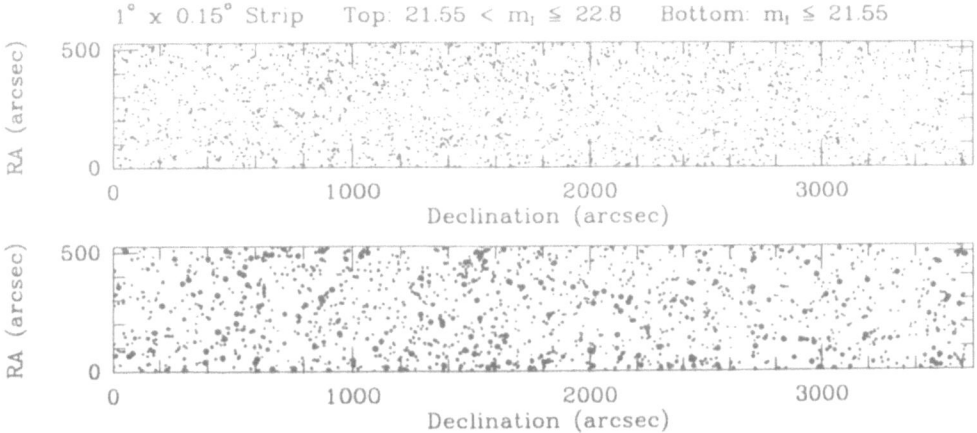

Figure 2. Galaxies in one strip of the new four-shooter cluster survey in the I band, plotted for two magnitude cuts. The area of the dots is roughly proportional to the brightness.

only) in one strip from the survey. It is quite evident that cluster-finding at faint levels is not an easy task.

The sad thing is that no such project is underway for a replacement for the Abell catalog. Such a task would involve the treatment of vast amounts of data, but a digitized data base already exists for a survey of somewhat more limted depth from the Space Telescope Science Institute guide star activity. It should be possible to reach the equivalent of Abell distance class 5 from these data, and the yellow passband used for this survey is better for comparison with the red data of higher-redshift surveys than the red Sky Survey band would be.

2. ACTIVE GALAXIES IN CLUSTERS

Since the pioneering work of Butcher and Oemler (1978), it has been recognized that the evolution of the population in clusters appears to be much more rapid than would have been expected, in the sense that even at moderate redshifts (a few tenths) the fraction of galaxies which have rest-frame colors typical of late-type systems is very much higher than in clusters at the present epoch. That work has been followed up spectroscopically by a number of workers, including Dressler and the author, (Dressler and Gunn 1982,1983, Dressler et al 1985, Gunn and Dressler 1988), Butcher and Oemler 1984), Lavery and Henry (1986), Couch and Sharples (1987), and Mellier (1988) (see also Soucail et al 1987). What all have found is a bit surprising; in some cases the blue galaxies do have spectra typical of late-type systems, but just as often they have spectra which are indicative of activity of varying degrees of violence. The active galaxies are of three kinds: a) Real Type I Seyferts, which are rare but much more common than at the present epoch. b) High-excitation narrow-line emission galaxies, often quite blue. An unknown fraction of these are probably Type II Seyferts, but most are probably not. They show very strong [OIII] $\lambda 5007$ and $\lambda 4959$, and strong [OII] $\lambda 3727$. The Balmer series is usually much weaker and even in some cases apparently absent. c) Galaxies with yellow colors, $0^m.2$-$0^m.3$ bluer in rest $B - V$ than the ellipticals, whose blue spectra are dominated by what appears to be the light of main-sequence A stars. Dressler and I have called these systems "E+A" or "post-starburst" galaxies, the latter appellation because it appears that the only reasonable explanation for the spectra is that the galaxy in question has undergone an extensive starburst involving about 20% of its mass (if the burst IMF is normal) of the order of 10^9 years ago. The combined rest-frame equivalent widths of Hβ and Hγ are of the order of 7 Å, to be compared with 2-3 Å for ordinary spirals of equivalent colors. These objects generally show no emission lines, in contrast with their equivalents found in the field at similar redshifts; we will return to that point later. Figures 3 and 4 show a composite of spectra with strong E+A characteristics and a set of objects with strong emission, all from the Dressler-Gunn sample.

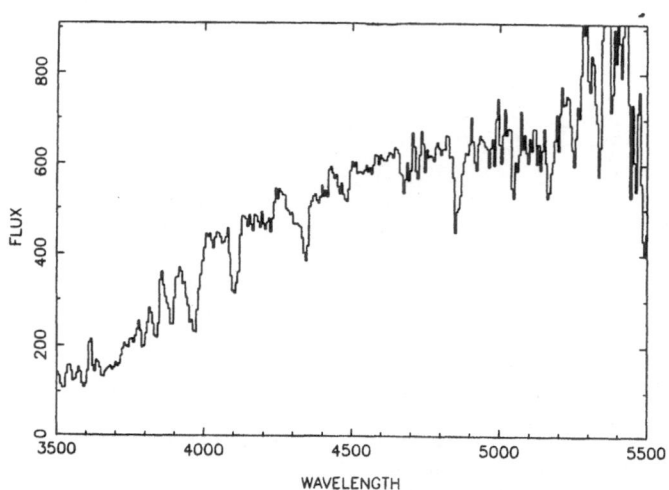

Figure 3. A composite spectrum from the intermediate-redshift sample of Dressler and Gunn of galaxies showing very strong E+A characteristics. Note the still quite strong magnesium feature at $\lambda 5200$ from the underlying old stellar population.

Dressler and I have studied most extensively a sample of 7 clusters with a mean redshift of 0.45 and a range of 0.38 to 0.55. We have a limited amount of data in three more distant clusters at redshifts of 0.76, 0.81, and 0.92, but the quality and quantity of these data are very much inferior to those in the lower redshift sample. This sample has been discussed in a preliminary way elsewhere (Gunn and Dressler 1988, Dressler 1987) and is currently being prepared for final publication. We will review it here most cursorily, and refer the interested to the previous preliminary discussion, and, for the patient, to the final paper. The data are summarized in Table I, in which Column 1 lists the cluster designation, 2 the mean redshift, 3 the number of spectra obtained, 4 the number of members among those spectra, 5, and 6, the number of active galaxies of the E+A type and the two emission-line types lumped together, 7 the fraction of active galaxies arrived at by the technique described in Gunn and Dressler (1988), 8 the velocity dispersion in $km\ s^{-1}$, and 9 the limiting r magnitude of the spectroscopy. The spectroscopic sample is in no way *complete* to that limit; it merely indicates the faintest objects for which we have spectra.

The 3C295 cluster is dynamically peculiar; it is not a very rich cluster, and yet has the highest velocity dispersion of the whole set. The 12 member redshifts are grouped in three small clumps, 5 at .450, 2 at .460, and 5 at .470, and with such small numbers it is quite impossible to guess what is going on. More work clearly needs to be done, and until it is we will exclude that cluster from our discussion about dynamics. The velocity dispersions are interesting. Both of

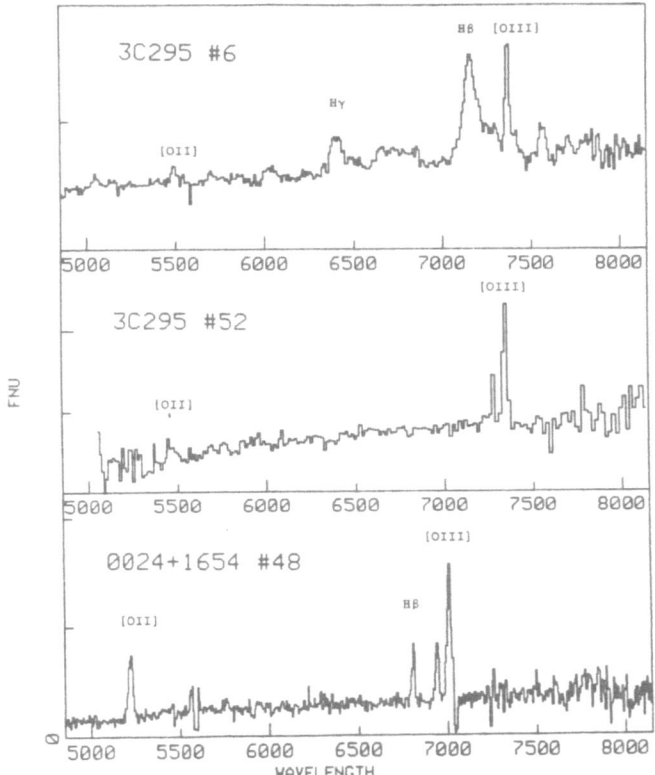

Figure 4. Three emission-line galaxies from the Dressler-Gunn sample.

TABLE 1: The z = 0.45 Sample

Cluster	$\bar{z}$	Sp	Mem	E+A	em	f_{active}	$\sigma(km\ s^{-1})$	r_{lim}
9HFCL27	0.378	27	16	3	1	27 ± 15	560	22.5
9HF$\alpha\beta$	0.391	36	29	2	10	27 ± 8	1600	21.5
Cl0024+24	0.407	48	30	6	1	22 ± 9	1350	22.0
3HFCL2	0.419	18	15	1	4	34 ± 17	590	22.0
3C295	0.467	22	12	3	3	39 ± 18	3000	22.0
16HF$\beta\beta$	0.540	31	22	3	5	37 ± 15	450	22.5
Cl0016+16	0.546	49	30	7	5	31 ± 10	1600	22.5

the serendipitously found clusters (0024,0016) have very high dispersions, as one might expect; their richness and central concentration called attention to them on material which was not specifically obtained for the purpose of surveying for clusters. Of the rest, all from the Gunn-Oke-Hoessel survey, only one (9h$\alpha\beta$) has a very large dispersion, and that cluster was included in the spectroscopic survey precisely because it is the richest cluster in the sample, and quite probably the richest (though not by any means the most centrally concentrated) cluster known. The others, all from the GOH survey, have dispersions typical of Abell richness 1 clusters, consistent with their populations, and we may feel reasonably sure that they are typical of moderately rich clusters at the epoch at which they are observed.

Though there appears from table 1 to be large variations in the relative proportions of the types of activity among the clusters in the sample, the differences are not statistically significant; to investigate further whether the differences are real and possibly correlated with other cluster properties will require a much more extensive sample, which will not be easy to come by. The mean level of activity, defined by the same spectroscopic criteria, is at least a factor of 4 higher than in similar clusters at the current epoch, and the incidence of the E+A phenomenon at least an order of magnitude higher (3 or 4 of the galaxies in the low-redshift Dressler and Shectman (1988) spectroscopic survey, which contains about 1000 cluster galaxies in an absolute magnitude range similar to that used in our sample, have *bona fide* E+A spectra.)

A striking property of the active galaxies which has been commented on by most workers in the field is their spatial distribution. They are quite definitely associated with the cluster but are very much less centrally concentrated than the "passive" (spectroscopically early-type) galaxies. Along with this, and fully as striking, is their velocity distribution, which in our sample shows a dispersion some 70% larger than the passive galaxies in the clusters.

3. THE MECHANISM FOR THE ACTIVITY

Starburst activity in galaxies at the present epoch is often, if not usually, associated with interaction (see, *e. g.* , Lonsdale et al 1984). Indeed, in the cluster sample of Lavery and Henry (1986) at redshift about 0.2, the evidence is strong that there as well the galaxies with strong emission are often interacting. (One would not expect the E+A galaxies to still be near their neighbor which caused the burst, since the cluster velocity dispersion would carry them far away over the 10^9 years since the burst). In our more distant sample, which has signal-to-noise in the imaging data not notably inferior to the data of Lavery and Henry, we find no evidence of interaction and a much higher level of activity. It is quite possible that a small fraction of the activity, consistent with the *numbers* seen in the Lavery-Henry sample, is caused by interaction, but the majority is apparently not.

Dressler and I have suggested that the trigger for the activity may well be the entrance of gas-rich galaxies for the first time into the high-temperature, high-pressure intracluster medium. This mechanism has several attractive features and not a few difficulties. Any suggested mechanism must account for the fact that the kind of activity seen at redshift 0.5 has essentially stopped at the current epoch. This is the single most difficult thing to explain, and no model put forward to date handles the question really satisfactorily. As we shall discuss below, however, recent advances in understanding how the hot intracluster medium develops may well provide a quantitative explanation under our "first infall" picture.

The "first infall" model does explain in a natural way the spatial and velocity distributions of the active galaxies. The mass infalling at any given time has a distribution which, because of transverse velocities acquired from substructure during the collapse, is never as centrally condensed as the old, already virialized core (Ryden and Gunn, 1987). From the observed radii (of the order of half a megaparsec), the velocities (projected speeds of order 1000 $km\ s^{-1}$) and the inferred lifetimes of somewhat less than 10^9 years, it would appear that we do not see the E+A galaxies spread out over their orbits, but that the phoenomenon lasts only long enough for us to see them near their (first) pericenter passage. At this phase, of course, they are travelling much faster than the average velocity dispersion, since the apocenter radius is typically several times that of the pericenter.

One difficulty which any picture faces may, in fact, provide an important part of the answer. The bursts, as mentioned, involve as much as 20% of the stellar mass of the parent galaxy. In the post-burst phase in which they exhibit strong E+A characteristics they are typically about half a magnitude to a magnitude brighter at V than they were before the burst, which places most of the ones we have studied one to two magnitudes fainter than L^* in their quiescent phase. Galaxies this bright at the present epoch have no more gas than this in total, and the preferentially early-type ones associated with the outskirts of clusters not nearly that much. Much of what they have–certainly the weakly bound outer parts of the distribution and the intercloud medium everywhere–will be lost to ram-pressure stripping as the galaxy enters the hot medium. Only the dense clouds will remain bound to the galaxy and be crushed by the hot, high-pressure intracluster gas, which typically has pressures 10 to 100 times higher than the interstellar pressures in the Galaxy. Unless, therefore, the mass function is very different from the one in the underlying galaxy, big galaxies are simply not gassy enough at the present epoch to show the kind of activity we see at $z = 0.5$, no matter what the trigger mechanism is. It is suggested, then, that one reason we do not see E+A galaxies today among bright cluster galaxies is a general and quite marked decline in gas content over the last third to half of the age of the universe.

This, by itself, would seem insufficient, given the statistical spread of gas content observed in galaxies. The clusters themselves are changing as well, however, and the two together might be sufficient to explain the precipitous drop in activity. It has not until recently been clear where the standing shock which separates the hot medium from the colder, more-or-less comoving medium falling in is, or whether, with a reasonable initial density fluctuation spectrum, a well-defined shock exists at all. Models with dark matter and gas have recently been constructed by Evrard (1988) using the soft-particle hydrodynamics (SPH) technique and Cold Dark Matter initial conditions (though the CDM spectrum is almost white for the mass range of interest, and the details of the spectrum are probably not very important.) What he finds is very interesting; the growth of subclustering in the infalling material does preheat the gas, so that the pressure averaged on spherical shells rises continuously but much more rapidly than the density outside some transition radius which may be associated with the "classical" shock radius. Over a short range in radius the behavior changes, as it would be expected to, to isothermality, and inside the pressure is essentially proportional to the density with a temperature slightly lower than the virial temperature for the cluster. Outside the "shock" the pressure rises roughly like r^{-5}; inside, like r^{-2}. The transition radius itself moves outward with a velocity about a quarter the velocity dispersion, and so for a cluster with a dispersion of 700 $km\ s^{-1}$, say, moves about a megaparsec since $z = 0.5$ if $h = 0.5$. In the clusters we have studied the active galaxies are contained with little central concentration inside a circle about 500 kpc in radius, which suggests that the transition radius is a somewhat bigger than that, and has probably almost trebled by the present epoch. It is thus now at a radius for these clusters of the order of two Mpc.

What does this mean for our picture of the triggering of the burst? When the transition radius is small, the pressure at the transition radius is high, since the pressure inside it goes like r^{-2} and is roughly constant at a given radius with time. Outside the transition radius the pressure rises quite steeply with time as the galaxy falls in, and if the transition radius is 700 kpc, the infall velocity 1400 $km\ s^{-1}$, the e-folding time for the pressure increase is about 1.5×10^8 yr, much shorter than the duration of the E+A phenomenon. At present, however, the timescale is of order three times as long because of the much larger shock radius, and the pressure at the shock hardly higher than typical interstellar pressures. The pressure gradient inside the shock is much shallower than outside, so that if the "burst" is triggered by the gentle rise of pressure inside the shock the timescale altogether is an order of magnitude longer. One then probably cannot describe the phenomenon as a burst, but instead as a prolonged enhanced period of star formation. This, together with the inferred relative paucity of gas, may well explain the absence of E+As in present-day clusters.

In addition to all of this, the mean pericenter distances for galaxies crossing the shock radius for the first time scales approximately like that radius, so the projected density against the central regions of the cluster will have decreased for a constant infall rate by an order of magnitude, and the infall rate itself is likely to have dropped by a factor of two or so.

What is the connection with the emission-line galaxies? We have always assumed that the emission-line objects are simply those in the burst phase itself; this phase must be of the order of a factor of ten shorter in duration than the E+A phase. The numbers of the emission-line objects is comparable to the E+As, not an order of magnitude smaller, but the brightness of the burst is probably an order of magnitude larger than the quiescent luminosity of the parent, to be compared to the factor of two or so during the E+A phase. Thus many faint systems which have faded out of our sample by the time they have become E+A's are visible during the burst phase. This explanation is easier to swallow when one thinks of a real shock with very short timescales for the pressure increase; the characteristic timescales derived above of a bit longer than 10^8 yr may be uncomfortably long, but detailed models will have to be made before strong conclusions can be drawn. It is also likely that a given parcel of gas has a much more violent history than the spherically averaged shell described above, and that must be taken into account as well.

It is interesting that the numbers and lifetimes are consistent with, and indeed almost force the view, that the E+A phenomenon is a one-time event. At the epoch we observe our sample the universe is perhaps 7 Gyr old. The E+A phenomenon lasts a bit less than 1 Gyr, and of the order of 10% of galaxies in the central regions of the cluster are E+A's, and are only a bit brighter during this phase. There thus seems to be no room for recurrence, which is another piece of evidence supporting a rather catastrophic mechanism.

Bright galaxies spectroscopically similar to E+As are found in the field at redshifts similar to our cluster sample, though with somewhat reduced frequency (Broadhurst et al 1988) and with one important spectroscopic difference—*viz.* , they *nearly always* have emission; the cluster E+As *never* have emission. This supports, I believe, the view that galaxies in general at these epochs were much more gas-rich than at present. Such a "pregnant" system can be triggered by any one of a large number of mechanisms, but the evidence is that when it happens in the field it is not *global*. When the A stars dominate the spectrum there is still gas and still star formation going on which is responsible for the emission lines. In the clusters, however, there is none; again, a global mechanism which suddenly precipitates star formation (in some of the gas) or loss (of the rest) of all the gas in a galaxy is required, and the shock/ram-stripping mechanism seems ideally suited to the task.

The Broadhurst et al (1988) field sample and their interpretation of their redshift distribution *vis-a-vis* the galaxy counts also supports the notion that galaxies were much gassier at relatively recent past epochs. They find that the data are most easily interpreted by a progressive steepening at earlier times of the faint-end slope of a Schechter-like luminosity function, and that the excess in the counts over a "no-evolution" model is due to galaxies with burst-like spectral characteristics. Again, galaxies like this are rare in the field today, and if this phenomenon persists to higher redshifts (their median redshift is only a bit bigger than 0.2) the most natural explanation would seem to be a progressive increase with look-back time of the luminosity at which typical galaxies can support bursts–which involves at the very least them being sufficiently gas-rich to do so.

4. PASSIVE GALAXIES AND THE EPOCH OF GALAXY FORMATION

The population synthesis models of Pickles and van der Kruit (1988) are probably the best available today, but have not been applied to high-quality data for redshifts bigger than 0.3 or so. For the question of age determination, it is probably as good or better to do simple differential comparisons between spectra of distant objects and nearby ellipticals. Figure 3 shows data of Oke, Hoessel, Schneider, and the author for the central elliptical in Cl132227+3027, at a redshift of 0.755. The relative blueness and weakening of spectral features is evident in comparison with the template spectrum (NGC 4881 in Coma). The synthesis models of Tinsley and Gunn (1976) can be used to derive the rate of change of B-V color and 4000-Å break strength D with age; the absolute values of the color and break strength are highly dependent on the details of the models, but the quantities $d(B - V)/d\log t$ and $d\log D/d\log t$ are not very, and one finds from those models that

$$\Delta(B - V) \sim 0.28 \log(t/t_0),$$

and

$$\frac{\Delta D}{D} \sim 0.57 \log(t/t0). \tag{1}$$

These quantities for the spectrum illustrated are about 0.15 and 0.30, respectively, with uncertainties of the order of 20% of the values. The ratio is correct, and both give ages of $0.29t_0$. If Ω is unity, the redshift corresponds to an age for the universe of $0.42t_0$. If at that time the galaxy was $0.29\ t_0$ old, the formation time for the bulk of the stars was $0.13t_0 \pm 0.04t_0$, which places the formation epoch between a redshift of 4.0 and 2.25. These values are substantially independent of the Hubble constant (the value of H_0 influences the constants in Equation 1, but not by very large amounts), and most of the uncertainties in the evolutionary models—but not, unfortunately, the large uncertainties in asymptotic giant-branch evolution for metal-rich intermediate-age populations; the models assume that the form and level of the giant branch changes almost not at all with age.

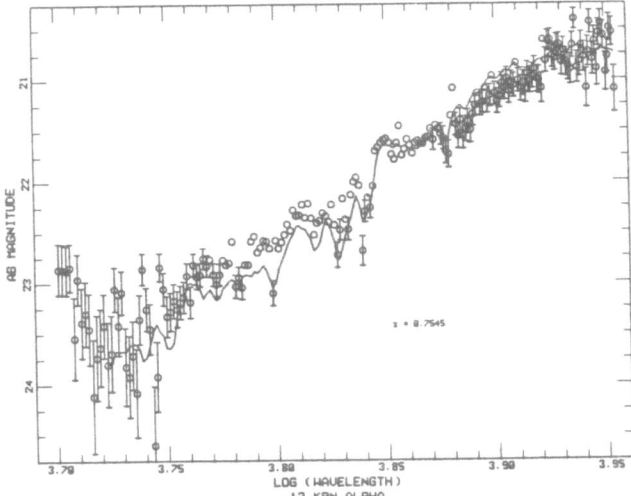

Figure 5. A 6000 second CCD spectrum of the brightest galaxy in Cl132227+3027, a rich, concentrated cluster in the GOH sample at z=0.755. The solid line is the redshifted spectrum of NGC4889 in Coma.

More and higher-quality spectra of objects at this and slightly higher redshifts will be forthcoming, and the observational situation will improve, but better models may not be easy to come by in the near future. In addition, the connection to the general question of the epoch of formation of *typical* galaxies is tenuous at best. Just as one is free to interpret the very high-redshift galaxies found associated with radio sources (Djorgovsky *et. al.* 1987, Chambers *et. al.* 1988, Lilly, this conference) as the result of the collapse of rare, very high-density peaks in the initial density distribution, so here as well. We *know* that the central regions of clusters are associated with high-amplitude, long-wave fluctuations (if, of course, the hierarchical clustering picture has anything at all to do with the formation of structure), so it would not be surprising if galaxy formation occurred somewhat earlier there than elsewhere—in any case, one would not expect these very atypical regions to be representative. Worse, the incompleteness in the cluster catalog is catastrophic at this redshift, so we have no real idea just how rare this environment is.

5. SUMMARY

It appears clear that clusters at moderate redshifts have populations which are rather different from those of present-day clusters, in the sense that a much higher fraction of galaxies show extensive and violent star-formation activity either currently or in their recent past. The numbers and timescales suggest that these bursts are associated with a single catastrophic event in the lifetimes of cluster galaxies which dumps a substantial fraction of the gas in the galaxy into stars and removes any residue. There is strong evidence that galaxies at the observed epochs were much more gas-rich than at present. Dressler and I have suggested that the triggering event for the burst is the first entrance of an infalling galaxy into the hot, shocked intracluster gas, and recent models of the behavior of the shock with time by Evrard make it possible to understand the dramatic decrease of this activity between $z = 0.5$ and the present.

The "passive" members of distant clusters can be used to estimate ages. One of our best spectra for a cluster at $z = 0.75$ has been crudely analyzed and yields a formation redshift of about 2.9 with considerable uncertainty. The conditions for the formation of brightest cluster galaxies are, however, not likely to be at all representative of those in the field, and conclusions about their epoch of formation, while highly interesting of themselves, are likely to shed little light on the question of when *typical* galaxies form.

The author would like to extend thanks to the Durham group for very kind hospitality before and during this meeting, and for many exciting and informative discussions about this and their own excellent work in this area, and to his many collaborators in the interlocking research efforts reviewed here. This research was supported by the National Science Foundation, the National Aeronautics and Space Administration, and the John D. and Catherine T. MacArthur Foundation.

References

Abell, G. O., 1958, *Ap. J. Suppl.* **3**, 211.

Broadhurst, T. J., Ellis, R. S., and Shanks, T., 1988, submitted to *M. N. R. A. S.*.

Butcher, H., and Oemler, A., 1978, *Ap. J.* **219**, 18.

Butcher, H., and Oemler, A., 1984, *Nature* **310**, 31.

Chambers, K., Miley, G. K., and van Breugel, W., 1988, preprint.

Couch, W. J., and Sharples R. M., 1987, *M. N. R. A. S.* **229**, 423.

Djorgovsky, S., Strauss, M. A., Perley, R. A., Spinrad, H., and McCarthy, P., 1987, *A. J.* **93**, 1318.

Dressler, A., 1980, *Ap. J.* **236**, 351.

Dressler, A., and Gunn, J. E., 1982, *Ap. J.* **263**, 533.

Dressler, A., and Gunn, J. E., 1983, *Ap. J.* **270**, 7.

Dressler, A., Gunn, J. E., and Schneider, D. P., 1985, *Ap. J.* **294**, 70.

Dressler, A., 1987, in *Nearly Normal Galaxies from the Planck Time to the Present*, S. M. Faber, ed., Springer-Verlag, New York, 265.

Dressler, A., and Shectman, S. A., 1988, *A. J.* **95**, 284.

Dressler, A., and Gunn, J. E., 1988, in *The Large Scale Structure of the Universe*, (IAU Symposium 130, Balaton, Hungary), J. Audouze, ed, Reidel, Dordrecht.

Evrard, A. E., 1988, preprint.

Gunn, J. E., Hoessel, J. G., and Oke, J. B., 1986, *Ap. J.* **306**, 30.

Gunn, J. E, Carr, M., Danielson, G. E., Lorenz, E., Lucinio, R.,Nenow, V., Schneider, D. P., Smith, J. D., Westphal, J. A., and Zimmerman, B. A., 1987, *Optical Engineering* **26**, 779.

Gunn, J. E. and Dressler, A., 1988, in *Towards Understanding Galaxies at Large Redshift*, R. G. Kron and A. Renzini, eds, Kluwer, Dordrecht, p. 227.

Lavery, R. J., and Henry, J. P., 1986, *Ap. J. (Lett.)* **304**, L5.

Lilley, S. J., 1989, these proceedings.

Lonsdale, C., Persson, S. E., and Matthews, K., 1984, *Ap. J.* **287**, 95.

Mellier, Y., 1988, in *Towards Understanding Galaxies at Large Redshift*, R. G. Kron and A. Renzini, eds, Kluwer, Dordrecht, p. 227.

Pickles, A. J. and van der Kruit, P. C., 1988, in *Towards Understanding Galaxies at Large Redshift*, R. G. Kron and A. Renzini, eds, Kluwer, Dordrecht, p. 29. See also Pickles, A. J., these proceedings.

Ryden, B. S., and Gunn, J. E., 1987, *Ap. J.* **318**, 15.

Shane, C. D., and Wirtanen, C. A., 1967, *Publ. Lick Obs.* **22**, 1.

Shectman, S. A., 1985, *Ap. J. (Suppl.)* **57**, 77.

Soucail, G., Mellier, Y., Fort, B., and Cailloux, M., 1988, *Ast. and Ap. Suppl.* **73** 471.

Tinsley, B. M, and Gunn, J. E., 1976, *Ap. J.* **203**, 52.

DISCUSSION:

SILK: The gas-rich progenitors that you required to account for the E + A galaxies you see in distant clusters could plausibly be disk galaxies. Is there any evidence for morphological peculiarities in high resolution images of these galaxies that distinguish them from normal E galaxies?

GUNN: We do not have enough spatial resolution yet. Remember that these objects are about a magnitude brighter than their quiescent state, and are likely to be small.

SHANKS: Results by Yee and collaborators suggest that low luminosity, radio-quiet QSOs with $0.5 < z < 0.75$ hardly ever exist in rich cluster environments. How much of a constraint does this type of observation place on the suggestion that the Seyfert "fraction" in rich clusters is increasing towards higher redshifts?

GUNN: I don't know. It is puzzling that when QSO's do show up at higher redshifts they appear to be central objects. All of our Seyferts are on the outskirts of the clusters.

FIELD: The spectra of E + A objects you showed have H absorption (consistent with A-stars) but no HII or [O II] emission, as would be expected from the HII regions around the accompanying O-B stars. How do you explain this?

GUNN: If the burst is sharply confined in time, there are no early- type stars by the time we see the objects as E + A galaxies. It is also possible that the shock event leaves no gas to excite.

EVOLUTION OF COMPACT GROUPS AND FORMATION OF ELLIPTICALS

Joshua E. Barnes
Institute for Advanced Study
Princeton, NJ 08540 USA

ABSTRACT. A numerical simulation of a group of six disk galaxies is described. This compact group evolves through multiple mergers on a timescale of only a few crossing times. Merger remnants are typically slowly-rotating triaxial systems with de Vaucouleur's-law luminosity profiles and structural parameters generally consistent with bright ellipticals. If the compact groups we observe are correctly interpreted as fleeting phases in the dynamical evolution of ordinary groups, the merger remnants they leave behind are examples of galaxies forming at $z \simeq 0$.

1. INTRODUCTION

Outside of the rich clusters, galaxies are not scattered at random but group into more or less distinct configurations. Most of these groups are relatively loose systems, with crossing times of 10^9 to 10^{10} y, a significant fraction of the age of the universe. Our own local group, for example, is an extended system only now collapsing out of the Hubble flow. Deeper surveys occasionally yield much more compact groups, with estimated crossing times as short as 10^8 y. These rather rare systems – Hickson (1982) lists ~ 100 in a sample extending to $z \simeq 0.05$ – contain a relatively large number of tidally disturbed galaxies.

The dynamical evolution of a compact, gravitationally bound group of galaxies is thus an interesting problem. Theoretical arguments indicate that such groups undergo radical restructuring in a few crossing times due to dynamical friction, tidal stripping, and galactic collisions. Numerical simulations by Carnevali *et al.* (1981), Barnes (1985) and others have qualitatively confirmed the theoretical predictions. The existing models, however, do not come all that close to representing the actual spatial and dynamical structure of a group of galaxies. With bigger supercomputers and better software this failing can be remedied.

2. NUMERICAL SIMULATION

Fig. 1 shows the initial conditions used in this calculation: a binary hierarchy with galaxies at exterior nodes. Clustering hierarchies have often been used to describe the distribution of galaxies, although the application to compact groups is something of an extrapolation. The six galaxies present initially are composite bulge/disk/halo systems with a 4:1 ratio of dark to luminous matter, similar to model 1 of Barnes (1988). Two of these galaxies have mass 2.5, while four are half-sized models scaled to have the same characteristic surface densities. All dimensional quantities are given in an arbitrary set of units with $G = 1$; scaling the larger disks roughly to our own Galaxy, the units of length, time, and mass are 28 kpc, 200 Myr, and 10^{11} M$_\odot$, respectively. A total of $N = 65536$ particles were used, of which only the 32768 representing luminous mass are plotted. At this time the dark mass is entirely concentrated in

179

C. S. Frenk et al. (eds.), The Epoch of Galaxy Formation, 179–183.

180

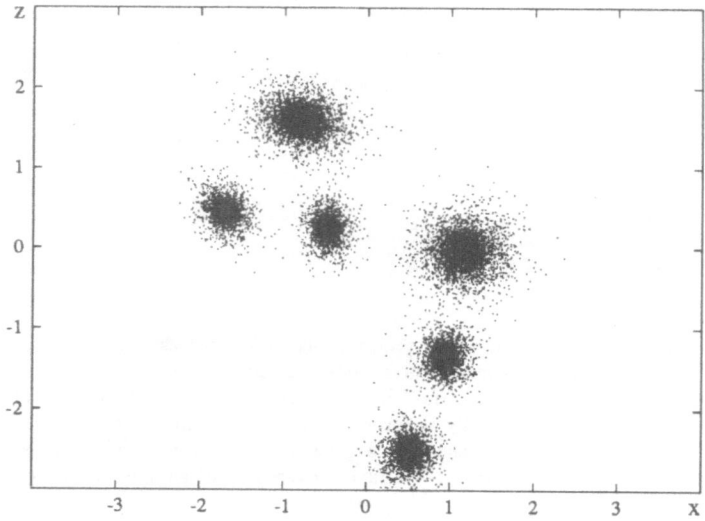

Fig. 1. Group model, initial conditions. Only luminous particles are plotted; the dark mass is all in individual galactic halos.

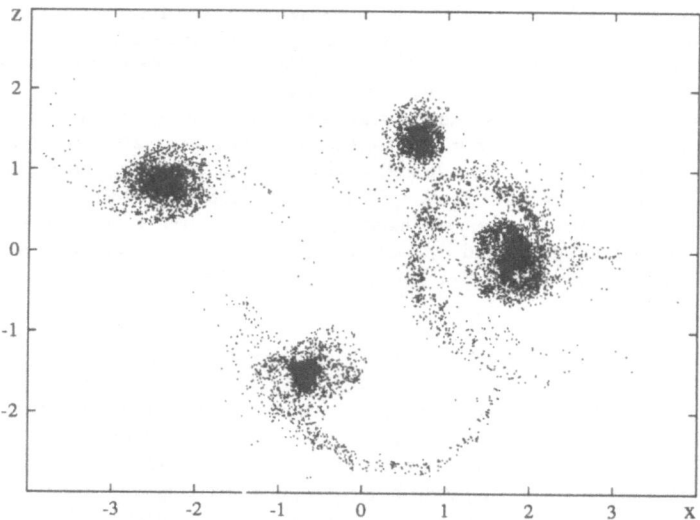

Fig. 2. Group model, time $t = 2$. The two merger remnants are marked by massive tidal tails.

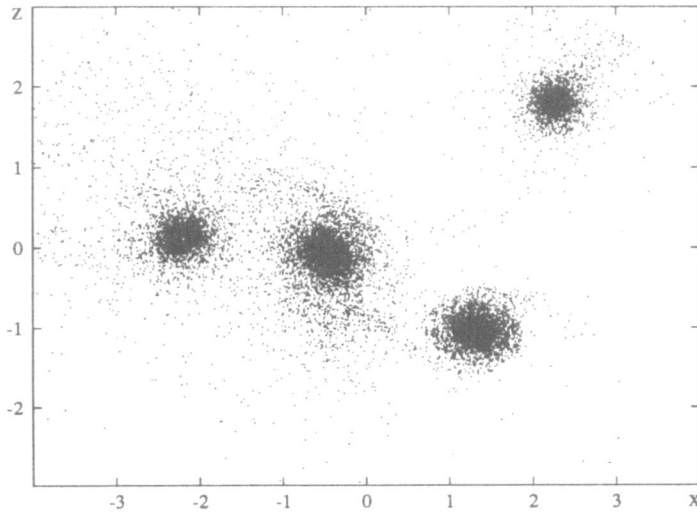

Fig. 3. Group model, time $t = 8$. The larger merger remnant has taken up residence at the center of the group.

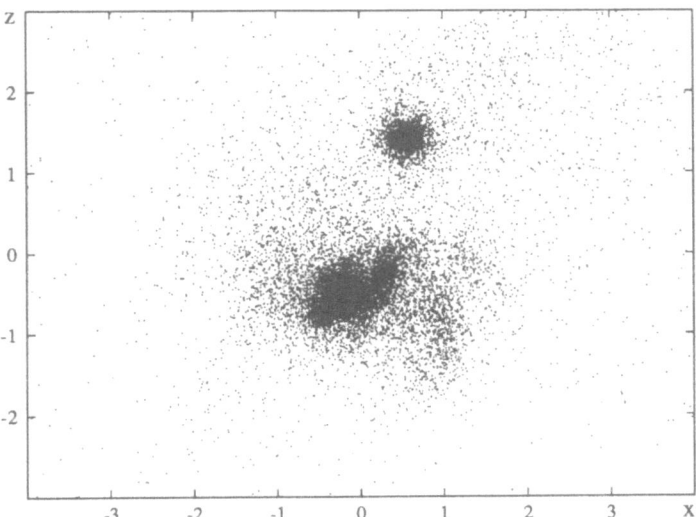

Fig. 4. Group model, time $t = 12$. The larger disk galaxy has just been accreted; the smaller one is slightly to the right of the central remnant.

dark halos around individual galaxies. Some compact groups have such low M/L ratios that little additional dark matter, beyond that bound to individual galaxies, can be present. But in general, it seems likely that groups formed by gravitational clustering would possess a common dark halo, not yet included in the initial conditions.

The evolution of the system was computed using a modified "tree" algorithm, running on a Cyber 205. This method provides force calculations with a relative accuracy of order 10^{-3}, quite sufficient for the job at hand. Forces were softened using the usual formula, with $\varepsilon = 0.015$, roughly 0.2 the exponential scale length of the smaller disks. Particle trajectories were calculated using a simple leap-frog with a time-step of $\Delta t = 1/128$.

By $t = 2$ time units, two mergers have occurred, leaving four distinct galaxies in the system. Fig. 2 shows the group at this time; the merger remnants are marked by their long tidal tails. The remnant at $x = -0.8$ involves the pair of small disk galaxies initially associated with the large spiral at $x = -2.3$ The remnant at $x = 2$ results from the merger of a small disk galaxy and a large one, and now comprises the most massive object in the group. The dark halos surrounding the merger remnants are much more extended than those around the original galaxies, reaching roughly as far as the luminous tails do, and even overlapping to some degree.

Around time $t = 6$, the two merger remnants undergo a moderately close encounter, which further distends their halos, and liberates a diffuse spray of luminous particles. Following this encounter, the more massive remnant takes up residence at the center of the group. By time $t = 8$, shown in fig. 3, the smaller remnant is at $x = -2.2$, enveloped by a cloud of tidal debris. The two surviving disk galaxies are as yet relatively undisturbed; the larger, at $x = 1.3$, is on a roughly circular orbit about the central remnant, while the smaller, at $x = 2.2$, is falling back into the group on an approximately radial orbit.

The disk galaxies are largely disrupted over the next four time units. This is the outcome of a series of close encounters, starting with a collision between the smaller disk galaxy and the central remnant at time $t = 10$, and followed by a passage of the smaller remnant at time $t = 10.5$. By this stage, however, the behavior of the system is poorly described in terms of two-galaxy encounters; interactions between between three or even all four galaxies are seen. At time $t = 12$, shown in fig. 4, the larger disk galaxy has been incorporated into the central remnant, while the the smaller disk galaxy, having been stripped of much of its luminous mass, is closely orbiting the center. Only the small remnant, at the top, is relatively clear of the central monster; after several more passages, it merges with the central object at time $t = 17.6$.

3. CONCLUSIONS

Starting from a hierarchy of galaxies, the group model presented here evolved into a system dominated by a central merger remnant, which then accreted the remaining members of the group. The whole process took of order one dynamical time, consistent with the earlier models of Barnes (1985). Earlier models, however, could say little about the nature of the remnants produced when compact groups decay. The present simulation, together with results presented by Barnes (1988), suggests a tentative answer: the endpoint of the evolution of a compact group is a dynamically ordinary elliptical galaxy.

References

Barnes, J., 1985. *MNRAS* **215**, 517.
Barnes, J., 1988. *Ap. J.* **331**, 699.
Carnevali, P., Cavalere, A. & Santangelo, P., 1981. *Ap. J.* **249**, 449.
Hickson, P., 1982. *Ap. J.* **255**, 382.

DISCUSSION:

PEEBLES: There are lots of groups of spirals with crossing times less than the Hubble time. Have you or someone else estimated the abundance of field elliptical galaxies to be expected if such groups were fusing into proto-ellipticals at a rate comparable to the crossing rate?

BARNES: At this point, I think there are no good data-bases from which to derive reliable estimates of merging rates - compact groups are so short-lived, and hence rare, that one would have to survey a large volume of space to obtain useful statistical results. One possible problem is that the net luminosity of the merger products may be rather high compared to the luminosity function of E galaxies, but fading as the stellar populations age may solve this.

SELLWOOD: Merging started so quickly in this simulation that it is hard to believe your initial conditions could ever have arisen in nature.

BARNES: One would like to start the simulations with less compact groups, and allow the orbits to decay until merging began. But we *do* see compact spiral-rich groups which look much like my initial conditions, so nature may not share your theoretical expectations. In any case, I would not jump to such a conclusion on the basis of only one simulation.

SILK: Metallicity is an important discriminant between ellipticals - there is an increase of metallicity with luminosity, and there are metallicity gradients within ellipticals. Do you think that your model is capable of accounting for these properties?

BARNES: By and large, merging preserves gradients present in the original galaxies - for example, the central parts of the merger remnants are dominated by particles in the bulges and inner-most parts of the original disks. If metallicity gradients are present in the disk galaxies, the merger remnants will also have gradients. I really haven't a clue about the metallicity-luminosity relation - that's outside the scope of the models.

KAISER: Does the half light radius turn out OK?

BARNES: The half-light radii of the merger products are quite consistent with E galaxies; for example, if the infalling disks have exponential scale lengths of $\alpha^{-1} \sim 3.5$ kpc, the effective radius of the remnant is roughly $r_e \lesssim 4$ kpc.

PROPERTIES OF GALAXY CLUSTERS ASSOCIATED WITH QUASARS

H. K. C. Yee
Département de Physique, Université de Montréal
C. P. 6128, Succursale A, Montréal, PQ, H3C 3J7, Canada

E. Ellingson
Steward Observatory, Tucson, AZ 85721, U. S. A.

R. F. Green
KPNO, NOAO, P.O. Box 26732, Tucson, AZ 85726, U. S. A.

C. J. Pritchet
Department of Physics and Astronomy
University of Victoria, Victoria, BC, Canada

1. Introduction

The existence of a relationship between environment and quasar activty has been demonstrated by many recent investigations. In particular, preliminary evidence of a time dependence of this correlation has been provided by Yee and Green (1987, hereafter YG87) who found that at $z > 0.5$, unlike at lower redshifts, radio-loud quasars do exist in the cores of rich clusters. Thus, the study of galaxy clusters harbouring quasars will provide us not only with insights into the triggering and fueling of quasar activity, but also with evidence for evolution in the properties of moderately high-redshift galaxy clusters. This paper presents some preliminary results from our study of galaxy clusters which were discovered by virtue of their association with quasars. Specifically, we present data on 4 clusters found around quasars at redshifts ranging from 0.2 to 0.65. $H_o = 50$ km/sec/Mpc and $q_o = 0.02$ are used throughout this paper.

2. Data

Imaging data for 4 clusters associated with quasars are presented. The observations, obtained through Gunn i, r, and g filters using CCD detectors, are summarized in columns 1 to 3 of Table 1. An image of the cluster associated with 3C 206 is shown in Figure 1. Spectroscopic data obtained from the KPNO 4m using multi-aperture and multi-slit techniques are also available. Detailed results for objects in the field of one of the quasars, 3C 206, will be presented.

C. S. Frenk et al. (eds.), The Epoch of Galaxy Formation, 185–189.

3. Results

3.1. Imaging

Using background galaxy count corrections, various properties of the clusters can be estimated from the imaging data. The richness of the clusters is quantitatively measured by the parameter B_{gq}, (see Table 1), the galaxy-quasar spatial covariance function amplitude, which is normalize for both luminosity and spatial distribution using the KE1 model in YG87. These amplitudes allow us to make an indirect comparison with the Abell richness class (Longair and Seldner 1979): PKS 0405-12 is equivalent to an Abell class 2 cluster, 3C 206 and 3C 263 are as rich as the average Abell 1 cluster, while PKS 0812+02 has a richness between that of class 1 and 0. 3C 206 is somewhat of an anomaly in that it is the only quasar in the sample of YG87 with $z < 0.5$ found situated in the core of a cluster of richness Abell class 1 or greater. Two clusters, 3C 206 and PKS 0405-12, have a large enough excess of galaxy counts to allow crude estimates of the luminosity function (LF) to be made. In both cases, the characteristic magnitudes of the Schechter LF fits are consistent with those obtained by YG87 for galaxies associated with quasars at these redshifts.

TABLE 1

Name	z	CCD Imaging	B_{gq}	Core Radius	f_b
3C 206	0.198	CTIO 4m, $i\ r\ g$	638 ± 197	37 kpc	0.14 ± 0.08
PKS 0812+02	0.402	CTIO 1.5m, $i\ r\ g$	550 ± 180	205	0.48 ± 0.16
PKS 0405-12	0.574	CTIO 4m, $i\ r\ g$	1250 ± 260	195	0.60 ± 0.15
3C 263	0.652	CFHT 3.6m, $i\ r$	850 ± 260	90	0.28 ± 0.18

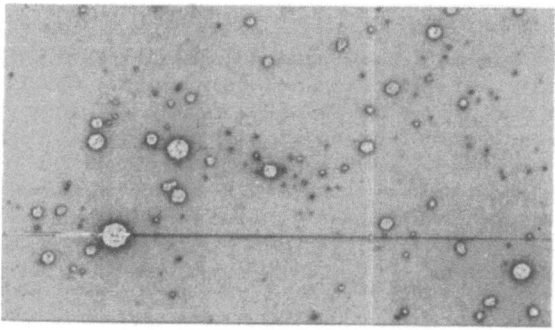

Figure 1. i image of the cluster associated with the quasar 3C 206. Note the extremely high galaxy density around the quasar which is situated in the center of the cluster.

A striking property which can be seen visually in the images of these clusters is the compactness and high galaxy density of the cluster cores. The radial profiles of the surface density of excess galaxies computed using concentric circles centred on the quasars are fitted to an empirical King profile. The resultant core radii are tabulated in column 5 of Table 1. The radii for the latter 3 clusters are about factors of 2 to 4 smaller than the average expected for Abell class 1 clusters (Dressler, 1980). For 3C 206, the lowest redshift object, the extremely high galaxy density and elongated structure in the proximity of the quasar produce a bad fit to the core profile, giving an abnormally small core radius. The compactness of the core, in general, may be one of the features most relevant to the formation of a quasar in the center of a rich cluster. It should be noted that this effect is probably even stronger in reality since the core radii are fitted without including the quasar as one of the galaxies.

Two- and three-color photometry and similar data from background galaxy counts allow us to estimate the fraction of blue galaxies in the clusters. A galaxy is classified as blue if it is bluer than a critical color, defined as 0.1 mag redder than the color of an Sbc galaxy at the redshift of the quasar. The blue fractions within a radius of 500 kpc of the quasar, f_b, are tabulated in column 6 of Table 1. Of these values, two, those of PKS 0405-12 and PKS 0812+02, are about a factor of two higher than that obtained for a sample of clusters by Butcher and Oemler (1985) at similar redshifts. An interesting point is that for all 4 clusters, there appears to be a significant number of blue galaxies in the central core of the clusters, a result not expected for ordinary dense clusters. Within a smaller radius of 200 kpc ($\approx$ the core radius for these clusters), there is actually a small increase in f_b compared to the larger area; whereas for "normal" clusters, a significant decrease is expected (Gunn and Dressler, 1988).

3.2. Spectroscopy

Three of the four quasars have at least one galaxy in the cluster with known redshift that is similar to that of the quasar. For a relatively bright blue galaxy close to PKS 0405-12, Marr and Spinrad (1985) found an emission line spectrum with a redshift identical to that of the quasar. Mulit-object spectroscopy of galaxies has been obtained at the KPNO 4m telescope for objects in the fields of 3C 206 and PKS 0812+02. For the latter field, only preliminary results have been derived. These indicate the existence of at least several galaxies at the same redshift as the quasar. For 3C 206, redshifts for 18 galaxies were obtained; 11 of these are found to have redshifts similar to that of the quasar, and are considered as members of a cluster associated with the quasar. An example of the spectra is shown in Fig. 2. The mean redshift of the cluster, including the quasar, is $z = 0.1971$. The quasar, with $z = 0.1976$, is close to, but not at, the exact dynamical center of the cluster. The line-of-sight velocity dispersion is 500 ± 110 km/sec, which is somewhat lower than the average of ≈ 800 km/sec expected for Abell class 1 clusters (Struble and Rood 1987). The spectra and colours of all but one of the 11 members which we observed spectroscopically are consistent with elliptical or early spiral galaxies. This result is consistent with the f_b estimated from photometry. The spectrum of the blue member galaxy (Fig. 2) shows [O II] $\lambda3727$ emission and Balmer absorption lines, typical of late-type galaxies. A more detailed investigation of the cluster associated with 3C 206 can be found in Ellingson, Yee and Green (1988).

4. Discussion

There are currently several models linking quasar activity with properties of the quasar environment. Galaxy interactions and mergers can act as perturbations which trigger the onset of feeding of the central engine (e.g., Roos, 1985). Within a rich cluster, the stripping action of the intracluster medium (ICM) will also have direct consequences for the sustenance of the AGN activity (Stocke and Perrenod, 1982), since it will deprive at least one component of the fuel for the nuclei.

The preliminary data on the clusters identified two properties which appear to support these ideas. Low relative velocities and the high spikes of galaxy density in the cores of the

188

clusters will result in effective gravitational interactions between cluster members and the host galaxy. The apparent existence of blue galaxies in the dense center of the clusters can be interpreted as an indication of inefficient gas stripping. The idea of a low-pressure ICM is further supported by the fact that all four quasars have classical double-lobed Faranoff-Riley Class (FR) II radio morphology. Normally, at low redshift, the lower power, more relaxed FR I morphology is expected to be associated with rich galaxy cluster environments (Longair and Seldner, 1979); whereas FR II morphology is found mostly within lower galaxy density, and hence lower intracluster medium density, regions.

Conversely, these scenarios support the interpretation that the change in the preferred environment of radio-loud quasars discovered by YG87 is a consequence of the evolution of the environment of rich clusters between the two epochs. As the cluster collapses and virializes, fewer effective gravitational interactions and a higher ICM density would be expected. This would result in a rapid decrease of the availability of fuel for the central engine in the galaxy at the center of a rich cluster, causing the dimming of the quasar and a change in the radio morphology.

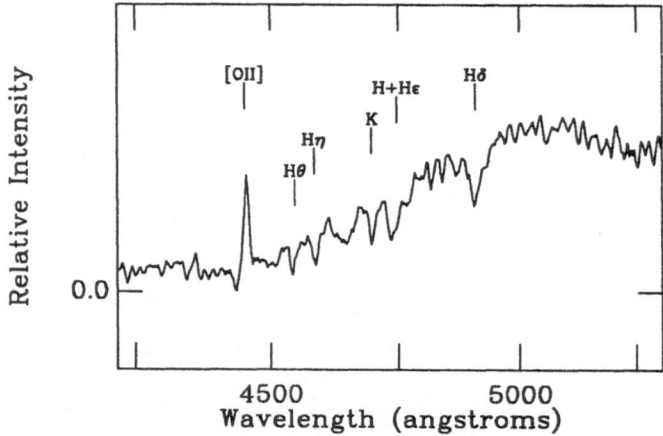

Figure 2. Spectrum of a blue cluster member of the cluster associated with 3C 206, showing typical high excitation signatures of [O II] emission and Balmer absorptions.

References

Butcher, H., and Oemler, A. 1978, *Ap. J.*, **226**,559.
Dressler, A. 1980, *Ap. J.*, **236**, 351.
Ellingson, E., Yee, H. K. C., and Green, R. F. 1988, *A. J.*, submitted.
Gunn, J. E., and Dressler, A. 1988, in the *Proceedings of the 5th Erice Workshop on Towards Understanding Galaxies at Large Redshift*, ed. R. G. Kron and A. Renzini, page 227.
Longair, M. S., and Seldner, M. 1979, *M. N. R. A. S.*, **189**, 433.
Marr, J., and Spinrad, H. 1985, *P. A. S. P.*, **97**, 684.
Roos, N. 1985, *A. A.*, **104**, 218.
Stocke, J. T., and Perenod, S. C. 1985, *Ap. J.*, **245**, 375.
Struble, M. F., and Rood, H. J. 1985, *Ap. J. Suppl.*, **63**, 681.
Yee, H. K. C., and Green, R. F. 1987, *Ap. J.*, **319**, 28 (YG87).

DISCUSSION:

ELLIS: The recent discovery of a spectacularly luminous IRAS galaxy at high redshift apparently associated with a cD in a rich cluster might give some support to your theory. Do you expect infrared emission from a dying QSO in a rich cluster?

YEE: I believe that currently some people favour IR emission during the early stage of QSO formation rather than the dying stage. However, this observation may corroborate the possibility that, unlike their present day counterparts, cD galaxies may well have a very active stage in their evolution. I would like to make two points concerning the possibility of quasars in the center of rich clusters as precursors to cD galaxies. One is that, in our images, at least visually, the quasar often appears to be at the center of the cluster; and if we replace the quasar by a cD galaxy, the picture would look "normal". Second, Hintzen and collaborators have been studying 3C 275.1, a radio-loud quasar in a very rich cluster. They interpreted the 100 kpc nebulosity surrounding the quasar as a cD galaxy.

PEACOCK: I believe your sample has such a strong redshift - radio power correlation that your $< B_{gq} >$- z relation could well reflect a correlation with power, not a true epoch dependence.

YEE: It is certainly a possibility that we cannot rule out completely at this time. A definitive test would be to obtain a large enough sample to test the dependence of radio power on environment. I hope our imaging survey of Parkes quasars recently completed at CTIO will resolve this. There are two other points I wish to raise. One is that our sample is also an apparent magnitude limited sample. Within individual redshift bins, we do not see a dependence of the level of optical activity with environment. Second, the strongest argument for a redshift dependence is in finding a difference in number density of quasars in rich clusters, a very difficult number to assess at this point. Simple-minded statistics at this point suggest that we should have seen several z < 0.5 radio loud quasars situated in Abell class 1 clusters, whereas we have possibly one (i.e. 3C206).

History of Star Formation in Normal Galaxies

A. J. Pickles
Kapteyn Astronomical Institute
Postbus 800
9700 AV Groningen, The Netherlands

ABSTRACT. Medium resolution spectroscopic observations have been obtained for the brighter galaxies in 6 clusters covering the redshift range $0.18 \leq z \leq 0.39$. These observations have been analysed by an evolutionary synthesis technique tied to the VandenBerg isochrone scale, to determine the trends with redshift of the stellar age and metallicity distributions. The results indicate that these systems are dominated by a generally old (9–12 Gyr), metal-rich population, with a small contribution from younger ($\sim$5 Gyr) material. The amount of this intermediate age component is found to increase slightly with redshift, and to be more pronounced for the fainter galaxies. The metallicity of the more recently formed stars appears to be low, indicating that this component may have formed from the accretion of relatively pristine material.

1. Introduction

This talk summarises the results of an observational study undertaken at ESO, La Silla, in collaboration with Dr. P. C. van der Kruit. The study was designed to obtain high quality, medium resolution optical spectra of a sample of bright galaxies in clusters up to the maximum redshift for which these spectra could be reliably synthesised. Our object was to obtain spectra of sufficient quality to permit good stellar population modelling by a synthesis technique, and to look for population trends within these clusters as a function of redshift and of luminosity.

The principle observational constraint on redshift limit is set by the redshift at which important synthesis features like the Mgb region at 5200 Å move to the observational region where increasing sky noise, declining apparent magnitude and declining detector sensitivity conspire to render measurement of this feature unreliable. The present limits of a study such as this are therefore about z=0.4, which however samples a sufficient fraction ($\sim$25%) of a Hubble time to enable population trends to be seen and studied.

C. S. Frenk et al. (eds.), The Epoch of Galaxy Formation, 191–204.

Our investigation was prompted by three lines of observational evidence, current in the mid 1980's, concerning stellar populations and their evolution in composite systems. These were i) The existence of a generally increasing fraction of blue galaxies in clusters as a function of redshift (Butcher and Oemler 1984), ii) the existence of a large colour range for galaxies at all redshifts $z \geq 0.2$ (Lilly and Longair 1984, Lebofsky and Eisenhardt 1986), and iii) the evidence for intermediate age components in nearby elliptical galaxies (O'Connell 1976, 1980, Pickles 1985, Rose 1985).

The brightest galaxies in nearby clusters are ellipticals which are red, generally have little gas or dust, and have light, colour and stellar velocity distributions reminiscent of old globular cluster systems. For these reasons they have generally been presumed to consist exclusively of old stars, and to have formed early and rapidly. Indeed the redshifted colours of present-day ellipticals form the red envelope of observed galaxy colours, and indicate that at least some bright galaxies concluded most of their star formation before the epoch $z=1$.

The bluer galaxies could in principle be spiral galaxies whose star formation rate was higher in the past, they could be galaxies undergoing morphological transition as a result of merger activity, or elliptical galaxies observed during or just after a delayed burst of star formation. These possibilities are directly relevant to the history of star formation in galaxies and to their colour and luminosity evolution, and hence to cosmology. Because of the faintness and compactness of galaxies at even modest redshifts however, it is difficult to measure their morphological type distributions for comparison with present-day counterparts, at least before the advent of deep, high resolution imaging with HST (but see also Thompson 1986).

Good quality spectra of the integrated light from these galaxies are now routinely available at a number of large telescopes however, and their analysis by spectral decomposition techniques offers probably the best present and future line of attack for studying the composition of these systems. Population synthesis can give a detailed breakdown of the distributions of stellar age and metallicity contributing to a composite spectrum. The scale and zero points are slightly model dependent, which can lead to disagreement between different approaches, but the technique should always be differentially reliable when carefully applied. The technique is therefore ideally suited to this analysis of the history of star formation, by comparison of similar galaxies at different lookback times.

2. Methodology

In the work described here, we are concerned to determine the age and metallicity distributions of the stellar populations in elliptical galaxies on some readily accessible scale. In particular, we wish to quantify the changes in these distributions with age or lookback time.

Because we wish to investigate the history of star formation in such systems by comparing galaxies at different redshifts, we choose to do this via the synthesis of medium resolution spectra. The spectra enable us to unambiguously determine cluster membership at large redshift, and restrict our sample to purely stellar systems with absorption line spectra. They also give the energy distribution over a relatively broad wavelength region in a form which does not requi:e recourse to uncertain K corrections, and give fairly detailed information on the mean metallicity from many sensitive features.

The differences with redshift for the brightest, cluster dominating galaxies are known to be slight however, either in terms of their broad-band energy distributions (Lilly and Longair 1984, Lebofsky and Eisenhardt 1986) or of their coarse spectra features such as the 4000 Å break (Hamilton 1985). This is not surprising given the short main sequence lifetime of the brighter blue stars (spectral type A or earlier) associated with recurrent or ongoing star formation, but implies that the traces of such activity will be found only in quite detailed analysis of many weaker spectral features.

We have chosen not to synthesise our observed spectra in terms of a flux library of stellar spectra. The method gives reliable stellar syntheses (O'Connell 1976, Pickles 1985), but does not directly give information on the age or metallicity distributions contained within the system. This information comes from comparing the synthesis results with stellar evolutionary theory, principally by comparison with theoretical isochrones. Of course, absolute age information in particular depends somewhat on the particular theoretical framework adopted (eg. Ciardullo and Demarque 1977, VandenBerg 1983, Green, Demarque and King 1987), and care must be taken when comparing different studies to determine the theoretical framework within which these results are expressed.

For systems which can be composite in metallicity and age simultaneously, interpretation can be difficult even in the case that direct CM diagrams are available, mainly because the critical main sequence turnoff region is widened and blurred (see figure 1). The blue or red edges of this distribution then refer to only a part of the population, and the problem is to determine how much.

The practitioners of synthesis techniques can learn a lot obviously from the excellent theoretical fits to CM data on Galactic globular clusters (eg. VandenBerg 1983, Stetson and Harris 1988). The narrow intrinsic colour dispersion of these systems confirms our expectations of mono-valued age and metallicity for these clusters.

Good CM data is also available for local group galaxies (Mould, Kristian and Da Costa 1983, 1984) and the Galactic centre (Terndrup 1987). The large intrinsic colour dispersion in these diagrams implies a significant spread of at least metallicity however, (Mould and Kristian 1986), and could easily encompass a large age spread aswell.

For this reason we choose to incorporate our knowledge of the isochrones directly into the flux library. We have chosen a set of isochrones which encompass the ranges of age and metallicity (particularly high metallicity) expected to be present, and for each isochrone we co-add appropriate stellar spectra in the proportions dictated by the isochrone tabulation to form a Model Isochrone Spectrum whose colours and line strengths are appropriate to the chosen spectrum (see below).

The method we use is therefore empirical in the sense that there are essentially no preconcieved age or metallicity restrictions which the solutions must satisfy. It is evolutionary in the sense that we directly derive fractional distributions in terms of metallicity and age, on a system which is internally and differentially reliable, and sufficiently well defined that it can be compared to other results with care. The optimising code operates without significant constraints, since the stellar constraints (relative stellar fractions) are already built into the Model Isochrone Spectra. Optimisation is achieved subject to a user specification of the relative weight to be given to spectral 'feature' or 'continuum' regions. Emphasising goodness of fit around specific features can alter the results slightly, but we use the same criteria always to preserve differential reliability. In an absolute sense, we prefer to give strong weight to important metallicity features, since these are a relatively small part of the total spectrum, but provide the only reliable discriminant between the competing colour effects of age and metallicity variations.

We generate alternate solutions (particularly solutions which are forced to be old) by using the same fit criteria, but restricting the library components allowed. This enables us to quantify the degredation suffered by non-optimal fits, and also to check for astrophysical compatability between the main sequence and giant branch solutions (see Pickles and van der Kruit 1988, for a discussion of this point).

3. Model Isochrone Spectra

For reasons of compatability with a standard system, which also includes high metallicities, we choose to use the isochrone tabulations of VandenBerg (1985 – V85) and VandenBerg and Laskarides (1987 – VL1987). The range of age and metallicity considered is listed below, where for simplicity we have adopted a helium abundance of $Y = 0.25$ throughout. The effects on the models of increasing the helium abundance in step with the mean metal abundance are slight, but would operate in the sense of making the metal-rich models slightly bluer.

VandenBerg Isochrone Parameters

Age (Gyr)	Z	[Fe/H]	mixing length ratio	Ref.
5,10,15	0.0017	-1.0	1.6	V85
5,10,15	0.0060	-0.5	1.6	V85
2,4,6,10,15	0.0169	0.0	1.6	V85
6,9,12,15	0.060	0.5	1.5	VL87

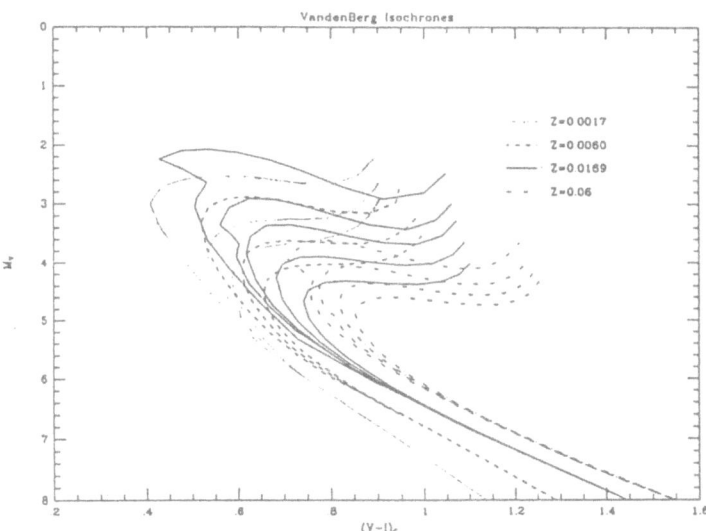

Figure 1: The isochrones from VandenBerg (1985) and VandenBerg and Laskarides (1987) which form the reference system used here are shown. Note that an observed broad CM diagram may encompass distributions in both age and metallicity

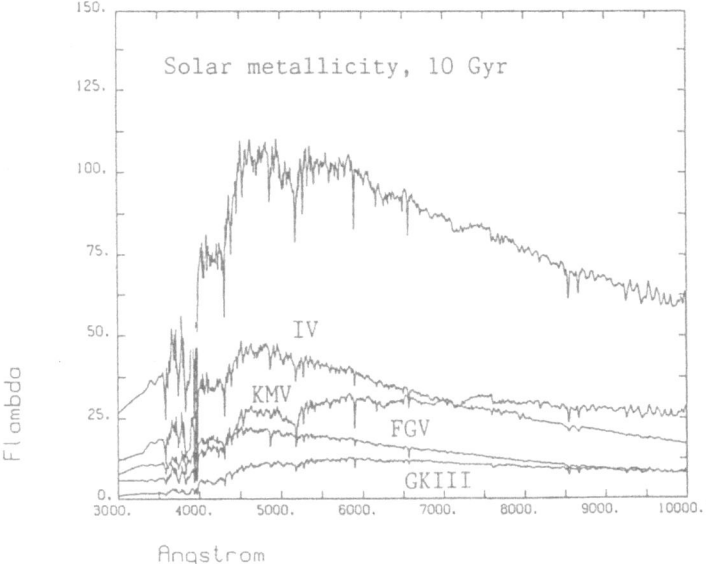

Figure 2: An example of a Model Isochrone Spectrum, formed by coadding stellar spectra in the proportions appropriate to a given isochrone. The fractional contributions from different stellar groups are also illustrated, but note that the resulting spectrum has only a partial contribution from giant branch groups (see text).

The selected isochrones are illustrated in figure 1, which serves to demonstrate how isochrones of several different ages as well as of different metallicities can be fitted to an observed CM diagram which shows intrinsic width (in both colour and luminosity). For these composite systems, with masses much larger than any single star formation cloud complex, there is no *a priori* reason to expect the stars to have the same age *or* metallicity, and hence a method of decoupling the effects of the two must be found, and preferably one which is easily related to a standard system.

The method of constructing the model isochrone spectra has been discussed before (Pickles and van der Kruit 1988), and will be described in full elsewhere (Pickles and van der Kruit, in preparation). An example of the resulting spectra is shown in figure 2, along with its stellar component breakdown.

It is important to note that the isochrone tabulations themselves reach only to the base of the giant branch (because of increasing theoretical uncertainties in the subsequent evolution). The model spectra shown here are complete to the same point, and hence lack a substantial fraction (up to 30 %) of the light at V which is expected to come from the later stages of giant branch evolution, aswell as the horizontal branch.

This is actually advantageous from the synthesis point of view, because the relevant stellar components can be included as free parameters in the flux library. The synthesis program itself can then determine empirically the contributions from these groups which best match the observed spectra, free from relatively uncertain theoretical constraints.

The metal-rich giant branch components include our EFOSC observations of stars in NGC6522, selected as metal-rich from Whitford and Rich (1983). These are strong-lined stars, which provide the best fit to the observed spectra of strong lined galaxies, and are the only stars which match detailed molecular features at 4650 Å for example.

The model isochrone spectra themselves include contributions from G and early K giants of appropriate metallicity, but are not complete on the giant branch. Metal-weak, solar and metal-rich giant branch components are included as free components in the flux library, and the synthesis fits are checked to see that the proportion of light coming from these components is compatible with the metallicity distribution derived from the MIS. The flux library also includes horizontal branch components as free parameters.

4. Synthesis of galaxies in medium redshift clusters

We have observed 6 clusters in the redshift range $0.18 \leq z \leq 0.39$, using the European Faint Object Spectrograph and Camera (EFOSC) on the 3.6m telescope at La Silla. Full observational details will be given elsewhere (Pickles and van der Kruit, in preparation), but the nature of the data and method of analysis is

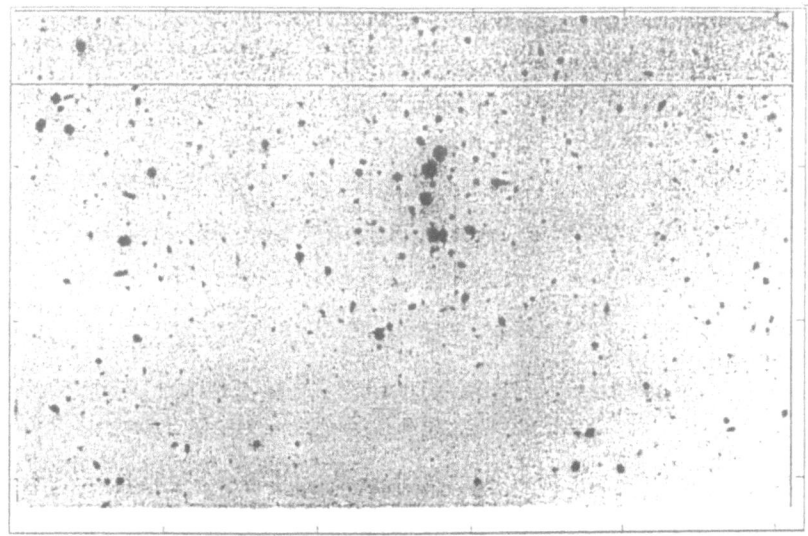

Figure 3: An R band CCD image of Abell 1525, obtained with EFOSC on the 3.6m telescope at La Silla. Spectroscopic observations have been obtained of the brightest cluster members

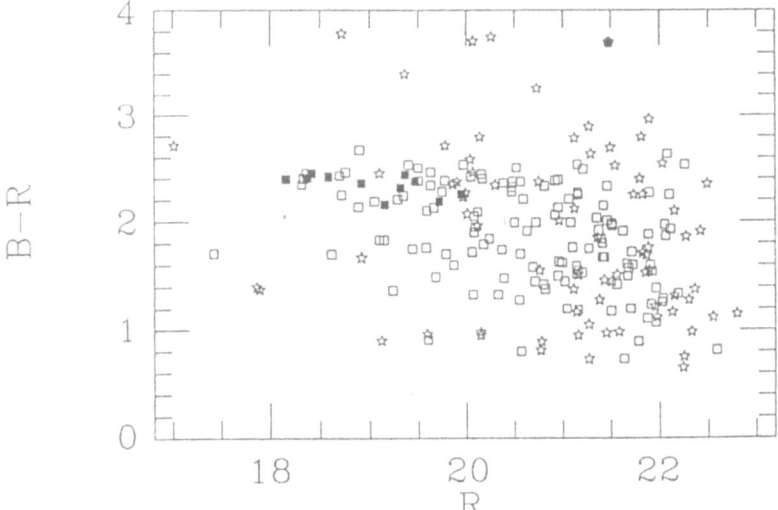

Figure 4: The colour-magnitude diagram obtained for A1525. Photometry in BVRI bands was used to check the flux calibrations of spectra, which were obtained for galaxies indicated as filled squares in the diagram. Spectra of the 5 brightest galaxies were co-added to form the 'bright' galaxy sample, the others form the 'faint' sample. Probable stars and objects determined to be non-cluster members are indicated by special symbols.

illustrated here.

Figure 3 shows an R band CCD image of one of the clusters, Abell 1525 at z=0.26, and figure 4 shows the colour magnitude diagram obtained for this cluster. We have obtained spectra of the filled objects in figure 4, which are mainly the brightest galaxies visible in the cluster centre in figure 3.

The spectra were extracted, wavelength and flux calibrated, and the flux calibration checked against the broad band colours obtained in direct mode. They were then dereddened and de-redshifted so that spectral synthesis could be performed in the rest frame.

The differences in colour and line strength are small for galaxies which fall near each other in the colour magnitude array, and we choose to define only two types in each cluster, and to increase signal to noise for each type by co-adding the spectra of galaxies of similar luminosity and colour.

The resulting spectra for A1525, illustrated in figure 5 together with their best synthesis fits, have resolution of 12 Å (2 pixels), and are well defined in the observed range 4000 – 9000 Å.

Spectra of the bright and faint combinations for A370 at z=0.37 are shown in figure 6, together with their best synthesis fits.

Cluster Age and Metallicity by Mass fraction

Cluster	Redshift	%V Light 2-6 (Gyr)	Age Group (Gyr) 2-6	Age Group (Gyr) 9-12	Age Group (Gyr) 15	$< Age >$ (Gyr)	$< [Fe/H] >$
Bright Group							
A1689	0.18	9	3	97	-	11.5	0.053
AC122	0.21	-	-	100	-	11.0	0.038
A2397	0.22	3	2	98	-	10.1	0.020
A1525	0.26	13	6	94	-	11.4	0.053
A370	0.37	-	-	99	-	11.4	0.047
Cl0024	0.39	11	9	91	-	10.3	0.031
Faint Group							
A1689	0.18	4	2	97	-	9.9	0.018
AC122	0.21	15	9	85	6	9.9	0.018
A2397	0.22	13	7	93	2	11.0	0.044
A1525	0.26	4	8	92	-	10.7	0.030
A370	0.37	24	17	83	-	10.1	0.036

The results of the synthesis are tabulated above, in terms of mass fractions contributed by the different age components. The mass fractions follow from the light fractions at V (which is what is actually fitted in the synthesis) and the model isochrone parameters, since the giant or horizontal branch components contribute negligible mass. The light contribution from the giant branch components is compatible with the model isochrone contributions in all cases, and the horizontal branch contribution is typically 1 – 2 % at V.

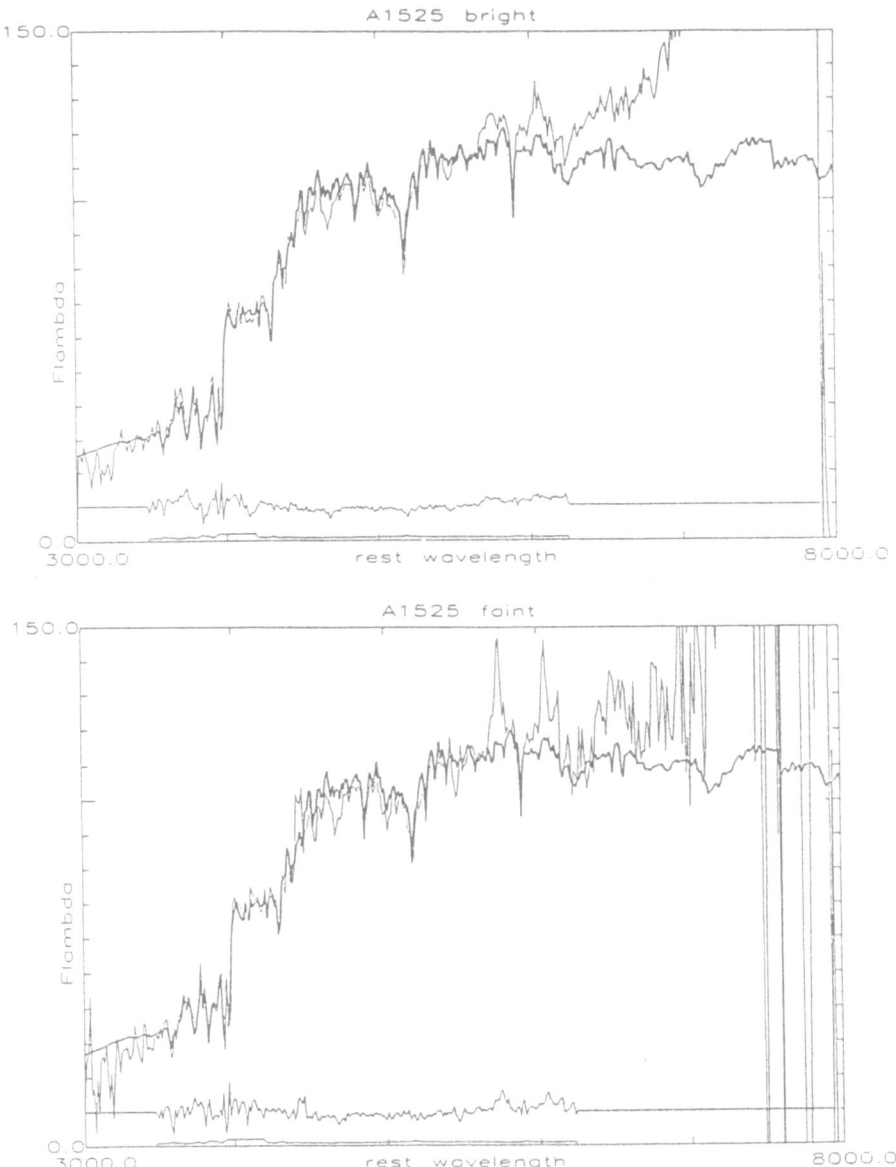

Figure 5: De-redshifted and co-added spectra of the bright and faint galaxy samples in A1525 (thin lines), together with their best synthesis fits (thick Lines). The trace at the bottom of each figure gives the error of the fit (in magnitudes), and indicates the wavelength region actually used in the synthesis

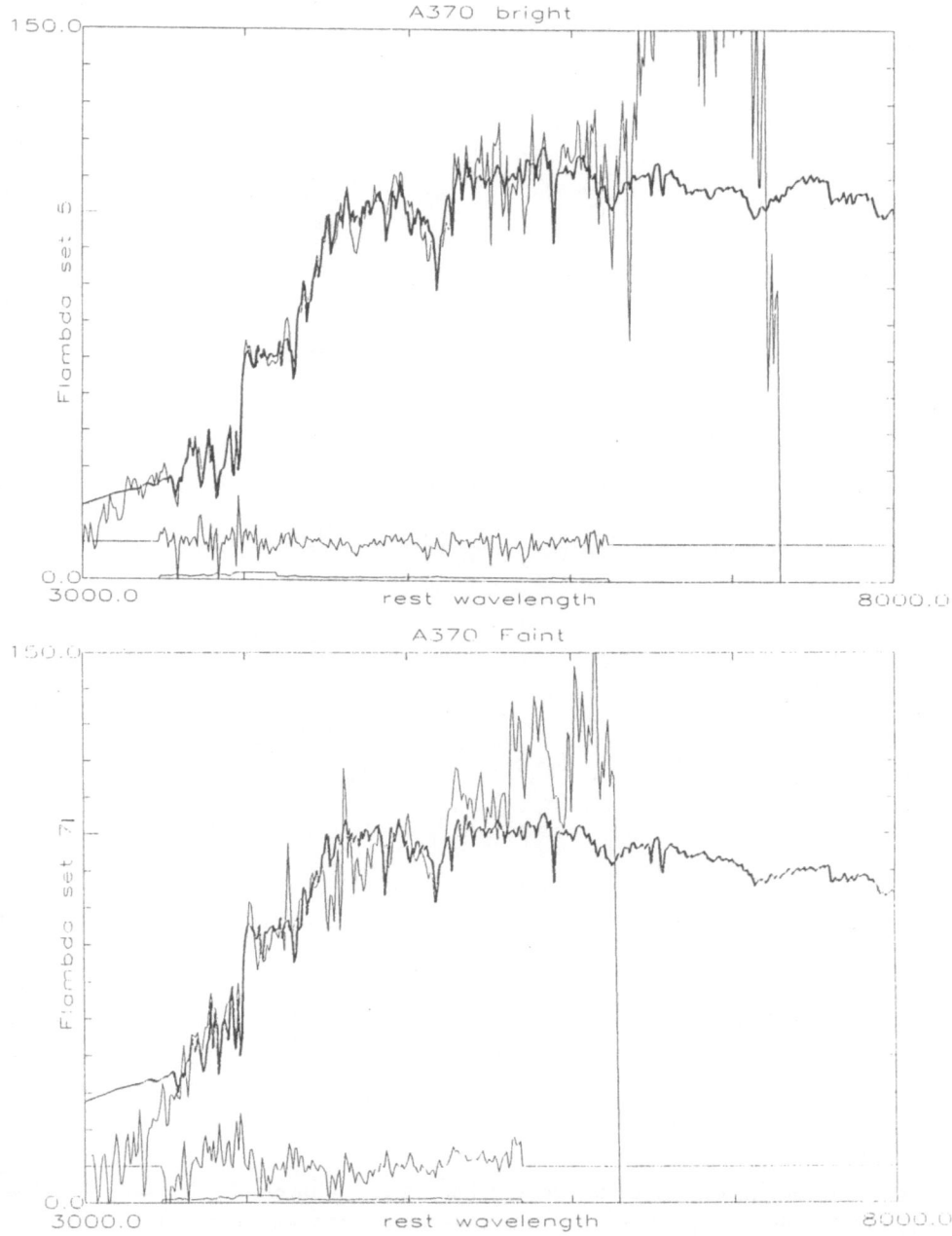

Figure 6: De-redshifted and co-added spectra of the bright and faint galaxy samples in A370 (thin lines), together with their best synthesis fits (thick Lines).

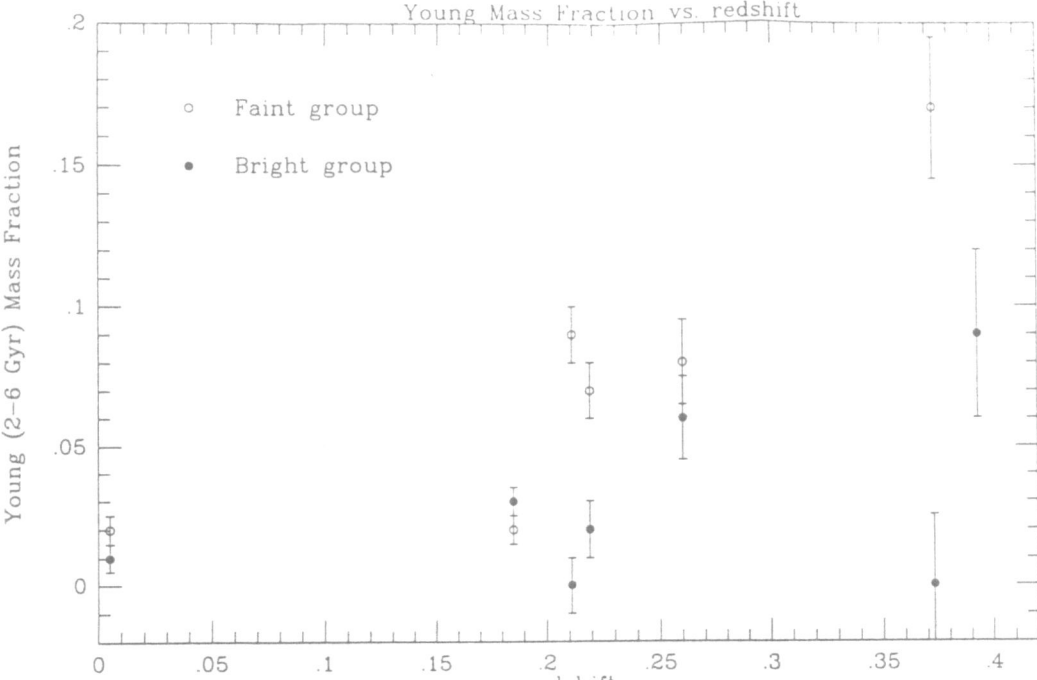

Figure 7: The mass fraction of the younger age component determined for the brightests and fainter galaxy combinations in each cluster as a function of red-shift. The points shown are the sum of all components in the age range 2 – 6 Gyr, although they are dominated by a 5 Gyr, metal-weak component in most cases. Error bars are estimated from the probable errors in the light fraction de-terminations, and the mass to light ratios.

5. Discussion

There are several points to note about the results presented here and shown graphically in figure 7, where the zero redshift results apply to the Fornax cluster (Pickles 1985).

- The 'Bright' galaxy sample comprises the brightest, cluster dominating galaxies in each cluster. They are all (with the exception of A2397) very strong lined, with mean metallicity 2 – 3 times solar.

- The 'faint' galaxy sample comprises galaxies which are typically 1 magnitude fainter than the bright sample, but these are still intrinsically bright galaxies, with mean metallicities substantially higher than solar. This fainter sample is intrinsically brighter than the 'starburst' galaxies described by Dressler and Gunn (1982, 1983) and Couch and Sharples (1987).

- The synthesis fits indicate that the majority of stars in each galaxy are 9 – 12 Gyr old. The fits can be forced to give older ages for this dominant component, but lead to an unphysical discrepancy between the mean metallicities on the main sequence and giant branches (Pickles and van der Kruit 1988). At face value these results indicate that the bulk of star formation occurred substantially later than the epoch of globular cluster formation, possibly favouring merger or accretion formation scenarios. These results do depend on the particular framework used however, and small scale changes in the age determinations are possible.

- The results indicate the presence of an increasing fraction of younger material with increasing redshift (and increasing lookback time, where the maximum redshift observed corresponds to a lookback time of about 25% of the age of the universe). This is expected if star formation continued (sporadically or slowly) for some Gyr after the bulk of star formation ceased. The errors at the highest redshifts (which are also the lowest quality spectra) are too large as yet to set significant constraints on the absolute epochs of star formation.

- The younger age components in the syntheses are exclusively metal-weak, indicating that star formation in these galaxies continued by accretion of metal-weak (pristine) material.

References

Butcher H. & Oemler A., 1984, *Astrophys. J.*, **285**, 436

Ciardullo R.B. & Demarque P., 1977, *Trans. Yale Univ. Obs.* **33**, 1

Couch W.J. & Sharples R.M., 1987, *Monthly Notices Roy. Astron. Soc.*, **229**, 423

Dressler A. & Gunn J.E., 1982, *Astrophys. J.*, **263**, 533

Dressler A. & Gunn J.E., 1983, *Astrophys. J.*, **270**, 7

Green E.M., Demarque P. & King C.R., 1987, Trans. Yale Univ. Obs., in preparation.

Hamilton D., 1985, *Astrophys. J.*, **297**, 371

Lebofsky M.J. & Eisenhardt P.R.M., 1986, *Astrophys. J.*, **300**, 151

Lilly S.J. & Longair M.S., 1984, *Monthly Notices Roy. Astron. Soc.*, **211**, 833

Mould J.R., Kristian J. & Da Costa G., 1983, *Astrophys. J.*, **270**, 471

Mould J.R., Kristian J. & Da Costa G., 1984, *Astrophys. J.*, **278**, 575

Mould J.R. & Kristian J., 1986, *Astrophys. J.*, **305**, 591

O'Connell R.W., 1976, *Astrophys. J.*, **206**, 370

O'Connell R.W., 1980, *Astrophys. J.*, **236**, 430

Pickles A.J., 1985, *Astrophys. J.*, **296**, 340

Pickles A.J. & van der Kruit P.C., 1988, *Interpreting Integrated Spectra*, in *Towards understanding Galaxies at Large Redshift*, R.G. Kron & A. Renzini (eds.)

Rose J.A., 1985, *Astron. J.*, **90**, 927

Stetson P.B. & Harris W.E., 1988, *Astron. J.*, **96**, 909

Terndrup D., 1987, Thesis, Caltech.

Thompson L. A., 1986, *Astrophys. J.*, **300**, 639

VandenBerg D.A., 1983, *Astrophys. J. Suppl.*, **51**, 29

VandenBerg D.A., 1985, *Astrophys. J. Suppl.*, **58**, 711

VandenBerg D.A. & Laskarides P.G., 1987, *Astrophys. J. Suppl.*, **64**, 103

Whitford A.E. & Rich R.M., 1983, *Astrophys. J.*, **274**, 723

DISCUSSION:

WHITE: If the hypothesis you wish to test is that ellipticals are made from populations of similar age to globular clusters, it seems that a direct test might be possible by attempting a synthesis using integrated spectra of globular clusters with a wide range of metallicities. Clusters are known in M31 with metallicities up to solar. Could they be used to carry out this test? What warnings should we take from observations that the spectra of M31 clusters appear to have some systematic differences (for example, in H line strength) with Galactic clusters?

PICKLES: The approach you suggest forms the basis of the synthesis models presented by Bica and Alloin, who use galactic clusters of "known" age and metallicity (up to a maximum which is close to solar). This could be extended to include M31 globulars - but would require them to have a reasonable metal-age calibration to be useful in synthesis i.e. to interpret composite systems in terms of components which are at least slightly better understood. The problem in synthesising ellipticals lies in their strong lines, indicating mean metallicities substantially above solar. As I've indicated, there are also features which appear there which are not present at all in solar metallicity stars. M31 globulars might be a good site for synthesis in terms of Galactic components however.

BRUZUAL: I think that predicting the UV in your synthetic spectrum will be interesting and provide constraints on the amount of recent star formation and on its metallicities.

PICKLES: I agree in principle, but think that models such as your own can be used to do this. Perhaps the most important point relevant to your question from this work is that the youngest significant population (in terms of mass) is usually a 5 Gyr, metal weak component, so the predicted UV flux level is not high.

JONES: I didn't quite appreciate your concept of "framework", particularly in relation to Brigitte's questions. Either you can fit a spectrum from 2000 Å - $2\mu m$ or you can't - it's not a question of the framework within which you interpret things. Can you, for example, fit the Ca IR triplet with your population synthesis?

PICKLES: (1) Brigitte questioned the differences in age between her models and those presented here. I think that the light contributed by different stellar groups in our respective models is really quite similar, but our derived ages are based on looking at different evolutionary frameworks to interpret this distribution. (2) We have been able to get very good fits up to $1\mu m$ (Pickles, *Astrophys. J.* 1985) and especially detailed fits for the whole of the near IR region including the Ca triplet, NaI (8190 Å) and FeII (9910 Å) — Carter, Visvanathan and Pickles (*Astrophys. J.* 1986) in nearby systems (Fornax cluster).

RECURRENT STAR FORMATION IN ELLIPTICAL GALAXIES

N. Arimoto
Institut fuer Theoretische Astrophysik
der Universitaet Heidelberg
Im Neuenheimer Feld 561, D-6900 Heidelberg 1, F.R.G.

ABSTRACT. A new galactic wind model is presented. Periodic star formations and winds are possible in giant elliptical galaxies after an initial wind. The model explains the paucity of gas in most ellipticals and predicts a significant contribution of intermediate-age stars into the present day galactic lights. During recurrent star formation phases ellipticals flare up, which could be observed as unusual blue objects in distant clusters of galaxies.

1. INTRODUCTION

Elliptical galaxies undoubtedly experienced active star formation in the early era. A recent supernova (SN)-driven wind model, done with Y.Yoshii (Arimoto and Yoshii, 1987; Yoshii and Arimoto, 1987; hereafter AY model) by using an evolutionary method of population synthesis, successfully reproduced the structual and chemical properties of elliptical galaxies, dwarf ellipticals, and globular clusters. AY model demonstrated convincingly that the spheroidal systems are really old – 15 10^9 yr – and certainly belong to a one parameter family of mass at their birth.

However, there are not a few evidences against an old age of the spheroidal systems. Many of static methods of population synthesis have shown non-negligible contribution of the intermediate age populations into total galactic lights of elliptical galaxies (eg., O'Connell, 1986; Pickles, 1985; Rose, 1985; Bica; 1988). Bica and Alloin (1988) have even suggested the existence of very young stars in several ellipticals which exhibit a sharp upturn in their UV spectra at $\lambda < 1500A$. Therefore, it would be quite interesting to ask whether it is possible to reconcil AY model with the static population synthesis results.

Arimoto and Yoshii (1987) just followed the chemical evolution of elliptical galaxies until the remaining gas was ejected completely in a wind. However, the nature of chemical evolution itself implies that the gas would accumulate again towards a galactic centre after a wind, because there is continuous mass loss from less massive stars which were born during an initial star forming phase. If all the gas remains at the centre, the present day gas fraction should amount to 20% and the heavy elements abundance should be as high as $2.5Z_\odot$ (Arimoto, 1989).

The observed distribution of the hydrogen mass-to-luminosity ratio of elliptical galaxies is $0.001 \le M_H/L_B(M_\odot/L_\odot) \le 0.3$ (Knapp et al., 1985). From the AY model of $M_G = 10^{12}M_\odot$ with $L_B = 10^{10.62}L_\odot$, Arimoto estimates an average gas accumulation rate as $dM_{gas}/dt = 14.3$ $M_\odot y^{-1}$. Then the time scale of gas storing up to $(M_H/L_B)_{max} = 0.3$ is $t_{acc} = 8.8$ 10^8 yr. This means, whatever the gas sweeping mechanism is, it should function at least each 10^9 yr. Parts of the gas observed in ellipticals might have an external origin (van Driel, 1987), which

C. S. Frenk et al. (eds.), The Epoch of Galaxy Formation, 205–209.

implies much smaller value of t_{acc}. Gas-rich ellipticals having $M_H/L_B = 0.1 - 0.3$, on the other hand, requir another constraint that the sweeping should not be continuous but be intermittent.

In a similar way to Sanders (1981), we have evaluated the cooling time scale of the stored gas. The results for $M_G = 10^{12}M_\odot$ are $t_{c,rad} \simeq 1.3 \ 10^7 \ yr$ and $t_{c,conv} \simeq 5.2 \ 10^8 \ yr$ if the solar abundances are assumed. Arimoto shows that the metallicity of the gas attains the maximum value $8Z_\odot$ immediately after a wind and is diluted down to $2.5Z_\odot$ by relatively metal-deficient materials from low mass stellar envelopes. In such a case of high metallicity, the cooling would proceed much more effectively than the present estimate (cf. Hensler, 1987). Therefore, it is very likely that the star formation takes place again in elliptical galaxies after an initial blustering wind.

2. RECURRENT STAR FORMATION

Extending AY wind model, we have followed a possible star formation in elliptical galaxies after a wind. The model has been computed for an elliptical of a mass $M_G = 2 \ 10^{12}M_\odot$ and an age $T_G = 15 \ 10^9 \ yr$ with a time step $10^7 \ yr$. The same IMF as AY model, $\phi(m) \propto m^{-0.95}$, is adopted. The SFR per unit mass is supposed to be proportional to the gas fraction $f_g(t)$:

$$C(t) = (\nu/\nu_o)f_g(t), \qquad \Omega_g(t) \geq E_{th}(t)$$

$$= 0, \qquad\qquad \Omega_g(t) < E_{th}(t)$$

where $\nu/\nu_o = 42$ with $\nu_o = 6.08 \ 10^{-18} \ s^{-1}$ (the SFR in the solar neighbourhood); $\Omega_g(t)$ and $E_{th}(t)$ are the binding energy of the gas and the available thermal energy, respectively. We assume that there remains no residual thermal energy in a system after winds. Details of AY model were given in Arimoto and Yoshii (1987). The advantage of this model is that it takes into account explicitly, in the population synthesis, the change in the stellar metallicity following the chemical evolution process (Arimoto and Yoshii, 1986).

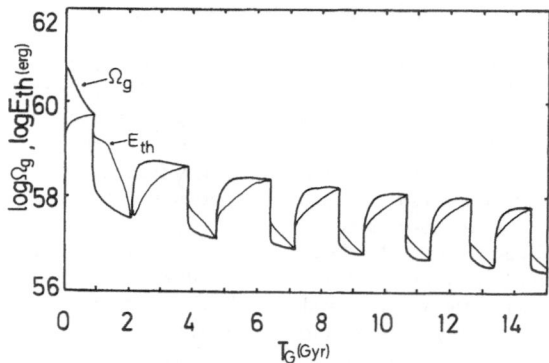

Figure.1-The theoretical evolution of the binding energy Ω_g (thick line) and the thermal energy E_{th} (thin line) of the wind model.

Figure 1 demonstrates clearly that the star formation occurs *recurrently* in a giant elliptical galaxy. At $t_{gw} \simeq 9 \ 10^8 \ yr$, a galactic wind takes place and expells the remaining gas completely from the system. The binding energy drops suddenly due to gas loss and also to an expansion of stellar system which always intends to keep a virial equilibrium. The star formation stops, while a wind blows successively owing to SN explosions of less massive progenitors until the thermal energy cools down to a level of the reduced binding energy. During this wind phase, the binding energy decreases gradually because of steady galactic mass loss. Once the thermal energy becomes smaller than the binding energy at $t \simeq 2 \ 10^9 \ yr$, a wind terminates and the star formation recuperates. The gas accumulates again and the binding energy augments. This second star formation phase continues until $t \simeq 4 \ 10^9 \ yr$, when newly exploding SN's supply sufficient thermal energy to induce another wind. In this way a cycle of the star formation and the steady wind repeats until the present epoch with an almost certain period of $2 \ 10^9 \ yr$.

Since there exists always gas supply from low mass stars, the gas mass fraction remains finite even during wind phases; the typical average is $f_g(t) \simeq 10^{-4}$. Once a wind stops, the gas fraction increases rapidly up to $f_g(t) \simeq 10^{-3}$ and remains nearly constant because the gas accumulation is balanced with the gas consumption due to the star formation.

To compare with observations it is more suitable to discuss the evolution of the gas mass in terms of a hydrogen mass-to-luminosity ratio, M_H/L_B, which is easy to observe, while the gas fraction is usually difficult because of uncertainties in estimating galactic total mass. Figure 2 shows the $log M_H/L_B$ vs T_G diagram. The present wind model explains quite naturally why most of elliptical galaxies have very little amount of gas ($M_H/L_B \leq 10^{-2}$). Figure 2 also suggests that the gas in several gas-rich ellipticals ($M_H/L_B > 10^{-2}$) might have an external origin.

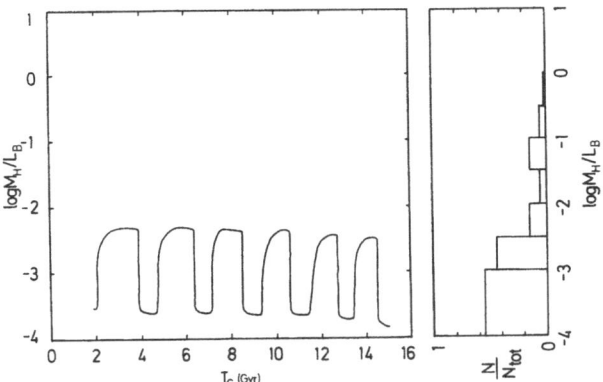

Figure.2-Left: The evolution of the theoretical hydrogen gas mass-to-luminosity ratio. Right: The observed distribution for elliptical galaxies taken from Knapp et al., (1985).

The gas metallicity is as high as $Z_g(t) \simeq 4Z_\odot$ and $1.6Z_\odot$ during star formation and wind phases, respectively. AY model introduced a luminosity weighted mean of stellar metallicity $[Fe/H]_\ell$. This luminosity mean is a useful observable quantity when the system is composed of different metal stars. When stars are being formed, $[Fe/H]_\ell \simeq +0.5$, mainly determined by young metal-rich OB stars. In wind phases, $[Fe/H]_\ell \simeq +0.3$, reflecting relatively low metallicity of old red giant branch stars. This suggests a possible way to study the nature of star bursts in distant elliptical galaxies. If these galaxies show a high value of $[Fe/H]_\ell$, equivalent to the line strength of metallic absorption lines, the origin of metal-rich gas should be regarded as internal. On the other hand, if $[Fe/H]_\ell < 0$, bursts should be supported by external metal-poor gas, perhaps due to galactic tidal interactions.

The most interesting effect of intermittent star formation appears in the photometric evolution. Figure 3 indicates how U-B and B-V colours are sensitive to the star formation. It should be noticed that a small amount of gas $(M_H/L_B \simeq 4~10^{-3})$, in the galactic nuclei, is sufficient to make a blue flash in colours. A galaxy becomes bluer by $\Delta(U-V) \simeq -0.6$, which is comparable with the blue colour excess of Butcher-Oemler galaxies (Butcher and Oemler, 1978). U-B and B-V colours at the present time are very similar to those of AY model. This means that the recurrent star formation has little effect on the present day photometric properties.

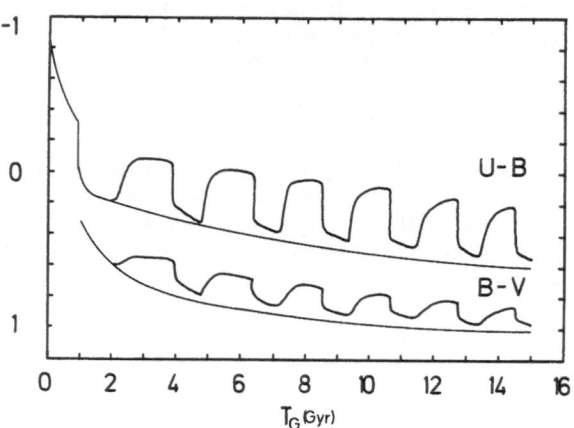

Figure.3-The evolution of the theoretical U-B and B-V colours of the present wind model (thick lines) and AY model (thin lines).

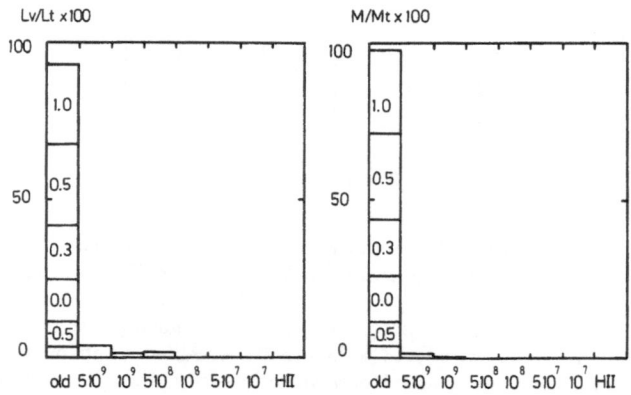

Figure.4-V-flux luminosity fractions (left) and mass fractions (right) of the present wind model. Metallicity fractions $[Z/Z_\odot]$ are indicated whenever possible.

Bica, Arimoto and Alloin (1988) have recently compared the population synthesis results for galaxy nuclei, done by Bica (1988), to those derived from the evolutionary method of Arimoto and Yoshii (1986). These authors interfaced the two different approaches by means of theoretical M/L_v ratio for single generation stellar systems computed for differing ages. By using the same single generation models, we transform the mass fractions of the different age and metalicity stars predicted by the present wind model into the V-flux fractions. Figure 4 should be compared directly to Fig.3 of Bica et al. (1988), which shows an empirical result for giant elliptical and S0 galaxies. Figure 4 demonstrates that the intermediate age ($10^9 - 5 \ 10^9 \ yr$) components born during the recurrent star formation phases contribute appreciably to the V-flux although their mass fractions are almost negligible. Thus the extended wind model gives an excellent agreement with the static population synthesis results.

3. CONCLUSIONS

We have computed a galactic wind model for giant elliptical galaxies by using the evolutionary method of population synthesis. After an initial wind, the recurrent star formation can take place at the galactic center. The model explains the paucity of gas in most giant elliptical galaxies and predicts non-negligible contribution of intermediate age populations into the V flux; thus is consistent with the static population synthesis results.

During the star formation phases, giant ellipticals flare up in blue colours by $\Delta(U - V) \simeq -0.6$ mag, although the amount of gas converted into stars is quite small. This suggests the nature of Butcher-Oemler galaxies in distant clusters of galaxies.

However, the global properties such as i) the colour-magnitude relation, ii) the metallicity-luminosity relation, iii) the metallicity dispersion, iv) the surface brightness-diameter relation, and v) the luminosity-velocity dispersion relation are essentially determined by the old stars formed before an initial wind (Arimoto and Yoshii, 1987; Matteucci and Tornambe, 1987).

Acknowledgements: We thank Drs. C.Balkowski and J.Koeppen for useful discussions. Financial support from the Deutsche Forschungsgemeinschaft (SFB 328) is gratefully acknowledged.

References

Arimoto, N. 1989, in "Evolutionary Phenomena in Galaxies", eds., J.Beckman, B.Pagel, Reidel
Arimoto, N., Yoshii, Y.: 1986, Astron. Astrophys. **164**, 260
Arimoto, N., Yoshii, Y.: 1987, Astron. Astrophys. **173**, 23
Bica, E.: 1988, Astron. Astrophys. **195**, 75
Bica, E., Alloin, D.: 1988, Astron. Astrphys. **192**, 98
Bica, E., Arimoto, N., Alloin, D.: 1988, Astron. Astrophys. **202**, 8
Butcher, H., Oemler, G.: 1978, Astrophys. J. **219**, 18
Hensler, G.: 1987, Mitt. Astro. Ges. **70**, 141
Knapp, G.R., Turner, E.L., Cunniffe, P.E.: 1985, Astron. J. **90**, 454
Matteucci, F., Tornambe, A.: 1987, Astron. Astrophys. **185**, 51
O'Connel, R.W.: 1986, in "Spectral Evolution of Galaxies", eds., C.Chiosi, A.Renzini, Reidel
Pickles, A.J.: 1985, Astrophys. J. **296**, 340
Rose, J.A.: 1985, Astron. J. **90**, 1927
Sanders, R.H.: 1981, Astrophys. J. **244**, 820
van Driel, B.: 1987, Ph.D. thesis, Kapteyn Laboratorium, Groningen
Yoshii, Y., Arimoto, N.: 1987, Astron. Astrophys. **188**, 13

Constraints on Galaxy Formation from Cosmic Background Radiations

J. Richard Bond
CIAR Cosmology Program
Canadian Institute for Theoretical Astrophysics
University of Toronto
Toronto, Ontario
M5S 1A1, Canada

ABSTRACT   In this paper, I show what the current data on distortions and anisotropies of the cosmic background radiation (CMB) imply for the epoch of galaxy formation in theories for which the primordial random density field is Gaussian, as in inflation-based models. The new upper limits on CMB anisotropies at $7'$ (OVRO) and $\gtrsim 7°$ (RELICT 1) impose strong constraints on the amplitude of the fluctuation spectrum, on scales $\sim (5-20)\,h^{-1}\mathrm{Mpc}$ and $\sim 300 - 6000\,h^{-1}\mathrm{Mpc}$, respectively (for flat models). For a specified initial spectrum, these limits can be expressed in terms of the biasing factor b_ρ, which also largely determines the galaxy formation redshift. For $\Omega = 1$ CDM models, $b_\rho \gtrsim 0.7$ for $\Omega_B = 0.1$ and $h = 0.5$. For non-zero Λ CDM models with vacuum energy $\Omega_{vac} = 0.8$, $b_\rho \gtrsim 1.6$. Neutrino-dominated models have a non-linear redshift $z_{nl} \lesssim 0.5$ from the OVRO results, a very strong constraint. The Berkeley-Nagoya CMB distortion, if confirmed, would be a direct probe of the pregalactic universe. Arguments against Compton cooling as the source are presented. For dust emission models, a high formation redshift, ~ 30, is indicated, although this depends upon how exotic and abundant the dust may be. A large energy release, an order of magnitude more than the blue light that the galaxies we see would release in a Hubble time, is required. Constraints on the clustering of the dust and the luminosity sources from the OVRO experiment, and how sub-mm telescopes can improve these, are also discussed.

1. THE ANISOTROPY DATA

The anisotropy upper limits on both large and small scales have improved dramatically over the past few years. The current best constraints come from the Caltech Owens Valley (OVRO) experiment at $7'$. and the RELICT 1 data at very

211

C. S. Frenk et al. (eds.), The Epoch of Galaxy Formation, 211–226.

large angles. There has also been a tantalizing report of anisotropy at $8°$ and $5°$ by the Tenerife collaboration, which is rather difficult to reconcile with the RELICT results for the standard Zeldovich spectrum theories such as the adiabatic cold dark matter model. In this section, we give a brief review of the status of these experiments.

1.1 Very Small Angle Anisotropies — VLA

Observations of *rms* anisotropies at sub-arcminute resolution have recently been obtained with the VLA (Fomolont *et al.* 1988, Martin and Partridge 1988): $(\Delta T/T)_{rms} \lesssim 2 \times 10^{-4}$ for $\theta_{fwhm} \sim 18'' - 60''$. Fomolont *et al.* think that the signal is due to faint radio sources, so these values should probably be treated as upper limits. The angular scale probed corresponds to unperturbed comoving galactic scales, hence it might be thought that these results could directly probe the spectral amplitude on these scales. However, the width of the last scattering surface for standard recombination models ($\sim 20\, h^{-1}$Mpc corresponding to an angle $\theta_c \sim 10'$) is sufficiently large that primary anisotropies are effectively erased on angular scales less than the coherence angle $\theta_c \gg 1'$. Thus the VLA observations may tell us about secondary anisotropies but not about primary ones.

1.2 Small Angle Anisotropies — OVRO

The strongest upper limits on primary anisotropies come from the OVRO experiment (Readhead *et al.* 1988). The Caltech experiment is a 3-beam one, with the CMB temperature of the central beam subtracted from the average of the CMB temperatures in arcs of $30° \equiv 2\varphi_R$ on either side of the central beam, a distance $7.15'$ away. The smearing *fwhm* angle is $1.8'$. The corresponding separations being probed on the last scattering surface are $12\,\Omega^{-1/2}\, h^{-1}$Mpc and $\sim 1\,\Omega^{-1/2}\, h^{-1}$Mpc. There were a total of 12 such patches in a circle separated by 2^h in right ascension offset by $1°$ from the North Celestial Pole. In the 7^h patch, a time dependent weak radio source was detected and that field was not included in the analysis. In the Readhead *et al.* preprint, only the data from 7 fields is included. The 95% confidence limit upper bound is $(\Delta T/T)_{rms} \lesssim 1.8 \times 10^{-5}$ from a χ^2 analysis. The maximum likelihood technique gives a similar result: the upper limit is quite robust, no matter what the analysis procedure. However, with only 7 fields included, this bound should be applied with some caution, in particular to theories with non-Gaussian radiation patterns (§3).

To compare the OVRO results with theoretical computations of Gaussian perturbation fields, this upper limit should be compared with the *rms* fluctuations $\langle (\Delta T/T)^2 \rangle^{1/2}$, where

$$\langle (\Delta T/T)^2 \rangle = 2(C_F(0) - C_F(\varpi)) - \frac{1}{2}(C_F(0) - C_F(2\varpi)) \quad (1a)$$

$$-\int_0^{\varphi_R} \frac{d\varphi}{\varphi_R} \left(1 - \frac{\varphi}{\varphi_R}\right) \left[(C_F(0) - C_F(2\varpi \sin\varphi)) + (C_F(2\varpi) - C_F(2\varpi \cos\varphi)) \right] \quad (1b)$$

$$\varpi \equiv 2\sin(\theta/2), \quad \theta = 7.15', \quad \varphi_R = 15°.$$

Here, $C_F(\theta)$ is the radiation correlation function smoothed over the beam profile.

The term (1b) is typically less than a 10% correction to the classic 3-beam experiment result, eq.(1a). Results are given in §2.2.

1.3 Intermediate Angle Anisotropies — Tenerife

The Tenerife team (Davies *et al.* 1987) reports a detection of anisotropy of $(\Delta T/T)_{rms} \sim 4 \times 10^{-5}$ with $\theta=8°$ and $\theta_{fwhm}=8°$. Preliminary analysis of their $\theta=5°$ data at the same wavelength confirms persistence of the signal in the same area of the sky. The crucial experiment will be to take observations at other frequencies to decide by its spectrum whether the signal is a galactic or extragalactic radio source. Their 4×10^{-5} result is the value of $C(0)^{1/2}$ if $C(\theta)$ has a Gaussian profile in θ. Therefore this result cannot strictly be applied to the Zeldovich spectra models of most interest here, but we can take it as indicative. Although intermediate angles are probably the best place to look for anisotropy from a theoretical viewpoint, this and other intermediate angle experiments that have been undertaken so far (*e.g.*, Melchiorri *et al.* 1981) do not have an optimal design, since the two 'illuminated' patches on the microwave sky are predicted to be highly correlated in most models. Minimizing θ_{fwhm} while maintaining θ at a few degrees is preferable for testing theories, although there are extreme experimental difficulties in such strategies. Nonetheless such experiments are planned, *e.g.*, by Lubin.

1.4 Large Angle Anisotropies — RELICT 1

The Soviet satellite experiment RELICT 1 (Strukov *et al.* 1987) flown in 1983 gives upper limits on very large scales a factor of 4 better than those previously given, assuming a Zeldovich spectra. For such scale invariant fluctuations, the angular power spectrum of the radiation pattern is $C_\ell = C_2 6[\ell(\ell+1)]^{-1}$ for small $\ell \lesssim 30$, where the multipole ℓ roughly probes the angular scale ℓ^{-1} radians, and $C_2^{1/2} = a_2$ is the quadrupole amplitude. The power per decade of angular wavenumber, $\ell^2 C_\ell$ is therefore flat, as expected from the scale invariance. Assuming this form for C_ℓ, Strukov *et al.* obtain a 95% confidence quadrupole limit of $C_2^{1/2} < 2.5 \times 10^{-5}$. The fluctuation wavelengths they probe overlap with those probed by the Tenerife experiment. I think the two results are mutually compatible for a general fluctuation spectrum, but they are apparently incompatible for scale invariant spectra. Dropping the scale invariant assumption considerably relaxes the RELICT upper limits, to $C_2^{1/2} < 5 \times 10^{-5}$, still a factor of two better than the upper limit obtained by Lubin and Villela (1986) using the combined Berkeley-Princeton balloon data. RELICT 2, scheduled to fly in 1992, should probe deeply the 10^{-6} territory, down to the level of the biased CDM model ($C_2^{1/2} = 10^{-5}/b_\rho$).

2. BIASING FACTOR CONSTRAINTS FROM ANISOTROPIES

It is worthwhile to emphasize that *all* current theories of structure formation need to invoke inflation, whether the theory relies on scalar field fluctuations, normal or superconducting cosmic strings, explosions, or isocurvature baryon perturbations. No other viable mechanism to ensure flatness and smoothness has been proposed. The strong implication is that our local patch of the universe (the region we can see within about 10^4 Mpc) is likely to have $\Omega \approx 1$, if possible vacuum energy (non-zero Λ) contributions are included in the tally. A weaker implication relies on the specific model of fluctuation generation. If the perturbations that grew into the observed cosmic structure were from quantum oscillation modes of scalar fields, then it is most likely that the perturbation field was initially Gaussian with a fluctuation spectrum that was scale invariant. Although this has focussed our attention on this specific case (as I do in §2.2—2.5 below), it should be recognized that disproof of initial scale invariance or initial Gaussian-ness by observations will not disprove inflation: scale invariance might be broken and/or the statistics might not be Gaussian (Kofman and Linde 1987, Grinstein and Wise 1987, Salopek, Bond and Bardeen 1988).

Primordial Gaussian perturbations lead to a Gaussian primary radiation pattern. A non-Gaussian primary radiation pattern is a signature of cosmic string theories (Bouchet, Bennett and Stebbins 1988), as well as of exotic inflation models. Indeed, the cleanest way to decide on the statistics of the primordial fluctuations would rely on primary CMB anisotropy observations (e.g., Bond and Efstathiou 1986), since one does not have to untangle the initial underlying correlations from the complex dynamical correlations that develop with nonlinearity. If we are lucky, the CMB anisotropies will reveal to us the full mass distribution when the universe was simpler, and galaxy formation and its epoch will be derivable, albeit with some difficulty, from a known initial state.

The initial scale-invariant fluctuation power spectrum $\mathcal{P}_\rho(k)$ (per decade of wavenumber) for the density in various species has $\mathcal{P}_\rho(k)$ proportional to k^{3+n} with $n = 1$ for adiabatic perturbations and $n = -3$ for isocurvature perturbations. The scale invariance for adiabatic perturbations means flatness of the (initial) gravitational potential power spectrum ($\mathcal{P}_\Phi \propto k^{-4}\mathcal{P}_\rho(k)$). Of course, linear evolution bends and filters the spectral shape as the waves 're-enter the horizon'. Locally the shape can be parameterized by a slope $n(k)$. On very large scales, both isocurvature and adiabatic perturbations have $n = 1$.

One overall parameter characterizing the amplitude of the linear spectrum is required. It has now become conventional to define a biasing factor b_ρ to do this (Bardeen *et al.* 1986). The technical definition I use for b_ρ is 'J_3-normalization':

$$\langle \frac{\Delta M}{M}(< r = 10\,\mathrm{h}^{-1}\mathrm{Mpc})\frac{\delta\rho}{\rho}(0)\rangle \equiv \frac{1}{b_\rho^2}\langle \frac{\Delta N_g}{N_g}(< r)\frac{\delta n_g}{n_g}(0)\rangle \equiv \frac{3J_{3g}(r)}{b_\rho^2 r^3} \approx \frac{0.81}{b_\rho^2}. \quad (2)$$

Unfortunately, $10\,\mathrm{h}^{-1}\mathrm{Mpc}$ is probably not large enough to ensure validity of the theorem eq.(2) relating the ΔM average evaluated using the linear density fluctuation spectrum to the observationally determined ΔN_g average which gives the quoted 0.81 value, obtained from the CfA redshift survey; and the data are not very good on much larger scales, where we expect the theorem should hold (for $b_\rho = 1$). Consequently we can at best take (2) as an agreed upon convention for spectrum normalization. Basically it amounts to fixing the rms linear fluctuations on a Gaussian smoothing 'cluster' scale of $5\,\mathrm{h}^{-1}\mathrm{Mpc}$ to be $(\delta\rho/\rho)_{rms}(\mathrm{cl}) \approx 0.75/b_\rho$. The other widely used normalization convention, $\langle [(\Delta M/M)(< 8\,\mathrm{h}^{-1}\mathrm{Mpc})]^2\rangle = 1/b_\rho^2$, agrees to within 10% with this J_3-normalization for non-pathological spectra.

In §2.1, I review the mechanisms which give rise to primary anisotropies, those that directly probe the linear fluctuation spectrum without relying on the uncertainties of nonlinear dissipative evolution. None of these directly probe galaxy scales, so derived constraints on the epoch of galaxy formation rely on extrapolation from longer wavelengths, therefore requiring an assumption for spectral shape, such as an initial power law. In §2.2—2.5, current constraints on popular Gaussian models are given. The most significant points are: how close we are to interesting $\Delta T/T$ territory for the CDM models, and how strong the constraints on massive neutrino models and, to a lesser extent, $\Lambda \neq 0$ models are. These constraints have been derived in collaboration with George Efstathiou. We are preparing a more detailed set of computations since the OVRO results are so strong.

2.1 Mechanisms of Primary Anisotropies

The anisotropy from fluctuations of wavenumber k for photons arriving from direction $-\hat{q}$ has a simple form if we assume the recombination history has a Gaussian differential visibility function in conformal time, centred about the $\gamma-$decoupling time τ_r with width σ_r, which is not a bad approximation (see Bond 1987, §6, for details):

$$\frac{\Delta T}{T}(k,\hat{q},\tau_0) \approx e^{-ik\cdot\hat{q}R_r}e^{-(k\cdot\hat{q})^2\sigma_r^2/2}\left[\Phi(k,\tau_r)/3 + \frac{1}{4}\delta_{\gamma eff}(k,\tau_r) + \hat{q}\cdot\vec{v}_{Beff}(k,\tau_r)\right]$$
(3)

Here, R_r is the comoving distance to last scattering, $\approx 6000\,h^{-1}\mathrm{Mpc}$ ($=\tau_0$) for $\Omega = 1$ universes; τ_0 is the current time. A clear separation of the main effects affecting primary anisotropies is evident from this expression.

2.1.1 Last Scattering Surface Damping: $\exp(-(k\cdot\hat{q})^2\sigma_r^2/2)$ describes the damping due to the fuzziness $\sigma_r \sim 10\,h^{-1}\mathrm{Mpc}$ of the last scattering surface at $z \sim 1000$. The associated angle $\theta_c \sim 10'$ is approximately the coherence angle of the radiation correlation function.

2.1.2 The Sachs Wolfe Source, $\Phi/3$: Here Φ is the perturbed gravitational potential. This effect dominates on large angular scales $\theta \gtrsim 2°$. Strictly speaking this result is only valid for an $\Omega = 1$ universe dominated by non-relativistic matter for which Φ is constant. The general result is a path integral of the metric along the line of sight.

2.1.3 Photon Density Source, $\delta_{\gamma eff}$: This term leads to the often misused '$\Delta T/T = \frac{1}{3}\delta\rho_B/\rho_B$' rule for adiabatic fluctuations. If the perturbation mode is isocurvature rather than adiabatic, the fluctuations are initially perturbations in the entropy δ_s, without accompanying curvature perturbations. The entropy is taken per CDM particle for isocurvature axion perturbations or per baryon for isocurvature baryon perturbations. The latter are the old isothermal perturbations popular in the seventies. For isocurvature models, $\delta_\gamma \propto \delta_s$ at low k, which can lead to severe large angle constraints if the spectra are scale invariant. At large angles, this term has the same C_ℓ-shape as the Sachs-Wolfe effect, but is a factor of 5 larger. See Efstathiou and Bond (1986) and Bond (1987) for discussions.

2.1.4 The Electron Velocity Source, v_{Beff}: The bulk flow of the electrons around photon decoupling results in an asymmetry in Thomson scattering which leads to the anisotropy from this source. However, the fuzziness of the last scattering surface washes out this effect, dominantly for waves with wavevectors parallel to the line of sight where destructive interference arises from opposing contributions from the troughs and crests that straddle the photon decoupling surface. (Since $\Delta T/T \propto \int \sigma_T \bar{n}_e v_e dt$ is linear in the perturbations, where σ_T is the Thomson cross section and $\bar{n}_e$ is the average electron density.) For normal recombination, this term typically dominates on small angles, from $2°$ down to a few arcminutes.

A universe that never recombines has a very fuzzy decoupling surface, with a width corresponding to $\theta_c \sim 5°$, compared with the value $\theta_c \sim 10'$ of standard recombination. It was thought that small angle anisotropies would be essentially erased below $\sim \theta_c$ in these cases. However, as Vishniac (1987) has pointed out, significant small scale anisotropy can develop from quadratic nonlinearities in the scattering ($\propto \int \sigma_T \delta n_e v_e dt$) which do not suffer from destructive interference, as different wave-modes are coupled. This effect leads to significant anisotropies for isocurvature baryon models, even though early reionization seems likely.

2.2 Constraints on Adiabatic Cold Dark Matter Models

$\Omega = 1$, $h = 0.5$, $\Omega_B = 1 - \Omega_{cdm}$ CDM models with biased galaxy formation (Davis et al. 1985, Bardeen et al. 1986) are not in conflict with any of the observations, even for Ω_B as high as 0.5, unless we suppose the Tenerife experiment is seeing primary CMB anisotropies. The RELICT 1 result implies $b_\rho \gtrsim .4$. The OVRO results are shown in the following Table:

Ω_B	.03	.1	.2	.5
$b_\rho >$	.5	.7	.9	1.6

A convenient translation of b_ρ to the *rms* density linear fluctuation amplitude on galaxy scale (*i.e.*, a Gaussian filter $= 0.35 \, h^{-1}$Mpc) at redshift z is $(\delta\rho/\rho)_{rms}$(gal) $\approx 3.3(1.5/b_\rho)(1 + z)$. In a Bardeen et al. (1986) biasing model, the redshift of halo collapse would then be roughly $1 + z_g = 3/b_\rho$. Recall that the Davis et al. simulations adopted $b_\rho = 2.5$.

To have the biased CDM models agree with the large Tenerife result (if it proves to be primordial) one must have extra power on $\sim 300 \, h^{-1}$Mpc scales (where the Sachs-Wolfe effect dominates) compared with what the Zeldovich spectrum gives. To agree with RELICT, there would have to be a subsequent downturn in the gravitational potential perturbation spectrum at $\sim 500 - 6000 \, h^{-1}$Mpc. To break scale invariance with such a mountain of extra power, rather precisely placed, is not very attractive in inflationary models, but can occur (Kofman and Linde 1986, Bardeen, Bond and Salopek 1987, Bond, Salopek and Bardeen 1988, Salopek, Bond and Bardeen 1988.) Often accompanying the extra power would be a non-Gaussian component to the primordial post-inflation perturbations, and hence to the radiation pattern (Salopek, Bond and Bardeen 1988). The extra power would also aid in giving larger cluster-cluster correlations and larger and more coherent streaming velocities, while still giving cluster-galaxy and galaxy-galaxy correlation functions in agreement with the data (which the scale invariant CDM model seems to fit).

2.3 Constraints on Vacuum-dominated CDM Models

Invoking a nonzero cosmological constant, parameterized here by $\Omega_{vac} \equiv \Lambda/(3H_0^2)$, is one of the few ways to make models with low densities in CDM and baryons -- which have nice large scale structure properties (Bardeen, Bond and Efstathiou 1987, Blumenthal, Dekel and Primack 1988, but see Bond and Couchman 1988) — compatible with inflation. Such models would require $b_\rho \sim 1$ to avoid an excessively early redshift of galaxy formation (Bardeen, Bond and Efstathiou 1987). Although Efstathiou and I have not exhaustively explored the parameter space of these models, it is clear from the following table that models with $\Omega_{vac} = 0.8$ and $\Omega_{cdm} + \Omega_B = 0.2$ are now largely ruled out by the Caltech experiment.

Ω_B	.03	.03	.1	.1
h	.75	.5	.75	.5
$b_\rho >$	1	1.9	1.6	4.2

Open CDM-dominated models are very strongly ruled out by the Caltech results. For example, $\Omega = 0.2$, $\Omega_B = 0.03$, $h = 0.5$ would require $b_\rho > 7$.

2.4 Constraints on Neutrino-dominated Models

We estimate an $\Omega = 1$, $h = 0.5$, $\Omega_B = 0.1$ ($m_\nu \approx 24$ ev) model would need to have a nonlinear redshift ($z_{nl} = (\delta\rho/\rho)_{rms}(\text{gal}) - 1$) below 0.5 to be compatible with the OVRO data. Only $\sim 1\%$ of the gas would have shocked by $z = 1$ if $z_{nl} = 0.5$ defines the normalization amplitude, making recent efforts to resurrect this model somewhat problematical. Efstathiou and I are currently exploring these constraints in more detail. The RELICT 1 constraint is less severe, $z_{nl} \lesssim 1.4$.

2.5 Constraints on Isocurvature Models

Anisotropies for isocurvature baryon models with initial spectral index $n = 0$ (Poisson seed model) or $n = -1$ (phenomenological) that were very popular in the 1970s have been discussed recently by e.g., Peebles (1987) and Efstathiou and Bond (1987). The isocurvature effect (§2.1.3) dominates, determining the large angle anisotropies. Efstathiou (1987) displays a nice graph illustrating the vast region in $\Omega_B - n$ space ruled out by a combination of the OVRO result and the Melchiorri et al. (1981) $6°$ upper limit of 5×10^{-5} as the fiducial small and intermediate angle constraints for universes that never recombined (as expected since collapse on the smallest scales should occur shortly after $z = 1000$). The $6°$ limit implies $n \gtrsim -2$ is required for $\Omega_B = 1$, and $n \gtrsim -1$ is required for $\Omega_B = 0.1$. The $7'$ limit (where the Vishniac effect of §2.1.4 dominates) implies $n \lesssim -1$ for $\Omega_B = 1$ and $n \gtrsim -1$ for $\Omega_B = 0.1$ (and up to $\Omega_B = 0.4$). The $\Omega_B = 0.1$, $-1 \lesssim n \lesssim 0$, models often discussed in the seventies are therefore not yet ruled out. If, however, the universe has normal recombination, all of these models are ruled out.

For isocurvature baryon perturbations to be compatible with inflation, we would require $\Omega = 1$, and, unless scale invariance is broken, $n = -3$ (a flat spectrum in δ_s). If scale invariance is broken, it is very difficult to arrange for a $-1 \lesssim n \lesssim 0$ power law over an extended region. If we allow ourselves complete freedom in shaping the potential energy surface of the scalar fields which drive inflation, however, essentially any spectrum, likely of non-Gaussian form, can result (Salopek, Bond and Bardeen 1988). To have Ω_B compatible with primordial nucleosynthesis, $\Omega_B \lesssim 0.2$ is required in standard models (although $\Omega_B = 1$ may be conceivable in variants, using quark—hadron phase transition inhomogeneities, e.g.. Applegate et al. 1987, or decaying particles, *e.g.*, Dimopoulos *et al.* 1988). In that case, the residual would have to be made up with vacuum energy. The constraints discussed above for the open models would be less stringent in these cases, although the $n = -3$ flat cases would still be strongly ruled out.

The isocurvature CDM model, with axions (or other ultra-weakly interacting bosons of tiny mass) forming the cold dark matter, has a similar story for large angle anisotropies as the baryon models (Efstathiou and Bond 1986). $\Omega = 1$ models with $n = -3$ can be ruled out by large angle anisotropies, even with biased galaxy formation, but larger n cannot be excluded. By contrast with the baryon case, the small angle fluctuations are not very large, even with standard recombination. For example, in an $h = 0.75$ $\Omega_B = 0.03$, $n = -3$ model, $\Delta T/T$ is $7 \times 10^{-6}/b_\rho$ for the OVRO experiment.

3. SECONDARY BACKGROUNDS AND THEIR ANISOTROPIES

Once a non-Gaussian component to the random density field develops as structures turn-around and collapse, non-Gaussian *secondary* anisotropies arise from Thomson scattering by nonlinear bulk-flow currents, from Compton cooling of inhomogeneous hot gas, and by redshifted emission from primeval dust. These will be the only significant fluctuations in structure formation theories based on explosions, whether supernova-driven (Hogan 1984, Vishniac and Ostriker 1985, Bond 1987), radiation-driven (Hogan and Kaiser 1983) or driven by the electromagnetic energy from superconducting strings (Ostriker, Thompson and Witten 1987). These will lead to non-Gaussian radiation patterns. Hot gas in collapsed groups and clusters will also give rise to a non-Gaussian component to the microwave sky. Interesting limits on the nonlinear redshift of the massive neutrino model (but not on the CDM model) result, as shown in Bond (1987). However, the OVRO constraint is much stronger, and covers roughly the same wavenumber region. In this section, we shall focus primarily on constraints from the distortions and anisotropies in dust emission models, since these probe a wavenumber region inaccessible by other CMB probes.

Secondary radiation backgrounds and their associated anisotropies may probe the epochs when structure first formed in the universe. Current constraints allow large energy releases. I adopt $\Omega_R(\lambda) = 4\pi\nu I_\nu/(\rho_{crit}c^2)$, the energy density as a fraction of closure energy density, as a measure of the radiation content in a waveband about wavelength λ. Here, I_ν is the usual radiation intensity. To set the scale, the total CMB energy is $\Omega_{cmb}h^2 = 25 \times 10^{-6}$, the blue light luminosity density from galaxies, integrated over a Hubble time, is $\Omega_{blue}h^2 \sim 4.6^{+1.9}_{-1.4} \times 10^{-7}$ (using the luminosity density $\mathcal{L}_{blue} = 1.9^{+0.8}_{-0.6} \times 10^8\,h\,L_\odot\,\text{Mpc}^{-3}$ of Efstathiou *et al.* 1988) and the X-ray background peak is $\Omega_R(30\,\text{kev})\,h^2 = 1.7 \times 10^{-8}$. Upper limits on $\Omega_R h^2$ are: $\sim 10^{-7}$ in the UV, $\sim 5 \times 10^{-7}$ in the optical, $\sim 10^{-5}$ in the near IR (with a possible detection at this level at 2.2μ still claimed by Matsumoto, Akiba and Murakami 1988), $\sim 10^{-5}$ at 100μ, and $\sim 5 \times 10^{-6}$ in the sub-mm region (with detections reported by Gush 1981 and now by Matsumoto *et al.* 1988).

3.1 The Berkeley-Nagoya Distortion

The Matsumoto *et al.* (1988; hereafter BN) distortion on the Wein side of the CMB has an excess radiation density $\Omega_R(\lambda)$ above that in a $2.74°K$ unperturbed CMB, $\Omega_{cmb}(\lambda)$, in their three lowest frequency channels given by:

λ	480μ	709μ	1160μ
$\Omega_R(\lambda)$	3.1 ± 0.3	5.3 ± 0.5	1.7 ± 0.6
$\Omega_{cmb}(\lambda)$	1.0	7.1	18

Single temperature dust models typically have difficulty getting the 1160μ energy low enough if they fit the other two channels; the problem is slightly less severe for more rapidly declining grain opacities (with λ). Nonetheless, they do fit better than nonrelativistic Compton cooling, which has a specific spectral shape characterized by the Compton y-parameter, $y = \int n_e \sigma_T (T_e/m_e)dt$. The best fit ($y = 0.016$) has an unperturbed CMB temperature of $2.83°K$ (giving a Rayleigh Jeans temperature of $2.74°K$). For this case $\Omega_R(\lambda)$ would of course differ from the table. The total energy in the distortion for dust models is typically 25%, and for Compton cooling it is 6%. For nonrelativistic decaying particle models, the energy input would have to be $\sim 20\%$ as well (Bond 1987). I refer the reader to Kawasaki and Sato (1987) for a discussion of the (in my view unlikely) possibility the distortion is direct redshifted emission from an exotic decaying particle of a few ev.

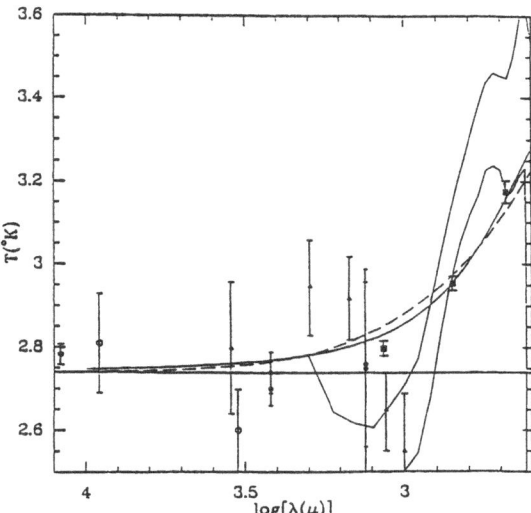

Figure 1: The current status of the CMB spectral data is plotted in terms of thermodynamic temperature, along with curves indicating the quality of a dust fit (with a λ^{-1} grain opacity, solid curve), and the $y = 0.016$ Compton cooling model (long-dashed curve). References to the Rayleigh-Jeans data are given in BCH2. The filled squares are the BN points, the filled triangles are the Peterson. Richards and Timusk (1985) points and the thin lines denote the 90% confidence limits of Gush (1981). The horizontal line is the best fit $2.74°K$ Rayleigh-Jeans temperature

of the CMB.

Of course, the difficulty of these sub-millimeter experiments invites caution, especially recalling the Wein distortion claims in the early seventies (Weiss 1980), and by Woody and Richards (1981) and Gush (1981), as well as the distortion constraints by Peterson, Richards and Timusk (1986) and, via dipole measurements, by deBernardis *et al.* (1985) and Halpern *et al.* (1988). COBE, scheduled for a 1989 launch, will be much more sensitive, capable of probing to $\Omega_R h^2 \sim 3 \times 10^{-8}$ over a spectral range $500 - 10^4 \mu$.

3.2 A Compton Cooling Distortion?

All late time energy injection models (*e.g.*, galaxy explosions) that could produce the large $\bar{y}$ required to fit the BN distortion would be expected to generate large (unobserved) Rayleigh Jeans anisotropies, $\Delta T/T \approx -2\Delta y$. To avoid this, it would be best to inject the energy at high redshift ($\gtrsim 50$) while the Universe is still thick to Compton scattering, and from a large number N of small scale sources, so that the anisotropies can be diminished both by diffusion and $1/\sqrt{N}$ effects. However, nuclear or accretion energy from stars or black holes is not effective since it primarily goes into radiation which, once the medium is ionized, propagates relatively freely to us. Although most of the heat pumped into the electrons by the ionization process Compton cools into the CMB, the energy is likely to be small compared with the direct redshifted radiation.

Early non-radiative energy injection does not suffer from this problem. One of the most intriguing possibilities for this is superconducting cosmic strings (Ostriker, Thompson and Witten 1986). Electromagnetic energy liberated below the plasma frequency of the pregalactic medium by SCS's would heat electrons which would then Compton cool, giving $\bar{y} \approx (0.3\text{-}5) \times 10^{-3} 10^6 G\mu$ (Ostriker and Thomson 1987), where μ is the string tension, provided at least 10% of the energy liberated by decaying strings is electromagnetic (with the remainder being in gravitational waves). Ostriker and Thompson show that the nature of the spectrum of string loops in the network imply anisotropies, generated predominantly at late times, will not be overly large: they estimate $(\Delta T/T)_{rms} \sim 2 \times 10^{-5}$ for switching angles $\theta > 4'$ without beam smearing for $G\mu = 10^{-6}$. Increasing $G\mu$ to give $\bar{y} = 0.016$ might then conflict with the OVRO limit, although more precise calculations will be required, especially since the radiation pattern would be non-Gaussian.

3.3 Primeval Dust Emission

Theoretical musings on the source of the 1979 Woody—Richards distortion centred primarily around dust models (Rowan-Robinson, Negroponte and Silk 1979, Wright 1982). Even before this, there was considerable speculation about whether dust emission could account for *all* of the CMB (*e.g.*, Layzer and Hively 1973, Rees 1978, Hoyle 1980, Wright 1982). However, to get the fairly good blackbody spectrum at cm wavelengths required fairly contrived circumstances (Hawkins and Wright 1988), including an alchemical mix of grain types, some with quite exotic properties (*e.g.*, iron conducting needles). In light of the 25% energy requirement of dust models of the BN data, I think we will have to take this possibility very seriously indeed if the distortion is confirmed by COBE.

Bond, Carr and Hogan (1986 [BCH1]) showed that if galaxies (even dwarfs) exist at $z \sim 10$ then they cover the sky, and if they are dust-laden then *all energy* from the near IR to the X gets absorbed and re-emitted in the far IR, with a peak wavelength $\lambda \sim 700\mu$ which is relatively model insensitive. McDowell (1986) and Negroponte (1986) came to similar conclusions. BCH1 showed that energy sources would necessarily deliver a distortion energy $\Omega_R/\Omega_{cmb} \sim 10^{-3}$, and that $\Omega_R/\Omega_{cmb} \sim 10^{-1}$ was plausible. In BCH1 and Bond (1987), we showed that the intensity fluctuations in the sub-mm range would typically be $\Delta I_\nu/I_\nu \sim 0.01 - 0.1$ from galaxies for a wide range of models.

In Bond, Carr and Hogan (1988 [BCH2]), we apply our theory to the BN data. This has pushed us into a different regime than we focussed on in BCH1; in that paper, we thought galaxies that would just cover the sky (optical depth ~ 1) would be the likely culprit. The BN data has led us to consider high energy inputs (§3.1), which are difficult to arrange for conventional metal-producing stars (BCH1, Lacey and Field 1988, BCH2), and high dust optical depths, which in turn indicate a high grain formation redshift. For example, at 700μ we find that the optical depth across the shell where the bulk of the emission takes place ranges from 0.03–0.3, for opacity laws from $\sim \lambda^0$ to $\sim \lambda^{-2}$, respectively. The redshift of the emission shell would then be

$$1 + z_d \gtrsim 29 \left[\frac{A_d \Omega_d h}{10^{-5}} \right]^{-0.4}, \qquad (4)$$

where A_d parameterizes the dust cross section. Technically, it is $A_d = \frac{4\pi}{3} \mathrm{Im} \left[\alpha_i^{e\,i} \right]$ in the long wavelength limit, where $\alpha_i^{e\,i}$ is the trace of the electric polarizability tensor of the grain. If A_d is wavelength dependent, which it is for every dust law but a $\sim \lambda^{-1}$ cross section, then it is to be evaluated at $700\mu/(1+z_d)$, making eq.(4) implicit. Over the relevant emission wavelengths, A_d is typically 0.1–1 for graphites and silicates, but could easily be ~ 100 for conducting iron whiskers. Assuming half the heavy elements are locked in grains, the dust abundance is

$$\Omega_d = 10^{-5} \frac{Z}{0.02} \frac{\Omega_B}{0.1} \frac{f_{B\,\mathrm{in\,gas}}}{0.1} \frac{f_{B\,\mathrm{collapsed}}}{0.1}, \qquad (5)$$

in terms of the metallicity Z, Ω_B, the fraction of the baryons B in gas, and in collapsed objects. For normal spiral galaxies, we might expect that Ω_d is $\sim 10^{-5}$ now with Population I dust abundances, and was at most 10^{-4} in the past if a large fraction of the metals were generated when the spirals were mostly gas. If dwarf galaxies with $M \sim 10^{6-9}\,M_\odot$ were to generate the metals, to avoid overpolluting the environment from which the halo of our Galaxy formed, we might allow at most Population II metals, hence again $\Omega_d \lesssim 10^{-4}$. If Z is to be below the minimum metallicity observed for Population II stars, $\sim 3 \times 10^{-6}$ (Norris *et al.* 1984), then the limit is much more severe, $\Omega_d \lesssim 10^{-6.5}$.

The freedom we have in Ω_d and A_d allows some freedom in z_d, which makes it difficult to rule models out. However, the plausible z_d's do seem to be too high for models based on late time pancaking, such as the neutrino-dominated or baryon-dominated adiabatic models. On the other hand, high z_d is absolutely natural for the isocurvature baryon-dominated models. Cosmic string scenarios also have structure forming early, accreting onto the seed string loops, providing ample opportunity for early dust and radiation production.

It is more difficult to constrain the CDM model. On a Gaussian smoothing scale of $R_{JBr} \approx 1.6\,h^{-1}$kpc, the reduced baryon Jeans wavenumber at recombination, the *rms* fluctuations are $\approx 22/b_\rho$. This scale would undergo conventional top hat collapse at a redshift $1 + z = 13/b_\rho$. To have molecular hydrogen form sufficiently to cool such objects, $R_{C,H_2} \sim 25\,h^{-1}$kpc is required, with *rms* fluctuations $\approx 10/b_\rho$. Thus, with the conventional biasing normalization used by White *et al.* (1984), $b_\rho = 2.5$, the smallest scale dwarfs form quite late (except for relatively rare 3σ peaks). Even with $b_\rho = 1.4$, as advocated by Bond (1987a,b) and Kaiser (1988), we still have difficulties having enough collapsed structure at high redshift to generate the required dust abundance; *e.g.*, only 10% of the universe would have collapsed by $z = 20$, predominantly in masses below $10^7\,M_\odot$, according to Press–Schechter (1974) theory. Of course one may rely on our ignorance of A_d to drop z_d to ~ 10, which the CDM model is more comfortable with. It is also possible that Ω_d was higher ($\sim 10^{-4}$) at high z, with the dust being subsequently swallowed by stars, as mentioned above. Nonlinear amplification of the primordial CDM short distance structure could also raise z_d. The $\Omega_{RT}/\Omega_{cmb} \sim 1/4$ result implies the energy sources must be exotic, and, if they are stellar or accretion sources, it is quite possible that they would have a profound effect on their environs through explosive or radiative injections of energy and radial momentum. Indeed the mock gravity model of Hogan and White (1987) suggests this may inevitably accompany such large dust emission. Broken scale invariance could also lead to more power at short scales.

BCH2 estimate anisotropies by assuming the dust and luminosity sources are distributed in collapsed objects with assumed profiles. The anisotropies then have a contribution from the Poisson fluctuations of the objects and from their clustering. In models such as the CDM theory, clustering dominates. In a simplified approximation to the detailed linearized radiative transfer equation solution, we show that the anisotropies can be written in terms of the fluctuations in dust optical depth and in the luminosity sources across the effective emission shell. The anisotropies are substantially diluted if many correlation lengths span the shell. If the dust is clumped in normal spiral galaxies or even in $\sim 10^{10}\,M_\odot$ dwarfs, the level would typically be above a percent in $\Delta I_\nu(\theta)/I_\nu$. However, if only very small dwarfs cause the distortion, the anisotropy level can also be much smaller.

However, interesting limits on the scale of dust or luminosity source correlations already follow from the OVRO experiment at 15000μ: for example, for λ^{-1} dust, BCH2 find the clustering length satisfies $r_0 \lesssim 0.2\,h^{-1}$Mpc; and $r_0 \lesssim 2\,h^{-1}$Mpc for all models we tried. (We assumed the 7-field OVRO limit represents a global limit; but see §1.2.) The truly interesting regime for the small dwarf models we are led to consider by the high z_d would require sub-millimeter telescopes. If they could achieve similar sensitivity levels to the OVRO experiment, they could push r_0 down to $\sim 4\,h^{-1}$kpc.

Acknowledgments: Thanks go to my collaborators George Efstathiou, who is the other half of the ongoing work on primary anisotropies reported in §2, and Bernard Carr and Craig Hogan, who are the other members of the dust trio that composed the tale of §3. The support of a Canadian Institute for Advanced Research Fellowship, a Sloan Foundation Fellowship and the NSERC of Canada is gratefully acknowledged.

4. REFERENCES

Applegate, J.H., Hogan, C.J. and Scherrer, R.J. 1987, *Phys. Rev.* D **35**, 1151.

Bardeen, J.M., Bond, J.R., Kaiser, N. and Szalay, A.S. 1986, *Ap. J.*, **304**, 15.

Bardeen, J.M., Bond, J.R. and Efstathiou, G. 1987, *Ap. J.*, **321**, 28.

Bardeen, J.M., Bond, J.R. and Salopek, D.S. 1987, in *Proceedings Second Canadian Conference on General Relativity and Relativistic Astrophysics*, eds. A. Coley, C. Dyer (Singapore: World Scientific).

Blumenthal, G., Dekel, A. and Primack, J., 1987, *Ap. J.* **326**, 539.

Bond, J.R., Carr, B.J. and Hogan, C.J. 1986, *Ap. J.* **306**, 428 [BCH1]; 1988, CITA Preprint [BCH2].

Bond, J.R. and Efstathiou, G. 1984, *Ap. J. Lett.* **285**, L45; 1987, *M.N.R.A.S.* **226**, 655.

Bond, J.R. 1987, in *The Early Universe*, Proc. NATO Summer School, Vancouver Is., Aug. 1986, ed., Unruh, W.G. (Dordrecht:Reidel).

Bond, J.R. 1987b, in *Cosmology and Particle Physics*, Proc. LBL Workshop, Berkeley CA 1986, ed. I. Hinchcliffe (Singapore: World Scientific).

Bond, J.R. 1988, in Proceedings I.A.U. Symposium 130, ed. Audouze, J. and Szalay, A.S. (Dordrecht: Reidel).

Bond, J.R., Salopek, D.S. and Bardeen, J.M. 1988, in *Proceedings Vatican Conference on Large Scale Velocities*, eds. G. Coyne, V. Rubin (Vatican: Pontifical Academy).

Bond, J.R. and Couchman, H.M.P. 1988, CITA preprint; Couchman, H.M.P. and Bond, J.R. 1988, in *Large Scale Structure and Motions in the Universe*, Proc. Trieste Conference, April 1988, 4 pages, ed. G. Giuricin *et al.*

Bouchet, F., Bennett, D.P. and Stebbins, A. 1988, *Nature* **335**, 410.

Davies, R.D., Lasenby, A.L., Watson, R.A., Daintree, E.J., Hopkins, J., Beckman, J., Sanchez-Almeida, J. and Rebolo, R. 1987, *Nature* **326**, 462.

Davis, M., Efstathiou, G., Frenk, C.S., and White, S.D.M. 1985, *Ap. J.* **292**, 371.

deBernardis, P., Masi, S., Malagoli, A. and Melchiorri, F. 1985 , *Ap. J.* **288**, 29.

Dimopoulos, S., Esmailzadeh, R., Hall, L. and Starkman, G. 1988, *Ap. J.*, in press.

Efstathiou, G. and Bond, J.R. 1986, *M.N.R.A.S.* **218**, 103; 1987, *M.N.R.A.S.* **227**, 33P.

Efstathiou, G., Ellis, R.S. and Peterson, B.A. 1988, *M.N.R.A.S.* **232**, 431.

Efstathiou, G. 1988, in *Proceedings Vatican Conference on Large Scale Velocities*, eds. G. Coyne, V. Rubin (Vatican: Pontifical Academy).

Fomolont, E.B., Kellerman, K.I., Anderson. M.C., Weistrop, D., Wall, J.V., Windhorst, R.A. and Kristian, J.A. 1988, *A. J.*, in press.

Grinstein, B. and Wise. M. 1987, *Ap. J.* **320**, 448.

Gush, H.P. 1981, *Phys.Rev. Lett.* **47**, 745.

Halpern, M., Benford, R., Meyer, S., Muehlner, D. and Weiss. R. 1988, *Ap.J.* **332**, 596.

Hawkins, I. and Wright, E.L. 1988, *Ap. J.* **324**, 46.

Hogan, C.J. 1984, *Ap. J. Lett.* **284**. L1.

Hogan, C. J. and Kaiser, N. 1983, *Ap. J.* **274**. 7.

Hogan, C.J. and White, S.D.M., 1986. *Nature.* **321**. 575.

Hoyle, F. 1980, *Steady State Cosmology Revisited.* (Cardiff Wales: University College Cardiff Press).

Kaiser, N. 1988, in *Proceedings Vatican Conference on Large Scale Velocities.* eds. G. Coyne, V. Rubin (Vatican: Pontifical Academy).

Kofman, L. A. and Linde A. D. 1987, *Nuc. Phys.* **B282**, 555.
Kawasaki, M. and Sato, K. 1987, *Pub. Astron. Soc. Japan*, **39**, 337.
Lacey, C. and Field, G. 1988, *Ap. J. Lett.* **330**, L1.
Layzer, D. and Hively, R. 1973, *Ap. J.* **179**, 361.
Lubin, P. and Villela, T. 1986, in B. F. Madore and R. B. Tully, eds. *Galaxy Distances and Deviations from the Hubble Flow* (Dordrecht: Reidel).
Martin, H. M. and Partridge, R. B., 1988, *Ap. J.* **324**, 794.
Matsumoto, T., Akiba, M. and Murakami, H. 1988, Preprint.
Matsumoto, T., Hayakawa, S., Matsuo, H., Murakami, H., Sato, S., Lange, A.E. and Richards, P.L. 1988, *Ap. J.*, **329**, 567. [BN]
McDowell, J. C. 1986, *M.N.R.A.S.* **223**, 763.
Melchiorri, F., Melchiorri, B.O., Ceccarelli, C. and Pietranera, L. 1981, *Ap. J. Lett.*, **250**, L1.
Negroponte, J. 1986, *M.N.R.A.S.* **222**, 19.
Norris *et al.* 1984, *M.N.R.A.S.* **222**, 19.
Ostriker, J.P., Thompson, C. and Witten, E. 1986, *Phys. Lett.* **B180**, 231.
Ostriker, J.P. and Thompson, C. 1987, *Ap. J. Lett.* **323**, L97.
Peebles, P.J.E., 1987, *Ap. J. Lett.* **277**, L1.
Peterson, J.B., Richards, P.L. and Timusk, T. 1985, *Phys. Rev. Lett.* **55**, 332.
Press, W.H. and Schechter, P. 1974, *Ap.J.* **187**, 425.
Readhead, A.C.S., Lawrence, C.R., Myers, S.T., Sargent, W.L.W., Hardebeck, H.E. and Moffet, A.T. 1988, Caltech preprint.
Rees, M.J. 1978, *Nature* **275**, 35.
Rowan-Robinson, M., Negroponte, J. and Silk, J. 1979, *Nature* **281**, 635.
Salopek, D.S., Bond, J.R. and Bardeen, J.M. 1988, CITA Preprint.
Strukov, I.A., Skulachev, D.P. and Klypin, A.A. 1987, in Proceedings I.A.U. Symposium 130, ed. Audouze, J. and Szalay, A.S. (Dordrecht: Reidel).
Vishniac, E.T. and Ostriker, J.P. 1985, *Societa Italiana di Fisica Conference Proc.* **1**, 137.
Vishniac, E.T. 1987, *Ap. J.*, **322**, 597.
Weiss, R. 1980, *Ann. Rev. Astron. Ap.* **18**, 489.
Woody, D.P. and Richards, P.L. 1981, *Ap.J.* **248**, 18.
Wright, E.L. 1982, *Ap. J.* **255**, 401.

DISCUSSION:

Peacock: I find the *a posteriori* rejection of one of the OVRO points disturbing. All these fields will contain some discrete sources; what one should do is map these first and subtract the smoothed effect of the small-scale data from the CMB experiment.

Bond: According to Readhead *et al.* the rejected field in question, at 7^h right ascension, has an 8σ detection of a known radio source which is time variable. Radio observations concurrent with the CMB measurements were not made, so the source amplitude cannot be subtracted out — your ideal strategy. They had little choice but to exclude this field. Otherwise their upper limit would be too sensitive to the subtraction scheme.

Silk: How do you reconcile your requirements for dust production in the early universe with the heavy element abundances in Lyman alpha absorption clouds and in the more extreme metal poor halo stars?

Bond: This depends in detail upon the model. The redshift of the dust depends upon the abundance of the grains. Enforcing $\Omega_d \lesssim 10^{-5}$, which is a reasonable number for either relatively uniformly distributed dust in Population II abundances ($Z \sim 10^{-4}$), or clumped dust in galactic scale objects with Population I abundances, fixes the (high) redshift of dust emission. Only the most metal poor stars would violate this, and I could always lower the overall abundance to accomodate them. This would require either raising the dust emission redshift or the absorption opacity in the redshifted millimeter band, *e.g.*, by invoking whiskers. Alternatively, I could rely on clumpiness of the pockets where dust forms. These dust pockets might preferentially avoid the low Z star forming regions in the protohalo or the low N_{HI} Lyman alpha absorption cloud forming regions in the IGM. The latter does not seem inconceivable in the CDM model, for the relevant part of the fluctuation spectrum for the $10^8 \, M_\odot$ clouds where the dust would form is the flicker noise region where the cloud-in-cloud problem is severe, and where small scale objects imbedded in larger scale overdense regions are the first to collapse. I can't say more since the gas dynamics of flicker noise spectra is very difficult to handle, either analytically or numerically. Nonetheless, I regard the metallicity constraint as our second most severe challenge, after the energetics one. Still I don't regard it as disastrous for the dust concept yet.

Peebles: Your point that small objects can form at high redshift in the standard CDM theory is well taken. I would note, however, that the space positions of these small objects are not strongly correlated with the large L_* galaxies that would form at low z, so you are in danger of filling the nearby voids with an unacceptably large number of dwarf galaxies. Can you give us an estimate of the expected distance to the nearest remnant of the generation of objects that would have produced the Berkeley-Nagoya anomaly in your theory?

Bond: Dating back at least to the famous Peebles and Dicke primeval globular cluster hypothesis, it has always been questionable why we haven't seen the $10^{6-8} \, M_\odot$ bricks which are supposed to build galactic structure, whether the theory is isocurvature baryon, adiabatic baryon, or CDM. Indeed, CDM may fare the best because of the tremendous cross-talk from scale to scale in the flicker noise region, leading to relatively high correlations and large accretion rates onto larger scale objects. Nonetheless, these small scale things will not form only in large scale dense environments, because the short distance power will tend to exceed the medium distance power (galactic scale) and swamp large distance power (cluster and void scale). Using the theory of Gaussian fields applied to the CDM spectrum, I would get a mean comoving $\sim 10^7\,M_\odot$ peak spacing of about 100 kpc, only some of which would be likely to generate dust. Some of the collapsed peaks would, of course, be here in the Milky Way, having been more likely to generate dust because they collapsed at an earlier time in an intermediate scale overdense region. I have no idea what the current state would be of the ones that have not accreted. I do expect that the gas would be gone because the binding energies would be so low; the dark matter would also be prone to tidal disruption in the Local Group. In other words I bet I could hide them, at least until I know better how to calculate in this region. I regard the metallicity constraint as a more severe problem than the proximity one. These little dwarfs would certainly be undetectable in the distant voids.

Field: Simon White and Craig Hogan proposed in *Nature* a while ago that mock gravity would work in circumstances which, in retrospect, are like those at $z \sim 10$ needed to explain the Matsumoto *et al.* submillimeter excess. I have been investigating this idea with a student at Harvard, Bogi Wang. We concur that the effect exists, although the physics is different in detail. The preliminary result is that the mock gravity would clump the dust and gas on the scale of $10^8\,M_\odot$ in baryons. Would future measurements of anisotropy, perhaps at submillimeter wavelengths, be able to test this?

Bond: It would certainly be nicer if larger scale structure than $10^8\,M_\odot$ could form. However, ground-based submillimeter telescopes such as JCMT might ultimately probe clustering levels of a few tens of kiloparsecs, below the range you are finding, if sensitivity levels similar to those at Rayleigh Jeans wavelengths are achievable in the 800μ window. Certainly the first round of experiments won't get there, but, with a lot of integration time, there is hope. (See BCH2 for a more detailed discussion.)

EXPLAINING THE NAGOYA-BERKELEY SUBMILLIMETRE BACKGROUND

B.J.Carr
School of Mathematical Sciences
Queen Mary College
Mile End Road
London E1 4NS

ABSTRACT. The most natural source of the Nagoya-Berkeley excess is pregalactic or protogalactic radiation reprocessed by dust. The sort of infrared galaxies observed at the present epoch could produce the energy in the excess but only if there was a large luminosity evolution. We therefore consider pregalactic radiation sources with absorption by dust which is either intergalactic or confined to galaxies which cover the sky. In this case, fitting the data requires that the reprocessing occur at a rather high redshift (z>30). It would be hard to generate enough dust at such an early epoch in the pancake or standard cold dark matter scenarios but it would be possible in a more general hierarchical clustering picture. Possible sources for the radiation would be VMOs (very massive stars whose black hole remnants may provide the dark matter in galactic halos), cosmic explosions or decaying elementary particles.

1. INTRODUCTION

Matsumoto *et al.* (1988) have reported a significant distortion in the spectrum of the microwave background radiation in the submillimetre waveband 400-700μ. This follows a rocket experiment by a team from Nagoya and Berkeley (NB) in February 1987. The form of the distortion is indicated in Figure 1. Measurements were made at six wavelengths: the 1160μ intensity confirms the usual microwave temperature, while the intensities at the three shortest wavelengths (102, 137 and 262μ) are well-explained by interstellar dust emission. The surprising results are the measurements at the two intermediate wavelengths (481 and 709μ); these are significantly in excess of the expected microwave flux, even though there is no obvious local source of radiation at these wavelengths. If this represents a cosmological background, it peaks at around 700μ and has a density

$$\Omega_R \simeq 5 \times 10^{-6} h^{-2} \simeq 0.2 \ \Omega_C \qquad (1)$$

in units of the critical value. Here h is the Hubble parameter in units

227

C. S. Frenk et al. (eds.), The Epoch of Galaxy Formation, 227–234.

of 100 km/s/Mpc and Ω_C is the density parameter associated with the microwave background itself. If one believes the error bars in Figure 1, this is a 13σ effect, so such a large density poses a severe challenge to the standard cosmological model.

Figure 1. The spectrum of the submillimetre background (Matsumoto et al. 1988). The fluxes obtained by Matsumoto et al. are shown by ● with the vertical error bars and the horizontal bars for the effective bandwidths. The results of other measurements are shown for comparison by 0 (Peterson et al. 1985), □ (Meyer and Jura 1985), ▉ (Crane et al. 1986), X (Smoot et al. 1985, 1987) and △ (Johnson and Wilkinson 1987). IRAS data at 60 and 100 μm are shown by ◆ .

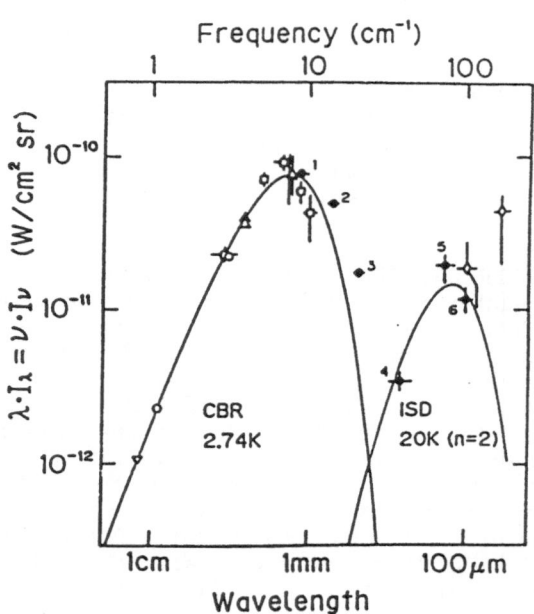

One can envisage two types of explanation for the excess. Firstly, it may represent *Compton distortion* due to the microwave photons having passed through hot ionized intergalactic gas. The gas would need to have a temperature of about 10^8K and the thermal bremmstrahlung radiation from such gas has already been invoked to explain the hard X-ray background (Guilbert & Fabian 1986). However, the Compton distortion model fits the data only if the Rayleigh-Jeans temperature of the microwave background is much smaller than usually assumed (Hayakawa *et al.* 1987). Secondly, the submillimetre excess may represent emission from cosmological dust, the dust itself having been heated by some radiation source. In this case, one can fit the observed distortion rather well (Bond *et al.* 1988; henceforth BCH), so this is probably the most satisfactory explanation. Both models face severe energetic demands: in the first, one has to heat up intergalactic gas, whereas in the second, one has to heat up the dust. The energy requirements are similar, although one needs somewhat more exotic models - such as cosmic explosions (Yoshioka & Ikeuchi 1987) or superconducting cosmic strings (Ostriker *et al.* 1986) - in the first case. We therefore focus on the dust model.

2. THE DUST MODEL

Even before the NB experiment, Bond *et al.* (1986) pointed out that, in many circumstances, one would expect pregalactic or protogalactic radiation to be reprocessed by cosmological dust to produce a background in the far-infrared to submillimetre range. They predicted that the spectrum should peak at a wavelength

$$\lambda_{peak} \simeq 700 \left[\frac{\Omega_R h^2}{10^{-6}}\right]^{-1/5} \left[\frac{r_d}{0.1\mu}\right]^{1/5} \left[\frac{1+z_d}{10}\right]^{1/5} \mu \qquad (2)$$

where r_d is the grain radius and z_d is the redshift at which the radiation is reprocessed. (Throughout this section we assume that the grain opacity scales inversely as the wavelength.) It is clear that, with Ω_R given by eqn (1) and with reasonable normalizations for r_d and z_d, λ_{peak} has the sort of value observed. The dust required may be either a smoothly distributed pregalactic component, in which case it can absorb the radiation for

$$1+z_d > 8 \left[\frac{\Omega_d h}{10^{-5}}\right]^{-2/3} \left[\frac{r_d}{0.1\mu}\right]^{2/3} \qquad (3)$$

where Ω_d is the dust density parameter, or it may be confined to galaxies themselves. In the latter case, the redshift of galaxy formation (z_G) must satisfy two independent conditions: firstly, it must exceed the value given by eqn (3) with Ω_d normalized to the density appropriate for galaxies; secondly, one needs the galaxies to cover the sky, which – for galaxies like our own – requires

$$1+z_G > 10 \left[\frac{R_G}{10kpc}\right]^{2/3} \left[\frac{\Omega_{GB}}{0.01}\right]^{-2/3} \qquad (4)$$

Here R_G is the radius and Ω_{GB} is the baryonic density associated with the galaxies.

It is not clear whether conditions (3) or (4) can be satisifed. One has no direct evidence for pregalactic dust (except perhaps from the NB distortion itself) but in any hierarchical clustering picture one would expect at least some pregalactic dust production. For example, one could easily envisage the dust produced by the first dwarf galaxies being blown out into intergalactic space because the gravitational potential of the dwarfs would be so small; the amount of dust required to satisfy condition (3) would only be tiny for large redshifts. Condition (4) is more worrisome but it should be noted that, in the cold dark matter (CDM) picture at least, the smaller galaxies would be expected to have a larger covering factor than the ones like our own. In fact, a detailed comparison with the data (BCH) shows that one must impose an even stronger condition than eqn (3). If the radiation is produced *continuously*, one needs

$$1+z_d > 30 \left[\frac{\Omega_d h}{10^{-5}}\right]^{-2/5} \qquad (5)$$

whereas, if it is produced in a *burst*, one needs

$$1+z_d > 60 \left[\frac{r_d}{0.1\mu}\right]^{-1} \tag{6}$$

Since one could hardly have a grain radius exceeding 0.1μ, the continuous scenario obviously makes the least stringent demands on z_d.

Let us first determine the dust density associated with galaxies. If we assume that the dust-to-gas ratio is half the metallicity, its density can be expressed as

$$\Omega_d \simeq 10^{-5} \left[\frac{Z}{0.02}\right] \left[\frac{f_g}{0.1}\right] \left[\frac{f_B}{0.1}\right] \left[\frac{\Omega_B}{0.1}\right] \tag{7}$$

Here Z is the metallicity (normalized to the solar value), f_g is the gas-to-baryon ratio, f_B is the fraction of baryons in bound objects, and Ω_B is the baryon density parameter (normalized to the value required by cosmological nucleosynthesis considerations). Both f_g and f_B are around 0.1 at present. Although f_g could have been larger in the past, permitting Ω_d to have been as large as 10^{-4}, this would require a considerable amount of subsequent dust destruction, so it seems more reasonable to take $\Omega_d \simeq 10^{-5}$ for galaxies. In this case, $\Omega_d \simeq 10^{-5}$ would also be an upper limit to the pregalactic dust abundance. Eqn (5) then requires that dust formation occurs at $z_d > 30$. This would already seem to be incompatible with the *pancake* picture, where the first objects do not form until $z<5$. In the *hierarchical clustering* picture, objects can form much earlier than this. However, in the CDM version the first bound objects form at $z \simeq 20/b$, where b is the bias factor (usually taken to be in the range 1.5-2.5). Thus the dust must derive from very rare high-σ objects and a detailed analysis shows that the value of f_B at $z \simeq 30$ is too small (BCH). Hence one is forced to some non-standard version of the hierarchical clustering picture, in which (for example) the spectrum of density fluctuations is less flat.

3. PROTOGALACTIC SOURCES

Perhaps the most conservative explanation for the NB excess would be to invoke the most prominent source of far-infrared radiation at the present epoch, the *infrared galaxies* discovered by IRAS. In this case, the considerations of Section 2 do not apply because the radiation is being absorbed by local dust rather than a cosmological background. The spectra of such galaxies is dominated by two components: (i) the emission from disks (i.e. starlight re-radiated by interstellar dust); and (ii) bursts of star formation heavily shrouded in dust. The first (cooler) component would be most naturally associated with the NB background, while the second might be associated with the 100μ background tentatively identified by Rowan-Robinson (1986). In principle, one can try to fit both backgrounds simultaneously. The background intensity at frequency ν can be written as

$$I_\nu \approx \overline{nP_\nu} \int \frac{P(\nu(1+z))}{P(\nu)} \, f(z) \, c \, dt \qquad (8)$$

Here $P(\nu)$ is the source spectrum (assumed fixed), $\overline{nP_\nu}$ is the luminosity density at frequency ν, and $f(z)$ is the luminosity evolution factor, assumed to have the form

$$f(z) = \exp\left[Q\left(1 - \frac{1}{1+z}\right)\right] \qquad (9)$$

The IRAS point counts imply $Q \lesssim 8$. An attempt to fit both the 700μ and 100μ backgrounds, taken from Rowan-Robinson & Carr (1988), is shown in Figure 2: with sufficiently strong evolution ($Q=8$), one can get the energy density required but the spectral fit is not very good. On the other hand, the 100μ background is less secure, so one may not need to be constrained by it. In this scenario, it would be natural to identify the IRAS galaxies as the source of the minimum metallicity associated with Population I stars.

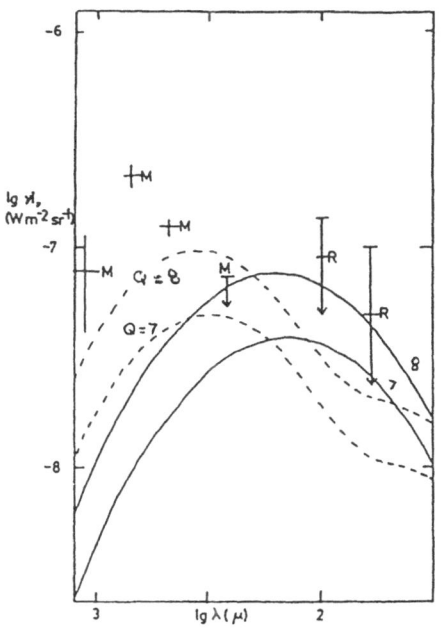

Figure 2. **Far infrared and submillimetre background. Data points labelled M are** from **Matsumoto et al. (1988).** Data **points labelled R are the IRAS isotropic component.** Solid **curves are predicted backgrounds for starburst component, broken curves for normal disc component, with luminosity evolution.**

4. PREGALACTIC SOURCES

The most natural pregalactic sources of the radiation which heats the dust grains would be *Population III stars* (Rowan-Robinson *et al.* 1979, McDowell 1986, Negroponte 1986). Such stars would be obligatory, for example, in the explosive or mock gravity theories of galaxy formation

(Hogan & White 1986). However, over most mass ranges, the stars which produce radiation would also produce metals. Only stars below 4 $M_\odot$ or above 200 $M_\odot$ may produce no metals at all. If we write the ratio of the metal yield to the radiation yield as R(M), we infer from eqn (1) that stars of mass M can produce the NB background only if they simultaneously generate a background metallicity

$$Z(M) \simeq 3 \times 10^{-3} \left[\frac{1+z_*}{10}\right] \left[\frac{\Omega_g h^2}{0.1}\right]^{-1} \left[\frac{R(M)}{250}\right] \quad (4M_\odot < M < 200M_\odot) \qquad (10)$$

Here z_* is the redshift at which the stars produce their light, which – from the considerations above – must exceed 10, Ω_g is the density of the gas which remains after pregalactic star formation, and R(M) is normalized to the value at which it is minimized (M≈10 $M_\odot$). It is clear that Z(M) necessarily exceeeds the metallicity associated with the poorest Population I stars (about 10^{-3}) throughout the range 4-200 $M_\odot$. On the other hand, stars below 4 $M_\odot$ have a main-sequence time of about $10^9 y$ and so necessarily burn out after z=10. We are therefore forced to invoke VMOs, i.e. stars larger than 200 $M_\odot$ which collapse entirely to black holes during their oxygen-burning phase.

Although VMOs do not seem to be forming prolifically at the present epoch, they may have done so at pregalactic eras. Indeed the black hole remnants of VMOs have already been invoked as candidates for the dark matter in galactic halos (Carr et al. 1984). In this case, the VMO density parameter Ω_* must be about 0.1 and they generate a radiation density

$$\Omega_R \simeq 4 \times 10^{-6} \left[\frac{1+z_*}{100}\right]^{-1} \left[\frac{\Omega_*}{0.1}\right] \qquad (11)$$

Thus we get exactly the density required providing $z_* \simeq 100$. This is rather too early for the CDM picture but it would be compatible with a more general form of the hierarchical clustering scenario (and we have seen that this may be required anyway). Lacey & Field (1988) have also concluded that one needs stars with an unconventional mass spectrum to produce the NR excess: using a more precise cosmological nucleosynthesis limit ($\Omega_g h^2 < 0.022$) and assuming a Scalo mass function for the stars, they show that Ω_R is too low by a factor of 5-8. It is interesting that pregalactic VMOs may be able to reionize the intergalactic medium (Carr et al. 1984) and thus solve the problem addressed by Shapiro (1989).

Another possible energy source is *black hole accretion*. Black holes of density Ω_{BH}, accreting at the Eddington limit for a Salpeter mass-doubling time and generating radiation with efficiency ϵ, would produce a radiation density (BCH)

$$\Omega_R \simeq 3 \times 10^{-4} \left[\frac{\epsilon}{0.1}\right]^{5/3} \left[\frac{\Omega_{BH}}{0.1}\right] h^{2/3} \qquad (12)$$

For example, if each galactic nucleus contains a $10^8 M_\odot$ black hole, then Ω_{BH} would be about 10^{-5}. However, in this case Ω_R would be too small even if ϵ were close to 1. Black holes in galactic halos would have a larger Ω_{BH} but they would be unlikely to accrete at the Eddington limit for a mass-doubling time (Carr 1981). If they accreted at the Bondi rate from a medium with the cosmological gas density Ω_g and temperature T at redshift z_*, they would generate a radiation density

$$\Omega_R \simeq 7 \times 10^{-7} h \left[\frac{\Omega_g}{0.1}\right] \left[\frac{\Omega_{BH}}{0.1}\right] \left[\frac{\epsilon}{0.1}\right] \left[\frac{M}{10^6 M_\odot}\right] \left[\frac{T}{10^4 K}\right]^{-3/2} \left[\frac{1+z_*}{10}\right]^{1/2} \quad (13)$$

Here we have normalized the black hole mass M to the maximum value consistent with dynamical constraints but, even in this case, Ω_R is too small unless ϵ is close to 1.

Another exotic explanation would be *decaying elementary particles* (Kawasaki & Sato 1986). If the particles decay radiatively at a redshift z_X and have mass m_X, then they should produce a background with density and peak wavelength

$$\Omega_R \simeq 5 \times 10^{-6} \Omega_X \left[\frac{1+z_X}{10^5}\right]^{-1}, \qquad \lambda_{peak} \simeq 120 \left[\frac{1+z_X}{10^5}\right] \left[\frac{m_X}{keV}\right]^{-1} \mu \quad (14)$$

Here Ω_X is the density which would be associated with the particles had they not decayed. We see that Ω_R has the required value if $\Omega_X \simeq (1+z_X)/10^5$. In order for the decays to generate the NB background directly, we would need to fine-tune m_X, but that would not be necessary if the radiation were reprocessed by dust.

REFERENCES

Bond,J.R., Carr,B.J. & Hogan,C.J., *Astrophys.J.* 306, 428 (1986).
Bond,J.R., Carr,B.J. & Hogan,C.J., preprint (1988).
Carr,B.J., *Mon.Not.R.Astr.Soc.* 194, 639 (1981).
Carr,B.J., Bond,J.R. & Arnett,W.D., *Astrophys.J.* 277, 445 (1984).
Guilbert,P.W. & Fabian,A.C., *Mon.Not.R.Astr.Soc.* 220, 439 (1986).
Hayakawa,S., Matsumoto,T., Matsuo,H., Murakami,H., Sato,S., Lange,A.E. & Richards,P.L., *Pub.Astr.Soc.Japan* 39, 941 (1987).
Hogan,C.J. & White,S.D.M., *Nature* 321, 575 (1986).
Kawasaki,M. & Sato,K., *Phys.Lett.B.* 169, 280 (1986).
Lacey,C.G. & Field,G.B., *Astrophys.J.Lett.* 330, L1 (1988).
Matsumoto,T., Hayakawa,S., Matsuo,H., Murakami,H., Sato, S., Lange,A.E. & Richards,P.L. *Astrophys.J.* 329, 567 (1988).
McDowell,J.C., *Mon.Not.R.Astr.Soc.* 223, 763 (1986).
Negroponte,J., *Mon.Not.R.Astr.Soc.* 222, 19 (1986).
Ostriker,J., Thompson,C. & Witten,E., *Phys.Lett.B.* 180, 231 (1986).
Rowan-Robinson,M., *Mon.Not.R.Astr.Soc.* 219, 737 (1986).
Rowan-Robinson,M. & Carr,B.J., in *The Post-Recombination Universe*, ed. A.Lasenby & N.Kaiser (Reidel, Dordrecht) (1988).
Rowan-Robinson,M., Negroponte,J. & Silk,J., *Nature* 281, 635 (1979).
Shapiro, P., this volume (1989).
Yoshioka,S. & Ikeuchi,S., *Astrophys.J.Lett.* 323, L7 (1987).

DISCUSSION:

SILK: There is an interesting consequence if the halo consists of VMO remnants. Any initial mass functions weighted towards VMO's would plausibly be expected to contain some stars of mass between 2 and 10 $M_\odot$. Their white dwarf remnants might be observable if existing proper motion surveys, which detect the onset of disk formation at the faint end of the white dwarf luminosity function could be extended by several magnitudes.

CARR: The number of white dwarfs expected depends on the mass spectrum of the stars. Only a small fraction of the mass can be in stars below about 200 $M_\odot$ (the mass above which one gets collapse to a black hole) else one would overproduce metals. One would not necessarily have any white dwarfs, but I agree that there would be interesting consequences if there were.

FIELD: My understanding is that small objects form first in the CDM scenario. Your proposal to produce the submillimetre excess with VMO's would be natural if VMO's form at z > 30. Is that to be expected in the CDM scenario?

CARR: When the first subgalactic objects form in the CDM picture depends on the bias factor b. For b $\simeq$ 1.5, only about 1% of the gas can have gone into bound objects by z $\simeq$ 30. Even if these objects fragment entirely into VMOs that is not quite enough to generate the submillimetre excess. One would have enough VMOs by z $\simeq$ 15 but the dust density would have to be rather large to absorb the VMO radiation then. Of course, these conclusions only apply if one adopts the specific spectrum of density fluctuations assumed in the CDM picture. There would be no problem in producing enough VMO's by z $\simeq$ 30 in a more general hierarchical clustering model.

FALL: It looks to me as if the cosmological density of dust you adopt, $\Omega_{dust} \sim 10^{-5}$, is very close to the current observational limits. At redshifts of 2-3, the density of neutral hydrogen in the damped Lyα systems in the Wolfe *et al* survey is $\Omega_{HI} \simeq 10^{-3}$. Pei and I find $(D/G)_{DLy\alpha} < 1/2\ (D/G)_{Gal}$ at the 95% confidence level, which gives $\Omega_{dust} \leq 10^{-5}$. There are some, less direct, limits on the dust-to-gas ratio in a few damped Lyα systems that would imply even lower limits on dust.

CARR: Wolfe's neutral hydrogen density is somewhat higher than you indicate, $\Omega_{HI} \simeq 0.002$ h^{-1}. If one combines that with the limit of yourself and Pei, and takes $(D/G)_{Gal} \simeq 0.01$, one gets $\Omega_D \simeq 10^{-5}\ h^{-1}$. As you say, that is close to the value we use. Stronger limits on Ω_D would require the dust to form at an even higher redshift. However, if one permits dust destruction, Ω_D could be as large as 10^{-4} for some period and one would then only require the dust to form at a redshift of 10.

SILK: There is an interesting consequence if the halo consists of VMO remnants. Any initial mass function weighted towards VMOs would plausibly be expected to contain some stars of mass between 2 and 10 $M_\odot$. Their white dwarf remnants might be observable if existing proper motion surveys, which detect the onset of disk formation at the faint end of the white dwarf luminosity function, could be extended by several magnitudes.

STAR FORMATION AND THE X-RAY BACKGROUND

R. E. GRIFFITHS
Space Telescope Science Institute
3700 San Martin Drive,
Baltimore MD 21218 USA

ABSTRACT. The potential contribution of massive x-ray binaries to the 3-20 kev X-ray background has largely been ignored heretofore. It is pointed out that such objects, in the wake of star formation, can contribute at least as much as quasars to the x-ray background, while solving the problems of spectral shape and sky surface density.

1. Introduction

It has been established observationally that at least 30% of the all-sky x-ray background (XRB) in the energy range 1-3 kev comes from active galactic nuclei in the redshift interval 0.4 to 1.2 (Griffiths et al., 1983, 1988), and it is quite plausible that AGN contribute most of the XRB in the 1-3 kev range (Morisawa and Takahara 1988). However, analysis of spatial fluctuations in the Einstein deep survey counts (Hamilton and Helfand, 1987; Barcons and Fabian, 1988) has indicated the presence of a relatively smooth component of the XRB with a corresponding surface density of discrete sources of at least several thousand per square degree.

Besides accounting for the smoothness of the XRB on both small and large scales, the greatest problem with the AGN interpretation of the whole of the XRB is the discrepancy between the XRB spectrum and that of nearby quasars and Seyfert galaxies. Schwartz and Tucker (1988), however, have shown that present data on AGN spectra are consistent with the XRB originating from quasars at redshifts of 1 to 2. These authors have shown how plausible quasar spectra can fit that of the x-ray background over the energy range 1 to 30 kev. To achieve the same end, Morisawa and Takahara (1988) invoke evolution of the AGN spectra at high redhsift to explain the XRB spectrum, finding, of course, that flat spectra are needed above redshifts of 2. Although there is evidence that the optical/x-ray flux ratio α_{ox} evolves, it is found to change with optical luminosity, and not with redshift (Avni 1987). There is therefore no evidence for evolution of the overall optical/x-ray spectra of quasars, or for the x-ray spectra per se, and the model of Morisawa and Takahara must again be considered an *ad hoc* presumption about AGN spectra. The arc-minute scale fluctuations observed by Hamilton and Helfand, however, are shown to be consistent with the expected number of AGN below the detection limit of the Einstein deep surveys. These AGN would presumably have optical magnitudes in the range 22 to 30, which may well number thousands per square degree.

Alternative sources of the XRB have been discussed in recent years: Daly (1987)

235

C. S. Frenk et al. (eds.), The Epoch of Galaxy Formation, 235–241.
© *1989 by Kluwer Academic Publishers.*

has proposed massive, hot condensates in the early universe, requiring vast energy input; Guilbert and Fabian (1986) have supported the hypothesis that the emission is bremsstrahlung from hot, intergalactic gas; and Leiter and Boldt (1982) proposed "precursor" quasars with flat x-ray spectra, for which there is no observational evidence thus far. Clusters of galaxies probably contribute at the level of at least 5–10%, or more according to Schaeffer and Silk (1988), but it is difficult for the predicted temperatures to account for the observed hard spectrum. Each of these alternative scenarios thus has severe observational or theoretical difficulties.

We consider here an alternative or additional source for the all-sky XRB, especially in the 3–20 kev energy range above that of the Einstein observatory, *viz.* the integrated x-ray emission from an early population of low-metallicity massive x-ray binaries (MXRB) in regions of star formation. Such sources are typified by SMC X-1, with less luminous examples in our own Galaxy. This possibility was first mentioned by Bookbinder *et al.* (1979). Earlier estimates of the contribution of "normal" galaxies to the XRB were at < 1% (Rowan-Robinson and Fabian 1975), unless the x-ray sources in normal galaxies evolve in the same way as radio sources, in which case they would acccount for most of it (Silk 1968, when Population I x-ray sources were unknown). Van Paradijs (1978) considered the contribution of "normal" galaxies to the XRB, setting a "firm" upper limit at the level of 25%, unless massive halos are present around galaxies, and their formation was accompanied by the production of MXRB. The assumptions made by van Paradijs and others pre-dated the observations of the Einstein Observatory and those of the IRAS all-sky survey. We therefore return to the question of the massive binaries, and make revised calculations based on more recent observations.

2. Population I MXRB in Low-Metallicity Systems

The Population I x-ray binaries were established as a class of x-ray emitters in the late 1960s and 1970s (see Rappaport and Joss, 1983, for a summary of their x-ray properties). MXRB have primaries of ~ 20 M_o, with accreting secondaries of 1 - 7 M_o. Their orbital periods are typically a few days, and their optical properties have been summarised by van Paradijs (1983). They start emitting x-rays about 10^7 years after the formation of the initial massive binary, and the epoch of x-ray emission is rather uncertain, but probably lasts for at least 10^5 to 10^6 years (see van den Heuvel 1983 for a review of the evolution of such systems). The presently active and known MXRB in our Galaxy number about 20 (van den Heuvel, 1983) with x-ray luminosities of $\sim 10^{38}$ ergs s^{-1} and typical x-ray temperatures >15 kev, spectra which make them potential contributors to the all-sky background, provided that the low-metallicity counterparts of such systems are sufficiently common in regions of star-formation at redshifts not exceeding ~ 1.

Our Galaxy does not contain any low-metallicity MXRB, so that sources like Cyg X-1 and Cen X-3 may be under-luminous examples of those to be found in star-forming galaxies. We have to look at least as far as the Local Group, where we find SMC X-1 and LMC X-4. The low metallicity of the SMC has been taken as the reason for the extraordinary x-ray output of the SMC - rivalling the total x-ray output of the Galaxy, with only one-tenth the mass (Clark *et al.* 1978). The O-stars in the SMC are also more luminous than O-stars in the Galaxy. Generally, low metal content means lower x-ray opacity in the accreting material, driving up the accretion rate and enhancing the production of hard x-rays. The high-energy tail of binaries like SMC X-1 may be especially important in considerations of the XRB.

Nearby galaxies include examples of super-luminous sources in M82 (Watson, Stanger and Griffiths 1984), M101 (Palumbo et al. 1981), and possibly NGC 253 (Fabbiano and Trinchieri, 1984). Long and Van Speybroeck (1983) drew attention to the compact binaries in "normal" galaxies, where it is not unusual to find individual sources with luminosities in excess of 10^{39} ergs s^{-1}—M101 has three such sources, M100 has a source emitting in excess of 10^{40} and M82 likewise has a source in an outlying arm with $L_x > 10^{40}$ ergs s^{-1}. Although not identified with massive x-ray binaries (note that this can be done with the HST in some cases), we make the plausible assumption here that such sources are members of the class typified by SMC X-1.

The lifetime of x-ray emission is probably enhanced in metal-poor MXRB: the supergiant branch on the Hayashi track is shifted to the blue, and the low metallicity means that core helium burning will occur on the blue side of the Hertzsprung gap; a longer time will be spent there and the accreting wind may also be stronger.

It is therefore interesting to ask what fraction of galaxies and associated low-metallicity dwarves would need to have, say, on average, 100–1000 low-metallicity MXRB emitting at 10^{39-40} ergs s^{-1} each, in order to contribute, say, 50% of the observed XRB at 5 kev? Integrating to $z = 1$, we find the answer to be roughly 10%.

The x-ray spectra of the MXRB are typically flat, i.e., they can be characterised by bremsstrahlung temperatures in excess of 10^8 K, or kt > 15 kev. Examples are reviewed in Rappaport and Joss (1983), where the ~ 20 kev cut-off is apparent for many, though not all, of the sources. If these sources are typical of those contributing to the XRB at 3–20 kev, then their average redshift cannot be much greater than ~ 1, with the dominant contribution coming from $z < 1$. Hard x-ray tails are by no means uncommon, however, amongst these sources.

In the absence of detailed estimates of the SFR, the IMF and the fraction of MXRB, the contribution, or limit thereon, of the massive binaries to the XRB can be estimated in the following ways:

3. MXRB and the X-Ray Background

3.1. PRELIMINARY ESTIMATES

Bookbinder et al. (1979) obtained a limit on the contribution of MXRB in star-forming regions by considering the x-ray to optical luminosity ratio of the whole SMC, $\sim 10^{-3}$. When normalised to the optical energy density in inter-galactic space, this resulted in a limit of $10^{-5.4}$ ev cm^{-3} for the energy density of x-rays from MXRB in systems like SMC X-1. The upper limit on this argument is provided by the ratio of x-ray to optical luminosities in a low-metallicity MXRB, which is typically of order 1–10, leading to a limiting energy density of $10^{-2.4}$ to $10^{-3.4}$ ev cm^{-3}, i.e., greatly exceeding the XRB energy density. This limit assumes that MXRB are extremely common, however, and scales down with the mass fraction of MXRB as a function of total mass in OB stars. If the latter fraction is only $\sim 1\%$, however, the contribution to the XRB is still significant.

Bookbinder *et al.* actually favored the x-ray emission from supernova-driven winds in star-forming regions: an example of this may have been observed in the diffuse emission observed along the minor axis of M82 (Watson, Stanger and Griffiths 1984), but the x-ray temperature of such gas is probably too low to explain the observed background spectrum. Note that the arguments of Giacconi and Zamorani (1987) also rule out a major contribution from optically thin sources of this kind, and instead support emission from optically-thick accretion disks, as in the MXRB.

3.2. INFRA-RED STARBURST GALAXIES

Young, Kleinmann and Allen (1988) have found evidence that infrared starburst galaxies are powered by massive, young stars in heavily obscured H II regions. Weedman (1986) constructed a luminosity function for starburst galaxies, based on the 60 μm IRAS counts, and then used a ratio of infrared to x-ray luminosity from a dozen objects which were observed both with IRAS and Einstein. This ratio was found to be $2.7 \, 10^{-8}$ from the sample of galaxies in Fabbiano, Feigelson and Zamorani (1981), chosen for their properties of unusual features generally arising from interactions (abnormality). Without invoking any evolution of starburst galaxies, Weedman found a contribution of at least 13% to the XRB at 2 kev. With moderate evolution proportional to $(1+z)^2$, he pointed out that the contribution to the XRB would be doubled. Furthermore, we note that the x-ray fluxes measured at 2 kev were uncorrected for absorption within the emitting galaxies, and the extrapolated fluxes and contribution above 3 kev would therefore be higher, on the assumption of MXRB spectra typical of those in the Local Group.

We note that the very faint radio source counterparts of Windhorst *et al.* (1985) have been explained as the emission from starburst/interacting galaxies, and the evolution of such objects may have an associated time constant of 25% of the Hubble time (Danese *et al.* 1987). Identifying the blue radio galaxies with star-forming galaxies and therefore with infra-red emission and associated x-ray emission from MXRB, the corresponding contribution to the 3–20 kev XRB could be at the level of about 50%.

More rapid evolution of the number counts of starbursting galaxies would actually produce x-ray emission exceeding the observed flux, and such evolution is also ruled out on the arguments presented in section (i) above, *i.e.*, the MXRB are optically luminous, with primaries radiating at $\sim$half the Eddington limit, where the primaries are about ten times as massive as the accreting compact x-ray sources.

It has also become apparent that nuclear activity may be related to starburst activity (Balzano, 1983). Ward (1988) has shown a correlation between x-ray luminosity and Brackett γ emission in starburst nuclei, supporting the hypothesis of x-ray emission arising from MXRB in the nuclear regions. Again, it would not be unreasonable for the evolution of such starburst nuclei to evolve as rapidly as quasars themselves.

3.3. H II (BLUE) GALAXIES

Ellis (1987) and others have observed excess blue emission in field galaxies at redshifts as low as 0.4, indicating the evolution of star-formation in galaxies at very moderate redshifts. The number of MXRB in these galaxies is expected to follow the size of the HII regions, or the number of O-stars. Such blue galaxies may be expected to have integrated x-ray emission of about 10^{41-42} ergs s^{-1}, and to number perhaps 20–30% of all galaxies at those redshifts. These objects may have properties similar to the HII galaxies described by Terlevich (1987).

Another way of estimating the MXRB contribution is to start with the luminosity function of line-emitting galaxies from the CfA redshift survey (Burg, 1987), and to make a similar estimate to that above, based on the observed x-ray to optical luminosity ratio of emission-line galaxies. A few examples of relevant objects are contained, in fact, in the Einstein deep surveys. Objects in U Min and in Pavo at $z \sim 0.4$ may be representative of this class, and may be distant examples of some of the narrow emission line galaxies found in the all-sky surveys of HEAO-1 and Ariel V, etc (Griffiths et al. 1979; Schnopper et al. 1978), although the latter were dominated by nuclear emission. The result of folding Burg's LF with the observed ratio of l_x/l_{opt} is that the contribution to the XRB is $\sim 30\%$ at 2 kev. With hard x-ray spectra, this fraction would be maintained or exceeded in the 3-20 kev range.

4. Conclusions and Constraints on Star-Formation

Star-forming galaxies may contribute substantially to the x-ray background in the 3–20 kev range, contributing at least $\sim 20\%$ and, plausibly $\sim 50\%$. The first examples of these sources may already have been found in the Einstein deep surveys (Griffiths 1988). The dominant contributiuon must then arise from relatively low redshifts, i.e., at $z < 1$, but depending on the high-energy tail of the MXRB. This interpretation of the XRB therefore supports active star formation at relatively recent epochs.

A difficulty with the prediction of x-ray emission from MXRB in star-forming systems is that much of the activity may take place in dwarf galaxies or satellite galaxies, by direct analogy with our own Galaxy and the MXRB in the LMC/SMC; or in galaxies of low surface brightness. In this context, we note that York et al (1986) have invoked a population of star-forming dwarf galaxies to explain QSO absorption line systems in intermediate and high-redshift galaxies.

Some starburst activity is also related to nuclear activity (Wilson et al. 1988), so that starburst/active nuclei, with established evolution, also contribute to the 3–20 kev XRB.

Barcons and Fabian (1987), and Bagoly, Meszaros and Meszaros (1988) have considered the constraints on possible contributors to the XRB that can be derived by considering the clustering properties of putative classes of sources, such as quasars. Whereas the contribution of quasars can in fact be limited to the level of about 20%, because of the observational evidence that the clustering of quasars is similar to that of clusters of galaxies, the same constraint does not hold for the star-forming regions of galaxies in general, and especially not to dwarf, irregular galaxies like the SMC. As Weedman (1986) has pointed out, the surface density of star-forming galaxies is also such that they easily satisfy the limits of fluctuations in the all-sky survey counts.

A prediction of the starburst origin of the XRB is, of course, that the starburst galaxies have hard x-ray spectra, similar to those of MXRB in the Local Group.

Verification of this will have to await AXAF, XMM, and other missions reaching x-ray energies of 8–10 kev. This hypothesis would also predict a possible correlation between the fluctuations in the Einstein deep survey counts and deep radio source counts.

5. References

Avni, Y., 1986, *Ann. New York Acad. Sci.*, **470**, 71.

Bagoly, Z., Meszaros, A. and Meszaros, P., 1988, preprint.

Balzano, V. A., 1983, *Ap. J.*, **268**, 602.

Barcons, X., and Fabian, A. C., 1988, *M.N.R.A.S.*, in press.

Barcons, X. and Fabian, A. C., 1989, *M.N.R.A.S.*, in press.

Bookbinder, J., Cowie, L. L., Krolik, J. H., Ostriker, J. P., and Rees, M. J., 1979, *Ap. J.*, **237**, 647.

Burg, R., 1987, Ph.D. Thesis, MIT.

Clark, G., Doxsey, R., Li, F., Jernigan, J. G., and van Paradijs, J., 1978, *Ap. J.*, **221**, L37.

Daly, R. A., 1987, *Ap. J.*, **322**, 20.

Danese, L., de Zotti, G., Franceschini, A., and Toffolati, L., 1987, *Ap. J.*, **318**, L15.

Fabian, A. C., 1988, in "The Post-Recombination Universe", NATO ASI Series C, vol. 240, eds. N. Kaiser and A. N. Lasenby, p. 51.

Fabbiano, G. and Trinchieri, G., 1984, *Ap. J.*, **286**, 491.

Fabbiano, G., Feigelson, E., and Zamorani, G., 1982, *Ap. J.*, **256**, 397.

Giacconi, R. and Zamorani, G., 1987, *Ap. J.*, **313**, 20.

Griffiths, R. E., Doxsey, R. E., Johnston, M. D., Schwartz, D. A., Schwarz, J., and Blades, J. C., 1979, *Ap. J.*, **230**, L21.

Griffiths, R. E., *et al.* 1983, *Ap. J.*, **269**, 375.

Griffiths, R. E., Tuohy, I. R., Brissenden, R. J. V., Ward, M. J., Murray, S. S., and Burg, R., 1988, in "The Post-Recombination Universe", NATO ASI Series C, vol. 240, eds. N. Kaiser and A. N. Lasenby, p. 91.

Guilbert, P. W., and Fabian, A. C., 1986, *M.N.R.A.S.*, **220**, 439.

Hamilton, T. T. and Helfand, D. J., 1987, *Ap. J.*, **318**, 93.

Ellis, R., 1987, Proc. IAU Symposium No. 124, "Observational Cosmology", eds. G. Burbidge and L. Z. Fang, p. 367.

Leiter, D., and Boldt, E., 1982, *Ap. J.*, **260**, 1.

Long, K. S. and van Speybroeck, L. P., 1983, in "Accretion-driven Stellar X-ray Sources", eds. W. H. G. Lewin and E. P. J. van den heuvel, p. 117.

Morisawa, K. and Takahara, F., 1989, *PASJ*, in press.

Palumbo, G. G. C., Maccacaro, T., Panagia, N. G., Vettolani, G., and Zamorani, G., 1981, *Ap. J.*, **247**, 484.

Persic, M., de Zotti, G., Danese, L., Palumbo, G., Franceschini, A., Boldt, E. A., and Marshall, F. E., 1988, preprint.

Rappaport, S. A. and Joss, P. C., 1983, in "Accretion-Driven Stellar X-ray Sources", eds. W. H. G. Lewin and E. P. J. van den Heuvel, p. 1.

Rowan-Robinson and Fabian, A. C., 1975, *M.N.R.A.S.*, **170**, 199.

Savonije, G. J., 1983, in "Accretion-driven Stellar X-ray Sources", eds. W. H. G. Lewin and E. P. J. van den heuvel, p. 343.

Schaeffer, R. and Silk, J., 1988, preprint.

Schnopper, H. W., Davis, M., Delvaille, J. P., Geller, M. J., and Huchra, J. P., 1978, *Nature*, **275**, 719.

Schwartz, D. A. and Tucker, W. H., 1988, *Ap. J.*, **332**, 157.

Silk, J., 1968, *Ap. J.*, **151**, L19.

Silk, J., 1987, in "Star Forming Regions", IAU Symp. 115, eds. M. Peimbert and J. Jugaku, p. 663.

Stewart, G. C., Fabian, A. C., Terlevich, R. J., and Hazard, C., 1982, *M.N.R.A.S.*, **200**, 61.

Terlevich, R., 1988, in "The Post-Recombination Universe", eds. N. Kaiser and A. N. Lasenby, NATO ASI Series C, vol. 240, p. 69.

van den Heuvel, E. P. J., 1983, in "Accretion-driven Stellar X-ray Sources", eds. W. H. G. Lewin and E. P. J. van den heuvel, p. 303.

van Paradijs, J., 1978, *Ap. J.*, **226**, 586.

van Paradijs, J., 1983, in "Accretion-driven Stellar X-ray Sources", eds. W. H. G. Lewin and E. P. J. van den Heuvel, p. 189.

Ward, M. J., 1988, *M.N.R.A.S.*, **231**, 1.

Watson, M. G. W., Stanger, V. and Griffiths, R. E., 1983, *Ap. J.*, **286**, 144.

Weedman, D. W., 1986, in "Star Formation in Galaxies", NASA CP-2466, ed. Carol J. Lonsdale, p. 351.

Wilson, A. S., 1988, *Astron. and Astrophys.*, in press.

Windhorst, R. A., Miley, G. K., Owen, F. N., Kron, R. G., and Koo, D. C., 1985, *Ap. J.*, **289**, 494.

York, D. C., Dopita, M., Green, R. and Bechtold, J., 1986, *Ap. J.*, **311**, 610.

Young, J. S., Kleinmann, S. G., and Allen, L. E., 1988, *Ap. J.*, **334**, L63.

Discussion:

Field:

Since the background has a temperature of 30 kev, don't you agree that it is impossible to compare the background with 30 kev sources of any redshift other than 0?

Griffiths:

After subtracting estimated contributions from AGN, etc., the spectrum no longer fits a 30 kev bremsstrahlung, but is, in fact, even harder. Some massive x-ray binaries have the requisite spectra, and could also tolerate being redshifted, but not to more than $z \sim 1$. MXRB probably do not contribute much to the XRB above ~ 20 kev.

Galaxy Formation and Biased Clustering

Nick Kaiser
Canadian Institute for Theoretical Astrophysics,
McLennan Physical Laboratories, Toronto, Ontario, M5S 1A1

Shaun Cole
Institute of Astronomy, Madingley Road, Cambridge, CB3 0HA

ABSTRACT.
We have used the Press-Schechter approximation to calculate the abundance and large-scale clustering of dark halos in the CDM cosmogony, and have explored two simple models for galaxy formation. In the first, which seems attractive for disk galaxies, we identify galaxies with dark halos forming at the present epoch. We find that halos with abundance like that of L_* galaxies have large scale clustering which is nearly unbiased, and the smaller halos are antibiased—quite a negative feature. In the second, the galaxies we see today are assumed to be a fossil remnant of halos which formed earlier. The stellar and dark matter velocity dispersions are assumed to be equal, and the luminosity for each halo is adjusted to give a tight and universal $L - V$ relation like that observed. We identify an epoch when the comoving number density of halos with a given luminosity is maximised, and we calculate how this number density is biased by long wavelength modes. The results here are more encouraging: All of the 'galaxies' in this model are positively biased. The enhancement of the light-to-mass ratios for rich clusters is substantial, though dependent on the normalisation. For a low normalisation the bias is strong enough to reconcile virial estimates of M/L for clusters with $\Omega = 1$.

1. Introduction

In this paper we use the approximation of Press and Schechter (1974) for the number density of dark halos as a function of mass and redshift. While approximating the halos as a two parameter (M, z) family is a major oversimplification, these parameters are sufficient to determine gross properties such as binding energies, pressures, ages etc. which we suspect are important in determining which halos contain luminous objects.

We are primarily interested in the (biased) clustering of the halos on large scales. To this end we use the 'peak-background split' approximation (Bardeen *et al.* (1986)), that is we approximate the effect of long wavelength perturbations as simply a modulation of the halo collapse times. This leads to a bias in the mass distribution function $n(M, a) \equiv \partial N(> M, a)/\partial M$, where $N(> M, a)$ is the Lagrangian number density of halos as a function of mass M and expansion factor a:

$$n'(M, a) = n(M, a') \quad \text{with} \quad a' = a(1 - \Delta_B/\Delta_{\text{obj}})). \qquad 1.1$$

243

C. S. Frenk et al. (eds.), The Epoch of Galaxy Formation, 243–256.

Here Δ_{obj} is the initial density contrast of the objects and Δ_B is the amplitude of the background perturbation. In this manner, we can calculate the linear response to a linear background perturbation Δ_B, and, for objects which have large bias, we can also calculate the non-linear response, while still remaining in the regime where $\Delta_B \lesssim 1$. This can be used to estimate the enhancement of the number of galaxies per unit mass in rich clusters, with which is of interest to compare the observed cluster light-to-mass ratios.

While it is fairly straightforward to calculate the instantaneous clustering bias for dark matter haloes, the application to real galaxies is much less clear. Naively, one might simply explore the M, z plane, find objects with number densities and masses (or velocity dispersions) commensurate with real galaxies and read off the bias factor or estimate the enhancement in the number density of galaxies per unit mass in clusters etc.. There are good reasons, however, to be suspicious of such results. One can certainly identify objects whose abundance would have been quite strongly enhanced in overdense regions at some epoch, but only because the analogous objects in underdense regions were due to collapse slightly later. It is therefore possible to contrive a 'theory for galaxy formation' which gives a substantial bias either by invoking some hypothetical feedback to switch off galaxy formation, or by postulating that the efficiency of star formation is a step function of the halo properties, with the position of the transition finely tuned to coincide with the edge of the Gaussian distribution. Here we wish to consider less radical and perhaps more plausible solutions to the problem.

A useful way to illustrate the problem we are faced with here is to consider the space time diagram for the virialised haloes (selected at a given density contrast): for a two dimensional space, this would resemble a forest of trees. At the uppermost levels (high redshift) we have thin branches representing sub-galactic objects. These merge together into progressively larger objects, and the clusters at the present epoch would be the thickest trunks at ground level. A theory for galaxy formation would tell us how to assign luminosity to these branches. We will explore two alternative models for galaxy formation: In the first we calculate the bias that would arise if galaxy formation is a steady ongoing process, and if, as suggested by numerical experiments, the properties of even the dense inner parts of halos are determined by the binding energy of the most recently collapsed material. This might be appropriate for the formation of disk galaxies, and in this case one can apply the instantaneous bias results fairly directly.

The alternative we will explore is to assume that the process of galaxy formation has largely terminated (and so this may perhaps be relevant to the formation of ellipticals and spheroids) and requires somewhat different calculational techniques. We will assume that some stars form in all of the branches of the tree according to some formula $L(M, z)$, and that the present day galaxies are fossil remnants of this earlier star forming activity. For this type of model one will obtain a positive bias if $\partial L/\partial z > 0$. We shall argue that if the stellar and dark matter velocity dispersions are equal then the small scatter and apparent universality of the $L - V$ relation do indeed imply a positive bias for galaxies in this model and we make quantitative predictions for the bias as a function of the number density of galaxies. While this model is empirically motivated and is relatively 'natural' in that no sharp threshold for galaxy formation is assumed, the physical cause of the bias remains uncertain. More unsettling still is the way that the inferred $L(M, z)$ function seems contrived to compress the intrinsically broad $M - V$ relation into a much narrower $L - V$ relation. Some possible modifications and possible physical basis are discussed.

2. Mass Spectrum and Instantaneous Bias Factor

The Press and Schechter (1974) approximation is

$$n(M, z, t)\, dM = \frac{2\rho_0}{\sqrt{2\pi}}\, \frac{t(1+z)}{M}\, \left(\frac{-1}{\sigma^2}\, \frac{\partial \sigma}{\partial M}\right)\, \exp\left(-\frac{t^2(1+z)^2}{2\sigma^2}\right)\, dM, \qquad (2.1)$$

where $\sigma(M)$ is the rms density fluctuation of the initial Gaussian density field when filtered on a mass scale M, and t is the threshold linear theory overdensity at which regions turnaround and collapse. We adopt $t = 1.68$ as applicable for simple spherical collapse, and use $\sigma(M)$ for CDM as calculated by Bond and Efstathiou (1984). The only free parameters are the normalisation, which is conventionally specified as σ_8, the linear theory rms mass fluctuation in spheres of radius $8h^{-1}$Mpc, and the Hubble parameter $H = 100\,h$ km s^{-1}Mpc^{-1} which determines the fundamental length scale in the problem. We shall use the spectrum appropriate for h = 0.5.

This mass function may be derived from a very simple picture for the formation of non-linear condensations: Regions which are initially overdense by a fraction $(\Delta\rho/\rho)_i$ turn around and form virialised objects with a final density contrast $(\Delta\rho/\rho)_f \simeq 200$ after the universe has expanded by a factor $(a_f/a_i) \simeq 1.68/(\Delta\rho/\rho)_i$. This approximation enables the final distribution of non-linear objects to be calculated from the statistics of the initial Gaussian density field. It is remarkable that this approximation derived from consideration of spherically symmetric perturbations seems to describe the N-body results quite well; $\Delta\rho/\rho \simeq 200$ does indeed seem to delineate the transition from infall to quasi-equilibrium, and the mass spectrum seems to agree well with results of group finding algorithms applied to numerical experiments (Efstathiou et al. 1988).

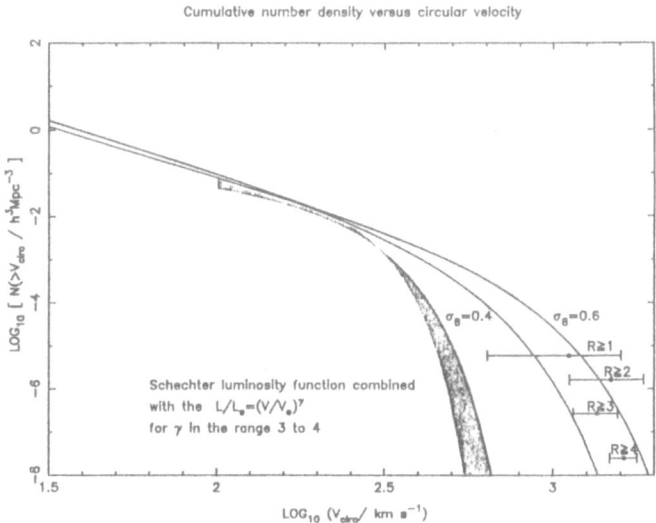

Figure 1. A comparison of the present day spectrum of dark matter halos as predicted

by the Press Schechter formalism, for normalisations of $\sigma_8 = 0.4$ and 0.6, with the Abell cluster and galaxy data.

In figure 1 we compare the predictions of this mass function with observations as follows: First we convert the mass distribution to a distribution in rotation velocity, as this is more directly observable. Then for Abell clusters we have determined $N(> V)$ using the Struble Rood (1987) compilation, and for galaxies we have converted the luminosity function to $N(> V)$ using the empirical Tully Fisher relation.

Comparing these distributions we see that with a normalisation in the range $\sigma_8 = 0.4 - 0.6$ the number densities of DM halos and galaxies agree quite well, though only over a small range of velocity dispersion. How the bright end cut-off in the galaxy luminosity function comes about is not fully understood. It may be associated with the inability of gas to cool in high temperature halos (Rees and Ostriker 1977). Empirically it seems that above some mass the theoretical curve should really be associated with the number density of groups and clusters of galaxies rather than individual galaxies. At the low velocity end there are increasingly more halos than galaxies of the same velocity dispersion and so the faint end slope is too steep (Schaeffer and Silk 1985). This is also seen in N-body CDM calculations.

From the mass function (equation 2.1) describing the evolving distribution of halos we can determine the bias parameter $b(M, z)$ by applying the peak-background split. In a region subject to a long-wave perturbation Δ_B the threshold for collapse will be reduced to $t = 1.68 - \Delta_B$, and consequently the number of halos of mass M at redshift z will be increased by a factor $n(M, z, 1.68 - \Delta_B)/n(M, z, 1.68)$. Thus the bias parameter $b(M, z)$, which is the ratio of the Eulerian perturbation in number density to the present density perturbation is given by,

$$b(M, z) = \frac{1}{(1 + z)} - \frac{1}{(1 + z)} \times \left. \frac{\partial \ln n(M, z, t)}{\partial t} \right|_{t=1.68} , \qquad (2.3)$$

or, using equation (2.1),

$$b(M, z) = \frac{1}{(1 + z)} - \frac{1}{1.68(1 + z)} + \frac{1.68(1 + z)}{\sigma^2(M)} . \qquad (2.3)$$

3. Galaxy Formation

We now consider two highly idealised models for the formation of galaxies. In the first we shall assume that galaxy formation is an ongoing process, with the luminosity tied to the properties of the most currently virialising material, so the clustering can be calculated by taking a slice through the distribution of halos at the present epoch. In the second we assume that galaxy formation on the whole has finished—what we see today are fossil remnants of the past history of the halo dendogram—and we calculate the biasing implied by identifying the stellar and dark matter velocity dispersions and making the 'galaxies' respect a universal and tight $L - V$ relation.

3.1. Late Galaxy Formation

A popular picture for formation of disk galaxies which has evolved over the years (notable contributions being the works of White and Rees 1978, Fall and Efstathiou 1980, Gunn 1982) can be outlined as follows: Collapsing perturbations violently relax to form dark matter haloes with density profiles $\rho(r) \propto r^{-2}$, while tidal fields induce angular momentum in these halos, with with spin parameters $\lambda \simeq 0.05 - 0.1$. Gas falling in is shock heated to the halo virial temperature, and adopts a hydrostatic equilibrium configuration with a density run like that of the dark halo. At later times gas from some radius within the currently virialising radius can cool in a Hubble time. This gas contracts quasistatically until rotationally supported, producing an approximately self gravitating rotationally supported disk with $V_{\rm rot} \simeq V_{\rm halo}$. It is encouraging that for a halo with $V_{\rm rot}$ equal to that of our galaxy and with $\rho_{\rm baryon} \simeq 0.1 \rho_{\rm tot}$, in accord with standard nucleosynthesis, the radius at which $t_{\rm cool} = t_H$ is ~ 100 kpc, so after collapse by a factor $\simeq 1/\lambda$ this might plausibly produce a disk of reasonable dimensions. Somewhat less encouraging is the $L - V$ relation predicted here. The mass-velocity relation is $M \propto V^3$, since the objects have the same density at their virialising radius, but a smaller fraction of the gas can cool in the hotter halos, so the predicted $\log L - \log V$ slope is always less than 3. Approximating the cooling time as $t_{\rm cool} \propto T^\beta \rho^{-1}$ one predicts $L \propto V^{(3-\beta)}$, or roughly $L \propto V^2$ for galactic temperatures. This is a shallower slope than the 'canonical' $L \propto V^4$, though not so far from the slope of $\simeq 2.7$ which is more relevant for optical luminosity (Tully and Fisher, 1977) Perhaps it is unreasonable to expect the theory to predict such details with great accuracy, and a positive point is that this type of model would at least predict that the galaxies form a 1-parameter family; a property not automatically shared by other models.

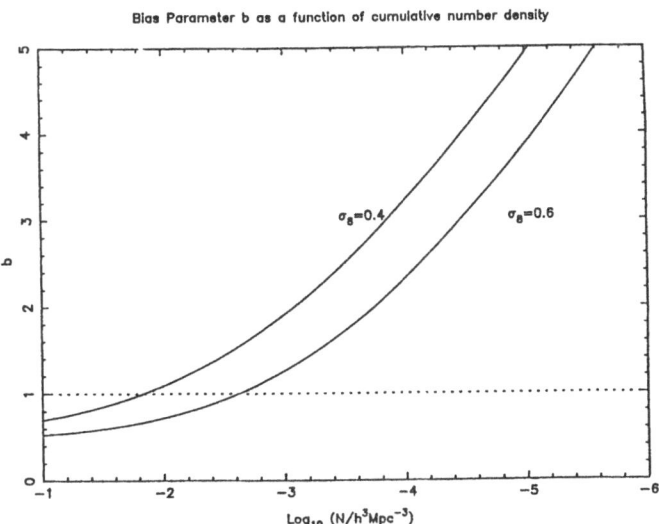

Bias Parameter b as a function of cumulative number density

Figure 2. The bias parameter b as a function of halo cumulative number density.

In this picture, galaxy formation would be an ongoing process; the number density of halos with a given rotation velocity is only slowly varying, and in any one of these halos the radius for which $t_{\mathrm{cool}} = t_H$, and therefore also the mass of cooled gas, is an increasing function of time. If so it would seem reasonable to associate the present distribution of halos with the galaxy distribution. If this is the case then the expected bias for these objects is simply given by equation 2.3 with $z = 0$. In figure 2 we have plotted the bias as a function of cumulative number density. We have chosen number density rather than velocity dispersion, for example, as this seems to be more robustly determined by the model. We see that for rare objects such as rich clusters which are on the steeply rising part of the mass function, figure 1, a strong positive bias is predicted. However for objects with number densities $n \simeq 10^{-2} h^3 \mathrm{Mpc}^{-3}$, as is typical of ordinary bright galaxies, no bias or even an anti-bias would be predicted unless a very low normalization were to be adopted.

We consider this prediction of such a low bias to be quite a negative feature of this otherwise physically well motivated scenario for galaxy formation. The bias parameter derived here applies strictly only to large scale structure, and with a late formation epoch it is difficult to extrapolate into the regime of currently non-linear clustering. Nonetheless, if one takes seriously the idea that the supply of gas for star formation is regulated by the condition that $t_{\mathrm{cool}} = t_H$, then on quite general grounds it is easy to see that even in the non-linear regime one expects to obtain a net antibias for this gas supply, and therefore, most naturally, for the net abundance of luminous stars. The problem is simply that in regions which are overdense on large scales, the characteristic virial temperatures of halos will surely be enhanced, while the density of the halos at virialisation is just a multiple of the global background density, and is therefore expected to be unchanged, hence the ratio t_{cool}/t_H will be increased.

One weakness of this model is that the halo properties at quite high density contrasts where the gas can currently cool are locked to the properties at larger radii where the new material is currently falling in. This assumption is motivated by N-body experiments, which lead one to attribute the non-linear profiles of clusters to relaxation. An alternative view is that the profiles are determined by the initial spectrum (e.g. Hoffman and Shaham 1985), with the dense inner parts of halos reflecting the binding energy of material which virialised at some earlier epoch, and which has remained dynamically intact since then. If this picture is closer to reality then other calculational techniques are called for.

3.2. Early galaxy formation

In this section we will take a rather different approach. This is designed to model the biasing that might develop if galaxy formation has effectively finished. Rather than assume some *ad hoc* threshold for galaxy formation, we let the luminosity generated in each branch of the space-time halo-dendogram be a continuous function of the halo properties, with this function chosen so that the 'galaxies' respect the observed relation between luminosity and velocity dispersion. Specifically, we assume that $V_{\mathrm{stars}} = V_{\mathrm{halo}}$ and that each halo is assigned a luminosity which is proportional to the 4th power of the stellar velocity dispersion. This rule is empirically motivated, and relatively 'natural' in that no sharp threshold is introduced.

At first sight it might seem that the empirical $L - V$ relations would be compatible with an unbiased 'mass-traces-light' model. After all, it is well known that in the CDM model, the spectral index is $n \simeq -2$ on galactic scales, and this results in a mean mass velocity relation $M_* \propto V_*^4$. Thus if we simply assumed that mass converts to light with equal efficiency everywhere then this would reproduce the observed $L - V$ relation, and the clustering of stellar luminosity would of course be unbiased. While it is true that the mean $\log L - \log V$ relation is acceptable, on closer inspection a couple of problems emerge. We will discuss these in some detail. The failure of the unbiased model on these grounds is instructive: It shows that under the assumptions given at least, some kind of clustering bias seems inevitable. It also prompts us to carefully examine the neccessity of these assumptions.

The first problem concerns the scatter in the mass-velocity relation. If we simply take the positive half of a Gaussian for the distribution of collapse redshifts at a given mass, and convert this to the distribution of the absolute magnitude estimator $M = 2.5 \log V^4$, this gives a predicted scatter of about 2.2 magnitudes. This is something of an overestimate since it includes perturbations which have yet to collapse. Truncating the distribution helps somewhat, but the revised estimate of about 1 magnitude is still considerably larger than that observed, and this presumably includes a substantial component from instrumental errors etc.. It would be possible to avoid this problem if, as envisaged by Faber (1982), the different morphological types are stratified along lines of constant ν, possibly because of systematic dependence of the angular momentum on ν. (Here ν denotes the initial amplitude of the perturbation in units of the rms.) It is important to realise that with this type of stratification, the various galaxy types would still be unbiased tracers of the mass, as we would expect long wavelength modes to have little effect on spins, or on other parameters depending on the local shape of the initial perturbation field. It now seems unlikely that the correlation between ν and angular momentum is sufficiently strong to effect the required stratification. Perhaps there is some other 'shape' determinant for morphology—even if this were the case, the model would still fall foul of a second, less escapable problem.

This second problem concerns the effect of long wavelength density perturbations. These will perturb the rotation velocities but not the luminosities (by assumption), so that in regions which are now overdense we would expect to see less luminous galaxies at a given velocity dispersion. This would cause the apparent Hubble constant to depend on the local density of the galaxies used as distance estimators, and would play havoc with studies of departures from pure Hubble flow. To get an idea of the strength of this effect consider a 'top-hat' perturbation of radius R, at distance r from the observer and with overdensity $\Delta_B \ll 1$. If the mean $L - V$ relation has slope 4, then the fractional error in the distance is

$$\frac{\Delta r}{r} = \frac{\Delta_B}{1.68(1 + z_f)},$$

where we assume that the halos containing the galaxies collapse, on average, at redshift z_f. On the other hand, the real peculiar velocity due to a spherical perturbation of radius R and amplitude Δ_B is $v_{\text{pec}} = (1/3)\Delta_B H R$, so the spurious 'peculiar velocity' $H \Delta r$ would overwhelm the true peculiar velocities for perturbations at distances $r > 1.68(1 + z_f)R/3$. Thus we see that in this hypothetical unbiased model, such systematic offsets in the $L - V$ relations would greatly exceed any real peculiar velocities, but this is apparently not seen.

Setting $L = V^4$ exactly obliviates both problems and also implies a bias. A straightforward way to see this is to consider an individual object and imagine subjecting this to a background perturbation Δ_B. From the virial theorem and the assumptions about virialisation, $L = V^4$ implies $L = M^{4/3}(1 + z_c)^2$. Thus an earlier collapse time leads to a higher L and therefore to a net luminosity bias.

$$L \rightarrow L' = (1 + \Delta_B/\Delta_{\text{obj}})^2 L,$$

A great advantage of the crude analytical approach adopted here is that one can see explicitly where the bias comes from, *vis* the highly non-trivial assumed dependence of L/M on collapse redshift. However empirically motivated and 'natural' the identification

$L \equiv V^4$, this theory still lacks a physical explanation. All we have shown is that assuming both $V_{\text{halo}} = V_{\text{stars}}$ and $L \propto M_{\text{halo}}$ looks unacceptable. In what follows we shall explore the consequences of retaining the former assumption, and therefore making variations in L/M reproduce a tight and universal $L - V$ relation.

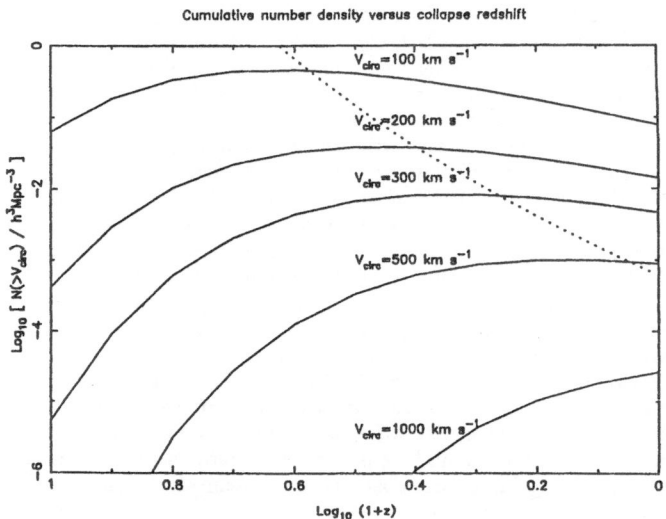

Figure 3. The cumulative number density of halos with circular velocities greater than V_{circ} as a function of $(1+z)$.

One can use the Press Schechter formalism to quantify the biasing. Figure 3 shows, as a function of redshift, the number density of halos $N(\geq V_{\text{circ}})$ with circular velocities greater than some value V_{circ}, and therefore with luminosities greater than $L_*\ (V_{\text{halo}}/V_*)^4$. Each member of this family of curves rises rapidly as the hierarchical growth proceeds, reaches a broad maximum and then gradually falls as these systems merge to form systems with higher velocity dispersions. With the normalization is $\sigma_8 = 0.6$ we find that halos with circular velocity $V_* = 300$ km s^{-1} reach their peak number density of 8×10^{-3} h^{3}Mpc^{-3} at a redshift of $z \simeq 1$. This agrees reasonably well with the present number density of galaxies of that velocity dispersion. Furthermore the mass and radius for such a condensation at redshift $z = 1$ are $M = 2.2 \times 10^{12}$ h$^{-1}M_\odot$ and $r = 100$ h^{-1}kpc, which seem reasonable for the formation of a typical L_* galaxy, if the baryons can dissipate their binding energy and collapse by about a factor ~ 10 in radius.

Having made this identification of the galaxy distribution and the distribution of the CDM halos we are in a position to address the question of biasing. If we consider adding a background perturbation $\Delta_B = \Delta_0/(1+z)$ to some region, then the effect is to cause each overdense region to collapse earlier, so the curves of figure 5 become displaced to the left; $(1+z)' = (1+z)(1+\Delta_B/1.68)$, but also relabelled; $V' = V(1+\Delta_B/1.68)^{1/2}$, since an earlier collapse time will lead to a higher velocity dispersion. This causes the peak number per unit mass at a given limiting velocity dispersion to have a strong non-linear dependence on the magnitude of the background perturbation. This is well fitted by a simple exponential $N_{\max}(\geq V) \propto \exp(\gamma(V)\Delta_0)$. On large scales, where Δ_0 is small, this converts to a present day bias given by $b = 1+\gamma$. Figure 4 shows how this bias parameter b depends on both the velocity dispersion V and normalisation σ_8, and indicates that in this model the rarer more luminous galaxies are more strongly clustered. A qualitative difference between this model and that considered in the previous section is that here one finds that galaxies of all luminosities are positively biased.

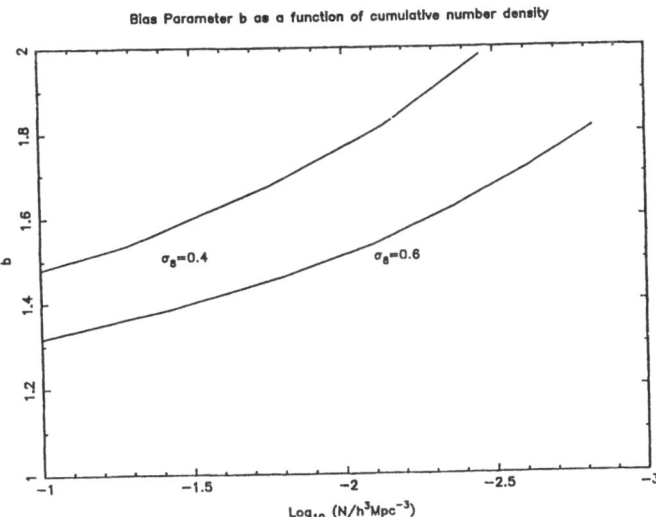

Figure 4. Bias factor b vs halo cumulative number density.

In this model galaxies form before clusters collapse, so we can also estimate the amount by which the L/M for clusters will be enhanced over the global value. We have chosen this statistic, rather than say the enhancement of the number of galaxies per unit mass, as the net luminosity is presumably insensitive to merging. If we identify as clusters the recently virialized systems whose mean initial overdensity when smoothed on a cluster mass scale equals 1.68, then the luminosity function in the clusters will be modified from the global value $\Phi(L)$, equation (2.3) to $\Phi_{\mathrm{clus}}(L) \propto \exp(\gamma(L))\Phi(L)$. Integrating over this new luminosity function using the computed values of γ, we find the enhancement of L/M in the clusters is about 2.2 for $\sigma_8 = 0.6$ and 4.2 for $\sigma_8 = 0.4$, with a slow variation with the lower luminosity limit.

Thus we see that in this model, the predicted bias for large-scale clustering is substantial, and is similar, though slightly less than that determined from analysis of dipoles of galaxy counts (Kaiser and Lahav 1988). The enhancement of L/M in clusters is also large. For the low normalisation at least, the bias is strong enough to reconcile the estimates of cluster mass-to-light ratios with closure density. For the high normalisation the bias is weaker, but this model is not necessarily excluded as this effect may be complemented by dynamical friction which could be efficient at further segregating light and mass in the non-linear environment provided by the cluster and its subunits (Hoffman *et al.* 1982, Barnes 1983, Evrard 1986, 1987).

3.3. Alternatives and possible physical basis

We have argued that the joint assumptions $V_{stars} = V_{halo}$ and $L \propto M$ have unacceptable consequences the scatter and universality of the luminosity-velocity dispersion relations. The simple model we have explored above keeps the former assumption and puts all the 'galaxies' exactly on a tight and universal $L - V$ relation by *fiat*. The observed scatter, while small, is not zero, nor is the relation for real galaxies known to be absolutely universal—all we know is that spurious 'peculiar velocities' do not seem to be very much larger than the real velocities we expect from the gravitational influence of observed large-scale clustering. There is therefore possibility to relax the simple assumptions made hitherto, with important consequences for the clustering bias.

One possibility is to stick with $V_{stars} = V_{halo}$, but adopt a more general form for $L(M, z)$. For instance, one might entertain models such as,

$$L = M^\alpha (1 + z_c)^\beta ,$$

in which α and β are constants (Kaiser, 1988a). The model of §3.2 is the special case $\alpha = 4/3$, $\beta = 2$. With this model we can still arrange that the mean $L - V$ relation has the form $L \propto V_{halo}^4$. If we approximate the initial power spectrum as a power-law with spectral index $n = -2$, as appropriate on the mass scale of galaxies, then this constraint requires

$$\alpha = (\beta + 6)/6 .$$

The model has one free parameter; β. For the unbiased model ($\beta = 0$) we argued that the scatter and systematic shifts in the $L - V$ relation were unacceptable. The model of the previous section has $\beta = 2$, and amply satisfies any observational constraint of this kind. The scatter about the relation is proportional to $|\beta - 2|$, so the observations would probably accomodate a β as large as 3. These 'high-β' models are particularly interesting since the clustering bias parameter b is proportional to β. Moreover, the enhancement of L/M in a cluster is, as we have seen, a strongly non-linear function of b, so with $\beta \simeq 3$ one would reconcile the virial estimates with $\Omega = 1$ even with a larger normalisation.

The high-β models also have very intriguing consequences for large-scale streaming studies. Just as in the unbiased model the prediction was that in overdense regions the galaxies at a given L would have higher rotation velocities, if $\beta > 2$ the converse applies. This has the effect of appearing to enhance the scale of peculiar motions (Kaiser, 1988b): If we are falling towards an attractor then we will underestimate the distances to galaxies in this overdense region. Consequently, the attractor itself will appear to be moving in the same direction as us, even if, in fact, it is at rest. If there is any problem with the amplitude or scale of large-scale motions—though the evidence at present (e.g. Kaiser and Lahav, 1988) for any such problem seems slight—then the beneficial effects of the high-β model are twofold: not only does this model permit a higher normalisation while still giving substantial enhancement of L/M in clusters, but it also predicts that the derived 'peculiar velocities' will be augmented by spurious systematic effects, and that the apparent coherence length for the streaming motions be enhanced.

So far we have said nothing about the physical mechanism that, in these models, adjusts the efficiency of star formation to maintain an acceptable $L - V$ relation. The L/M dependence on mass is quite weak; what these models require is that an earlier collapse time means substantially more stars, but why? An earlier collapse will increase the density, the virial temperature, and the pressure. One piece of physics that many would suspect plays an important role is the collisional cooling of the gas. Higher density certainly tends to increase the cooling rate, but the increase in temperature nearly balances this, so the ratio of cooling time to dynamical time is hardly altered by a long wave swell—contours of t_{cool}/t_{dyn} are roughly parallel to lines of constant mass in the M, z plane—so this mechanism is not the biasing agent (though the cooling criterion may well be important in setting the high luminosity cut-off in the luminosity function). More promising is the idea that one can form more stars per unit mass of gas in a deeper potential wells. Such a dependence has been suggested by Larson (1974), and the idea has more recently been revived by Dekel and Silk (1986) in the context of CDM. Such a dependence might plausibly arise if supernovae are efficient at expelling gas from shallow potential wells. For this regulation to be important for ordinary bright galaxies requires more energy and/or better coupling to the gas than assumed by Dekel and Silk, but much less than assumed by proponents of the explosive scenario for galaxy formation (Ostriker and Cowie, 1981; Ikeuchi, 1981). A related possibility is that the increased pressure might modify the initial stellar mass function by reducing the Jeans length. Ashman and Carr (1988) have suggested that such an effect might result in *high* mass-to-light ratios in high pressure systems. While a very large increase in pressure might suppress the formation of moderate mass stars, for the rather modest variation in pressure envisaged here it seems more likely that the effect would be to suppress predominantly the higher mass supernovae progenitors, so, if these act as regulators we would expect *lower* mass-to-light ratio at high pressure, and therefore a positive bias. Yet another possibility is that speeding up the collapse allows more stars to form before the supernovae explode.

Finally, one might consider relaxing the $V_{stars} = V_{halo}$ assumption. The fluctuations in light-to-mass ratio that seem required if we keep to this appear, in retrospect, somewhat contrived. In the model of §3.2 there has to be some kind of conspiracy to put the galaxies born in halos with a broad $M - V$ relation back on a tight $L - V$ relation. It is almost as if galaxies had been carefully designed to give astronomers good distance estimates! Perhaps what this is telling us is that the $L - V$ relations have even less connection to the parent halo properties than we have hitherto assumed. There is some observational constraint on how much the ratio of stellar to dark velocities can vary, but given that with an empirical $L \propto V^4$ relation, a 20% variation in V would give a factor 2 change in L, there seems to be no real observational support for the strict equality we have assumed. It is easy to see how one can construct a model for galaxy formation with stronger bias than those considered above. One could, for instance, hypothesise that V_{stars} is a function of V_{halo}, with V_{stars}/V_{halo} an increasing function of V_{halo} and let $L = V_{stars}^4$ as before. This 'theory' automatically satisfies the observed $L - V$ relations. The calculation of the biasing proceeds much as before; long waves modulate V_{halo}, but this now converts to amplified fluctuations in V_{stars} and hence in light-to-mass ratio. These considerations lead us into a dangerous area of largely untestable speculation. One qualitative prediction that is perhaps worth mentioning is that postulating V_{stars}/V_{halo} to be an increasing function of V_{halo} would have a beneficial effect on the faint end slope of the luminosity function.

4. Summary and Conclusions

The Press-Schechter distribution, while a rather rough approximation compared to N-body experiments, provides a very useful medium for considering the biasing of astrophysical object. We consider the lack of exactitude in our treatment of the gravitational clustering of dark matter to be the least of our problems. We are clearly a long way from an *a priori* theory for the formation of galaxies; the best one can reasonably hope to do is to construct more or less physically reasonable rules for assigning luminosity to halos. The predicted clustering bias then plays an important role in testing these hypotheses. If we are correct in our understanding of the interplay between the different wavelength modes, then nature has provided us with a set of experiments where the initial conditions for galaxy formation were systematically varied by long-wave perturbations. If we are inclined to believe that the universe has critical density then the apparent strong bias of galaxy formation towards dense environments surely provides an important clue to help us unravel the physics involved.

We have looked at two very idealised models for galaxy formation. At one extreme, we associated halos with galaxies by assuming that the stellar velocity dispersion at radii ~ 10 kpc is identical to that of the most recently virialised dark matter at radii of several hundred kpc. One advantage of this model is that one would quite naturally expect the galaxies to form an essentially one-parameter family. However, we found that the lack of bias for the present epoch halos with abundance like that of bright galaxies to be too small. In the second model we explored, galaxy formation was assumed to have essentially finished by the present. The velocity of stars and dark particles were identified, and luminosities assigned according to a strict equality between L and V^4. This model, with its continuous dependence of luminosity on halo properties seems to us to be at least a small advance towards a potentially realistic theory from previous models which invoked a very sharp threshold to separate 'galaxies' from 'failed galaxies'. The biasing predicted in this model seems much more promising. For suitable normalisation, at least, we found that the predicted enhancement of light-to-mass ratios in clusters and other dense environments is sufficient to reconcile the hypothesis $\Omega = 1$ with virial analysis.

NK received support from NSERC and the Canadian Institute for Advanced Research. SC acknowledges a SERC studentship and the hospitality of CITA.

References

Ashman, K. and Carr, B., (1988) in proceedings of NATO ASI *"The Post-Recombination Universe"*, (Dordrecht:Reidel, eds N. Kaiser and A. Lasenby)

Bardeen, J. M., Bond, J. R., Kaiser, N. and Szalay, A. S., (1986) *Astrophys. J.*, **304**, 15.

Barnes, J., (1983) *Mon. Not. R. astr. Soc.*, **203**, 223.

Boyle, B. J., *et al.* (1988) *"Evolution of Large Scale Structures in the Universe"* (IAU Symp. 130; eds. J. Audouze and A Szalay; Reidel, Dordrecht)

Bond, J. R. and Efstathiou, G. (1984) *Astrophys. J. Lett.*, **285**, L45.

Dekel, A. and Silk, J., (1986) *Astrophys. J.*, **303**, 39.

Efstathiou, G., Frenk, C. S., White, S. D. M. and Davis. M., (1988). *Astrophys. J.* in press.

Evrard, A.E., (1986) *Astrophys. J.*, **310**, 1.

Evrard, A.E., (1987) *Astrophys. J.*, **316**, 36.

Faber, S.M., 1982. in *"Astrophysical Cosmology"* (Vatican City Pontifical Academy ed. H. A. Brück *et al.*)

Fall, S. M. and Efstathiou, G., (1980) *Mon. Not. R. astr. Soc.*, **193**, 189.

Gunn, J.E. (1982) in *"Astrophysical Cosmology"* (Vatican City Pontifical Academy ed. H. A. Brück *et al.*)

Hoffman, Y., Shaham, J. and Shaviv, G., (1982) *Astrophys. J.*, **263**, 413.

Hoffman, Y. and Shaham, J., (1985) *Astrophys. J.*, **297**, 16.

Ikeuchi, S. (1981) *Publ. Astron. Soc. Japan*, **33**, 211.

Kaiser, N., (1984) *Astrophys. J. Lett.*, **284**, L9.

Kaiser, N., (1988a) in *"Evolution of Large Scale Structures in the Universe"* (IAU Symp. 130; eds. J. Audouze and A Szalay; Reidel, Dordrecht)

Kaiser, N., (1988b) in *"Large-Scale Structure and Motions in the Universe"* ICTP, Trieste, April 1988

Kaiser, N., and Lahav, O., (1988) in *"Large scale motions in the universe"* (Proceedings of the Vatican study week)

Larson, R. B., (1974) *Mon. Not. R. astr. Soc.*, **169**, 229.

Ostriker, J. P. and Cowie, L. L., (1981) *Astrophys. J. Lett.*, **243**, L127.

Press, W. H. and Schechter, P., (1974) *Astrophys. J.*, **187**, 425.

Rees, M. J. and Ostriker, J. P., (1977) *Mon. Not. R. astr. Soc.*, **179**, 541.

Schaeffer, R. and Silk, J. (1985) *Astrophys. J.*, **292**, 319.

Struble, M. F. and Rood, H. J. (1987) *Astrophys. J. Suppl.*, **63**, 543.

Tully, R. B. and Fisher, J.R. (1977) *Astron. Astrophys.*, **54**, 661.

White, S. D. M. and Rees, M. J., (1978) *Mon. Not. R. astr. Soc.*, **183**, 341.

DISCUSSION:

WHITE: The reason you tend to find large scatter in predicted Tully–Fisher relations is because your halo models depend on two parameters (mass and peak height, circular velocity and formation epoch, or ...). Now, it seems unlikely that the halos of observed disk galaxies have been substantially modified since the disks were formed. Further, the present halos in numerical model, are essentially a one-parameter family; they may be characterised by circular velocity alone in the denser regions from which gas must presumably cool to make disks. As a result it seems natural to get a one parameter family of disks in which luminosity would be tightly correlated with circular velocity.

KAISER: The model I was trying to shoot down was one in which the galaxies we see today are some kind of fossil remnant of the 2-dimensional distribution of halos, and in which the luminosity just equals the mass of the halo – the problem being that mass and velocity dispersion are only weakly correlated in Gaussian models. The model you alluded to is quite different. I'd agree that in this kind of scheme, in which galaxy formation would be an ongoing process, one sees a slice through the halo distribution so one might expect a tight Tully Fisher relation for the resulting 1-parameter family. As I argued in my talk though, the problems here (at least if one takes the cooling criterion seriously) are that one would seem to get the wrong TF slope, and that one would find a net luminosity antibias. One can get round the former problem by invoking more physics (supernovae or whatever), and this would also give a more positive bias, but the analytic results suggest that there would still be a potential problem in the predicted antibias for small galaxies.

EVRARD: Perhaps your burial of Faber's ideas that ellipticals are 3σ perturbations and spirals are 2σ ones is a bit premature. If you consider E's as forming above fixed $\delta(M)$ (where $\delta(M)$ $=\nu_E\ \sigma(M)$), then the effect of a cluster bias will be to make 3σ objects in the field equivalent to, say, 2σ objects in clusters. The main effect of this will be to add *scatter* to the L – v relation in clusters rather than vary dramatically the zero point of the relation.

KAISER: The idea I was referring to was that the dimensionless spin parameter is the discriminant for morphological type. Now it may be that λ is anticorrelated with the height of a peak *relative to its neighbours*, in which case ellipticals might well correspond to those peaks above a threshold ν times some local rms fluctuation amplitude. I think it is important however, to distinguish this type of local threshold from a threshold measured in units of the global rms such as you refer to. I agree with your comments regarding such a global threshold, but as you will have gathered from my talk, I am somewhat disillusioned with models which invoke such a threshold simply in order to effect the desired biasing without any good physical basis.

PEEBLES: I agree that your estimate of the bias parameter, b $\sim$ 1.5, has many advantages, but, as you know, I think it suffers from the great disadvantage of disagreeing with the biasing wanted to reconcile most dynamical estimates of the mean mass density with that of the Einstein-de Sitter model. The cleanest test seems to be the Local Supercluster. Could you quote, for the record, suggested values for the Virgocentric velocity and galaxy density contrast that would be consistent with $\Omega = 1$ and b $=$ 1.5 and not observationally unreasonable?

KAISER: I'm not sure that I agree that the Local Supercluster provides the cleanest test, though it should of course be taken seriously. Problems in determining the density contrast include determining the background density, very difficult with the RSA redshift catalogue, and correcting for peculiar velocities. Determining the infall is no easy task either. I made some simple spherical models of the LSC (*MNRAS*, **227**, 1) and found that b $=$ 1 and an infall velocity of 350 km s^{-1} looked quite acceptable. The true density contrast is only about 80% for this infall, but the apparent density contrast is much greater. With b $=$ 1.5, a very slightly lower infall would also be OK. I realise that there is some evidence pointing towards considerably lower infall velocities, and if this holds up then we have a problem. My main worry about these calculations is the use of a spherical model centred on the Virgo cluster core. There seems to be no good reason why the outskirts of the supercluster – which dominate the acceleration – should be concentric with the high density cluster. It may be that fixing the centre in this way is forcing an unrealistic model, and this may well compromise the reliability of the results. As far as I know there has been no serious consideration of this aspect of the problem, but it is worth emphasising that in the somewhat analogous test applied to the deeper angular dipoles, but in which no modelling of the shape of the perturbation is imposed, the results do seem to be compatible with a b value similar to that determined here.

GALAXY CLUSTERS AND THE EPOCH OF GALAXY FORMATION

Carlos S. Frenk
Physics Department
University of Durham
England

ABSTRACT. Existing data sets can be used to constrain the amplitude of primordial density fluctuations and hence the epoch of galaxy formation. Velocity dispersions, x-ray temperatures and mass-to-light ratios of galaxy clusters are particularly sensitive diagnostics. The expected values of these quantities in the cold dark matter cosmogony are calculated. Comparison of the model predictions with the velocity dispersion data is hampered by projection effects inherent in the two-dimensional nature of cluster catalogues. The model predictions agree with the x-ray temperature and (M/L) data for a value of the biasing parameter $b \simeq 2 - 2.5$. With this normalisation the cold dark matter cosmogony predicts a recent epoch of galaxy formation with much of the activity occurring at $z \lesssim 2.5$.

1. Introduction

In hierarchical theories of gravitational clustering, the epoch of galaxy formation is determined by the amplitude of fluctuations in the *density* distribution on galactic scales. These fluctuations need not be replicated in the *galaxy* distribution. Indeed, if the mean cosmic density approaches the closure value, galaxies must be more clustered than the dark mass (Efstathiou *et. al.* 1988a). The segregation between galaxies and mass may be quantified by a biasing parameter which we define as $b(r) = \sigma_{gal}/\sigma_\rho$ where $\sigma_{gal}(r)$ and $\sigma_\rho(r)$ are the fluctuations in the galaxy and density distributions respectively filtered with a "top hat" window function of radius r.

A variety of observational data now exists which can be used to constrain separately the values of σ_ρ and b on certain scales, independently of any particular cosmogonic model. This should be distinguished from other data sets which test a specific form of the fluctuation spectrum over a range of scales. Table 1 lists some of the "observables" and the quantity to which they are most sensitive. Note that some of these "observables" probe different scales. For example, the *rms* peculiar velocities, $< v_{pec}^2 >^{1/2}$, of galaxy pairs are determined by σ_ρ on a scale of $\sim 2h^{-1}$ Mpc whereas large scale streaming motions reflect the value of σ_ρ on a scale of $\sim 20h^{-1}$ Mpc. (Here and below h denotes the Hubble constant in units of 100 km s^{-1}Mpc^{-1}).

The best studied and in many ways the most successful model of galaxy formation to date is the cold dark matter (CDM) cosmogony. Its basic premises are that the Universe contains a closure density of weakly interacting elementary particles and that primordial fluctuations are adiabatic and scale invariant. Its only free parameter is the normalisation of the primordial fluctuation spectrum which must be determined empirically. In the CDM model this is equivalent to a determination of either σ_ρ or b on any one scale. This is because the parameter b is

C. S. Frenk et al. (eds.), The Epoch of Galaxy Formation, 257–264.

Table 1. Diagnostics for σ_ρ, b and the fluctuation spectrum

σ_ρ	b	fluctuation spectrum
$< v_{pec}^2 >^{1/2}$ of galaxy pairs $< v_{pec}^2 >^{1/2}$ of clusters internal $< V^2 >^{1/2}$ of clusters T_x of cluster gas large scale streaming	$\Omega_{cosmic\ virial\ theorem}$ (M/L) of clusters v_{infall} towards Virgo	$\xi_{cluster-cluster}$ $\xi_{galaxy-cluster}$ abundance of galaxies natural bias

approximately independent of scale since the predicted mass and galaxy autocorrelation functions turn out to have approximately the same slope (Davis $et.$ $al.$, 1985). Futhermore, since $\sigma_{gal}(8h^{-1}Mpc)$ is measured to be $\simeq 1$, then $b \simeq 1/\sigma_\rho(8h^{-1}Mpc)$. Any of the diagnostics listed in the first two columns of Table 1 can be used to fix the normalisation and, of course, if the theory is a correct description, they should all give the same answer. Note that a given epoch of formation $(1 + z)$ scales as b^{-1} while velocities scale as $b^{-1/2}$.

From the values of $< v_{pec}^2 >^{1/2}$ for galaxy pairs obtained from N-body simulations, Davis $et.$ $al.$ (1985) found $b = 2.5$. From the galaxy autocorrelation function Bardeen $et.$ $al.$ (1986), inferred $b = 1.7$ whereas an analysis of large scale streaming motions led Kaiser and Lahav (1988) to advocate $b \sim (1.4 - 1.7)$. With the exception of this last study for which, unfortunately, the observational data is most controversial, the other estimates of b were based on fairly rough arguments which were appropriate in the early days of the CDM model. This model has now been developed to such an extent that a careful re-examination of its normalisation (and thus of the predicted epoch of galaxy formation) is called for. In this talk I will consider the use of data on galaxy clusters for this purpose. Although much of the discussion that follows will make specific reference to the CDM model many of the arguments and results can be readily extended to other hierachical clustering models.

2. Galaxy clusters as probes of the large-scale density field

Rich galaxy clusters are rare objects. In hierarchical clustering theories they grow from density fluctuations in the tail of the distribution, usually assumed to be a Gaussian random process. Properties such as their abundance or mass-to-light ratios are therefore very sensitive to the value of σ_ρ on large scales. To exploit this feature we require the ability to count clusters in a complete and homogenous way. Here we will consider counting clusters according to their velocity dispersion and to the temperature, T_x, of their x-ray gas.

The information contained in the abundance of clusters should be clearly distinguished from that contained in their mass-to-light ratios. These measure the relative degree to which galaxies and mass are overrepresented in clusters, ie the product $b\,\Omega$ where b is the biasing parameter (on a cluster scale) and Ω is the cosmological density parameter. Only when one assumes a specific form for the spectrum of fluctuations is the determination of b equivalent to a measurement of σ_ρ.

Within any particular model it is often possible to calculate the expected abundance and properties of galaxy clusters. In general, the number of assumptions and thus the degree of uncertainty increases as one proceeds from the calculation of the distribution of velocity dispersions to that of the distribution of x-ray temperatures to that of the (M/L) ratios. Paradoxically, the uncertainty in the observations increases in exactly the opposite order.

Perhaps the best observationally determined quantities are the (M/L) ratios. Assuming that galaxies $within$ $clusters$ trace the mass, one finds a mean $(M/L) \sim 0.25$ in units of the (M/L) ratio required to close the Universe and a surprisingly small scatter (Colless 1988, Geller 1988).

Taking into account both cluster to cluster variations and uncertainties in the mean cosmic luminosity density (Efstathiou *et. al.* 1988b) the uncertainty in (M/L) is ~ 0.1. Although in a somewhat model dependent way, the x-ray temperatures of the intracluster gas can be measured with about 20% accuracy in clusters with good x-ray data (*cf*, for example, Edge and Stewart, in preparation). The problem here is that the set of clusters with reliable measurements is not a complete sample so the distribution of T_x must be derived in an indirect way. Finally, and perhaps contrary to one's naive expectation, the true distribution of internal velocity dispersions is the least well determined datum. Clusters selected, as Abell clusters are, from 2-D data may be severely contaminated by projection effects which would lead, in general, to a gross overestimate of the high velocity dispersion tail of the distribution. To summarize, it is perhaps disappointing that those properties which are relatively "clean" from a theoretical point of view are observationally "dirty", and viceversa.

3. Results

The results discussed here are part of a collaboration with G. Efstathiou, M. Davis and S. White. In order to study the statistical properties of galaxy clusters and their dependence on the normalisation of the fluctuation spectrum we performed a set of 5 N-body simulations from CDM initial conditions. Each simulation represents a cube of size 360 Mpc containing 262144 particles each of mass $1.2 \times 10^{13} M_\odot$. (Here and below I assume $h = 0.5$.) The numerical techniques are described in detail by Efstathiou *et. al.* (1985). The simulations were continued past the epoch which we have identified with the present day in our previous work (*cf* Davis *et. al.* 1985, Frenk *et. al.* 1985, 1988, White *et. al.* 1987a, 1987b) and which requires $b = 2.5$. Identifying a different epoch in the simulations with the present corresponds to having $b = 2.5/a$, where a is the expansion factor from our fiducial "present". Results were analyzed for values of $b = 3.3, 2.5, 2.0, 1.6, 1.3$ and 1.0, the last value corresponding to a model in which the galaxies trace the mass.

The sites of galaxy clusters in the simulations were identified by picking out high density regions using a friends-of-friends algorithm. Particles within an "Abell radius" of 3 Mpc were used to calculate cluster properties. I now discuss in order of increasing theoretical uncertainty the three properties of interest here:

3.1. THE DISTRIBUTION OF VELOCITY DISPERSIONS

Velocity dispersions can be measured easily and accurately in our simulations which have a spatial resolution of ~ 200 kpc. The velocity dispersions are determined purely by the depth of the gravitational potential well so they can be estimated with no reference to galaxies other than the assumption that they are test particles. The cumulative distributions of line-of-sight velocity dispersions, $< V_{l.o.s}^2 >^{1/2}$, obtained for different values of b are shown in Figure 1a, normalized to the mean density of Abell clusters of richness class $R > 0$, $n_{Abell} = 7.5 \times 10^{-7}$ Mpc^{-3}. The predicted distributions have been convolved with a Gaussian of width $0.2 < V_{l.o.s}^2 >^{1/2}$ in order to mimic the internal errors in the estimates for real clusters. (This procedure tends to populate the high velocity tail of the distributions.) The dashed line in the figure shows the corresponding distribution for 72 Abell clusters with velocity dispersions measured from at least 10 galaxies (Moore and Frenk, in preparation.)

At first sight the comparison between the model predictions and the data is rather disappointing. The shape of the observed distribution differs significantly from the model for all values of b. One might anticipate, however, that this comparison is suspect because our clusters were defined as true three-dimensional objects whereas Abell clusters were identified on projection. A preliminary examination of the projection effects in an "Abell catalogue" constructed

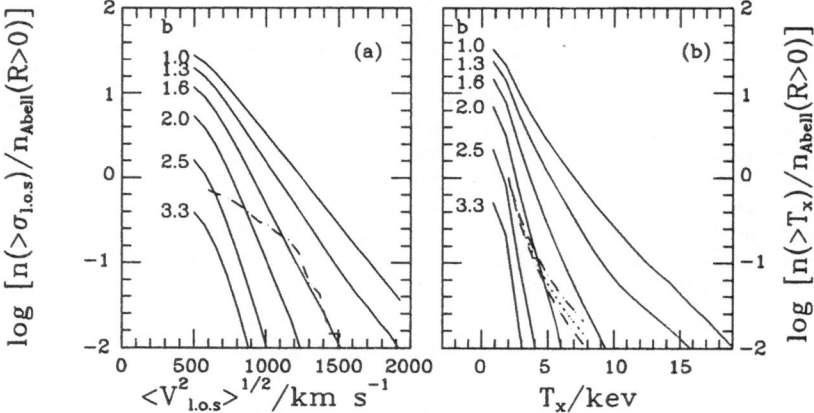

Figure 1.(a) Cumulative distribution of velocity dispersions of galaxy clusters. The solid lines give the predictions of the CDM model for various values of the biasing parameter; the dashed line represents the observational data. The number densities are normalised to $n_{Abell}(R > 0) = 7.5 \times 10^{-7}$ Mpc^{-3}. (b) Cumulative distribution of x-ray temperatures of galaxy clusters. Symbols and notation are as in (a); the error range for the data is described in the text.

from our simulations shows that these effects are indeed quite dramatic. On projection clusters with very large velocity dispersions are found but they invariably turn out to be made up of widely spaced small clumps strung along the line of sight. Constructing a projected cluster catalogue, however, requires the identification of "galaxies" in the simulations and this destroys the attractive theoretical simplicity which the problem appeared to have originally. Our preliminary results, which will be detailed in a forthcoming paper, indicate that the comparison in Figure 1(a) is meaningless and that no cosmological inferences can be made from the existence of clusters with very large apparent velocity dispersions unless projection effects are taken into account. This conclusion is reinforced by the examination of the x-ray properties of clusters to which I now turn.

3.2. THE DISTRIBUTION OF X-RAY TEMPERATURES

To study the properties of the intracluster gas, hydrodynamic effects must be considered in addition to gravity. This introduces an additional level of complexity and corresponding uncertainty in the theoretical predictions. The first treatment of the problem in the context of the CDM cosmogony was recently carried out by Evrard (in preparation, *cf* also this volume) who combined our standard N-body techniques with a "smooth particle hydrodynamics" approach. Assuming that cooling may be neglected, he finds that the infall and shock-heating of the gas as the cluster collapses leads to an approximately isothermal final distribution for the gas within the Abell radius with little segregation from the dark matter. The ratio of the specific energies of the gas and dark matter within the Abell radius is constant to within 10% and has the value $\beta \equiv <V_{l.o.s}^2>/[kT_x(\mu m_p)^{-1}] = 1.25$, where k is Boltzmann's constant, μ is the mean molecular weight and m_p the proton mass. This result is very convenient since it allows us to infer T_x directly from our simulations, which contain no gas, by simply measuring $< V_{l.o.s}^2 >$ and multiplying by the appropiate factor. The resulting cumulative distributions, again convolved with a Gaussian of width $0.2\,T_x$ to mimic the effect of internal measurement errors, are shown in Figure 1b.

Obtaining the observed distribution of T_x is not straightforward because of the unknown selection criteria in the sample of clusters with determinations of T_x. One can proceed through an indirect route as follows. The $HEAO$ 1 satellite carried out a complete flux-limited x-ray survey of a large area of sky which allows an estimate of the cluster x-ray luminosity function in the range $2.4 \times 10^{43} < L_x/ergs\ s^{-1} < 2.8 \times 10^{45}$ (Piccionotti $et.$ $al.$ 1982). There are also determinations of x-ray luminosities and temperatures for a heterogeneous sample of clusters from this and other satellites. L_x and T_x exhibit a very good correlation which can be folded into the luminosity function to obtain the temperature distribution. So long as the luminosity function is unbiased, the inferred T_x distribution will also be unbiased regardless of the selection effects in the sample with measured T_x. The most recent compilation which includes many new $EXOSAT$ results is that of Edge and Stewart (in preparation) who find $L_x = (7 \pm 1) \times 10^{42}\,T_x^{2.7\pm0.2}$ for $T_x > 2$ kev. Figure 1b shows the cumulative temperature distribution derived from this relation and from the luminosity function mentioned above. The main source of internal error is the scatter in the L_x-T_x relation; the dashed-dot lines correspond to the range allowed by the quoted errors in this relation.

Figure 1b shows that the observed distribution at abundances greater than $\sim 0.05\ n_{Abell}$ agrees quite well with the model predictions for $b \simeq 2 - 2.5$. At lower abundances the data show a tail of high temperature clusters which is not produced in the models. This discrepancy could reflect a shortcoming of the model, either of Evrard's hydrodynamic calculations, or of the CDM fluctuation spectrum on large scales. On the other hand, it could also be due to the small number of clusters in the tail which, furthermore, is quite sensitive to observational errors in the temperature determinations. Since the x-ray luminosity function is based on only 30 clusters, the shape of the observed temperature distribution for $T_x > 5\ kev$ is determined by only 1.75 clusters. Doubling the 20% errors assumed in the figure, makes the tail of the distribution consistent with the model predictions for $b \simeq 2$. Unfortunately, a more accurate determination of the bright end of the luminosity function will have to await the next generation of x-ray satellites. From his own simulations Evrard concluded that $b \sim 1.5$ partly because his simulations did not include the statistical distribution of clumps expected in the CDM model and partly because he considered exclusively the tail of the distribution.

If the ratio of the specific energies of the gas and the dark matter is approximately the same for all clusters as in Evrard's models, then the distributions of $< V_{l.o.s}^2 >$ and T_x should be approximately the same apart from the coordinate transformation given by the constant β defined above. The model predictions, of course, follow this expectation. The $observed$ distributions, on the other hand, do not, as can be seen from a comparison of Figures 1a and 1b. The distribution of T_x falls off steeply whereas that of $< V_{l.o.s}^2 >^{1/2}$ declines gently at first and cuts off only at large values. There are two possible interpretations of this discrepancy: one, that β is not a constant in real clusters; the other, that the observed distribution of $< V_{l.o.s}^2 >^{1/2}$ is severely affected by projection effects in Abell's catalogue. This latter interpretation is consistent with the arguments of the previous subsection.

3.3. MASS-TO-LIGHT RATIOS

The calculation of luminosities introduces a further level of complexity beyond the relatively straightforward treatment of the mass provided by the N-body simulations. Fortunately, much effort has been devoted to this problem in the context of the CDM cosmogony; this is encapsulated in the "high peak model" for biasing which has been applied to simulations by Davis $et.$ $al.$ (1985) and developed analytically by Bardeen $et.$ $al.$ (1986). In this model "galaxies" are identified with the high peaks of the suitably filtered linear density field. Some justification for this approximation is provided by the close association between high peaks and the sites of formation of galactic halos in N-body simulations (Frenk *et. al.* 1988). Using the analytic machinery developed by Bardeen *et. al.* it is possible to calculate the expected number of "galaxies" even in simulations which, like ours, do not have enough resolution to locate the high peaks directly. The average luminosity of the "galaxies" is obtained by requiring the mean luminosity density of the model to be equal to its observed value. This procedure is described in detail by White *et. al.* (1987a).

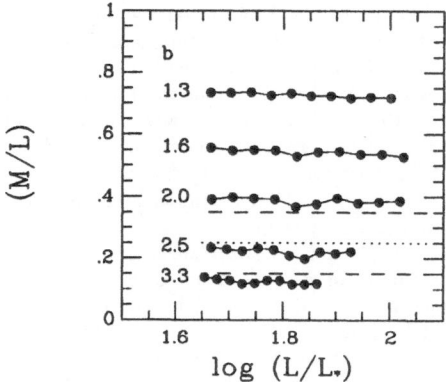

Figure 2. Mass-to-light ratios of galaxy clusters as a function of luminosity. Filled circles give the predictions of the CDM model for various values of the biasing parameter. The dotted and dashed lines represent the mean observed value and its standard deviation respectively. (M/L) is in units of the value required for closure and L is in units of $L_* = 3.9 \times 10^{10} L_\odot$

The predicted mass-to-light ratios of clusters as a function of luminosity are shown in Figure 2 for the various values of the biasing parameter b. (M/L) decreases very slowly with L. The mean observed value is shown by a dotted line and its 1σ range by the dashed lines. The data are consistent with $b \simeq 2 - 3$; smaller values are strongly ruled out. From the observational point of view this test provides the strongest constraint on b. The theoretical predictions are also reliable if one is prepared to accept the high peak model which, from other applications, would seem to be a good first approximation.

4. Discussion and conclusions

The distributions of velocity dispersions and x-ray temperatures of galaxy clusters can be used to place useful constraints on the amplitude, σ_ρ, of large-scale density fluctuations. The mass-to-light ratios constrain the combination $b\,\Omega$ where b is the biasing parameter. In the CDM cosmogony, σ_ρ and b are simply related so the problem can be rephrased as a determination of b from all three observables. Most of the arguments are applicable in general to hierarchical clustering theories in which the primordial density field is a Gaussian random process. In this case, the distinction between the quantities that the different observables measure should be kept in mind.

The simplest property to predict is the distribution of velocity dispersions of true three-dimensional clumps. This, however, cannot be compared directly with the observational data which may suffer from severe projection effects due to the two-dimensional nature of cluster catalogues. The temperature distribution of the x-ray emitting gas in clusters is not affected by projection effects, but the theoretical predictions must now rely on hydrodynamical models. The strongest constraint comes from the (M/L) ratios which are reasonably well determined for many clusters. In this case the theoretical predictions require an assumption about the relative distributions of galaxies and dark matter such as that given by the "high peak model".

It is pleasing that the two useful diagnostics, the distribution of T_x and the mean value of (M/L) lead to very similar answers, $b \simeq 2 - 2.5$. This result agrees with the conclusions from our previous N-body work, but disagrees with the value $b \sim 1.5$ estimated by Kaiser and Lahav (1988) from an analysis of large scale streaming motions. This disagreement confirms that large bulk velocities on large scales are inconsistent with the standard CDM model. Another potential difficulty is the excess, at the 5% level, of observed clusters with high values of T_x which are not present in the models. The existing data, however, are insufficient to assess how serious this difficulty really is. Elsewhere in this volume Carlberg and Couchman also argue in favour of a low value of b. It is difficult to see how their conclusion can be reconciled with the observed (M/L) ratios of clusters.

The normalization of the CDM spectrum derived here has an immediate application to the main topic of this workshop. It confirms the previous conclusion from N-body work that galaxy formation in the CDM cosmogony is a recent and protracted affair that continues until the present day. In this cosmogony the Universe at redshifts $z < 2$ is extremely dynamic with the bulk of the stars and galaxies being formed during this time. Many of the contributions in this volume would seem to suggest that this is not a drawback of the model but one of its attractive features.

Acknowledgements. I thank Marc Davis, George Efstathiou and Simon White for allowing me to discuss our results prior to joint publication. I also thank Gus Evrard and Jim Gunn for valuable discussions and Alistair Edge and Gordon Stewart for access to their unpublished x-ray data. The simulations described here were carried out using a grant form the NSF San Diego Supercomputer Center. This work was supported by grants from SERC and the Nuffield Foundation.

References

Bardeen, J.M., Bond, J.R., Kaiser, N. and Szalay, A.S. (1986). *Astrophys. J.*, **304**, 15.

Colless, M. (1988). *Ph.D. Thesis*, Cambridge University.

Davis, M., Efstathiou, G., Frenk, C.S. and White, S.D.M. (1985). *Astrophys. J.*, **292**, 371.

Efstathiou, G., Davis, M. Frenk, C.S. and White, S.D.M. (1985). *Astrophys. J. Suppl.*, **57**, 241.

Efstathiou, G., Ellis, R.S. and Peterson, B.A. (1988b). *Mon. Not. R. astron. Soc.*, **232**, 431.

Efstathiou, G., Frenk, C.S., White, S.D.M. and Davis, M. (1988a). *Mon. Not. R. astron. Soc.*, in press.

Frenk, C.S., White, S.D.M., Efstathiou, G. and Davis, M. (1985). *Nature* **317**, 595.

Frenk, C.S., White, S.D.M., Davis, M. and Efstathiou, G. (1988). *Astrophys. J.*, **327**, 507.

Geller, M. (1988). In *Large scale structures in the Universe*, Saas-Fee lectures, eds. L. Martinet and M. Mayor, (Geneva: Geneva Observatory), 69.

Kaiser, N. and Lahav, O. (1988). *Mon. Not. R. astron. Soc.*, in press.

Piccionotti, G., Mushotzky, R.F., Boldt, E.A., Holt, S.S, Marshall, F.E., Serlemistos, P.J. and Shafer, R.A. (1982). *Astrophys. J.*, **253**, 485.

White, S.D.M., Frenk, C.S., Davis, M. and Efstathiou, G. (1987a). *Astrophys. J.*, **313**, 505.

White, S.D.M., Davis, M., Efstathiou, G. and Frenk, C.S. (1987b). *Nature* **330**, 451.

DISCUSSION:

GUNN: I think the fact that the M/L of galaxies in clusters is almost certainly 2-3 times that in the field will influence your determination of the assignment of light to peaks, and will result in even higher values of b.

FRENK: That is right. The luminosity assigned to each peak is determined by requiring the mean luminosity density in the model to be equal to its observed value with no allowance for environmental effects. The size of the correction would depend on the morphological mix of galaxies in clusters compared to that in the samples used to estimate the observed mean luminosity density. This is not easy to quantify but I suspect it would be a relatively small effect, certainly in the direction that you point out.

MELOTT: You report the resolution of your simulation as 200 kpc. However, the particle separation is such that each one represents about 10^{13} solar masses. This is not a very good approximation to a nearly continuous density field. How do you know that this does not strongly affect your velocity dispersions? The test cases reported in Efstathiou *et al* are too weakly evolved to show this effect.

FRENK: The mean number of particles per cluster in the simulations ranges from several tens at early times to a few hundred at late times. This is sufficient to estimate velocity dispersions reliably. This is demonstrated by the fact that clusters in an earlier set of simulations (White *et al* 1987) had only about half the number of particles of the present ones, but very similar velocity dispersions.

EXPLORING ORIGINS FOR THE HUBBLE SEQUENCE

August E. Evrard
Institute of Astronomy, Madingley Road, Cambridge CB3-0HA, UK
and
Astronomy Department, University of California, Berkeley, CA 94720

ABSTRACT. An analytic formalism which allows calculation of separate mass functions for disk, spheroid and dwarf galaxies is applied to the cold dark matter model. Dwarfs are assumed to form within halos with virial temperatures below a critical value $T_c \simeq 10^{5.4}$K. Two sets of criteria are used to differentiate spheroids from disks — one based on cooling timescales and another based on initial perturbation height. The cooling timescale model fails to reproduce simple observed properties of the galaxy distribution. The model based on perturbation height is relatively successful and predicts a "natural" morphological bias which reproduces the observed morphology-density effect in clusters. This model suggests that the transition from the disk-forming to spheroid-forming regimes occurs at a critical *pressure* which is *independent of environment*.

1. Motivation

Although we lack a genuine physical understanding for the origin of galaxy morphology, there are some widely-held ideas on what processes *broadly* separate the Hubble types. For example, dwarfs are thought to originate in potential wells which are too shallow to contain gas heated by a first generation of supernovae (Dekel and Silk 1986). The implication is that "normal" (*i.e.* non-dwarf) elliptical, lenticular, and spiral galaxies are systems with virial temperature above some critical value T_c.

The early simulations of Larson (1975 and references therein) led him to suggest a physically compelling distinction between the formation of disk and spheroidal systems. He argued for the star formation rate as the critical element in determining whether or not a significant disk structure could be formed. If stars were formed rapidly during the collapse of a protogalaxy, consuming a large gas fraction in the process, the collapse of the galaxy would involve mainly violent relaxation of a cold stellar component. As N-body simulations have shown, the end product is likely to resemble an elliptical galaxy. If the initial star formation rate was low, then the gaseous collapse would preferentially result in the formation of a disk.

A second, empirically motivated distinction between spirals and ellipticals was suggested by Faber (1982) who pointed out that the observed Tully-Fischer and Faber-Jackson relations ($L \propto v^4$) would arise naturally in clustering from an initial scale-free fluctuation spectrum with spectral index $n = -2$ ($P(k) \propto k^n$). Blumenthal *et al.* (1984) noted that this is approximately the slope of the cold dark matter spectrum on galactic scales. They also pointed out that the observed narrow scatter and trend of decreasing velocity from early to late systems at a given luminosity would require that ellipticals arise from $\sim 3\sigma$ fluctuations while spirals are perhaps

C. S. Frenk et al. (eds.), The Epoch of Galaxy Formation, 265–269.

$\sim 2\sigma$ objects. Further impetus for this model comes from biased galaxy formation — the idea that galaxies form at $\gtrsim 2.5\sigma$ peaks in the density field (Bardeen *et al.* 1986). This biasing is *necessary* in an $\Omega = 1$ cold dark matter model if it is to reproduce the observed kinematics of the galaxy distribution on scales of $1 - 10\,h_{50}^{-1}$ Mpc (Davis *et al.* 1985).

The purpose of this work is mainly to test the two sets of ideas suggested by Larson and Faber for differentiating disk from spheroidal systems. More details can be found in a paper submitted for publication (Evrard 1989). Since quantitative predictions require an assumed cosmogony, we consider the specific case of adiabatic fluctuations in a CDM universe with cosmological parameters $\Omega = 1, \Omega_b = 0.1$ and $H_o = 50\,h_{50}$ km s^{-1} Mpc^{-1}. Two normalizations are considered — the BBKS bias $b = 1.7$ and the DEFW bias $b = 2.5$, where $b = 1/\sigma_o(16\,h_{50}^{-1}$ Mpc).

2. Morphological Mass Functions

The Press-Schechter (1974) formula is used to detemine the mass function of dark matter halos at a given redshift $n_H(M, z)$. The virial densities and temperatures of the gas within the halos are inferred from a simple spherically symmetric collapse model. Cooling timescales are determined using the primoridial cooling curve of Fall and Rees (1985).

The aim is to divide the population of halos among the Hubble types dwarf (Dw), spheroidal (E), and disk (Sp) given the defining physical criteria listed in the following table.

Model	Label	Dw	E	Sp
τ-ratio	τ_r	$T < T_c$	$\tau_{cool} < \tau_{dyn}$	$\tau_{dyn} < \tau_{cool} < \tau_H$
ν-threshold	ν_t^I	$T < T_c$	$\nu > \nu_E, \ \tau_{cool} < \tau_{dyn}$	$\nu_{Sp} < \nu < \nu_E, \ \tau_{cool} < \tau_{dyn}$
ν-threshold	ν_t^{II}	$T < T_c$	$\nu > \nu_E, \ \tau_{cool} < \tau_H$	$\nu_{Sp} < \nu < \nu_E, \ \tau_{cool} < \tau_H$

The τ-ratio model is based on the ideas of Larson while the ν-threshold model reflects those of Faber. A critical temperature $T_c = 10^{5.4}$ K is used for all models. Two versions of the ν-threshold model differ in allowing cooling on a dynamical time (ν_t^I) or a Hubble time (ν_t^{II}).

The analytic collapse model gives $\rho(M, \delta)$ and $T(M, \delta)$, the virial density and temperature in terms of initial overdensity δ and mass M. This allows us to rewrite the defining physical criteria above in terms of mass M and normalized pertubation height $\nu \equiv \delta/\sigma(M)$, where $\sigma(M)$ is the mass variance on scale M. For example, the temperature bound for dwarfs $T < T_c$ translates into an upper bound on perturbation height $\nu < \nu_{Dw}(M) = (T_c/2.3 \times 10^5 \text{K})M^{-2/3}\sigma^{-1}(M)$. Because the probability distribution of initial overdensities is *normal* in ν, we can then take the number density of galaxies of type X to be equal to the number density of halos times the fraction of collapsed perturbations which had values of ν which satisfy the conditions for that particular galaxy type

$$n_X(M, z) = n_H(M, z)\ f_X(M). \tag{1}$$

The forms of $f_X(M)$ for dwarfs, disks, and spheroids involve various combinations of complimentary error functions (see Evrard 1989).

3. Results

Figure 1 shows the comoving number density and relative abundance of disk and spheroidal systems as a function of redshift in each of the models. Note that the 'snapshot' style of this analysis does not allow us to define the present observed galaxy population — to do this would require a time operator capable of remembering individual halo histories. The horizontal lines in Figure 1 indicate the observed number density of galaxies brighter than $L_*/30$ and the observed

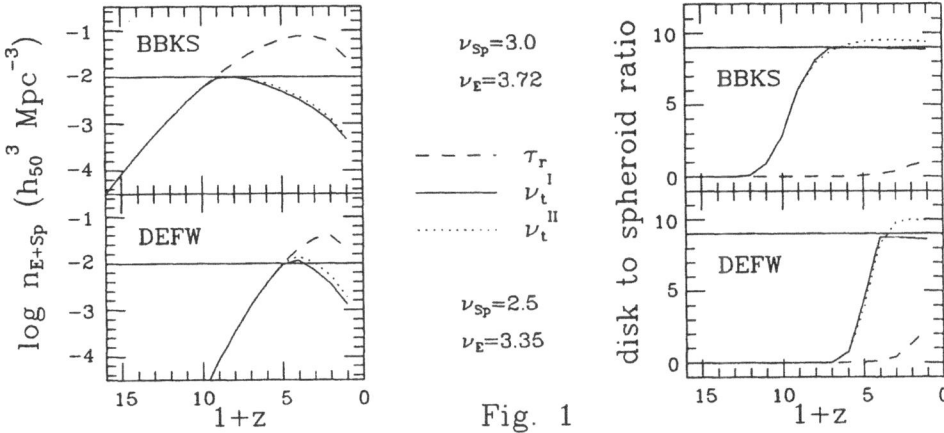

Fig. 1

10% global elliptical fraction. The τ-ratio model predicts an overabundace of halos capable of hosting normal galaxies at recent redshifts. More seriously, it predicts that *ellipticals should be at least as abundant as disk systems*, seriously in conflict with observations.

The ν-threshold model has two adjustable parameters, ν_E and ν_{Sp}, which can be tuned to yield the correct maximum galaxy number density and abundance ratio. The resulting values are $\nu_{Sp} = 3.0$, $\nu_E = 3.72$ (BBKS) and $\nu_{Sp} = 2.5$, $\nu_E = 3.35$ (DEFW). The first success for this model comes from examining the predicted mass functions shown in Figure 2 for snapshots taken at $z = 7$ (dashed lines), $z = 3$ (solid), and $z = 0$ (dotted). (Only the BBKS results are shown, the DEFW functions are similar.) The ν-threshold model is in good qualitative agreement with the work of Bingelli (1987), who observed that the luminosity functions of non-dwarf ellipticals and disk systems (Sp's and S0's) were similar in both shape and range, being roughly bell-shaped two decades wide in L. With appropriate choice of cooling intermediate between a dynamical and Hubble time, it is clear that the ν-threshold model can reproduce this observation. On the other hand, the τ-ratio model predicts that spirals should be more massive than ellipticals, typically by a factor ~ 5.

In searching for a more physical interpretation of the ν-threshold model, we note that for CDM the virial pressure at $M \sim 10^{12}$ M$_\odot$ scales as

$$P_{vir} \propto \delta^4 M^{2/3} \sim \nu^4.$$

The critical values ν_E, ν_{Sp} thus define **critical presures** P_E, P_{Sp}. The question is: Are these pressures meaningful? One way to address this question is to *bias* the galaxy population (*i.e.* imbed galaxies within protoclusters) and ask if the observed cluster population abundances are reproduced *from the same critical values P_E and P_{Sp} inferred globally*. Figure 3 shows the ratio of disks to spheroids for the global population (light line) and for the galaxies biased in 3.5σ clusters of mass 10^{15} M$_\odot$ (medium line) and 10^{14} M$_\odot$ (heavy line). The horizontal dotted line shows the mean observed morphology-density effect in clusters — roughly an enhancement from 10% ellipticals in the field to 20% in dense groups and clusters.

The observed morphology-density enhancement is reproduced in the ν-threshold model using the same critical pressures inferred globally and biasing expected from cluster core or dense group masses. The shapes of the mass functions are little affected by the bias, in agreement with the observations of Bingelli who finds similar galaxy luminosity functions in the field and Virgo. Only the mix of galaxy types changes with density.

268

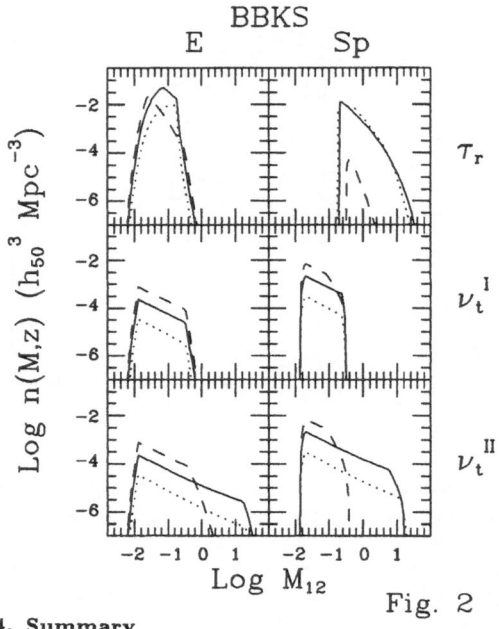

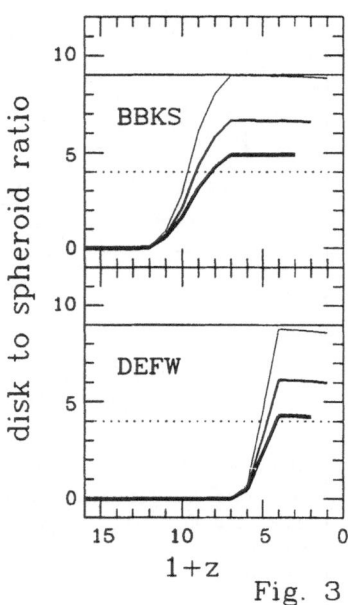

BBKS

E   Sp

τ_r

ν_t^I

ν_t^{II}

Log $n(M,z)$ (h_{50}^3 Mpc^{-3})

Log M_{12}

Fig. 2

disk to spheroid ratio

BBKS

DEFW

$1+z$

Fig. 3

4. Summary

Although physically appealing, using cooling timescales to distinguish disk from spheroidal systems leads to the prediction of ellipticals being more numerous and less massive than spirals, a situation in direct contrast with reality. The model in which ellipticals form from rare high density peaks and disk systems from less rare perturbations matches well many observations of the galaxy distribution. One *physical* interpretation is of a **pressure-sensitive** morphology. The morphology-density effect arises 'naturally' from the enhancement in virial pressure caused by the background cluster perturbation. As a final note, from Figure 1 we see that "THE EPOCH OF GALAXY FORMATION", defined as the epoch in which half the dark matter halos capable of hosting normal galaxies were in place, is $z_g \simeq 9$ for the BBKS normalization and $z_g \simeq 4$ for the DEFW bias. Adding an interval of 4×10^8 yr for the first generation of stars would result in the "epoch of star formation" being $z_* \simeq 5$ (BBKS) and $z_* \simeq 3$ (DEFW).

References

Bardeen, J.M., Bond, J.R., Kaiser, N. and Szalay, A.S. 1986, Ap. J., **304**, 15. (BBKS)
Bingelli, B. 1987, in *Nearly Normal Galaxies*, ed. S.M. Faber, (New York: Springer Verlag).
Blumenthal, G.R., Faber, S.M., Primack, J.R., and Rees, M.J. 1984, *Nature* , **311**, 517.
Davis, M., Efstathiou, G., Frenk, C.S. and White, S.D.M. 1985, Ap. J., **292**, 371. (DEFW)
Dekel, A. and Silk, J. 1986, Ap. J., **303**, 39.
Evrard, A.E. 1989, *Ap. J.*, submitted.
Faber, S.M. 1982, in *Astrophysical Cosmology*, ed. H.A. Bruck, G.V. Coyne and M.S. Longair,
       (Vatican:Pontificia Academia Scientiarium), p. 191.
Fall, M. and Rees, M.J. 1985, *Ap. J.*, **298**, 18.
Larson, R.B. 1975, *M.N.R.A.S.*, **173**, 671.
Press, W.H. and Schechter, P. 1974, *Ap. J.*, **187**, 425.

DISCUSSION:

FRENK: Couldn't the same criticism that was made yesterday to the stellar population synthesis models be made here also? Namely, that you are assuming that the morphology of galaxies seen today was the same in the past. How would evolution in the morphology (such as spirals turning into ellipticals) affect your conclusion?

EVRARD: You'll notice that I haven't strictly said anything about the present galaxy distribution expected from this simple model. That would require a time evolution operator to combine the snap-shot calculations I've done into an observed present-day distribution. I'm not quite sure how to do that with any confidence right now, as it involves knowing the detailed dynamical history each halo has had. What I've hoped to do, by taking snap-shots at different z, is to exhibit the possible range of behaviour in each of the models.

GALAXY MERGERS IN A CDM MODEL

R. G. Carlberg and H. M. P. Couchman
Dept. of Astronomy and C. I. T. A.
University of Toronto
Toronto, Ontario
Canada, M5S 1A1

ABSTRACT. A particle code is used to follow the evolution of a Cold Dark Matter spectrum in an $\Omega = 1$ background, treating half of the 524288 particles as a gas that can cool and turn into stars. The number density of galaxies as a function of redshift is well fit by the BBKS peaks statistics, using a sharp threshold having a normalized overdensity $\nu_t = 1.44(1+z)/\sigma_0(M)$, where $\sigma_0(M)$ is the linear extrapolation of the initial RMS density fluctuation on mass scale M. The dark halo number density-redshift relation is fit with the same ν_t using the volume averaging Press-Schechter model. The merger rates of the model galaxies are compared to observational estimates.

1. INTRODUCTION

The important distinction between dark matter and galaxies is that dissipation reduces the merger cross-section of galaxies such that they separate away from the ongoing process of clustering and merging of dark halos. A simple dissipative model is used in this paper to study the formation of galaxies in a CDM universe, and to show that many of their statistical properties are different from those of the dark halos.

2. NUMERICAL METHODS

The statistical properties of galaxies are largely a result of the collapse of density peaks, so that it is essential that each galaxy be composed of several particles, rather than initially identifying a single particle with a galaxy. In these simulations we have a total N of 524288, half of which are gas with $\Omega_b = 1/11$, giving each a mass of $1.54 \times 10^9 M_\odot$ in a 40 Mpc cube (all lengths are quoted with $H_0 = 50 \, \mathrm{km \, s^{-1} \, Mpc^{-1}}$). The Cold Dark Matter spectrum, as given in Bond and Efstathiou (1984), is added to a uniform random distribution of particles. At the expansion that we identify as the current epoch the linear extrapolation of the spectrum gives a calculated mass variance in 16 Mpc top hat spheres of 0.51 (direct measurement finds 0.54) and ultimately leads to our experiments being slightly underevolved.

The gravitational accelerations of the particles on one another are calculated with a PPPM code. The gas dynamics algorithm introduces scale breaking into the gas component at an appropriate mass scale, and is not really intended to follow the details of galaxy formation. When the gas particles collide, there are three possible outcomes, no cooling, instantaneous

271

C. S. Frenk et al. (eds.), The Epoch of Galaxy Formation, 271–274.

complete cooling, or star formation. The outcomes are regulated by the density and temperature of the gas, as set by a simplified version of the cooling curve (Raymond, Cox and Smith 1976). That is, if the density exceeds a critical value of $10^{-2}\,\mathrm{cm}^{-3}$ and the collision velocity along the line of centers does not exceed a few hundred $\mathrm{km\,s}^{-1}$, the gas instantaneously is cooled to a temperature of $10^4\mathrm{K}$. Otherwise the collision is completely adiabatic and the velocities along the line of centers are simply reversed. If the density exceeds $5 \times 10^3\,\mathrm{cm}^{-3}$ the particles turn into stars and are subsequently collisionless.

We performed a series of tests, raising and lowering the critical temperatures by factors of 4, and the densities by a factor of 8, to test for sensitivity to the gas parameters. The variations can make significant changes to the numbers and masses of the individual galaxies, but the galaxy correlation length is restricted to variations of approximately 10% over this range of gas parameters.

3. STAR FORMATION

The redshift of peak star formation is near 4.5, corresponding to the time when the mass scale of dwarf galaxies, that correspond to two and three particles in these simulations, are collapsing. The model predicts a current epoch average star formation rate per galaxy of $dM/dt \simeq 0.2\,\mathrm{M_\odot\,yr}^{-1}$. At a redshift of 4 the star formation rate is equivalent to $105\,\mathrm{M_\odot\,yr}^{-1}$, averaged over the 1000 galaxies that eventually form, although no large galaxies are present at a redshift 4. It must be emphasized that the way in which stars form determines the properties of the resulting galaxies. In this particular scheme gas turns into stars as soon as the gas density exceeds some threshold, providing that the temperature is not too high. In this case we expect star formation to occur as quickly as small ($10^{10}\,\mathrm{M_\odot}$ minimum here) bound units collapse, and the epoch of star formation will be significantly earlier than the epoch of formation of large galaxies. Because clustering bias diminishes with time, early star formation will tend to favour a low bias.

4. GALAXIES AND MERGERS

Galaxies are defined as clumps of star particles within a specified maximum distance of each other. In the 40 Mpc box a link length of 0.1 grid units leaves only 11883 of the 89224 star particles as individuals, and is adopted as the standard value for the galaxies. Galaxies are required to contain at least 5 star particles, scaling to a minimum luminosity of $M_B = -18.5$ (using an assumed stellar $M/L = 2$). In a 40 Mpc box model, there are 985 detected galaxies, with a mean mass of 10 particles and a mean half-mass radius of 0.069 grid units, or 43 kpc. In a 80 Mpc box a link length of 0.1 grid unit leaves 7314 single stars, and finds 3644 galaxies containing at least 3 particles, corresponding to a minimum luminosity of -20.2. The mean radius of these galaxies is 0.049 units, or 61 kpc.

To estimate the current epoch merger rate in these data we use the criterion that some minimum fraction, f, of the mass of a pre-existing galaxy be incorporated into a galaxy at a lower redshift. We have chosen $f = 0.3$, that is, if anything in excess of 30% of the mass of a pre-existing galaxy is given to another galaxy, it is counted as a merger. We find that the merger rate per luminous galaxy in the model remains nearly constant up to a redshift of 1 at approximately 0.020 Gyr^{-1} per galaxy. Raising the required retention fraction, f, to 60% decreases the number of merged galaxies between z=0 and z=0.16 to 24 out of 985, or 0.009 Gyr^{-1} at the current epoch. Nearly 30% of the galaxies ($M_B \leq -18.5$) appear between a redshift of 1.34 and 0.81, largely as a consequence of the merging of small galaxies building up larger ones.

Toomre (1977) estimated that 11 pairs out of a total of approximately 4000 NGC galaxies are certainly destined to merge, and that the galaxies remain in an obviously disturbed state for 0.5 Gyr, implying a merger rate of 0.005 Gyr^{-1}, a factor of 4 *less* than in the simulation data. A similar fraction of interacting galaxies is found in the Arp and Madore (1986) atlas, where they find that 12.6% of their 6445 catalogued galaxies are "interacting doubles". The 812 interacting doubles represent 1.04% of the 77835 galaxies examined, a factor of 3.7 times higher than Toomre's (1977) 0.28%, although the Arp and Madore (1986) catalogue includes interacting doubles to much wider separations than Toomre would have included in his list of doomed galaxies.

The rate of increase of the mean mass of the model galaxies is dominated by small galaxies falling into larger ones, simply because larger galaxies generally have larger halos, and most of the galaxies are small galaxies. Bahcall and Tremaine (1988) combine the Malin and Carter (1983) counts of shell galaxies with the shell formation studies of Malin, Quinn, and Carter (1983) to estimate that the fractional rate of mass gain from small mergers onto a large elliptical is $d\ln(M)/dt = 0.017$ Gyr^{-1}. The model data indicates a fractional mass gain rate of 0.049 Gyr^{-1}, of which 0.002 Gyr^{-1} is a result of residual star formation, and 0.023 Gyr^{-1} is from massive galaxies merging, leaving 0.022 Gyr^{-1} as a consequence of low mass mergers, only 50% more than the observational estimate.

We conclude that the current observational estimates of mass gain through mergers are 1.5-3 times lower than the values inferred for the current epoch in the models. The discrepancy may be partly due to conservative observational estimates, but the model estimates are affected by the limited resolution and the low number of particles in the model galaxies artificially increasing the merger cross-sections.

5. NUMBER DENSITY-REDSHIFT RELATIONS

The Press-Schechter theory predicts $N(M,z)dM$, the number of collapsed objects between mass M and $M + dM$, for self-similar clustering. The fraction of the mass in collapsed objects of mass scale M is given as $f(M) = \text{erfc}[\delta_c(1+z)/(\sqrt{2}\sigma_0(M))]$, where δ_c is the normalizing factor giving the size of the density fluctuation that results in a dark halo. Here $\sigma_0(M)$ is the linear extrapolation to redshift zero of the RMS mass density fluctuation after filtering the initial density field with a Gaussian filter. For the purposes of comparison to our N-body data we evaluate the expected number of objects in mass intervals from M to $2M$.

An alternate description for the number redshift relation is the statistics of peaks presented by Bardeen, *et al.* (1986). Integrating their peak density equation from a critical threshold overdensity $\nu_t = \delta_c(1+z)/\sigma_0(M)$ to infinity gives a prediction of the number density-redshift relation.

Figure 1a shows that the number density-redshift relation for the stars and galaxies are well fit using the peaks theory using $\delta_c = 1.44$, and Figure 1b shows that the dark halo data are well fit by the Press-Schechter theory, using the same δ_c. This parameter is accurate to approximately 10%.

The peaks theory predicts the number density-redshift data for stars and galaxies with impressive accuracy, and the Press-Schechter theory works equally well for the dissipationless halos. The absence of a merging description in the peaks theory is not a serious deficiency when a comparison is made to this simulation, since only 30% of the galaxies have detectable mergers, and the number of galaxies never declined. On the other hand, the peaks theory cannot be successfully applied to the halos because the initially abundant galactic mass halos significantly decrease in numbers as they merge together.

274

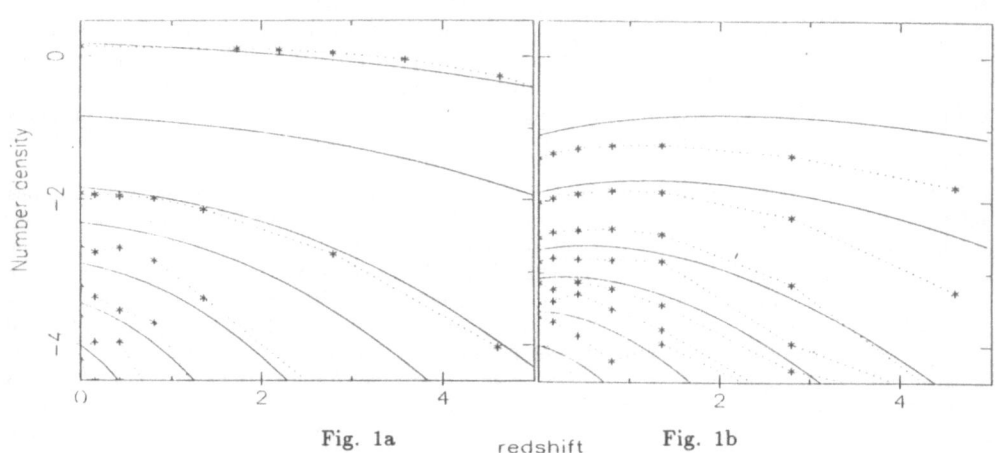

GAUSSIAN PEAKS THEORY   PRESS–SCHECHTER

Fig. 1a   redshift   Fig. 1b

6. DISCUSSION AND CONCLUSIONS

The main result is that our dissipative model successfully breaks galaxy clustering away from the clustering of dissipationless halos. In spite of the complex manner by which galaxies are formed, the statistical properties of the galaxies are well described with the BBKS statistics of non-merging density peaks. The Press-Schechter theory is an equally useful description of the constantly merging dark halos. A common value for the threshold overdensity, found here to be $\delta_c = 1.44$, fits both galaxies and halos. This implies that on the mass scale of galaxies, $10^{12}\,M_\odot$, the current epoch overdensity that is contributing to galaxy formation corresponds to a relatively low overdensity, $\nu_t = 0.65$, leading to a low clustering bias. The main reason for the low peaks in galaxies is that the redshift of star formation is much earlier ($z_* = 5$) than the redshift of the appearance of the bright galaxies, $z_{gal} \simeq 1.5$. Galactic mass gain through merging in the model is roughly in accord with other theoretical estimates, but is about a factor of 1.5-3 times higher than observational estimates.

References

Arp, H. A., and Madore, B. F. 1986, *A Catalogue of Southern Peculiar Galaxies and Associations* (Cambridge University Press: Cambridge).
Bahcall, S. and Tremaine, S. D. 1988, *Ap. J. (Letters)*, **326**, L1.
Bardeen, J. M., Bond, J. R., Kaiser, N., and Szalay, A. S. 1986, *Ap. J.*, **304**, 15.
Bond, J. R., and Efstathiou, G. E. 1984, *Ap. J. (Letters)*, **285**, L45.
Malin, D. F., and Carter, D. 1983, *Ap. J.*, **274**, 534.
Malin, D. F., Quinn, P. J., and Graham, J. A. 1983, *Ap. J. (Letters)*, **272**, L5.
Press, W. H., and Schechter, P. 1974, *Ap. J.*, **187**, 425.
Raymond, J. C., Cox, D. P., and Smith, B. W. 1976, *Ap. J. (Letters)*, **204**, L290.
Toomre, A. 1977, in *Evolution of Galaxies and Stellar Populations*, ed. B. M. Tinsley and R. B. Larson, (Yale Observatory: New Haven) p. 401.

CLUSTERING AND DYNAMICS IN A DISSIPATIVE CDM N-BODY SIMULATION

H.M.P. Couchman
Canadian Institute for Theoretical Astrophysics,

R.G. Carlberg
Department of Astronomy,
University of Toronto, Toronto, ON M5S 1A1, Canada

ABSTRACT. An N-body code is used to follow the evolution of an $\Omega = 1$, $h = 0.5$ Cold Dark Matter spectrum in cubes of 40, 80 and 200 Mpc on a side. We model a universe consisting of collisionless dark matter and a dissipative baryonic component. Cooling enables gas particles to congregate into small groups which are then identified as galaxies. The the bias present in the model, at the time identified as the present, is relatively small; $b = 1.3 \pm 0.2$. The pairwise velocities on small scales are a factor of two smaller in the galaxies than in the dark matter. Estimates of the mass density of the Universe derived from the Cosmic Virial Theorem and from virial analysis of groups gives $\Omega \sim 0.2$.

INTRODUCTION

The standard $\Omega = 1$ Cold Dark Matter model may be reconciled with $\Omega \simeq 0.2$ inferred from the motions and clustering of galaxies only if galaxies formed in such a way that they reflect only a fraction of the total matter. There have been a variety of estimates of the amount of bias present in the galaxy distribution. Davis *et al.* (1985) labelled particles in a N-body model as galaxies only if they exceeded an overdensity threshold 2.5 times the RMS fluctuation amplitude on galaxy scales as measured in the initial fluctuations. This gives an effective bias of around 2.05 (Lilje and Efstathiou 1988). A somewhat lower bias, b=1.7, was derived using the statistics of peaks (Bardeen *et al.* 1986). An important constraint that requires a small bias for the CDM model to be viable is the large-scale streaming velocities of galaxies observed by Lynden-Bell *et al.* (1988) who conclude that $b = 1$ is compatible with the data but that $b = 2$ or higher is improbable.

This paper presents the results of a series of N-body simulations which include collisionless dark matter together with a gaseous component which can dissipate and form galaxies. Experiments were run in cubic boxes of three different linear scales, 40, 80 and 200 Mpc. The simulations were evolved with 524288 particles using a PPPM code with periodic boundary conditions. Half of the particles were treated as dark matter the other half as gas. Each gas particle was assigned as mass of one tenth of a dark matter particle. By running experiments on three different scales we were able to examine the influence of the CDM spectrum on both large scale structure and on galaxy scales. Details of the numerical techniques involved and of the properties of the galaxies formed are discussed by Carlberg and Couchman (this volume).

TWO-POINT CORRELATIONS AND BIAS

Bias — the degree to which the clustering of galaxies differs from that of the dark matter — may be measured in a variety of ways. Here we adopt the definition $b = \sqrt{\xi_{gg}/\xi_{\rho\rho}}$. In this case b is a function of radial separation r. An alternative that has been used is $b = \sigma_N(8h^{-1}\,\text{Mpc})/\sigma_\rho(8h^{-1}\,\text{Mpc})$ where σ_N is the RMS variation in the number of galaxies on that scale. These two definitions are essentially equivalent in linear theory. Fluctuation spectra are commonly normalized to the present-

C. S. Frenk et al. (eds.), The Epoch of Galaxy Formation, 275–279.

day clustering by calculating the RMS variation of the number of galaxies in a sphere of radius r. This is observed to be unity at a radius of $8h^{-1}$ Mpc. For this reason some authors have taken the bias to be simply the reciprocal of the linear extrapolation of the RMS overdensity of the dark matter on this scale. This may cause confusion, however, when comparing estimates of the bias if $\sigma_\rho(8h^{-1}\,\text{Mpc}) \neq 1$ (see below).

In the 40 Mpc box each gas particle weighs $1.54 \times 10^9\,M_\odot$. A galaxy with 20 particles in this box has a luminosity which corresponds to $M_B \simeq -20.0$. Thus even for the smallest box a galaxy is composed of only a few particles. A problem arises, however, on the longest wavelengths in the 40 Mpc box since the correlation length of galaxies is observed to be ~ 10 Mpc and scales greater than this will not be evolved correctly by the simulation method. In view of these restrictions the 80 Mpc box was used to set the normalization to the present epoch. This choice gives good resolution on galaxy scales whilst ensuring that the longest waves in the box are evolving linearly. Galaxies are identified by linking together particles that have successfully cooled and formed 'stars' as described in Carlberg and Couchman (1988). In the 80 Mpc box a link length of 0.1 grid units was used where the linear size of the box was 64 grid units. This procedure give 3644 galaxies containing at least 3 particles corresponding to a minimum luminosity equivalent to $M_B = -20.2$. This assumes an $M/L \simeq 2$ in the baryonic component or 22 overall. The mean size of these galaxies was 0.049 grid units or 61 kpc.

The two-point correlation functions for the galaxies and dark matter are shown in Figure 1. For the three particle and greater galaxies $\sigma_N(16\text{Mpc}) = 0.69$ suggesting that the simulation is somewhat underevolved unless the Davis and Peebles (1983) normalization applies to galaxies brighter than $M_B = -20.2$. (Galaxies with $n \geq 8$ have $\sigma_N(16\text{Mpc}) = 0.92$.) Since the luminosity function in the simulation has relatively more galaxies at low luminosities than is observed the precise comparison is uncertain. At this epoch $\sigma_\rho(16\text{Mpc}) = 0.54$. These figures suggest a bias value $b \sim 1.3$ using the preferred definition adopted above. Alternatively; there are 961 galaxies with $n \geq 5$. This better corresponds with the space density of galaxies brighter than $M_B = -20.2$ as found by Felten (1985). This would imply an $M/L \sim 4$. For these galaxies $\sigma_N(16\,\text{Mpc}) = 0.8$ implying a bias of $b \simeq 1.5$. In this case the value of σ_N suggests that the simulation is underevolved by about 20%. Since bias tends to decrease with time as lower peaks collapse and the dark matter clustering evolves towards that of the galaxies, continued evolution will reduce the bias below the values given above. These results suggest that the true bias at the epoch when $\sigma_N(16\,\text{Mpc}) = 1$ is in the range $b = 1.3 \pm 0.2$.

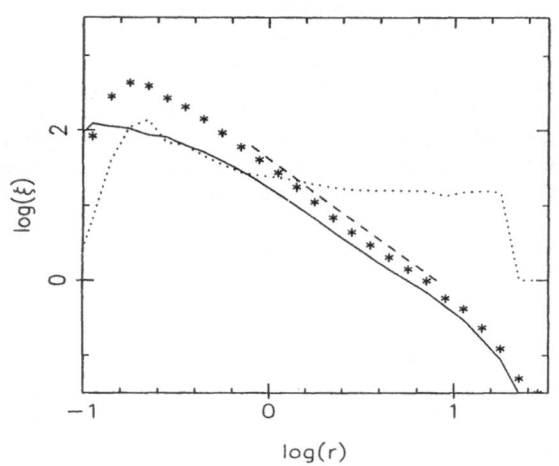

Figure 1. The two-point correlation function of the galaxies in the simulation containing $n \geq 3$ particles (asterisks) together with that for the dark matter (solid line) in the 80 Mpc box. The dashed line is the observed correlation of galaxies. The dotted line is the bias value on a linear scale ($b = \sqrt{\xi_{gg}/\xi_{\rho\rho}}$). It should be noted that the bias is relatively constant over a wide range of scales. The radial scale is expressed in grid units.

In order to compare the results of these simulations with those obtained by Davis et al. (1985),

White *et al.* (1987) we found dark haloes in our simulations which correspond to their measured haloes. Our experiments are somewhat more evolved than theirs: they find $\sigma_\rho(16\,\mathrm{Mpc}) = 0.4$ at the present epoch. A redshift of 0.43 was chosen in our experiments for comparison, as this best corresponds to the amplitude of their spectrum. At this epoch we find $\sigma_\rho(16\,\mathrm{Mpc}) = 0.36$. A link length of 0.2 grid units gives 446 haloes having more than 20 particles. These haloes have a mean implied circular velocity of $240\,\mathrm{km\,s^{-1}}$ and represent a population similar to the $V_c \geq 200\,\mathrm{km\,s^{-1}}$ haloes of White *et al.* (1987). These were found to have $b \simeq 2$ in excellent agreement with the results of Lilje and Efstathiou (1988). We may conclude from this that these experiments reproduce those of Davis *et al.* to $\simeq 20\%$. The primary new feature of our experiments is to include a simple dynamical prescription of star formation that marks the galaxies. To attain the observed clustering amplitude the experiments with these galaxies must be evolved so that the dark matter is quite strongly clustered, giving rise to a low bias estimate.

DYNAMICAL ESTIMATES OF Ω

It is commonly supposed that a high value of bias must be present in the galaxy distribution for an $\Omega = 1$ universe to be compatible with low dynamical estimates of Ω. We find that the implied matter density inferred from the cosmic virial theorem and from group virial analyses is compatible with $\Omega \simeq 0.2$ because of the presence of a large bias in the velocities of galaxies when compared with those of the dark matter.

Figure 2 shows the parwise velocities of the galaxies and dark matter in the 40 Mpc box. It can be seen that at small separations the pairwise velocities of the galaxies are a factor of about two smaller than those of the dark matter with the velocity ellipsoid of the galaxies being somewhat more radial. At larger separations the pairwise velocitiy rises towards that of the dark matter. Similar results hold for the dark matter halo population.

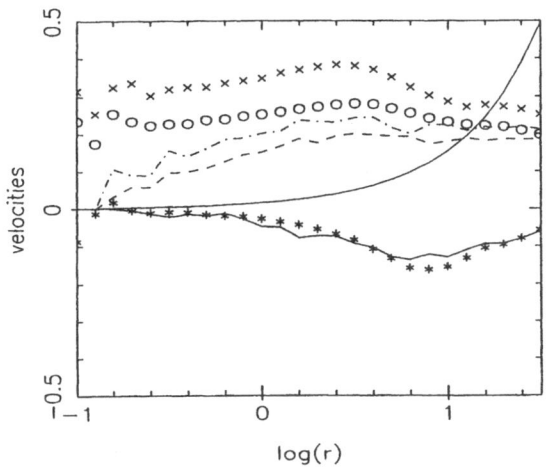

Figure 2. The pairwise velocities of the galaxies in the 40 Mpc box (lines) and of the dark matter (symbols). In each case the following quantities are shown: $\langle v_r \rangle$ solid line and asterisks; σ_r dashed line and circles; σ_T dash-dot line and crosses. The thin solid line is the Hubble velocity at that separation. The units of velocity are expressed in terms of the Hubble velocity across the box. The radial separation is given in grid units.

A preliminary application of the Cosmic Virial Theorem gives $\Omega \sim 0.2$–0.3. A further estimate of Ω was made by constructing a group catalogue for comparison with the Huchra and Geller (1982) CfA group catalogue. A link length of 2.5 Mpc was used in the 80 Mpc box to find groups. There were 251 groups with three or more members, this leaves 20% of galaxies as singles and 11% as binaries. These are similar to the estimates of Nolthenius and White (1987). The median harmonic radius of the groups is 1.35 Mpc, median velocity dispersion is $147\,\mathrm{km\,s^{-1}}$, and the median ratio of the virial mass estimate to the total mass is $\Omega = 0.18$. The largest group has 238 members a harmonic radius of 2.7 Mpc, a velocity dispersion of $764\,\mathrm{km\,s^{-1}}$, and an estimated virial mass of $5.4 \times 10^{14}\,M_\odot$. Groups found in the 40 Mpc box give similar results. Thus, even for the low bias measured in these experiments, the value of Ω is seriously underestimated.

CONCLUSIONS

The most important feature of these experiments is that the dissipated and dissipationless components have significantly different histories. In particular the galaxies and dark matter haloes have very different merger rates. Haloes effectively fall together once and merge whereas galaxies survive 2–4 times longer on average (Carlberg and Couchman 1988). The results is that galaxies do not trace dark matter haloes particulary well. It is this feature together with the more evolved dark matter distribution of these experiments that lead to a low bias and results differing from those of Davis *et al.* (1985).

A low bias $b \sim 1.3$ has many benefits for Cold Dark Matter cosmologies. In particular observed large-scale streaming velocities are more compatible with CDM and the epoch of galaxy formation occurs at redshifts > 1.

The marked difference in velocity between the galaxies and dark matter may be understood heuristically as follows. Firstly galaxies tend to form near the peaks of the density distribution and these have smaller velocities than the typical point. Secondly the dissipated stellar component lies near the bottom of a dark matter potental well which may have a much higher dispersion. As the distribution evolves the velocities will grow under the action of gravity whilst mergers will tend to reduce relative velocities at the expense of internal velocity dispersion. After the epoch of rapid merging, $z \sim 1.5$, galaxy velocities increase together with the dark matter velocities as the clustering evolves.

Although we believe that the bias estimate of this paper is based on a reasonable set of rules for the formation of galaxies (see Carlberg and Couchman 1988), the low bias is largely a consequence of the redshift of star formation being much earlier, $z_* \simeq 5$, than the redshift of the appearance of bright galaxies, $z_{gal} \simeq 1.5$. A scheme that retarded star formation until much heavier objects were turning around would be likely to give a higher bias.

REFERENCES

Bardeen, J.M., Bond, J.R., Kaiser, N. and Szalay, A.S. 1986. *Astrophys. J.*, **304**, 15.

Carlberg, R.G. and Couchman, H.M.P., 1988. *Astrophys. J.*, submitted.

Davis, M., Efstathiou, G.E., Frenk, C.S. and White, S.D.M., 1985. *Astrophys. J.*, **292**, 371.

Felten, J.E., 1985. *Comments Astrophys.*, **11**, 2.

Huchra, J.P. and Geller, M.J., 1982. *Astrophys. J.*, **257**, 423.

Lilje, P. and Efstathiou, G.E., 1988. *Mon. Not. R. astr. Soc.*, submitted.

Lynden-Bell, D., Faber, S.M., Burstein, D., Davies, R.L., Dressler, A., Terlevich, R.J. and Wegner, G., 1988. *Ap. J.*, **326**, 1988.

Nolthenius, R. and White, S.D.M, 1987. *Mon. Not. R. astr. Soc.*, **225**, 505.

White, S.D.M., Davis, M., Efstathiou, G. and Frenk, C.S. 1987. *Nature*, **330**, 451.

DISCUSSION:

SHAPIRO: From your graphs of $\xi(r)$, it appears that the largest wavelength that fits within your smallest box has gone nonlinear. How can you justify the assumption of periodic boundary conditions for your simulation in this case?

COUCHMAN: The longest waves in the 40 Mpc box are indeed becoming nonlinear and will therefore not be evolved correctly by the adopted simulation method. For this reason we use the 80 Mpc box, in which this problem does not arise, to set the normalisation.

THE GALAXY LUMINOSITY FUNCTION: AN ALTERNATIVE TO THE PRESS AND SCHECHTER TECHNIQUE

E. Martínez-González & J.L. Sanz
Departamento de Física Moderna
Universidad de Cantabria
39005 Santander, Spain

ABSTRACT. We discuss a new model of galaxy formation, based on the biased scenario, where each galaxy morphology is characterized by several parameters in the initial conditions. The main result is the morphological luminosity function that is compared with observations of E+SO and S+Im galaxies. We also obtain the epoch of galaxy formation in this model.

1. INTRODUCTION

Observations of the luminosity function (LF) for all galaxies have been carried out by many authors using different galaxy samples. The main characteristic of these observations is that the LF is well represented by the Schechter function $\Phi(L) \propto L^{\alpha} \exp(-L/L_*)$, with the values $\alpha \approx -1$ and $L_* \approx 10^{10} L_{\odot}$ (see Felten 1985, Efstathiou 1987). On the other hand, studies of the LF for each Hubble type show bell-like shapes. This last behaviour has been found by Sandage et al. (1985) for galaxies in the Virgo cluster and by Thompson and Gregory (1980) for the Coma cluster.

In order to explain those observations Press and Schechter (1974) proposed a model where the seeds of galaxies are identified with the initial regions of the random field that are able to collapse by the present time. Later on, Epstein (1984) refined the Press and Schechter model for a Poisson model solving the cloud in cloud problem. While these attempts to get the LF involve all the galaxies, here we consider the LF for each morphological type. The main idea of our approach is to mimic the complex fenomena that take part in the process of galaxy formation by applying a high frecuency filter on the initial conditions for each morphology. Even though we do not know anyting about the possible shape of this filter (we will choose a Gaussian one for simplicity) we think that it is not going to play a crucial role in the final determination of the LF. The steps necesary to calculate the LF for each galaxy type are given in the next section.

2. THE MORPHOLOGICAL LUMINOSITY FUNCTION

A detailed derivation of the LF is given elsewhere (Martínez-Gonález and Sanz 1988). Here we will breafly sunmmarize the main points.

C. S. Frenk et al. (eds.), The Epoch of Galaxy Formation, 281–284.

We assume a Gaussian random field for the initial density fluctuations and consider that each morphological type of galaxies (E, S, ...) comes from the seed regions with density fluctuation above some global level ν of the field filtered on an appropiate scale R_f. We then can calculate the probability density of intervals $P(l)$ along any arbitrary line crossing these seed regions. From $P(l)$ it is possible to obtain the radii probability density of these regions $P(R)$, assuming previously that they are spherical. While this last assumption is streactly valid for high values of ν, it has been proved by 2D numerical simulations that it is still a good approximation for $\nu \approx 2$ (Appel 1987).

To get the mass function $P(M)$ from $P(R)$ we can easily do it by using the relation $M = 4\pi/3 R^3 \rho_b$ bettwen the mass of a seed region and its radius (ρ_b is the background density). Finally, by assuming a constant M/L ratio, L being the luminosity, we can obtain the LF.

3. COMPARISON WITH OBSERVATIONS

Two free parameters characterize each galaxy morphology: the threshold ν and the filtering scale R_f (or equivalently the mean luminosity $\langle L \rangle$). We assume the constant mass to light ratios $(M/L)_E = 80h$ and $(M/L)_S = 30h$ for ellipticals and spirals respectively (Faber 1981). Results for the case of scale-free spectra are given in Martínez–González and Sanz (1987). In the figures below we show the case of a cold dark matter model with adiabatic fluctuations and compare it with the observations of Sandage et al. (1985) for E+S0 and S+Im galaxies. For the free parameters we have chosen $\nu = 2.5$ and $R_f = 0.3h^{-1}Mpc$ for E+S0 and $\nu = 2$ and $R_f = 0.125h^{-1}Mpc$ for S+Im. From the figures we see that the observations can be reproduced by these particular choices of the model.

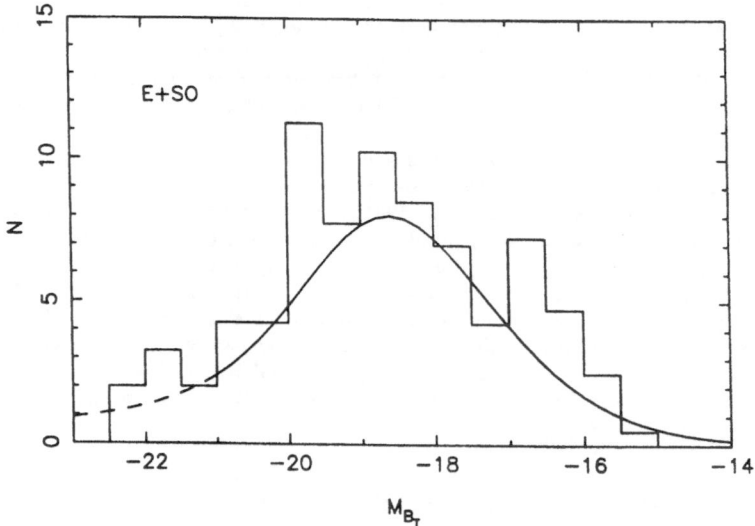

Figure 1. Luminosity distribution for E+S0 galaxies. The predictions of the cold dark matter model (solid line) are compared with observations from Sandage *et. al.* 1985 (histogram).

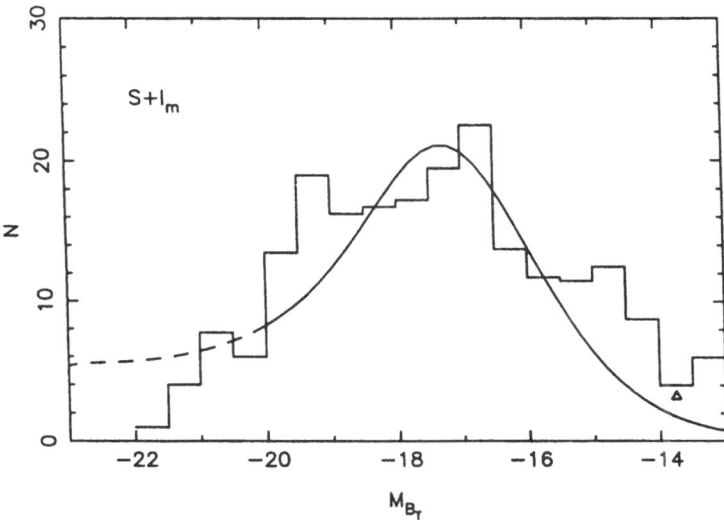

Figure 2. As Figure 1 for S+Im galaxies.

4. THE EPOCH OF COLLAPSE

Assuming that the time of collapse z_c for a seed region is given by the spherical model, we have

$$z_c = \frac{\nu \sigma_0}{\delta_c} - 1$$

where $\delta_c = 1.68$ and σ_0 is the r.m.s. fluctuation at present. For the case of E+SO galaxies all of them have collapsed at a redshift $z_c = 6$ and for S+Im this redshift is $z_c = 6.8$. The relative number of collapsed objects at redshift z, n/n_0 is given by the equation

$$\frac{n}{n_0} = \left[\frac{\text{erfc}\left(\frac{\delta_c}{\sqrt{2}\sigma_0}(1 + z_c)\right)}{\text{erfc}\left(\frac{\delta_c}{\sqrt{2}\sigma_0}(1 + z)\right)} \right]^2 \exp\left(-\frac{\delta_c^2}{2\sigma_0^2}[(1 + z)^2 - (1 + z_c)^2]\right)$$

In the figures below we show this evolution.

284

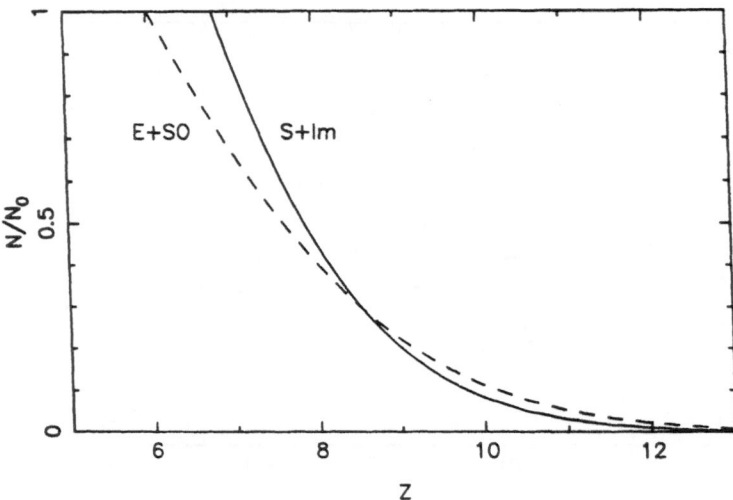

Figure 3. Fraction of collapsed objects as a function of redshift predicted in the cold dark matter model.

References

Appel, L., 1987, private communication.
Efstathiou, G., Ellis, R.S., and Peterson, B.A., 1988, *M.N.R.A.S.*, **232**, 431
Epstein, R., 1984, *Ap. J.*, **281**,545
Faber, S.M., 1981, Proceedings of the Study week on *Cosmolgy and Fundamental Physics*, Pont. Acad. Scient., Scripta Varia, **48**, 191
Felten, J.E., 1985, *Comments on Astrophys.*, **11**, 53
Martínez–González, E. and Sanz, J.L., 1988, Proceedings of the NATO Advanced Study Institute *The Post-Recombination Universe*, ed. Kaiser N. and Lasenby A.N.. Kluwer
Martínez–González, E. and Sanz, J.L., 1988, *Ap. J.*, in press
Press, W.H. and Schechter, P., 1974, *Ap. J.*, **187**, 425
Sandage, A., Binggeli, B. and Tamman, G.A., 1985, *A. J.*, **90**, 1759
Thompson, L.A. and Gregory, S.A., 1980, *Ap. J.*, **242**, 1

DISCUSSION:

BONOMETTO: Does the large mass cut-off in the number density have anything to do with the balance between collapse and cooling time? Or does this come only from the exponential dependence of ν on number densities?

MARTINEZ: The large mass cut-off is due to the fact that you find far fewer big objects than small objects above the threshold. It is clear that it must be related to some physical mechanism such as cooling time in relation to collapse time, and that we think it must have been, in some way, taken into account by the use of the filter acting on the initial conditions.

GALACTIC EVOLUTION AND GLOBAL STAR FORMATION

Joseph Silk
University of California at Berkeley
Astronomy Department
Berkeley, California 94720

Rosemary F. G. Wyse
The Johns Hopkins University
Department of Physics and Astronomy
Baltimore, Maryland 21218

ABSTRACT. The basic ingredients of a model for galactic star formation are reviewed and incorporated into a simple scheme for galactic evolution. A unified model in which the star formation rate is proportional to the ratio of gas surface density to local dynamical time–scale describes star formation in starburst galaxies, protogalactic disks, and protogalactic spheroids. The possible role of active galactic nuclei in triggering early star formation provides an alternative evolutionary scheme.

1. Introduction

Knowledge of the star formation rate throughout the universe as a function of space and time is the primary goal of both observers and theorists who work in the area of galaxy formation. Armed with this knowledge, one would be able to unveil the epoch of protogalaxy collapse. The difficulty which arises and is hindering progress in this direction is the lack of an adequate theory of star formation. This review will discuss the uncertainties inherent in prescriptions for the rate of star formation, and will describe simplistic schemes for global star formation in the context of galactic evolution models.

The basic ingredients of a global star formation model are the initial mass function, the star formation rate, and the star formation efficiency. These are defined and described in §2. The following section (§3) debates the role of disks: need star formation necessarily occur in galactic disks? §4 discusses a theoretical prescription for star formation in disk galaxies, and §5 applies it to starbursts and early galactic evolution. We develop the cosmological context for protogalactic star formation, and describe the probable origin of morphological differences between galaxies. Active galactic nuclei (AGN) could also be important triggers of star formation, and §6 is devoted to exploring the role of AGN in galactic evolution. A final section summarizes our results.

C. S. Frenk et al. (eds.), The Epoch of Galaxy Formation, 285–308.

2. Fundamentals of Global Star Formation

Critical star formation parameters that must be specified in order to prescribe a galactic evolution model include the stellar initial mass function, which is the mass distribution of all stars ever formed, the rate of star formation, which we define as the instantaneous rate of mass consumed by star formation, and the efficiency of star formation, a measure of the total mass of stars that a gas cloud can form prior to gas exhaustion – thus we take into account both the returned fraction per generation of stars and the star formation per dynamical time.

2.1 INITIAL MASS FUNCTION (IMF)

2.1.1. *The IMF in Normal Galaxies.* The IMF is not well known even in normal galaxies. The most detailed data exist for the solar neighbourhood over the range $\sim 0.1 M_\odot$ to $\sim 100 M_\odot$, as reviewed by Scalo (1987). For the short-lived (compared to the age of the galaxy) massive stars as well as at the lowest masses, there is a considerable uncertainty in the conversion from luminosity to mass function, which requires knowledge of the star formation history. In addition, direct estimates of the IMF from star counts of luminous stars have widely differing logarithmic slopes, spanning the range $-1.0 \geq \Gamma \geq -2.2$, where Γ is defined by

$$\frac{dN(M)}{d\ln M} \propto M^\Gamma. \tag{1}$$

Observations of nearby galaxies require that for stellar masses greater than $\sim 2 M_\odot$, $\Gamma \approx -1.5(\pm 0.5)$ with uncertainty arising from incompleteness, reddening corrections, and theoretical evolutionary tracks. In the LMC, the IMF is indistinguishable from the solar neighbourhood IMF. In the outer regions of the LMC, there are HII regions where far UV observations indicate an apparent truncation in the IMF above $\sim 30 M_\odot$ (Lequeux 1987). Only massive stars appear to have formed in the spiral arms of M81 and M83, although in the latter case, there is evidence for truncation above $\sim 15 M_\odot$ (Jensen *et al.* 1981). Data on several galactic open clusters are suggestive of a turn-over below $\sim 1 M_\odot$, although this result is not definitive. There is the well–known phenomenon of T associations, which demonstrate that low mass star formation is not inevitably accompanied by *contemporary* massive star formation. A further complication is that the IMF may be time–dependent over short timescales: star–forming regions appear to form progressively more massive stars. The solar neighbourhood IMF does have features near 1.2 $M_\odot$ and $3 M_\odot$ that are suggestive of prior bursts of star formation according to Scalo (1987), although they are easily understood if the IMF has not been invariant, and in particular, may have become progressively more deficient in low mass stars with each succeeding burst. Barry (1988) has found evidence for local peaks in the number-age histograms of nearby G-dwarfs (ages derived from chromospheric activity), at ages of $\sim 10^8$ yr and $\sim 5 \times 10^9$ yr which may also support the idea of bursts of stars formation in the disk. The chemical evolution of such a model has not been tested in detail.

2.1.2 *The IMF in Starburst Galaxies.* Studies of several starburst galaxies have concluded that the IMF is deficient in low mass stars, with truncation below $M_l = 2 - 3M_\odot$. The evidence was originally based on the large 2 μm luminosities of M82 and NGC 253, which require many old supergiants to have evolved within the age of the burst ($\sim 3 \times 10^7$ yr) inferred from modelling $(M/L)_{bol}$, broad band colours, and Hα line strength (Rieke *et al.* 1980). The case of M82 has been especially well studied, and it is clear that the starburst is contained within the central 600pc. More recent data reviewed by Rieke (1988) on the near infrared line spectrum constrains the temperature of the exciting stars, setting an upper limit to their mass of $\sim 25M_\odot$. The CO absorption bands from cool stars set a lower bound on the duration of the starburst. High resolution millimetre–wave observations are used to obtain a reddening–free estimate of the ionizing photon flux, and molecular line studies yield estimates of total mass (from rotation curve modelling) and gas mass in the starburst region. One has to produce a far infrared luminosity of $\sim 4 \times 10^{10} L_\odot$ from a gas supply of $\sim 10^8 M_\odot$. Formation of almost exclusively massive stars provides the only simple explanation. The observed supernova rate of $1/(2\text{--}20)$ yr^{-1}, and evidence for a supernova–driven x–ray wind, provide supporting evidence for a high massive star formation rate in the starburst region (Kronberg 1988). The apparent star formation efficiency (SFE) is large for any reasonable IMF: e.g. with $10 < M < 25M_\odot$, the SFE is ~ 30 percent (Rieke 1988). However allowance for gas return ($R \approx 0.9$) brings the global SFE down to the nominal range (SFE ≈ 3 percent) for giant molecular clouds in our galaxy.

Independent arguments for other starburst galaxies have arrived at similar conclusions. The weakness of the Balmer series absorption lines in the spectra of the interacting systems Mk 171 and IC 2153 leads one to infer a high value of the lower mass cutoff – $M_\ell \sim 10M_\odot$ (Augarde and Lequeux 1985). Similar values of $M_\ell \sim 4M_\odot$ and $\sim 10M_\odot$ are derived for the two blue compact dwarfs I Zw 36 (Viallefond and Thuan 1983) and ESO338 - IG04 (Bergvall and Jörsäter 1988), respectively based on their large Hβ equivalent widths. Further indirect arguments that favor large M_ℓ have been summarized by Scalo (1987): these include the derived short gas-exhaustion time–scales, the extremely low ($\lesssim 0.01$) bolometric mass-to-light ratios observed in extreme cases, the persistence of blue light relative to Hα emission expected for a low M_ℓ- IMF that is not found in interacting galaxy samples, and the low values of C/O measured in several blue compact dwarfs and dwarf irregulars.

2.1.3 *The Primordial IMF.* Did the primordial IMF differ from the solar neighbourhood IMF? All we can offer are indirect arguments, of varying degrees of plausibility. There are solar–mass stars of exceedingly low metallicity: the most metal–poor is a $0.4M_\odot$ dwarf with [Fe/H] ≈ -5.6 (Liebert *et al.* 1986). The precursor stars which synthesized the heavy elements in Population II stars must have included many stars of mass $\sim 20 - 50M_\odot$ in order to account most simply for the observed abundances of oxygen and other elements.

Primordial massive stars could have been responsible for early ionization of the intergalactic medium. It is debatable whether quasars could provide sufficient ionizing photons to account for the ionization level inferred from the Gunn–Peterson test; the calculation is rather dependent on ill-constrained parameters but the most plausible values for these suggest that some other ionizing source is required. However, efficient formation of stars with masses above $\sim 20M_\odot$ amounting to a mass fraction, relative to the critical cosmological density $\Omega_* \lesssim 0.005$, comparable to the amount of matter in galactic disks, would suffice, although these stars would have generated a substantial fraction of the Population

I heavy element abundance (Shapiro, these proceedings). A much larger mass fraction in massive stars, $\Omega_* \gtrsim 0.1$, is necessary if primordial stars are to produce the excess energy observed in the sub–millimeter spectral distortion of the cosmic microwave background (Bond, these proceedings). These stars must exist at $z > 10$ in order for there to be sufficient optical depth in the dust (also presumed to be produced by these stars) to absorb the stellar photons, and most of their processed ejecta must be undetectable since we do not observe a sufficiently high level of chemical enrichment anywhere (assuming that the stars are not VMOs which collapse to black holes), and perhaps hidden in the form of dark baryonic matter outside the luminous regions of galaxies.

It is not implausible that dark halos consist of baryonic matter in the form of stellar relics. The standard primordial nucleosynthesis constraint from ^{6}Li, ^{2}H and ^{3}He sets a lower bound on Ω_b of ~ 0.02, which exceeds the luminous baryonic matter density by a factor ~ 4. Isothermal (r^{-2}) dark halos could therefore extend, on the average, for several optical radii and consist of dark baryonic matter. This dark matter could consist of substellar objects (of planetary mass) or massive ($\gtrsim 200 M_\odot$) black holes that efficiently swallow any processed debris: it cannot consist of matter originally with an IMF containing stars anywhere in the conventional range from 0.1 to 100 $M_\odot$, otherwise either excessive light would be radiated or excessive processing of protogalactic matter would have occurred.

2.1.4 *The Theoretical IMF.* Theoretical prescriptions for the IMF are *post hoc*. If one had to predict the mass of a star, it would be $\sim 10^{-2} M_\odot$ (the minimum opacity-limited Jeans mass) or $\sim 10^4 - 10^5 M_\odot$ (the magnetic Jeans mass for an interstellar cloud). Neither limit has direct relevance to the primordial IMF, nor for that matter, to the conventional IMF. Since the thermal Jeans mass varies as $T^2 P^{-1/2}$, it has been argued that high pressures (as inferred for some cooling flows) favor very low mass star formation: however, the far stronger dependence on temperature, the absolute value of which is poorly determined, renders this argument meaningless. A more quantitative reason for the irrelevance of substantial pressure variations in different regions on the IMF is that $P \approx$ constant in galactic molecular clouds and cloud cores, under the single assumption that these regions are virialized over a range of scales that spans regions where both massive and exclusively low mass stars are forming (Larson 1981, Myers 1983, Dame *et al.* 1986). The number of Jupiters forming in a cooling flow can be predicted about as well as the number of angels able to fit on the point of a needle. It is equally absurd to argue, at least on the basis of any star formation theory, that $10^6 M_\odot$ VMOs form efficiently out of primordial clouds, as opposed to, say, conventional clouds or galactic nuclei.

The Jeans mass itself may have little bearing on the IMF: it is at best a lower bound on fragment masses. Processes such as accretion and coalescence increase fragment size, while angular momentum is a key limiting factor: the net outcome is unpredictable. We can make reasonable statements about the masses of potentially star–forming gas clouds, but the theory of star formation in conventional molecular clouds is entirely driven by the observed characteristics of the gas and of the young stars observed therein. Since we observe neither primordial clouds nor primordial protostars, we are at somewhat of a theoretical impasse in attempting to extend our theories of current star formation. The only sensible approach is to utilize what indirect evidence there is, which suggests that primordial stars (see above) and primordial clouds (see below) spanned a similar range to those observed today.

2.2 THE RATE OF STAR FORMATION (SFR)

2.2.1 *Normal Galaxies.* The star formation history of a typical spiral or irregular galaxy is crudely measured from its Lyman continuum flux, which probes star formation over the past $10^7 - 10^8$ yr, blue luminosity, sensitive to the past $\sim 10^9$ yr, and total stellar mass, which yields the mean SFR (Gallagher, Hunter and Tutukov 1984). Normal galaxies are characterized by a nearly constant SFR, although some irregulars have a slowly rising SFR. Within the crude time-resolution available, it is difficult in the former case to distinguish a uniform SFR from a discontinuous one, consisting of a series of bursts (where a burst is defined to be a greatly elevated, by $\gtrsim 10$, SFR) of suitable duty cycle. If the present SFR greatly exceeds the mean value, as it does for some rare irregular galaxies, a burst of star formation is evidently in progress. The duration of the burst can be modeled: typically, we infer a duration of $\sim 3 \times 10^7 - 10^8$ yr, over which time up to $\sim 10\%$ of the stellar content of the galaxy may be generated. The statistics of starbursts can be inferred from IRAS–selected samples, and for Hα studies of spiral samples, to imply that most disk galaxies could have undergone several starbursts. This idea can be tested in detail by the elemental abundance information available for galactic stars (Wyse and Gilmore 1988).

2.2.2 *Starburst Galaxies.* The luminosity function of IRAS galaxies provides a measure of the frequency of starburst systems. In a normal spiral, such as our own galaxy, $L_{IR}/L_B \sim 0.5$, whereas in extreme starburst galaxies, this ratio increases to $\gtrsim 100$. About 1% of luminous galaxies have L_{IR}/L_B enhanced by a factor of ~ 10, consistent with the hypothesis that most ordinary galaxies have undergone at least one starburst of duration $\lesssim 10^8$ yr. The starburst phenomena is associated with galaxy interactions, especially at high luminosities: the fraction of mergers or close pairs increases from 25% at modest luminosity ($L_{IR} \sim 3 \times 10^{10}$ $L_\odot$) to 100% for $L_{IR} > 10^{12}$ $L_\odot$ (Scoville 1988). There are indications that bars are associated with the weaker starbursts, and that active nuclei provide a substantial part of the infrared flux in the most luminous starburst galaxies. Most bursts in interacting systems are confined to the nuclear regions ($\lesssim 1$ kpc), while intense disk star formation activity, seen via the emission, accompanies strong nuclear emission in some of the strongest starburst galaxies. The star formation rate is found to be enhanced per unit mass of gas, as measured by L_{IR}/M_{H_2}, in starburst systems, but has a broad dispersion.

2.2.3 *Protogalaxies.* The star formation rate during the first $\sim 10^9$ yr in the protogalaxy must have been substantially higher than the present or time–averaged value in normal spirals. The observed elemental abundance distributions of metal-poor ([Fe/H] $\lesssim -1$) stars suggest that the formation phase of the protospheroid took $\lesssim 10^9$ years (Wyse and Gilmore 1988) while the smooth $r^{1/4}$ light profiles observed for many spheroids seem to require that the metal-poor spheroids form stars on of order a free-fall time, or a few $\times 10^8$ yr (van Albada 1982). This latter argument applied to a luminous elliptical galaxy predicts early star formation rates of $\sim 10^3 M_\odot$ yr^{-1}. The thin-disk age, estimated from the white dwarf luminosity function and theoretical cooling curves or ages of open clusters, exceeds 10 Gyr, and may be several Gyr less than the spheroid age, in agreement with theoretical arguments: a slow dissipative contraction time, relative to the dynamical time, is necessary to form a flattened stellar population.

2.3 THE STAR FORMATION EFFICIENCY (SFE)

2.3.1 *Normal Galaxies.* The local efficiency of star formation will be defined as the mass fraction of stars and long–lived remnants formed from a gas cloud within its lifetime. The net efficiency allows for return of gas that is ejected from stars via winds, planetary nebulae, and supernova explosions. The returned fraction (R) is about 1/3 for a solar neighbourhood IMF, and may be as large as ~ 0.9 if the IMF is truncated below several solar masses.

The SFE is measured in giant molecular clouds in our galaxy to be about 2%. One can also compute a global SFE value defined over a dynamical time–scale: with a star formation rate of 3 $M_\odot$ yr^{-1} and a gas content of 6 $\times 10^9$ $M_\odot$, we infer a time–scale for star formation with a standard (solar neighbourhood) IMF of 3×10^9 yr, yielding a mean SFE of 3% for a dynamical time of $\sim 10^8$ yr. Similar star formation time–scales and SFE's are inferred for many spiral galaxies. This leads to an apparent coincidence (*cf.* Sandage 1986): gas exhaustion would seem to be imminent, since the age of the disk is probably between 8 and 12 $\times 10^9$ yr, according to different estimates. If the disk age is near the lower limit of the allowed range, then the observed gas fraction of about 10% of the disk is consistent with $\sim 2 - 3$ e–foldings of star formation in a simple closed box model with SFR proportional to gas surface density and a standard IMF. An older disk may require a bimodal IMF (Güsten and Mezger 1983; Larson 1986; Sandage 1986), or gas infall to replenish the gas reservoir.

2.3.2 *Starburst Galaxies.* There is little doubt that a high SFE is required in starburst galaxies. The observed rate of massive star formation, from the infrared observations, combined with the molecular hydrogen mass inferred from CO flux measurements, suggests that strongly interacting or merging galaxies, which describe the category of galaxies with high far infrared–to–CO ratios, have gas consumption time–scales that are up to an order of magnitude shorter than in galactic giant molecular clouds. Values for a sample of interacting galaxies are $L_{FIR} \sim 2 \times 10^{11}$ $L_\odot$, $L_{FIR}/L_B = 6$, and $M(H_2) = 3 \times 10^9$ $M_\odot$. Adopting an IMF with $\Gamma = -1.35$, Gallagher and Hunter (1987) find that the rate of formation of stars above $2 M_\odot$ is $\sim 2 \times 10^{-10}$ $L_{FIR}(L_\odot)$ $M_\odot$ yr^{-1}. The mean gas-consumption time–scale is $\sim 1 \times 10^8$ yr with no allowance for gas return, these systems including Mrk 231, Mrk 273, NGC 1614, Arp 193, and Arp 220 as extreme examples (Solomon and Sage 1988). To obtain the SFE, we require comparison of the gas-consumption time with the local dynamical time–scale. If the starburst is occurring within the innermost kpc of the galaxy, as appears to be the case for Arp 220, the estimated dynamical time–scale is only $\sim (1-2) \times 10^7$ yr, implying an SFE of about 10% even if we make no allowance for any low mass star formation. Allowance for gas return, however, reduces the net SFE to near the galactic value.

2.3.3 *Theoretical Arguments.* Presumably similar physical mechanisms are involved for star formation in starburst galaxies as occur in our own galaxy. One may simply have to appeal to more extreme parameters to account for the enhanced SFR (and SFE if necessary), and variations in the IMF. These mechanisms are predominantly the following processes. Large–scale density waves in self–gravitating, gas–rich galactic stellar disks drive molecular cloud aggregation over a dynamical time–scale (Kwan and Valdes 1987). Smaller clouds are magnetically supported (Myers and Goodman 1988). The gas cools efficiently, so that when cloud masses are enhanced, cloud collapse is initiated. Generic collapse results in formation of compressed sheets, which are unstable to fragmentation. The resulting fragments have excessive angular momentum to form stars: magnetic torquing is the most efficient means of transporting angular momentum and producing stellar mass accretion disks and protostars. Outflows from convectively unstable contracting protostars as well as winds from massive stars self-regulate the star formation efficiency in molecular clouds, which are eventually dispersed by the ionization fronts driven by OB stars (e.g., Larson 1988a).

The global disk instabilities may be enhanced by cold gas injection, since addition of a cold, dissipative component enhances the instability of a marginally unstable disk (Sellwood and Carlberg 1984). Radial gas flows, induced by bars, tidal interactions and mergers, are plausibly expected to enhance the rate of cloud formation and coagulation throughout the galactic disk, and, presumably, the rate of star formation (Combes 1988). Once sufficient gas accumulates in the inner kpc of a galaxy that a self-gravitating gas disk forms, numerical simulations suggest that star formation becomes strongly enhanced by the ensuing non-axisymmetric disk instabilities. It seems plausible to speculate that the enhanced viscosity in such a marginally unstable disk drives gas flows that are capable of growing and feeding a central black hole, thereby causally coupling nuclear activity to an intense starburst phase (Lin et al. 1988). A note of caution is called for, however: it may be that the nuclear activity triggers the starburst (see §6 below).

3. Disks as Precursors to Star Formation

Protogalactic disks may not be inevitable prerequisites for efficient star formation from the theorist's perspective, but we shall argue that circumstantial evidence would bring about a conviction before any reasonable jury.

Firstly, a cold self-gravitating disk is unstable to fragmentation, whereas an initially uniform, collapsing spheroid is stable. Only by imposing sizeable initial fluctuations on scales above the initial Jeans length can one hope to initiate fragmentation in a generic collapse, prior to formation of a thin disk or sheet. Of course, one cannot exclude this possibility from the point of view of initial conditions, but one would have to carefully tune the star formation parameters in order to produce the desired high efficiency.

Secondly, star formation is observed to occur almost uniquely in galactic disks. Certainly, global star formation is only found in disks, and starburst galaxies are invariably disk galaxies, albeit, in some cases, undergoing strong interactions.

Thirdly, mergers of disks plausibly result in elliptical galaxy formation (Toomre 1977). There is evidence of both an observational and theoretical nature that the violent relaxation accompanying mergers yields de Vaucouleurs profiles in the final stellar distributions (van Albada 1982; McGlynn 1982). Moreover, close inspection of many ellipticals reveals evidence for past mergers, in the form of outlying shells of stars and of dynamical subsystems (Quinn and Hernquist 1987). This suggests that most, if not all, ellipticals underwent mergers in the past, some several gigayears ago.

Fourthly, mergers are only likely to occur in encounters when galaxy relative velocities are low, of order the internal stellar velocity dispersion. This excludes rich clusters as being an environment where merging occurs, and tells us that galaxy groups are the ideal environment for mergers. Groups are disk galaxy-dominated: again, suggesting that merging gas-rich disks are the precursors to ellipticals. In fact, the observed groups generally have too long a galaxy merger time-scale to efficiently form ellipticals. It is their compact counterparts that are the culprits, and most of these are presumably destined to have been incorporated into the great clusters of galaxies. The rare isolated compact goups are indeed observed to be spiral–deficient relative to the field (Hickson 1982), and are presumed to be relatively short–lived, perhaps being destined to form cD galaxies (Barnes 1985, Merritt 1985).

Fifthly, a dissipative collapse is required, not merely to account for rotationally supported disks, but for important features of spheroids, such as metallicity gradients and the correlation of metallicity with luminosity. Dissipative collapses generally result in disk formation, and an ensuing merger will not efficiently erase metallicity correlations: even gradients survive. Indeed, a transient pancake is also a generic feature of cold, non-dissipative collapses.

Of course, these arguments do not guarantee that protodisks are precursors to protospheroids. One could well imagine that the basic star forming units in a protogalaxy are clouds of globular cluster mass (Larson 1988b). An objection is that there is no evidence that this could be an efficient mode of star formation, whether observational or theoretical. Indeed, theoretical reasoning would suggest that the high velocity cloud collisions endemic to a collapsing spheroidal distribution would be a disruptive rather than a catalytic influence on the SFE. There is, finally, the intriguing suggestion that the highest star formation rates observed in extreme starbursts approach 10^3 $M_\odot$ yr^{-1}, coincidentally precisely what one infers for formation of a luminous spheroid. This is a dangerous comparison, however: the starbursts may form predominantly massive stars, whereas the protospheroid evidently formed at least a comparable number of low mass and massive stars depending on the adopted value of R. A bimodal IMF, with roughly comparable masses in low and high mass stars, might be acceptable both for protospheroids and for extreme starbursts.

4. Prescription for Global Star Formation

Star formation depends in detail on many parameters. It is clear that such effects as feedback, metallicity, magnetic fields, rotation, ionization, molecular chemistry, etc. must all play a role in regulating star formation. However, as with other complex nonlinear phenomena, one may hope that a simple phenomenological description may suffice to describe the key features.

4.1 AN EMPIRICAL LAW FOR THE STAR FORMATION RATE

We propose (Wyse and Silk 1989) the following modified Schmidt-law (*cf.* Schmidt 1959; 1963) for studying star formation in disks, as a function of position and of time:

$$\Psi(r,t) = \varepsilon\Omega(r)\mu_{HI}^n(r,t); \qquad 1 \lesssim n \lesssim 2, \tag{2}$$

where μ_{HI} is the surface density of HI, $\Omega(r)$ is the local angular frequency and ε is proportional to the efficiency of star formation, defined above. The dependence on $\Omega(r)$ allows the SFR, $\Psi(r,t)$, to depend on rotation and shear. It represents, to within a constant, the pattern frequency in a grand design spiral or the local shear rate in a flocculent spiral: in either case, $\Omega(r)$ is a measure of the rate at which density waves entrain gas clouds. We have assumed that SFR is simply proportional to the *atomic* gas surface density. Wyse (1986) showed that a law analogous to equation (1) for the formation rate of H_2 provides a good fit to the sharply-peaked H_2 distribution in most disk galaxies, although since the HI distribution is fairly constant and $\Omega(r)$ is steeply declining with radius, the power of μ_{HI} is not tightly constrained. The similarity between the H_2 distribution and that of young stars suggests that the two are causally related. The currently popular view is that when a molecular cloud grows sufficiently massive, via accretion, for example in the spiral arms, it becomes unstable and forms stars more prolifically than was previously possible in individual quasi–stable subclumps. Indeed, Rana and Wilkinson (1986) analysed various recent estimates for the run of SFR and gas densities with radius in our Milky Way galaxy and concluded that the most consistent result was that the SFR and the molecular hydrogen surface density were indeed related linearly, while the SFR and the atomic gas show no strong correlation. The association of OB star formation with spiral arms suggests that the spiral pattern velocity as well as the gas density is likely to be a determining factor in regulating the SFR, as noted by Oort (1974).

The models presented here are natural extensions of contemporary research into galactic evolution, having features in common with several alternative pictures. The parameterisation of SFR depending on gas density is a common feature of models of chemical evolution for the solar neighbourhood, such as those described by Lynden-Bell (1975), Talbot and Arnett (1975), Vader and de Jong (1981), Güsten and Metzger (1983), Lacey and Fall (1983, 1985), and Rana and Wilkinson (1986, 1987), with the component of gas being variously the atomic gas, the molecular gas, and the sum of the two. A consensus from these models is that an additional feature is needed to explain abundance gradients in disk galaxies along with simultaneously producing the radial profiles of gas and star formation. Talbot and Arnett (1975) introduced a metallicity dependence in the SFR, as did Rana and Wilkinson; Lacey and Fall (1985) allowed gas to flow radially; Tinsley and Larson (1978) found metallicity gradients to be a feature of the dynamical collapse models of Larson (1976), while Talbot (1980) and Güsten and Metzger (1983) appealed to a radial variation of the stellar initial mass function. Here we show that the radial dependence of the angular frequency provides an alternative (or perhaps additional) mechanism, as expected from early work on density wave theory (*cf.* Jensen, Strom and Strom 1976) but as we show here, not restricted to that class of model.

Perhaps a more fundamental approach to star formation involves study of the stability of a rotating self–gravitating disk of gas and stars to both axisymmetric and non–axisymmetric perturbations. A critical parameter that governs the local stability of such a disk is the Toomre Q–parameter (*e.g.* Toomre 1981),

$$Q = \sigma_g \kappa / \pi G \mu,$$

where κ is the epicyclic frequency ($\kappa \simeq 2\Omega$ if $\Omega \propto r^{-1}$), σ_g is the velocity dispersion of the disk, and μ is the surface density. For marginal stability, one requires $1 \lesssim Q \lesssim 2$, a condition that may be satisfied in spiral galaxies: in this case, local gravitational instabilities are subjected to swing amplification and account for the observed density waves. A self–gravitating gas–rich disk may be sufficiently dissipative to maintain Q near unity, so that swing amplification should be an especially strong generator of massive star–forming molecular clouds. Suppose that the dissipation rate is able to maintain $Q \sim 1$: then the growth rate for density waves is $\sim \Omega$, and the molecular cloud formation rate, and consequently, the SFR, is

$$\Psi(r,t) = \varepsilon \Omega(r) \mu(r,t), \tag{3}$$

as above with $n = 1$. We have oversimplified the above discussion primarily by ignoring the fact that a disk consists of stars and gas, which itself consists of several components, and using just 'μ' as a generic surface density. However, one would still expect an effective Q to describe the effects of self–gravity on the gas component (Jog and Solomon 1984) which may affect the effective value of n.

Note that equation (2) refers to the instantaneous SFR, which disregards the complication that gas may either be recycled through many generations of stars, or locked up in low-mass stars or remnants. To incorporate this, we can write $\varepsilon = \text{SFE}/(1\text{-R})$, where the star formation efficiency, as before, refers to the mass in long–lived stars and remnants. The rate at which mass is permanently removed from the gas component of galaxies will, in the simplest approximation, be proportional to the SFR, the constant of proportionality depending on the stellar IMF, as described in the next section.

4.2 TESTS OF THE STAR FORMATION LAW

The above prescription for molecular cloud and star formation in galactic disks has to meet various tests. We have already mentioned its success in accounting for the present-day radial distribution of H_2 and young, massive stars. Further tests of our proposed prescription for the SFR arise when we consider the implied radial dependence of the chemical evolution of galactic disks. The rotation curve of a typical disk, with $\Omega(r) \sim$ constant at $r \lesssim 1$ kpc, and $\Omega(r) \propto r^{-1}$ at $r \geq 3$ kpc, naturally gives rise to an increased SFR in the inner disk that results in enhanced stellar processing and a gas metallicity approaching the nucleosynthetic yield, whereas the outer disk remains relatively gas–rich and metal-poor. One can quantify these arguments by making the following assumptions : we will adopt the Instantaneous Recycling Approximation (IRA), which describes the return of (enriched) gas to the interstellar medium by the returned fraction, R, assumed constant, such that the rate at which matter is permanently locked up into stars, denoted here by $\dot{\mu}_*(r,t)$, is simply $(1-R)\Psi(r,t)$. We will further assume that our disk evolves as a closed system, with no radial flows, and thus the chemical evolution follows that of the 'simple, closed-box' model, which is known to predict too many long-lived stars of low metallicity for the observed solar neighbourhood metallicity distribution, independent of star formation rate, if one assumes zero initial metallicity (the 'G-dwarf problem' cf. van den Bergh 1962; Pagel and Patchett 1975). However, such models can provide a good approximation to the solar neighbourhood G-dwarf distribution provided that one adopts an initial metallicity of $0.1 \lesssim Z_\odot \lesssim 0.25$ (Audouze and Tinsley 1976), a value which may arise naturally during the formation of the thick disk and halo components of the Galaxy (Gilmore and Wyse 1986).

The further assumption that as the disk evolves, atomic gas is transformed into stars and molecular hydrogen – i.e. molecular clouds are not entirely dissociated when stars form, leading to a net creation of molecular gas at a rate simply proportional to the rate of depletion of HI – allows one to derive a simple set of analytic equations for the radial distribution and temporal evolution of the HI gas content, the SFR and the metallicity, for each of $n = 1$ and $n = 2$. The shapes of the resulting metallicity gradients are in good agreement with the observed gradients in metallicity (oxygen, produced by short-lived massive stars; Diaz and Tosi 1984; Tosi and Diaz 1985) of HII regions in several external spiral galaxies. The present models predict gradients that are independent of the nature of the spiral structure in a given galaxy, and thus are consistent with the fact that metallicity gradients of equal amplitude are observed for galaxies of very different spiral-arm characteristics – M51, a classic 'grand-design' spiral, and NGC 2403, a 'flocculent' spiral are examples (Diaz and Tosi). This contrasts with models that seek to explain the existence of gradients with, for example, radial flows that arise in the velocity field of a quasi-steady density wave. Irregular galaxies, which are typically in a state of near rigid-body rotation, are not predicted to develop any gradient in SFR or, therefore, in metallicity, which is also in accord with the observations.

One can also calculate the predicted history of metal enrichment at any radius. Estimates of the age-metallicity relation exist for stars in the solar neighbourhood (Twarog 1980; Carlberg et al. 1984) and for clusters in the anti-center direction and in the LMC (Geisler 1988). The fit to the solar-neighbourhood observations (which have a large spread) is satisfactory for both $n = 1, 2$, and the rates of depletion of gas – and thus the variations in predicted SFR over the age of the disk – are consistent with the (rather poorly-defined) constraints from, e.g. , continuity of the IMF (Scalo 1987). The LMC/anti-center predictions may require a lowering of the star formation efficiency, even after taking account of the lower initial gas surface density (relative to the solar neighbourhood). This 'lowering of the efficiency' most probably reflects the fact that gas is likely to flow out of these regions due to supernova heating.

The B–V and U–B gradients of these models can be derived according to two complementary prescriptions. For $n = 1$, the SFR at each radius decays exponentially with time, analogous to Bruzual's (1983) 'μ–models', but with a different e–folding time for each radius. One can thus simply add together different Bruzual models to synthesize a disk; this yields good agreement both with the actual values of the colours in the solar neighbourhood, and with the slope of the colour gradients (remembering that Bruzual's models lack any chemical evolution, consisting of solar metallicity stars everywhere, and so will underestimate the colour gradients). One can also use the prescription of Larson and Tinsley (1978) to relate B−V and U−B colours to the recent SFR relative to the past average SFR, for either $n = 1$ or $n = 2$. We have compared the predicted B−V gradients at age 10 Gyr for our Milky Way $n = 1$ model using Bruzual's colours with the Larson-Tinsley colour prescription and, for comparison, the predictions of the Villumsen, Gunn and Casertano (1988) infall model for galactic evolution, at 10.2 Gyr. Our model is somewhat redder than the Villumsen et al. model, and is in good agreement with the Bahcall-Soneira (1980) solar-neighbourhood colour (B−V = 0.68) at age 10 Gyr, and with van der Kruit's (1986) analysis of the Pioneer data (B−V = 0.84 ± 0.15) at age 15 Gyr.

Estimation of *global* parameters such as SFR obviously allows us to utilize larger samples of galaxies for comparison with observations. Donas et al. (1987) obtained UV fluxes (from a balloon-borne telescope) for a large sample of spiral galaxies spanning a range of Hubble types and gas contents, and derived estimates of the current SFR in these galaxies (note that this involves extrapolation over all stellar masses through an adopted IMF, and corrections for dust extinction). Data on a variety of Hubble types supports our proposed $\Omega(r)$ dependence. The gas depletion time–scale is found to increase markedly towards later Hubble types, median values rising from $\sim$ 2Gyr for Sbc galaxies to $\sim$ 7Gyr for Sdm galaxies (Donas et al. 1987). The gas surface density does not vary much with Hubble type, and the decrease in rotation rate provides a natural explanation of the lower SFR, and consequently higher gas fraction, observed in late Hubble types. An intermediate value of n $(3/2)$ is favored by the far UV data, but one should not take this too seriously because of uncertain extinction corrections.

5. A Minimal Model for Protogalactic Star Formation

Now that our simplest model for star formation, defined by equation (3), meets a variety of tests in our galaxy and in nearby galaxies, we propose to apply it in a more general context. The parameter ε equals the ratio of local rotation time–scale to "star formation" time–scale: μ_{gas}/Ψ, in other words $\varepsilon = $ SFE/(1-R). We generalize our interpretation of SFE by the following: we allow ε to vary with galactic environment and we consider $\Omega(r)^{-1}$ to measure the local dynamical time, rather than simply rotation period. For example, we anticipate that more efficient star formation (larger ε) should occur as a consequence of a merger, as suggested by the observations. Theoretically, we have in mind the consequence of a self-gravitating gas disk which develops following gas infall, perhaps via radial inflows, triggered by a close tidal interaction or a merger of a gas-rich dwarf with a disk galaxy, leading to formation of a central bar. Such dynamical interactions should generate disk instabilities that produce transient density waves which enhance the star formation efficiency. In addition, the increase in velocity dispersion implied by the presumed near constancy of Q as μ increases implies that the mass spectrum of fragmentation may be affected: one is likely to increase the minimum fragment mass, perhaps raising M_ℓ, and reducing the locked-up fraction in low mass stars.

The rotation rate $\Omega(r)$ will also be interpreted more literally as the local dynamical time-scale: our star formation prescription (1) will be applied to more general systems than a cold (centrifugally-supported), self-gravitating disk. This generalization is one we do not attempt to justify other than qualitatively: the local dynamical time-scale controls the rate of accumulation of the molecular gas that feeds star formation, and must, therefore, be a critical factor in controlling the SFR.

5.1 THE MODELS

Two prescriptions for protogalactic star formation will be studied. In our protodisk model, we inject gas, increasing μ_{HI} by a factor of 2, but leave $\varepsilon \equiv$ SFE/(1-R) unchanged in a disk that already has been forming stars for several Gyr. The gas injection simulates the effect of infall or accretion of a gas-rich dwarf, and provides either a single delayed burst of star formation, or a series of bursts if the gas is injected in several smaller increments. The rationale for not enhancing the SFE is simply that disks process gas with a low SFE, remaining gas-rich for several Gyr. We assume further that the gas (HI) surface density is enhanced equally over the face of the disk, rather than being confined to the central regions (as in the more powerful 'starburst galaxies').

Our second model is appropriate to a protospheroid. Again, we commence with a disk quiescently forming stars for several Gyr, and then inject a considerable amount of gas, equal to the original mass, but only in the *central* regions. This allows the gas to be dominant and form a self-gravitating disk. One now expects the SFR to be considerably enhanced, and we increase ε by an order of magnitude. This choice is justified conceptually by the *a posteriori* knowledge that ellipticals and spheroids must have formed stars within one, or at most a few, dynamical time-scales, representing a considerable enhancement in the disk SFE, which itself corresponds to ~ 50 dynamical time-scales. Also, since we assume that the starburst is concentrated in the central regions, the increase of $\Omega(r)$ towards the center will by itself increase the SFE.

Both models involve, and require, formation of low mass stars, as well as massive stars. There may be, in addition, an exclusively massive-star-forming mode, as appears to be the case in extreme starburst galaxies. Our prescription is trivially generalized to include a bimodal IMF if such a refinement is required by adjusting R, since $\varepsilon \equiv SFE/(1\text{-}R)$.

Quiescent formation of stars, or at least, star formation with low efficiency, is natural in a hierarchical model for the evolution of large–scale structure prior to the collapse of massive systems corresponding to luminous galaxies. Massive star formation is highly disruptive of gas–rich shallow potential wells. Only when the escape velocity exceeds ~ 100 kms^{-1} can one efficiently retain supernova ejecta and suppress gas loss via a strong supernova-driven wind (Larson 1974; Dekel and Silk 1986). As the hierarchy evolves, more and more mass accretes, and more and more massive subsystems develop. With an initial gaussian fluctuation spectrum that has considerable large–scale power, as in the cold dark matter theory, one may distinguish two alternative evolution models in which mergers of density peaks play different roles. An isolated fluctuation in a low density region may accrete many small fluctuations but will not run into density peaks of comparable scale. However, a density peak in a region modulated by a large–scale fluctuation, where a protocluster is destined to develop, will be liable to undergo a merger with a comparable scale fluctuation while the hierarchy is progressing from group to cluster scales. We identify the former case with a protodisk at low redshift undergoing mild infall, and the latter with a protoelliptical at high redshift forming by violent mergers (*cf.* Zurek and Quinn 1988; Frenk *et al.* 1985).

5.2 MODEL RESULTS

To set the models into a cosmological context we shall adopt $\Omega_o = 1$, and $H_o = 50$km s^{-1}Mpc^{-1}, with the epoch of first star formation ('galaxy formation') at a redshift of 4, or ~ 2 Gyr after the Big Bang. The age (defined as the time since first star formation) of models corresponding to present-day galaxies is then ~ 11.5 Gyr. The epoch of 'galaxy formation' should really also be allowed to vary; such models are in preparation. The properties of the models which we shall study consist of the photometric evolution, the evolution of the gas content, and the chemical evolution. As in §4, we shall derive colours from the model star formation history using Larson and Tinsley's prescription, setting the derived colours to the red limit of $B-V = 1.0$, $U-B = 0.64$ for relative star formation rates below their calibration, and extrapolating to the blue for high relative star formation rates. (Our models are similar in spirit to those of Larson and Tinsley.) We need to improve our treatment of chemical evolution since here we will be in the low gas-fraction regime, where the IRA (used in §4) breaks down. We have therefore implemented the Talbot and Arnett (1971) semi-equilibrium approximation when the IRA gas-fraction drops too low – of order a few percent.

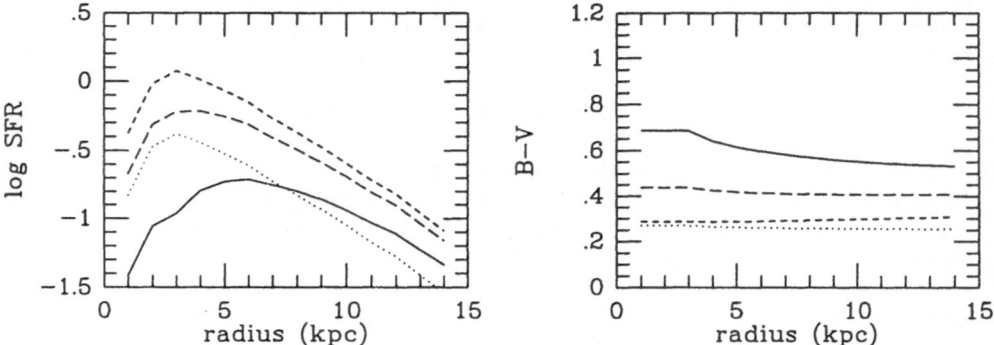

Figure 1: SFR (in $M_\odot/kpc^{-2}yr^{-1}$) and B–V colour at various times for the disk-galaxy model. Dotted curve represents age 1 Gyr, short-dashed represents 2 Gyr, the long-dashed represents 5 Gyr and the solid curve is for 10 Gyr.

Consider, by way of illustration, an initially exponential disk, with a flat rotation curve, which is forming stars at a rate comparable to the disk of the Milky Way Galaxy, until some time, say 2 Gyr after the onset of first star formation, when the HI surface density is instantaneously increased by some amount, the fractional increase being the same at all radii. The driving of SFR by HI gas content (equation [1]) leads to a corresponding increase in SFR, which consumes the gas, and causes the SFR to decay with time. An example of the evolution of such a system is shown in Figure 1, where we plot the radial dependencies of the SFR and B−V colour at various times, for a model with $n = 1$, and gas infall occurring at an age of 2 Gyr (or a redshift of $1 + z \sim 1.6$). The curves shown are for ages of the oldest stars of 1 Gyr (prior to gas infall; dotted line), 2 Gyr (just after gas infall; short-dashed line), 5 Gyr (long dashed line) and 10 Gyr (solid line). Note that the star formation rate changes most in the central regions, and this is reflected in the colour gradient since the colours are derived essentially from the ratio of present SFR to past average SFR. The chemical evolution of this model is shown in Figure 2, where we plot metallicity against age at two different radii – 3 kpc (dashed curve), where the rotation curve has been assumed to turn over, and 8 kpc (solid curve), a solar-neighbourhood environment. The metallicity initially increases, then is diluted when the metal-free gas is accreted. However, it recovers quickly. The glitch in the 3 kpc curve, at ages $\sim$ 7.5 Gyr is caused by the transition from IRA to the Talbot and Arnett formalism. The 8 kpc curve requires no such adjustment. (Note that the 'age' here is the same as the 'age' in Figure 1 and refers to the oldest stars formed, not the age of stars of that metallicity as is more often the case for this diagram.) Disk galaxies that formed in this manner would not differ from observed galaxies. We are currently exploring parameter space, and investigating $n = 2$ models, where the relationship between gas and SFR is such that the models remain blue for a longer period after gas accretion.

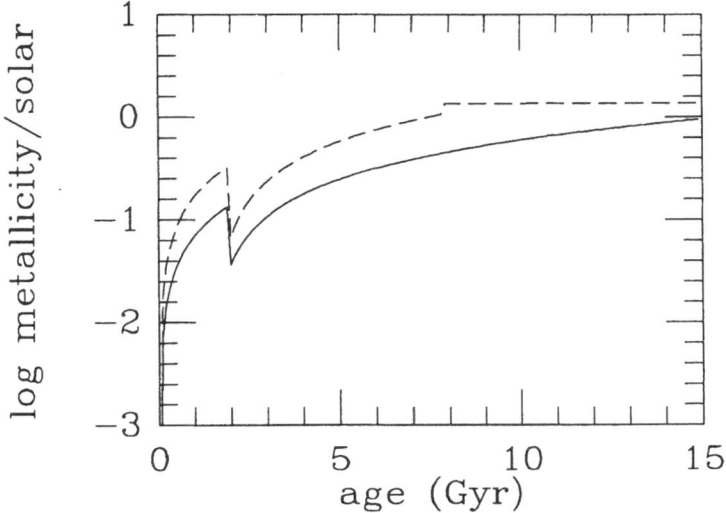

Figure 2: Metallicity evolution for two radii. The dashed curve is for 3 kpc, and the solid curve for 8 kpc.

The 'starburst' models for spheroid formation have enhanced star formation only in the central regions, where Ω is approximately constant so that in the current models there will be no spatial gradients. An $n = 1$ model, with an amount of gas equal to the total mass added at 2 Gyr, and the SFE subsequently enhanced by an order of magnitude, has the colour and metallicity evolution shown in Figure 3. The 'starburst' central region reddens and becomes metal-rich very quickly, mimicking giant elliptical galaxies and starburst nuclei.

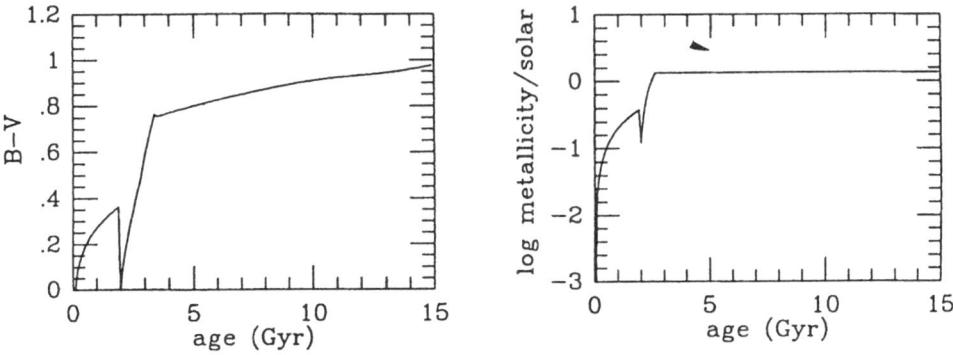

Figure 3: B−V colour and metallicity evolution of the central 'starburst' regions or protospheroid.

6. Active Galactic Nuclei and Global Star Formation

Circumstantial evidence links AGN and starbursts. Minkowski's object is the closest example of a starburst driven by a jet. Extreme starbursts show spectral evidence of an underlying AGN, or even a shrouded quasar. At high redshift, luminous radio galaxies are associated with intense, extended Lyman alpha emission, presumably due to excitation by many massive stars. In the latter two examples, it is not apparent whether the AGN or the radio source has provoked the star formation episode. Observational evidence (Spinrad, these proceedings) provides some indirect arguments. For example the optical emission is preferentially orientated along the radio jet at high redshift, whereas at low redshift, one generally finds the radio jet directed along the minor axis of a luminous elliptical galaxy. Thus it is tempting to infer that the radio emission has triggered the optical emission. Further clues which support this hypothesis are that the radio sources with associated optical emission at high redshift are steep spectrum sources, and are asymmetrical, with the foreshortening occurring in the direction of the optical emission. Radio jets at high redshift show systematic bending and asymmetries, and enhanced depolarization.

6.1 RADIO JETS AND STAR FORMATION

These observations are all consistent with the ideas that the jets are directed into the cloud medium characteristic of a protogalaxy. The pressure required to deflect a jet is of the order of the jet ram pressure; which is $P_R \sim (L_j/\pi r_b^2 c) \sim 10^6 k_B (L_j/10^{44}$ erg s$^{-1})(1$ kpc$/r_b)^{-2}$cm^{-3}K for typical jet parameters (radius r_b, jet luminosity L_j) over a protogalactic scale. The inferred pressure is similar to that expected in a protogalaxy ($n \sim 1$ cm^{-3}, T $\sim 10^6$ K), where $\sim 10^{11}$ $M_\odot$ of gas is within a radius of ~ 10 kpc and supported by random motions of ~ 100 km s^{-1}, equivalent to a virial temperature of $\sim 10^6$ K. Although the cloud random motions dissipate (to ~ 10km s^{-1}) during the formation of a protogalactic disk, the effective pressure is unchanged. The amount of gas entrained at the jet working surface and through the bow shock that separates the shocked jet particles from the protostellar gas is about $3\pi \rho r_b^2 v_b$, where v_b is the beam velocity and ρ is the mean density of protogalactic gas, or 210 $n(v_b/10^2$km s$^{-1})(r_b/1$ kpc$)^2 M_\odot$yr^{-1}. This clearly provides sufficient gas flow to fuel a powerful burst of star formation. Note that independently of any jet, one infers that in order to sustain a star formation rate as high as $\dot{M} \sim 1000 M_\odot$ yr^{-1}, we require a gas flow velocity, assuming a simple spherical geometry with a gravitationally induced flow, of $\sim (G\dot{M})^{1/3} \sim 200$ km s^{-1} even if the star formation were completely efficient (Larson 1988a). However, this is the typical flow velocity expected in a spheroidal potential characteristic of a massive galaxy, and the radial flow velocity induced in a disk would be considerably less. Evidently, the large flow velocity induced by the jet in the entrained gas may be necessary if star formation is to be moderately inefficient. Of course, we have not demonstrated that the entrained gas clouds will survive the jet interaction. A necessary condition is that the self-gravity of a cloud should exceed the ram pressure, which is equivalent to requiring the cloud surface density (if isothermal) to satisfy $\mu > (8\pi P_R/G)^{1/2} \sim 1000(L_j/10^{44}$ erg s$^{-1})^{1/2}(1$ kpc$/r_b)^{-1}$ $M_\odot$ pc^{-2}. This is about an order of magnitude larger than observed for galactic molecular clouds. If protogalactic molecular clouds are on the verge of gravitational instability, as is the case for conventional giant molecular clouds, the fact that the protogalactic pressure P_g balances P_R, inferred from jet bending, means that $\mu \sim (8\pi P_g/G)^{1/2}$, as is the case for galactic clouds.

There is direct evidence in the nearby starburst galaxy M82 from the far infrared [CI] and molecular line emission that the interstellar pressure is much higher than in our galaxy (e.g., Lugten et al. 1986; Carlstrom 1988). We also note that there is a remarkably tight correlation between star formation rate, as monitored by the IRAS flux, and the 6cm radio flux from starburst galaxies. Again, it is not inconceivable that this correlation arises from nuclear radio emission, although supernova remnants produced by massive stars must also play a role. The broad Lyman alpha lines, up to ~ 1000 kms^{-1} in width, observed in the diffuse emission associated with the high redshift radio galaxies, may be attributable to scattering of resonance photons in a dust–free gas envelope that has a velocity dispersion of this order.

6.2 RADIO JETS AND MAGNETIC FIELDS IN PROTOGALAXIES

If the radio jet is responsible for triggering star formation, there is a further noteworthy implication. The jet delivers relativistic particles and magnetic flux from the AGN to the boundary of the protogalaxy. The magnetic flux is of considerable interest, because a protogalaxy is usually presumed to be flux–free, in the absence of a primordial magnetic field. The nuclear field itself originates in the AGN, which is surrounded by an accretion disk, an ideal site for a turbulent dynamo. The nuclear magnetic field should build up to equipartition strength, saturating at

$$H_{nucl} \sim M_{nucl}^{-1} \, c^4 G^{-3/2} \left(r_{Sch}/r_{nucl}\right)^{3/2} \sim 10^{11} \left(r_{Sch}/r_{nucl}\right)^{3/2} M_8^{-1} \quad \text{gauss},$$

where $M_{nucl} \equiv M_8(10^8 M_\odot)$ is the mass of the nucleus, r_{Sch} is the Schwarzschild radius of the dark halo, and r_{nucl} is the radius of the accretion disk. This seeds an interstellar field

$$H_{ism} \sim H_{nucl} \left(r_{nucl}/r_g\right)^2 \sim 10^{-7} \quad \text{gauss}$$

over $r_g \sim 10$ kpc. Such a seed field is adequate to drive a protogalactic dynamo (Pudritz and Silk 1988). Magnetic fields are a crucial factor in surmounting the angular momentum barrier of conventional star formation, and the magnetic flux may be an important link between the radio jet and the ensuing star formation efficiency. Strong magnetic fields are inferred from Faraday rotation measurements to be present in the damped Lyman alpha absorption clouds that are putative protogalactic disks at $z = 2 - 3$ (Wolfe, these proceedings): an early phase of radio activity may have generated them. Star formation in such clouds is relatively quiescent, but the presence of heavy elements indicates that considerable enrichment and star formation must already have occurred at higher redshift.

6.3 AGN AND PROTOGALACTIC STAR FORMATION

Finally, we speculate on the connection between AGN–induced star formation and the protogalactic star formation generic to protogalaxies, mergers, and galactic disks that we previously described. In any bottom–up scheme for galaxy formation, galactic nuclei are most likely to form at high redshift ($z \gtrsim 10$). Rare, dense objects of mass $\sim 10^9 M_\odot$ or even $10^{10} M_\odot$ collapse first and subsequently accrete the bulk of the protogalaxy. The cold dark matter model, the isocurvature baryon–dominated model, and the cosmic string model for galaxy formation all allow, and, indeed require, early formation of galactic nuclei. Physical conditions in the early universe actually favor collapse to very high density.

Consider a protogalactic gas cloud, of radius ~ 100 kpc, and mean initial gas density ~ 0.001 cm^{-3} for a baryonic mass of $10^{11} M_\odot$. The gas is likely to be nonuniform, perhaps in clouds with random motions of $\lesssim 100$ km s^{-1}, and rotation is low (dimensionless angular momentum parameter $\lambda_i \sim 0.05$). Attainment of a high central density depends both on how efficiently the gas dissipates the random motions and on the initial level of the random motions, since both dissipation and violent relaxation play important roles. Following Silk (1988b) one may examine two limiting cases: a) dissipation in a dark halo means that rotational support regulates the final disk surface density to $\mu_f \sim \lambda_i^{-2}$; b) initial random motions are also of order $v/\sigma \sim \lambda_i$, and violent relaxation results in a final core density $\rho_f \sim \lambda_i^{-6}$. Evidently for formation of a dense nucleus, it is the dynamical effects that set the initial density of the nucleus, before mass–loss or two–body relaxation effects intervene to control the final evolution of a massive central object. Nevertheless, one still envisages the gas as consisting of a distribution of clouds of finite cross–section, and even in this case, however, it will be the efficiency of cloud–cloud collisions in dissipating cloud random motions that determine the final density of the nucleus. At $z>10$, Compton cooling guarantees that the gas initially remains isothermal, regardless of the relative velocity of cloud–cloud collisions. When Compton cooling is unimportant, atomic cooling processes do not suffice to cool protogalactic gas efficiently, and dissipation will, therefore, not be very effective. This will limit the maximum density attainable in the galactic core, where cloud–cloud collisions become inevitable. To demonstrate this, we note that in a collapsing protogalaxy, with typical parameters cloud velocity dispersion $\sigma \equiv \sigma_{100}(100$ km s$^{-1})$, cloud mass $M \equiv M_8(10^8 M_\odot)$, and mean halo gas density $n \equiv n_3(0.001 \text{cm}^{-3})$, the cloud density will be regulated by ram pressure of diffuse halo gas to be $\sim 0.1\sigma_{100}^2 n_3 \text{cm}^{-3}$, with cloud radius $\sim 2000 \, M_8^{1/3} \sigma_{100}^{-2/3} n_3^{-1/3}$ pc. Following a cloud–cloud collision, we ask whether the cloud is strongly radiative: this is a necessary condition for cool gas to survive and eventually condense into the protogalactic core. The Compton cooling time–scale of postshock gas, t_{Comp}, the bremsstrahlung cooling time–scale, t_{brem} and the time for the shock to traverse the cloud, t_{shock}, are given by

$$t_{Comp} = 8 \times 10^{19}(1+z)^{-4} \text{ s},$$

$$t_{brem} = 2 \times 10^{15}\sigma_{100}^{-1}n_3^{-1} \text{ s},$$

$$\text{and } t_{shock} = 2 \times 10^{15}M_8^{1/3}n_3^{-1/3}\sigma_{100}^{-5/3} \text{ s}.$$

Evidently t_{brem} is of order t_{shock} for typical protogalaxy parameters, whereas $t_{Comp} < t_{shock}$ at $1 + z \gtrsim 10$. Hence Compton cooling efficiently dissipates random energy of cloud motions released by cloud–cloud collisions at a redshift where the first structures are inferred to be forming in current models for galaxy formation (*cf.* Gunn 1982).

The clouds will survive collisions and fragment to form stars, thereby removing the gas dissipation, provided that the clouds become self–gravitating and collapse: for this to occur, the cloud surface density must satisfy $\mu > (8P_g/\pi G)^{1/2} \equiv 4M_8^{1/3}\sigma_{100}^{4/3}n_3^{2/3}M_\odot\text{pc}^{-2}$. However, postshock bremsstrahlung cooling is only effective over a column $\mu_{cool} \sim 10\sigma_{100}^2 M_\odot\text{pc}^{-2}$, and is of marginal significance in aiding the cloud to grow by coalescence in order to become unstable. Compton cooling plays a critical role in reducing μ_{cool} by an order of magnitude or more at $z > 10$, thereby allowing efficient cloud growth and ensuing collapse. Perhaps the redshift range 10–100 is the era of formation of dense galactic nuclei: in a low–Ω universe, possibly containing primordial isocurvature density perturbations, this is a natural possibility.

7. Summary

Despite huge uncertainties in the key parameters of star formation, namely the stellar initial mass function and the star formation efficiency, it is nevertheless possible to construct a simple model for the rate of star formation that meets all the usual tests with flying colors. In essence, the model asserts that

$$\text{SFR} = \frac{\text{SFE}}{1-\text{R}}\text{M}_{\text{gas}}\Omega(\text{r}),$$

where $\Omega(r)$ is the local dynamical rate (shear or rotation) and with the SFE adopted from studies of star formation in galactic molecular clouds, where it is about 3 percent. Enhanced massive star formation, interpreted as a bimodal IMF, sets the return fraction $\text{R} \approx 0.9$. This appears to be occurring in current epoch starbursts, and theoretical explanations are not hard to concoct (Silk 1989). All that remains to obtain an extreme starburst is to increase M_{gas} and $\Omega(r)$, each of which may plausibly be enhanced by an order of magnitude following a merger within the inner kpc of the galactic disk. This model describes starbursts, and is equally applicable to protogalactic star formation, where we envisage that mergers of gas–rich subsystems trigger vigorous early star formation. In this situation, the IMF cannot have differed too much from the local IMF, low mass stars having formed, although the SFR of massive stars may have been enhanced by an order–of–magnitude, as in Wyse and Silk (1987). Formation of a protospheroid and of a protodisk is indistinguishable from current star formation in extreme and intermediate starbursts, if suitably sustained over several episodes at least for protodisk formation. Mergers driving the protogalaxy phase involve gas–rich sub–systems, and are especially conducive to fueling the inner regions of a forming galaxy with gas. Within the inner kpc, the gas surface density in a protogalaxy must be the ratio of that in stars ($\sim 100M_\odot pc^{-2}$) to the SFE, or $\gtrsim 1000M_\odot$. One measures a molecular gas surface density of precisely this magnitude in extreme starburst systems such as Arp 220.

Non–axisymmetric gravitational instabilities play an important role once the gas is self–gravitating in amplifying density waves and in transferring angular momentum outward in the differentially rotating galaxy disk (Toomre 1981). The enhanced gas streaming provides a prolific supply of gas into the nuclear region on kpc scales (Norman 1988). It is plausible that the radial flow of gas is induced on much smaller scales, again by formation of a marginally unstable self–gravitating disk in which the viscosity is sufficient to permit formation and fueling of a central black hole (Lin *et al.* 1988). Nuclear activity may be an inevitable evolutionary consequence of a strong starburst, perhaps mediated by formation of a massive central star cluster (Scoville 1988). However an alternative possibility, argued here, is that the formation of the AGN may actually precede the starburst phase. This would happen in a cold–dark–matter model, or indeed in any bottom up scenario, in which galaxies formed "inside–out" (Silk and Szalay 1987). Dense clouds containing $10^9 - 10^{10} M_\odot$ condense at high redshift ($z > 10$) from rare fluctuations, and may provide a favorable environment for formation of dense nuclei and perhaps massive black holes. Compton cooling ensures that the *initial* collapse is dissipative and cold, and hence likely to evolve to high density. Only much later ($z \lesssim 5$) is the nucleus surrounded by the late–condensing, common fluctuations that are the source of the bulk of the protogalaxy mass. Mergers of these gaseous sub–condensations fuel the nucleus, and a plausible speculation is that nuclear activity, specifically a jet, triggers violent star formation in the more–or–less quiescently condensing protogalactic disk and halo. The characteristic properties of the high redshift galaxies being discovered via their radio activity emerge naturally in such a model. The major uncertainty is that one has no idea how common such a phenomenon is. Radio–selected samples of high redshift galaxies are not simply related to an unbiased sample, without knowledge of the radio luminosity function, which at $z > 2$ is not known. There are of course, low redshift analogs of jet–induced star formation, as well as the circumstantial evidence that associates a substantial fraction of the emission from an extreme starburst galaxy with an active nucleus.

Jet–induced star formation could lead to a radically different view of protogalaxy evolution. A gas–rich protogalaxy is a fertile, and highly unstable, environment where star formation can readily be induced. Gas clouds, as in our own galaxy, are on the verge of gravitational collapse, and their dissipative nature means that once collapse is induced, star formation is likely to be a run–away process. Protogalactic star formation may be self–induced, as appears to be likely for isolated disks, or merger–induced, the current favorite for galaxies in compact groups and clusters, or modulated by nuclear activity. The ongoing deep surveys promise to greatly increase available samples of high redshift galaxies. We may hope that these data will guide us towards discerning the dominant mode of protogalactic star formation and evolution.

Acknowledgements

We are grateful to many colleagues for helpful discussions and comments, including W. van Breugel, S.M. Fall, C.G. Lacey, C. McKee, R. Pudritz, P. McCarthy, and H. Spinrad. This research has been supported in part at Berkeley by grants from NASA and NSF.

References

Audouze, J. and Tinsley, B.M. 1976, *Ann. Rev. Astr. Ap.*, **14**, 43.

Augarde, R. and Lequeux, J. 1985, *Astr. Ap.*, **147**, 273.

Barnes, J. 1985, *M.N.R.A.S.*, **215**, 517.

Barry, D.C. 1988. Preprint.

Bahcall, J.N. and Soneira, R. 1980, *Ap. J. Suppl.*, **44**, 73.

Bergvall, N. and Jörsäter, S. 1988, *Nature*, **331**, 589.

Bruzual, G. 1983, *Ap. J.*, **273**, 105.

Carlberg, R.G., Dawson, P.C., Hsu, T. and Vandenberg, D.A. 1984, *Ap. J.*, **294**, 674.

Carlstrom, J. 1988. In *Galactic and Extragalactic Star Formation*, eds. R. Pudritz and M. Fich (Dordrecht: Kluwer), 571.

Combes, F. 1988. In *Galactic and Extragalactic Star Formation*, eds. R. Pudritz and M. Fich (Dordrecht: Kluwer), 475.

Dame, T. *et al.* 1986, *Ap. J.*, **309**, 892.

Dekel, A. and Silk, J. 1986, *Ap. J.*, **303**, 39.

Diaz, A. and Tosi, M. 1984, *M.N.R.A.S.*, **208**, 365.

Donas, J., Deharveng, J.M., Laget, M., Milliard, B. and Huguenin, D. 1987, *Astr. Ap.*, **180**, 12.

Frenk, C. *et al.* 1985, *Nature*, **317**, 595.

Gallagher, J.S. and Hunter, D.A. 1987. In *Star Formation in Galaxies*, ed. C.J. Lonsdale (Washington, D.C.: NASA), 167.

Gallagher, J.S., Hunter, D.A. and Tutukov, A.V. 1984, *Ap. J.*, **284**, 544.

Geisler, D. 1987, *A. J.*, **94**, 84.

Gilmore, G. and Wyse, R.F.G. 1986, *Nature*, **322**, 806.

Gunn, J.E. 1982. In *Astrophysical Cosmology*, eds. H.A. Brück, G.V. Coyne, M.S. Longair (Vatican: Pontificia Academia Scientiarum), 233.

Güsten, R. and Metzger, P.G. 1983, *Vistas in Astronomy*, **26**, 159.

Hickson, P. 1982, *Ap. J.*, **255**, 382.

Jensen, E.B., Strom, K.M. and Strom, S.E 1976, *Ap. J.*, **209**, 748.

Jensen, E.B., Talbot, R.J. and Dufour, R.J. 1981, *Ap. J.*, **243**, 716.

Jog, C. and Solomon, P.M. 1984, *Ap. J.*, **276**, 127.

Kronberg, P. 1988. In *Galactic and Extragalactic Star Formation*, eds. R. Pudritz and M. Fich (Dordrecht: Kluwer), 391.

Kwan, J. and Valdes, F. 1987, *Ap. J.*, **315**, 92.

Lacey, C.G. and Fall, S.M. 1983, *M.N.R.A.S.*, **204**, 791.

Lacey, C.G. and Fall, S.M. 1985, *Ap. J.*, **290**, 154.

Larson, R.B. 1974, *M.N.R.A.S.*, **169**, 229.

Larson, R.B. 1976, *M.N.R.A.S.*, **176**, 31.

Larson, R.B. 1981, *M.N.R.A.S*, **194**, 809.

Larson, R.B. 1986, *M.N.R.A.S.*, **218**, 409.

Larson, R.B. 1988a. In *Galactic and Extragalactic Star Formation*, eds. R. Pudritz and M. Fich (Dordrecht: Kluwer), 459.

Larson, R.B. 1988b. In *Globular Cluster Systems in Galaxies*, eds. J.E. Grindlay and A.G. Davis Philip (Dordrecht: Reidel), 311.

306

Larson, R.B. and Tinsley, B.M. 1978, *Ap. J.*, **219**, 46.

Lequeux, J. 1987. In *Starbursts and Galaxy Evolution*, ed. T. Montmerle (Gif–sur–Yvette: Editions Frontières), 59.

Liebert, J. *et al.* 1986, *Ap. J.*, **300**, 314.

Lin, D., Pringle, J. and Rees, M. 1988, *Ap. J.*, **328**, 103.

Lugten, J.B. *et al.* 1988, *Ap. J.*, **311**, L51.

Lynden-Bell, D. 1975, *Vistas in Astr.*, **19**, 299.

McGlynn, T. 1984, *Ap. J.*, **281**, 13.

Merritt, D. 1985, *Ap. J*, **289**, 18.

Myers, P. 1983, *Ap. J.*, **270**, 105.

Myers, P.C. and Goodman, A.P. 1988, *Ap. J.*, **326**, L27.

Norman, C.A. 1988. In *Galactic and Extragalactic Star Formation*, ed. R. Pudritz and M. Fich (Dordrecht: Kluwer), 495.

Oort, J.H. 1974. In *The Formation and Dynamics of Galaxies*, IAU Symposium **58**, ed. J.R. Shakeshaft (Dordrecht: Reidel), 375.

Pagel, B.E.J. and Patchett, B.E. 1975, *M.N.R.A.S.*, **172**, 13.

Pudritz, R. and Silk, J. 1988, *Ap. J.*, in press.

Quinn, P. and Hernquist, L 1987, *Ap. J.*, **312**, 1.

Rana, N.C. and Wilkinson, D.A. 1986, *M.N.R.A.S.*, **218**, 497.

Rana, N.C. and Wilkinson, D.A. 1987, *M.N.R.A.S.*, **226**, 395.

Rieke, G. *et al.* 1980, *Ap. J.*, **238**, 24.

Rieke, G. 1988. In *Galactic and Extragalactic Star Formation*, eds. R. Pudritz and M. Fich (Dordrecht: Kluwer), 561.

Sandage, A. 1986, *Astr. Ap.*, **161**, 89.

Scalo, J. 1987. In *Starbursts and Galaxy Evolution*, eds. T. Montmerle (Gif–sur–Yvette: Editions Frontières), 445.

Schmidt, M. 1959, *Ap. J.*, **129**, 243.

Schmidt, M. 1963, *Ap. J.*, **137**, 758.

Scoville, M. 1988. In *Galactic and Extragalactic Star Formation*, ed. R. Pudritz and M. Fich (Dordrecht: Kluwer), 541.

Sellwood, J.A. and Carlberg, R.G. 1984, *Ap. J.*, **282**, 61.

Silk, J. 1988. In *Galactic and Extragalactic Star Formation*, eds. R. Pudritz and M. Fich (Dordrecht: Kluwer), 503.

Silk, J. 1988b. In *Comets to Cosmology*, ed. A. Lawrence (Berlin: Springer–Verlag), 297.

Silk, J. 1989. In *The Physics and Chemistry of Interstellar Molecular Clouds*, ed. G.Winnewisser (Heidelberg: Springer–Verlag), in press.

Silk, J. and Szalay, A. 1987, *Ap. J.*, **323**, L107.

Solomon, P. M. and Sage, L. J. 1988, *Ap. J.*, in press.

Talbot, R.J. 1980, *Ap. J.*, **235**, 821.

Talbot, R.J. and Arnett, W.D. 1971, *Ap. J.*, **170**, 409.

Talbot, R.J. and Arnett, W.D. 1975, *Ap. J.*, **197**, 159.

Toomre, A. 1977, in *The Evolution of Galaxies and Stellar Populations*, eds. B.M. Tinsley and R.B. Larson (New Haven: Yale University), 401.

Toomre, A. 1981, in *The Structure and Evolution of Normal Galaxies*, eds. S.M. Fall and D. Lynden-Bell (Cambridge: Cambridge University), 111.

Tosi, M. and Diaz, A. 1985, *M.N.R.A.S.*, **217**, 571.

Twarog, B.A. 1980, *Ap. J.*, **242**, 242.
Vader, J.P. and de Jong, T. 1981, *Astr. Ap.*, **100**, 124.
van Albada, T. 1982, *M.N.R.A.S.*, **201**, 939.
van den Bergh, S. 1962, *A. J.*, **67**, 486.
van der Kruit, P. 1986, *Astr. Ap.*, **157**, 230.
Viallefond, F. and Thuan, T.X. 1983, *Ap. J.*, **269**, 444.
Villumsen, J., Gunn, J.E. and Casertano, S. 1988. Preprint.
Wyse, R.F.G. 1986, *Ap. J. (Letters)*, **311**, L41.
Wyse, R.F.G. and Gilmore, G. 1988, *A. J.*, **95**, 1404.
Wyse, R.F.G. and Silk, J. 1987, *Ap. J.*, **313**, L11.
Wyse, R.F.G. and Silk, J. 1989. *Ap. J.*, April 15 issue.
Young, J.S. 1986. In *Star Forming Regions*, IAU Symposium **115**, eds. M. Peimbert and J. Jugaku (Dordrecht: Reidel).
Young, J.S. and Scoville, N.Z. 1982a, *Ap. J.*, **258**, 467.
Young, J.S. and Scoville, N.Z. 1982b, *Ap. J. (Letters)*, **260**, L11.
Zurek, W. and Quinn, P. 1988. Preprint.

DISCUSSION:

BOWER: In your model, you suggest that the formation of a compact nucleus allows the onset of star formation because of the magnetic fields it generates. How do you suggest that this seed compact object would form?

SILK: Various models of structure formation, including cold dark matter and cosmic strings, suggest that masses of $\sim 10^9$ $M_\odot$ could condense out at an early epoch, certainly by $z \sim 10$. I suspect compact object formation is favoured at large redshift, because Compton cooling guarantees very cold initial conditions at the onset of collapse. Supersonic motion, should dissipate efficiently by shocks, and a sufficiently cold initial collapse should guarantee attainment of a high final density. Perhaps the resulting self gravitating gaseous disk would be the site for a dynamo, and turbulent diffusion, gravitationally or magnetically induced, would allow formation of a central compact object.

JONES: Can magnetic fields in galaxies really be generated by expulsion of field from sources (nuclei, planetary nebulae, ...)? The problem is to get the global topology correct and at the same time be consistent with the global Faraday rotation pattern.

SILK: Expulsion provides the seed. Supershells, generated by multiple supernova explosions, combined with differential rotation of the protogalactic disk, yield a large-scale field. Whether this would be consistent with the global field pattern I cannot say: certainly, this provides a plausible mechanism for the origin of the random component of the interstellar field.

PICKLES: Your models show ellipticals forming either *after* some dissipation by a merging process, or possibly by entraining matter in a beam model associated with an active nucleus. In both cases however, the star formation rates necessary to form giant ellipticals in a short time (less than a collapse time) seem to require > 1000 $M_\odot$ yr^{-1} ($\times$ 10^9 yr $\sim 10^{12}$ $M_\odot$). My understanding is that most star formation rates in starbursts are 100 - 200 $M_\odot$ yr^{-1} and are probably short lived. Does this not therefore require a more prolonged star formation process in E's? - possibly over several $\times$ 10^9 yrs?

SILK: Extreme starburst objects have inferred star formation rates of 500 $M_\odot$ yr^{-1} or even 1000 $M_\odot$ yr^{-1} . However, there is no evidence that solar mass stars form in such starbursts: in fact, there are arguments against this. I agree that the starburst phenomenon, which is certainly short-lived ($< 10^8$ yr) may have little in common with elliptical formation: rather the starburst analogue may correspond to the early phase of massive star formation that enriched the protogalactic gas.

FALL: One of the standard tests of chemical evolution models for the galactic disk is a comparison with the distribution of metallicities of G dwarfs in the solar neighbourbood. It is well known that with a constant initial mass function, one needs either infall or a non-zero initial metallicity of at least 0.1 $Z_\odot$ and, more likely, 0.3 $Z_\odot$. I wonder how the "G-dwarf problem" is solved by your models and whether the solution also has implications for the metallicity gradient within the galactic disk.

SILK: The general class of models that I described automatically yields a metallicity gradient in a differentially rotating disk. Solution of the "G-dwarf problem" requires additional input: my preferred solution is a finite critical metallicity of about 0.2 $Z_\odot$ in the thin disk, a value which actually arises via multicomponent models that incorporate a halo and thick disk. This solution does not affect the metallicity gradient: other proposed schemes involving *ad hoc* variations in the infall rate or radial inflow would independently generate metallicity gradients.

FIELD: You referred briefly to the interstellar magnetic field strength B as one of the many parameters affecting the rate of star formation. We know that B is important in the Galaxy in removing angular momentum as molecular-cloud cores collapse to stars. Presumably angular momentum is a problem in forming stars in young galaxies. But it takes several Gyr to build up a galactic B according to the Parker dynamo model, so would you agree there should be a several Gyr delay in forming stars in a disk galaxy?

SILK: I agree, provided that a galactic-scale field is necessary. However, to add a note of caution, smaller scale dynamos on the scale of primeval clouds or accretion disks around VMO's, AGN's, or rare primordial stars may produce enough magnetic flux to overcome the general angular momentum barrier to star formation. And perhaps a distinct mechanism, such as tidal torquing, might have been effective at transferring angular momentum in protogalactic star-forming clouds. No doubt, this would result in an IMF that differed from that observed today, but I doubt if this possibility could be ruled out.

GALAXY FORMATION AND COOLING FLOWS

Keith M. Ashman
School of Mathematical Sciences
Queen Mary College
Mile End Road
LONDON E1 4NS

ABSTRACT.   We argue that much of the dark matter in galactic halos consists of clusters of jupiters formed in pregalactic and protogalactic cooling flows.   We also suggest that the formation of globular clusters in the Fall–Rees model may be accompanied by the formation of dark clusters.

1. Baryonic Dark Matter – Clusters Of Jupiters

It is well known that standard primordial nucleosynthesis calculations imply that the baryonic density of the Universe falls short of the critical density by at least an order of magnitude (Yang *et al.* 1984).   However, unless the value of the Hubble parameter H_0 is very high, these calculations also suggest that most baryons in the Universe are dark.   A low value of H_0 around 50 $kms^{-1}Mpc^{-1}$ allows the Universe to be older than globular clusters and implies a baryonic density Ω_b of about 0.1.   This raises the possibility that dark galactic halos, which have a comparable density, may be baryonic.   The cluster dark matter would include the disrupted remnants of these halos, although non–baryonic matter, which is required anyway if $\Omega = 1$, may also contribute.

The *form* of baryonic dark matter in galactic halos has been discussed at length (Carr *et al.* 1984).   It appears that the only possible candidates are the black hole remnants of massive stars or jupiters [objects with masses below the hydrogen burning limit of around $0.08M_\odot$].   It has been argued that both options are unlikely since most studies of star forming regions indicate that such objects are scarce (Hegyi & Olive 1986).   However, we will argue that under suitable conditions jupiters *can* form in cosmologically significant numbers as a result of cooling flows.

There is some tentative observational evidence that dark galactic halos contain *clusters* of jupiters.   Firstly, Carr & Lacey (1987) have suggested dark clusters, with masses around $10^6 M_\odot$, can explain the heating of the Galactic disc.   This modifies an idea proposed by Lacey & Ostriker (1985) who used supermassive black holes to provide the heating.   However, black holes get dragged into the Galactic nucleus by dynamical friction leading to an excessive build up of material.   Clusters overcome this problem since they are disrupted before reaching the Galactic centre if their radius exceeds about 1pc.   However, for the clusters to survive and produce heating down to sufficiently small Galactocentric distances, their radii must be less than about 10pc.   For the clusters to avoid evaporation this constrains the constituents to have masses below a few $M_\odot$.   Thus the favoured objects in the Carr–Lacey clusters are jupiters.

Other evidence for dark clusters includes the claim by Sommer–Larsen & Christensen (1987) of the detection of a cluster in the Galactic halo with a mass

309

C. S. Frenk et al. (eds.), The Epoch of Galaxy Formation, 309–313.

around $10^6 M_\odot$ and high mass–to–light ratio. Gravitational lensing considerations have also been employed to investigate the dark matter distribution in galactic halos. Subramanian & Chitre (1987) have studied the quasar MG2016+112 which is lensed by an intervening galaxy. The intensity ratio in different wavebands suggests that the dark matter is clumped on scales between $3 \times 10^4 M_\odot$ and $3 \times 10^7 M_\odot$. Further, Nottale (1986) has shown that the flaring behaviour of 0846+51W1 can be explained by microlensing by a jupiter in the halo of an intervening galaxy.

2. Cluster Cooling Flows

X–ray observations indicate that most clusters of galaxies contain hot intracluster gas with a temperature of around 10^8K (Sarazin 1986). In many cases the cooling time t_c in the centre of the gas distribution is less than the Hubble time t_H, so that the gas is expected to cool and flow inwards, driven by the pressure of the outlying uncooled gas. Since the dynamical time of the gas t_f is less than t_c the flows are quasi–static, so that the pressure in the cooling region is maintained. Other observational evidence has been reviewed by Fabian (1987). Another important characteristic of cluster cooling flows is that they are invariably centred on massive cD galaxies.

The mass flow rate $\dot{M}$ in cluster cooling flows can be obtained from the X–ray luminosity. This yields typical values of $10^2 M_\odot yr^{-1}$ and exceptional ones of $10^3 M_\odot yr^{-1}$ (Johnstone *et al.* 1987). Observations of luminosity as a function of radial distance r within the flow indicate that the flow rate increases with radius so that mass is deposited throughout the flow. The natural interpretation is that the cooling gas is forming stars. However, if stars are formed with a standard initial mass function (IMF), then the accreting galaxies should be bluer and brighter than observed. The colours and luminosities of these galaxies can be reproduced if typically 1% of gas forms stars with a standard IMF and the remaining gas forms dark matter (Johnstone *et al.* 1987). The dark objects must be either low–mass stars or jupiters. Arnaud & Gilmore (1986) find that the average stellar mass in cooling flows must be below about $0.2 M_\odot$, so while low–mass stars are not ruled out, jupiters are favoured.

This view of the intracluster gas is not universally accepted. A common criticism is that any heating mechanism that operates in cooling flows will reduce $\dot{M}$ and could thereby permit a standard IMF. Such a view seems to stem from a belief in the universality of the standard IMF, but is difficult to sustain. The most convincing evidence against these heating arguments is that cooling gas *is* observed through spectroscopy, and that the amount of gas at relatively low temperature is just that expected from the observed $\dot{M}$ (Fabian 1987).

Models of cooling flows using a standard IMF seem to create more problems than they solve. However, the physical mechanism for the origin of the cooling flow IMF is uncertain. One suggestion is that the high pressures in cooling flows reduces the Jeans mass and hence the stellar fragment mass. Pressures in cooling flows are typically 10^5–$10^6 cm^{-3}$K, about 10^2 times greater than that in star forming regions in our Galaxy, so that the Jeans mass is lower by an order of magnitude. Although there is uncertainty in the origin of the cooling flow IMF, it should be recalled that the origin of the standard IMF is also poorly understood. Further, there are other situations, such as starburst galaxies and some globular clusters, where the IMF differs from the standard one (Scalo 1986).

While cluster cooling flows add weight to the view that at least some of the dark matter in the Universe is baryonic, they cannot directly solve any of the dark matter problems since they are restricted to the cores of clusters of galaxies. However, we now argue that analagous flows at earlier epochs may have far–reaching consequences.

3. Pregalactic And Protogalactic Cooling Flows

To establish whether conditions analogous to cluster cooling flows occur at earlier epochs we require a model for the formation of large–scale structure. We concentrate on the hierarchical clustering scenario since the situation is fairly simple. The detailed calculations have been described by Ashman & Carr (1988) who also consider cooling flows in the pancake and explosion scenarios.

Although there are several variants of the hierarchical clustering scenario, the common feature is that structure is built from the "bottom up". The first objects to condense out of the Hubble flow are small, with larger structures being formed through gravitational clustering. Overdense clouds that collapse and virialise can only form stars if the gas can cool on a Hubble time, otherwise the cloud is disrupted by the collapse of the next stage in the hierarchy. Since cooling flows are quasi–static, we need to find regions within cooling clouds in which locally $t_H > t_c > t_f$.

We model the density profile $\rho(r)$ of the virialised cloud by a constant density core to $r = a$ and a power law fall–off for $r > a$, where a is the core radius. We truncate this density profile at a radius R. Assuming that the cloud is isothermal, so that any temperature dependence of t_c can be neglected, we have that $t_c \propto \rho^{-1}$ and $t_f \propto \rho^{-1/2}$. Virialised clouds thus fall into three categories. (i) $t_c(r) < t_f(r)$ everywhere: this correponds to the region within the cooling curve of Rees & Ostriker (1977) and Silk (1977), but unlike cluster cooling flows such clouds are not quasi–static. (ii) $t_c(r) > t_f(r)$ everywhere: this is analogous to cluster cooling flows, but the requirement $t_c(r) < t_H$ means that only a small fraction of the cloud can cool. This is precisely what is observed in cluster cooling flows where the vast majority of the gas does not participate in the flow. (iii) $t_c(a) < t_f(a)$, $t_c(R) > t_f(R)$: in this case t_c is less than t_f at small radii, but beyond some radius it exceeds t_f so that the flow becomes quasi–static. The fraction of the gas participating in this flow is maximised when the mean cooling time and free–fall time are comparable; that is, when the cloud lies close to the Rees–Ostriker–Silk cooling curve (i.e. the line $t_c = t_f$).

In any particular version of the hierarchical clustering scenario one can specify the mass of clouds virialising at a given redshift by a trajectory M(z). Where M(z) crosses the cooling curve we therefore anticipate cooling flows and, if the pressure is high enough, the formation of jupiters. Ashman & Carr (1988) show that there are two cooling flow epochs and hence two mass–scales on which dark matter can be produced. The values depend on the form of M(z), but in the biased cold dark matter model they are $M \sim 10^6 M_\odot$ $z \sim 30$ (if H_2 is the dominant coolant) and $M \sim 10^{11} M_\odot$ $z \sim 10$. The pressures in the high mass case are comparable to those in cluster cooling flows. The situation in the low–mass case is less clear since the results are sensitive to the dominant cooling mechanism, but under certain conditions pressures can be high enough for jupiter formation. The situation is rather better in other hierarchical clustering pictures since small mass scales tend to bind earlier than in the CDM picture and hence at higher pressures.

If the low–mass cooling flows are dominant, galactic halos should contain dark clusters of jupiters with masses around $10^6 M_\odot$. This conclusion also holds in the explosion and pancake scenarios where, provided H_2 is the dominant cooling mechanism, the scale of clumping of the jupiters is also around $10^6 M_\odot$. If dark matter is formed directly on the high–mass scale, one might anticipate that the dark matter should be clumped in $10^{11} M_\odot$ objects. However, as we now discuss, jupiters may be formed under a wider range of conditions and high–mass cooling flows may also form $10^6 M_\odot$ clusters of jupiters.

4. Cooling Flows In The Fall-Rees Scenario

One problem with drawing an analogy between cluster cooling flows and conditions at earlier epochs is that the characteristics of cooling flows responsible for the formation of jupiters are uncertain. We now argue that the conditions considered by Fall & Rees (1985) in a model of globular cluster formation may include the relevant features. The Fall-Rees model assumes that the collapsing gas of a protogalaxy undergoes a thermal instability which leads to cool clouds embedded in a hot gas. In the absence of metals and molecular coolants the clouds cease cooling at around 10^4K at which point the Jeans mass is about $10^6 M_\odot$; hence the identification with the precursors of globular clusters. The situation of star formation occuring in cool clouds embedded in hot gas is highly reminiscent of cooling flows; in particular, the hot component will preserve the pressure in the cool clouds.

However, the Fall-Rees picture is concerned with the formation of visible objects, whereas we have argued that such conditions should produce dark matter. One possible resolution of this apparent contradiction is that cooling in protogalaxies continues via H_2 to 10^2K. The characteristic mass scale of globular clusters is lost, but by considering the detailed cooling history of the clouds it is possible to construct another mass scale around $10^6 M_\odot$ that distinguishes the formation of dark clusters and globular clusters (Ashman 1988). The idea is that clouds initially cool so quickly that they go out of pressure equilibrium with the surrounding gas. If star formation occurs before these clouds return to pressure equilibrium, then we argue that globular clusters are formed (since the pressure is low). The remaining clouds form stars at the high pressures characteristic of cooling flows and hence form dark clusters.

The Fall-Rees model concerns protogalaxies in which $t_c < t_f$, but the same thermal instability occurs in the high-mass $t_c \sim t_f$ case considered above, as well as in cluster cooling flows. Jupiters may therefore form in clusters in these cases as well.

References

Arnaud, K.A. & Gilmore, G., 1986. *Mon. Not. R. astr. Soc.*, **220**, 759.
Ashman, K.M., 1988. Preprint.
Ashman, K.M. & Carr, B.J., 1988. *Mon. Not. R. astr. Soc.*, in press.
Carr, B.J., Bond, J.R. & Arnett, W.D., 1984. *Astrophys. J.*, **277**, 445.
Carr, B.J. & Lacey, C.G., 1987. *Astrophys. J.*, **316**, 23.
Fabian, A.C., 1987. In *Proceedings Of The 22nd Moriond Meeting*, ed. Thuan, T.X., Montmerle, T. & Tran Thanh Van, J., (Editions Frontieres, France).
Fall, S.M. & Rees, M.J., 1985. *Astrophys. J.*, **298**, 18.
Hegyi, D.J. & Olive, K.A., 1986. *Astrophys. J.*, **303**, 56.
Johnstone, R.M., Fabian, A.C. & Nulsen, P.E.J., 1987. *Mon. Not. R. astr. Soc.*, **224**, 75.
Lacey, C.G. & Ostriker, J.P., 1985. *Astrophys. J.*, **299**, 633.
Nottale, L., 1986. *Astron. Astrophys.* **157**, 383.
Rees, M.J. & Ostriker, J.P., 1977. *Mon. Not. R. astr. Soc.*, **179**, 541.
Sarazin, C.L., 1986. *Rev. Mod. Phys.* **58**, 1.
Scalo, J.M., 1986. *Fund. Cos. Phys.*, **11**, 1.
Silk, J., 1977. *Astrophys. J.*, **211**, 638.
Sommer-Larsen & Christensen 1987. *Mon. Not. R. astr. Soc.*, **225**, 499.
Subramanian, K. & Chitre, S.M., 1987. *Astrophys. J.*, **313**, 13.
Yang, J., Turner, M.S., Steigman, G., Schramm, D.N., & Olive, K.A., 1984. *Astrophys. J.*, **281**, 493.

DISCUSSION:

THOMAS: The high mass PPCFs are most like the conventional cooling flows we see in groups and clusters today. I have produced some preliminary models which suggest that these could form giant elliptical galaxies.

ASHMAN: I agree. The gas temperature and density in this case are within an order of magnitude or so of cluster cooling flows.

THE EPOCH OF GALAXY FORMATION IN EXPLOSION MODELS

Christopher Thompson*
Joseph Henry Laboratories
Princeton University
Princeton NJ 08544 U.S.A.

ABSTRACT. We consider a class of models for the formation of large scale structure, in which each void is generated by a single giant explosion. The voids must form at a redshift $15 - 25$, for several independent reasons.
  1) Gas which is shock heated to a temperature $\sim 5 \cdot 10^7$ °K can cool sufficiently to fragment and form galaxies only when Compton cooling is efficient.
  2) The local velocity field is too chaotic unless the peculiar velocities of the expanding shells of galaxies are damped in the Hubble flow to less than ~ 150 km s^{-1}.
  3) The redistribution of baryons onto shells excites growing modes in the velocity field of wavelength comparable to the size of the voids. We require the RMS peculiar velocity to exceed ~ 500 km s^{-1}.
  4) If the voids form too early, then dark matter interior to the shells catches up with the shells, and galaxies and dark matter are no longer segregated.
Galaxies form later than the voids. The epoch at which the luminous component of galaxies is assembled depends on the rate at which the shells fragment. In the superconducting string model, galactic halos are assembled first via accretion of exotic dark matter onto loops. Nonetheless, in any explosion model a significant fraction of the dark matter in galaxies and clusters is *baryonic*. We argue that recent observations of a high-frequency excess in the CBR spectrum, if substantiated, provide evidence for enormous energy release *before* galaxy formation.

1. Introduction

Recent surveys of the distribution of galaxies have revealed structures which have surprising regularity to the eye (de Lapparent *et al.* 1986; Giovanelli *et al.* 1987). Galaxies appear to lie on sheets which (at least partially) enclose evacuated holes. Moreover, dynamical measurements of the density parameter Ω (see, e.g., Peebles 1986) yield a value $\Omega_{dyn} = 0.1 - 0.3$. These observations may be given a very simple, although hardly unique, interpretation: a significant fraction of baryons have been evacuated from the voids onto sheets via the localized release of energy. The amount of energy required is enormous. To form a blast of radius R at time t (redshift z) one requires, very roughly, $E \sim \Omega_S \rho R^5 t^{-2} = 2 \cdot 10^{64}$ erg $(1+z) \, \Delta^{-5} \left(H_0 R_0 / 1500 \text{ km s}^{-1} \right)^5 \left(\Omega_S h_{50} / 0.1 \right)$. Here Ω_S is the fraction of matter swept up by the blast, and Δ is the proportional increase in its comoving radius from time t to the present. The energy required to form a void of a fixed present diameter grows with z. We do not specify the energy source, except to say that it must be *exotic*. In the original explosion model (Ostriker and Cowie 1981) the energy source was

---

* Present address: Theoretical Astrophysics, Caltech 130-33, Pasadena CA 91125, U.S.A.

C. S. Frenk et al. (eds.), The Epoch of Galaxy Formation, 315–319.
© *1989 by Kluwer Academic Publishers.*

taken to be an active galaxy, with output $E \sim 10^{61}$ erg. As a result, galaxies were predicted to form on shells much smaller than the observed voids.

We also note the following. Once the baryons are segregated on shells, they make up a significant fraction of the local mass density. *This implies that a significant fraction of the dark matter observed in galaxies or clusters of galaxies is baryonic.* Indeed, *all* the dark matter is baryonic in an explosion model where the universe is open and $\Omega_{baryon} = \Omega_{dyn} = \Omega$.

2. Cooling of the Shocked Gas

In any explosion model for large scale structure, Compton cooling is essential (see also Section 4). The temperature of the gas behind a shock moving at peculiar velocity ΔV is $T_{PS} = 5 \cdot 10^7$ °K $(\Delta V/2000$ km s$^{-1})^2$. We ask whether gas which is shocked at redshift z_{shock} can Compton cool to $T \lesssim 2 \cdot 10^5$ °K, at which point atomic cooling takes over. Note that Compton cooling cuts off very quickly: the cooling time is $t \cdot [(1+z)/(1+z_{IC})]^{-5/2}$ where $1+z_{IC} = 7.7\, h_{50}^{-2/5}$. Gas can cool only if it is shocked above a well-defined redshift z_{cool}, which depends on ΔV:

$$1+z_{shock} > 1+z_{cool} = (1+z_{IC})\left[\frac{5}{3}\ln\left(\frac{T_{PS}}{2\cdot 10^5\,°K}\right)\right]^{2/5} = 12.5\, h_{50}^{-2/5}\left[2.0+\ln\left(\frac{\Delta V}{1000\,\text{km s}^{-1}}\right)\right]^{2/5}.$$
(1)

(We assume an Einstein-de Sitter cosmology.) Note that z_{cool} is roughly twice z_{IC} for $\Delta V = 1000 - 2000$ km s^{-1}, because the gas must cool at least a factor 100 in temperature.

To proceed further, we pick a specific model for the blasts. The classic $R \propto t^{4/5}$ adiabatic cosmological blast solution (e.g. Bertschinger 1983) is of little use because it assumes that *all* the ambient matter is swept up *and* $\Omega = 1$. If energy is released at a constant rate, as it is to a first approximation from a superconducting string, then ΔV is constant and $R = 3\Delta V t$. If gas fragments immediately upon cooling, then its peculiar velocity decays and it now lies a distance $H_0 R_0 = 1400\left([1 + z_{shock}]/20\right)^{-1/2}(\Delta/1.5)(\Delta V/2000\,\text{km s}^{-1})$ km s^{-1} from the blast's center. Requiring that $H_0 R_0 = 1500$ km s^{-1} and using (1), we find $\Delta V \simeq 2000\,(\Delta/1.5)^{-1}$ km s^{-1}.

3. The Local Velocity Field

The local velocity field follows a Hubble law quite closely (e.g. Sandage 1986), yet our galaxy moves at ~ 600 km s^{-1} with respect to the CBR rest frame. This means that galaxies in the Local Supercluster move essentially in unison, as confirmed by recent surveys (e.g. Dressler *et al.* 1987). The Local Supercluster has a natural interpretation as a ridge swept up by an explosion, and one is tempted, at first sight, to associate this collective motion with the residual expansion of the ridge. Recently Peebles (1988) has argued convincingly *against* a hydrodynamical origin for the the local velocity field: many galaxies lie well off the Supergalactic Plane, and these galaxies would be expected to have significant velocities with respect to galaxies on the Plane. Moreover, the local velocity vector lies in the Plane.

Indeed, the expansion of the shells makes a sufficiently small contribution to the local velocity field, if the voids form at $z \gtrsim 15$. The peculiar velocity of an isolated shell of galaxies of low surface density ($\Omega_S \ll 1$) decays to the present value $\Delta V_0 \simeq (1 + z_{shock})^{-1}\Delta V = 130\,[(1 + z_{shock})/15]^{-1}(\Delta V/2000\,\text{km s}^{-1})$. (One might associate this small residual velocity with the local anomaly discussed by Faber and Burstein 1988.) In comparison, the RMS peculiar velocity of galaxies closer than $H_0 R_0 = 1000$ km s^{-1} is ~ 150 km s^{-1} (Peebles 1988). One can check that the velocity dispersion of the galaxies in a shell is acceptably small if $\Omega_S \lesssim 0.2$.

In sum: the peculiar velocity of an expanding spherical shell is essentially a *pure decaying mode* in the limit $\Omega_S \ll 1$. The velocity field of the matter left behind in the shell is a mixture of decaying and growing modes. Eventually this "dark" matter catches up with the shell. Note that our first statement is true only for a *uniform, spherical* shell, surrounded by a uniform medium. Once we introduce interactions between expanding shells, we also introduce growing modes into the galaxy velocity field. The amplitude of these growing modes depends on the extent to which the individual bubbles depart from spherical symmetry. Thus hydrodynamical interactions between shells tend to enhance the relative importance of the growing modes.

The present RMS amplitude ΔV_0 of the growing modes depends on the epoch z_{void} when the voids were formed. At $z = z_{void}$ we have $\Delta V = \varepsilon \cdot \Omega_S H(z_{void}) R(z_{void})$, where the dimensionless parameter ε depends on the detailed distribution of matter on the shells, and $R(z_{void})$ is the physical diameter of a typical void. At present, ΔV is larger by a factor $(1 + z_{void})^{1/2}$. Normalizing to the present radius of the voids $R_0 = (1 + z_{void}) R(z_{void})$, we obtain $\Delta V_0 = \varepsilon (1 + z_{void}) \Omega_S H_0 R_0$. In particular, $\varepsilon = 0.16$ in the case where matter is distributed uniformly on a cubic grid of spacing $2R_0$. (This overestimates the departure from spherical symmetry of individual bubbles, but underestimates the concentration of mass toward the vertices where shells intersect.) *We conclude that RMS peculiar velocity equals 500 km s^{-1} if the voids form at $z_{void} \simeq (1.5 - 2)\Omega_S^{-1}$.* It is difficult to see how the coherence length of the velocity field could be much larger than the radius of a bubble. This small coherence length is potentially a serious problem with explosion models.

The requirement that dark matter and light matter remain segregated yields an upper bound on z_{void}. This bound is somewhat greater than $z_{void} = \Omega_S^{-1}$, but its exact value depends on complicated details such as the rate at which the shells fragment. (Of course, the bound does not exist if $\Omega = \Omega_{baryon} = \Omega_{dyn} \sim 0.1$.) Consider again the case where the blast is driven by a constant energy source. So long as the source is turned on, the blast grows at constant velocity, and the dark matter is left behind. We are interested in what happens after the source turns off (at time t_{off}). Before the dark matter catches up (at time t_{catch}), the peculiar velocity of the shell decays away and the shell reaches the comoving radius $R = \Delta \cdot (t/t_{off})^{2/3} R(t_{off})$. By this point, a fraction δ of the total energy $E = \dot{E} t_{off}$ has been transferred to the dark matter. Equating R with the radius $1.885(G\delta E)^{1/5} t^{4/5}$ (Bertschinger 1985) of an expanding hole in the dark matter, we obtain $\left(t_{catch}/t_{off} \right)^{2/3} = 0.17 \left(\Delta^5 / \delta \right) \Omega_S^{-1}$. Note the strong dependence on Δ. If the shell remains pressure driven and continues to sweep up mass after $t = t_{off}$, then $\delta = 1$, $\Delta = 1.5$ and $(t_{catch}/t_{off})^{2/3} = 1.4 \Omega_S^{-1}$ (*cf.* Thompson *et al.* 1988). If at some point the shell fragments and ceases to sweep up mass, then Δ is somewhat larger and the dark matter takes longer to catch up. Thus it is only marginally possible to match the amplitude of the local velocity field, and to keep dark matter and light matter segregated.

4. Epoch of the Formation of Galaxies and Lyman α Clouds

Cooling cosmological blast waves are subject both to gravitational and Rayleigh-Taylor instabilities (Vishniac 1983). These instabilities grow very slowly, unless the thin shell of gas behind the shock is very cold. Their maximum growth rate exceeds the Hubble expansion rate only if the gas is colder than $3 \cdot 10^4$ °K $(\Omega_S / 0.1)^2 (\Delta V / 2000$ km s$^{-1})^2$. The corresponding fragments have mass $10^{10} h_{50}^{-1} M_\odot ([1 + z]/15)^{-3/2} (\Omega_S / 0.1)^3 (\Delta V / 2000$ km s$^{-1})^2$. This is a reasonable mass for a young spheroid or elliptical if $\Omega_S \sim 0.1$. To build larger bound objects, perturbations of relatively large amplitude must be applied to the blast. These can be provided by collisions between shells or, in the superconducting string model, by accretion of dark matter onto loops before the formation of the voids. Without entering this thorny subject, we simply note that in an explosion model, one expects spheroidal components to assemble early, at $z \gtrsim 15$.

At $z \lesssim 15$ residual gas which has not been bound into galaxies is shock heated to high temperatures and cannot cool. The resulting Jeans mass is very large, and young galaxies accrete this phase quasi-statically, in a cooling flow. We conjecture that the delayed accretion of this phase through a large collapse factor leads to the formation of spiral disks.

Cold gas in pressure equilibrium with a blast has a small (three-dimensional) Jeans mass, $M_J(\text{cold}) \simeq 3 \cdot 10^8 M_\odot (1 + z)^{-3/2} (T / 10^4 $°K$)^2 (\Delta V / 2000$ km s$^{-1})^{-1}$. We identify clouds lighter than M_J(cold) with Ly-α absorption line systems. In this view, Ly-α clouds observed at $z \sim 2-3$ are remnants of blast waves formed at an earlier epoch when Compton cooling was efficient.

Very early galaxy formation is a potential problem in the superconducting string model, since space is filled by (relatively small) blast waves at $z \sim 10^3$. However, galaxy formation is suppressed at high redshift, for a number of reasons. For example, in the string model, the size of a blast wave is a well-defined function of epoch. The column density through a blast is $N = 10^{19}(1+z)^{3/2}(\Omega_{baryon}h_{50}/0.1)(\Delta V/2000\,\mathrm{km\,s^{-1}})$. At $z \gtrsim 10$, N exceeds the column density of gas through a galaxy. A young galaxy is disrupted by a passing blast wave at $z \gg 10$.

5. Compton Distortion of the CBR: Heating of the Early Universe

The high frequency excess in the CBR reported by Matsumoto et $al.$ (1988) has a shape similar to that produced by inverse Compton scattering off heated electrons. However, such a large Compton distortion ($\sim 5 - 10\%$ of the energy in the CBR) cannot be produced by explosions occuring at $z \lesssim 10^2$, because the thermal energy of a cosmological blast wave is only $1 - 2\%$ of its total energy (e.g. Bertschinger 1983). Indeed, if the thermal energy were much larger, the resulting angular fluctuations in the CBR would easily exceed observational bounds on the scale of several arc-minutes (Lawrence et $al.$ 1988). At $z \gtrsim 10^2$, essentially all the energy in a blast wave is transferred directly to the CBR via Compton drag. If the blasts reionize the universe (as expected if enough energy is injected to produce the Compton distortion), then fluctuations at a scale of a few degrees or less are smoothed out by multiple scattering above a redshift $\sim 60\,(\Omega_{baryon}h_{50}/0.1)^{-2/3}$. In the superconducting string model (Ostriker et $al.$ 1986; Thompson et $al.$ 1988) such an early generation of explosions occurs automatically, since decaying loops heat the universe continually. If the parameters of the model are fixed so that loops decaying at $z \sim 15 - 25$ generate blast waves as large and as numerous as the observed voids, then the Compton distortion generated by loops decaying at $100 < z < 10^5$ is similar in amplitude to the observed distortion (Ostriker and Thompson 1988).

I thank E. Bertschinger, J. Peebles, and especially J. Ostriker for stimulating conversations.

References

Bertschinger, E. 1983, $Ap.$ $J.$, **268**, 17.

Bertschinger, E. 1985, $Ap.$ $J.$ $Suppl.$, **58**, 1.

de Lapparent, V., Geller, M. J., and Huchra, J. P. 1986, $Ap.$ $J.$ $(Letters)$, **302**, L1.

Dressler, A., Faber, S. M., Burstein, D., Davies, R. L., Lynden-Bell, D., Terlevich, R., and Wegner, G. 1987, $Ap.$ $J.$ $(Letters)$, **313**, L37.

Faber, S. M., and Burstein, D. 1988, $Proc.$ $Pont.$ $Acad.$ $Sci.$ $Study$ $Week$ 27, $Large$-$Scale$ $Motions$ in the $Universe.$

Giovanelli, R., Haynes, M. P., Myers, S. T., and Roth, J. 1986, A. J., **92**, 250.

Lawrence, C. R., Readhead, A. C. S., and Myers, S. T. 1988, The $Post$-$Recombination$ $Universe$, ed. Kaiser, N. and Lasenby, A. N. (Dordrecht: Kluwer).

Ostriker, J. P., and Cowie, L. L. 1981, $Ap.$ $J.$ $(Letters)$, **243**, L127.

Ostriker, J. P., Thompson, C., and Witten, E. 1986, $Phys.$ $Lett.$, **180**, 231.

Ostriker, J. P., and Thompson, C. 1987, $Ap.$ $J.$ $(Letters)$, **323**, L97.

Peebles, P. J. E. 1986, $Nature$, **321**, 27.

Peebles, P. J. E. 1988, $Ap.$ $J.$, **332**, 17.

Sandage, A. 1986, $Ap.$ $J.$, **307**, 1.

Thompson, C., Witten, E., and Ostriker, J. P. 1988, $Rev.$ $Mod.$ $Phys.$, submitted.

Vishniac, E. T. 1983, Ap $J.$, **274**, 152.

DISCUSSION:

CARR: In predicting the velocity dispersion of galaxies in the explosion scenario, it makes a big difference whether the shells fragment before or after colliding. In the latter case, the gas can undergo considerable streaming before fragmenting. To which situation do your calculations pertain?

THOMPSON: The velocity field of the galaxies does depend sensitively on whether or not the shells suffer hydrodynamical interactions, that is, on whether or not they interpenetrate. I calculated the r.m.s. peculiar velocity of the dark matter assuming that the shells fragment on colliding. I doubt that the gas could be accelerated by hydrodynamical forces too long after the epoch $(1 + z \sim 15)$ that cooling becomes inefficient; as I described, it would be difficult to understand the observed smoothness of the local Hubble flow.

SHAPIRO: The shocks in the explosion model which radiatively cool will, in the absence of a very large ionizing radiation background, cool rapidly through $10^4 K$ to a much lower temperature of order 10^2 K by H_2 molecule formation and cooling. This is particularly likely to occur for the shocks which cool at high redshift $z \sim 10\text{-}20$ such as you discuss, since there is little reason to expect the high ionizing radiation levels required to suppress molecule formation at these redshifts. Such a change in the final temperature of postshock cooling would reduce the minimum unstable gravitational fragment mass by a factor of 10^4 from the values you have calculated by assuming the postshock gas cools only to $10^4 K$. What effect would this have on your conclusions?

THOMPSON: The column density through one Jeans length of cold material in pressure equilibrium with the blast, is independent of the temperature of the cold material. In the presence of a primeval magnetic field of present strength $\sim 10^{10}$ G (as required in the superconducting string model), the magnetic pressure balances the thermal pressure of the blast when the filaments cool to $10^4 K$. Further cooling would not, therefore, change my estimates of the Jeans mass of the cold material. The UV background inferred from the QSO proximity effect appears to be sufficient to inhibit molecular cooling in the cold filaments of $z = 2\text{-}3$ (we associate these with the Lyα clouds).

RECENT RESULTS ON COSMIC STRINGS AND GALAXY FORMATION

Robert H. Brandenberger
Department of Physics
Brown University
Providence, R.I. 02912, USA

ABSTRACT. The origin of galactic angular momentum via tidal torquing and the evolution of the mass function of galactic halos are discussed in the context of the cosmic string model.

1. Cosmic Strings and Galaxy Formation

Cosmic strings[1] are linear topological defects which form during a phase transition in the very early universe - typically about $10^{-35}s$ after the big bang. They arise in many but not all gauge theories in which the original symmetry group is spontaneously broken. Cosmic strings are thus lines of trapped energy with unusual dimensions. Their energy per unit length μ is about 10^{15} tons/cm and their width is about 10^{-22} of the radius of a hydrogen atom.

The result of detailed investigations of the formation and evolution of the network of cosmic strings is that at any time t there is of the order one infinite string crossing each horizon volume plus a distribution of loops given by

$$n(R,t) = v\ R^{-4} \left(\frac{z(R)}{z(t)} \right)^{-3}. \tag{1}$$

Here R is the radius of the loop, $z(t)$ is the redshift at time t, ν is a constant of the order 1 and $n(R,t)dR$ is the number density of loops in the radius interval $[R, R+dR]$.

The basic premise of the cosmic string model of formation of structure is that cosmic string loops form accretion seeds for structures such as galaxies and clusters. Accretion begins at the time t_{eq} of equal matter and radiation. Small loops seed galaxies, larger loops seed clusters. In more quantitative terms: objects with mean separation d are seeded by loops of radius $R(d)$ where $R(d)$ is chosen such that the number density of loops with radius $\geq R(d)$ equals the observed number density d^{-3}. Up to this point, the analysis is independent of μ and of the dark matter content of the universe. μ can be determined by demanding that the mass of a cluster agree with observations.[2] The result is $G\mu \sim 10^{-6}$ which translates into a scale of symmetry breaking of about 10^{16} GeV in the underlying gauge theory.

C. S. Frenk et al. (eds.), The Epoch of Galaxy Formation, 321–325.
© 1989 by Kluwer Academic Publishers.

The cosmic string theory of galaxy formation differs significantly from other currently popular models, for instance the "cold dark matter model" (CDM). In the latter the primordial energy density perturbations can be viewed as plane waves of all wavelengths superimposed with random phases. In the cosmic string model the initial density distribution is a collection of point perturbations of various amplitude superimposed on a homogeneous background.

The cosmic string model predicts a two point correlation function of clusters[3] which is independent of any parameters in the model, and the predictions match with observations within the observational error bars. Since $G\mu$ is fixed by demanding that the theory give the correct cluster mass, there are no more free parameters to adjust when studying galaxy formation. The predictions depend only on whether one has hot or cold dark matter. In the following sections I shall summarize some recent work on galaxy formation with cosmic strings, performed in collaboration with E. P. S. Shellard.[4]

2. Angular Momentum from Tidal Torquing

Tidal torquing as the source of angular momentum of galaxies has been studied in the context of the CDM model by many authors. Here I shall show that in the cosmic string model tidal torquing can generate enough angular momentum to explain observations, provided the dark matter is cold. The mechanism is well known: If galaxies are elliptical rather than spherical, the gravitational forces between nearest neighbors generate torques and hence angular momenta. In the cosmic string theory the eccentricity ϵ of a galaxy is generated by accretion onto a moving loop[5] and turns out to be about 1. The distance between nearest neighbor galaxies is set by the galaxy correlation function.

The angular momentum $\underline{L}$ is obtained by integrating the torque from t_{eq} to the present time t_0. The integral is dominated by the time t_f when all free gas has accreted onto some string loop and gives

$$L \simeq \frac{2}{3} \, G \, t_f \, m(t_f) \, \hat{m}(t_f) \, R^2(t_f) \, d^{-3}(t_f) \tag{2}$$

where $\hat{m}$ and R are mass and virialized radius of the galaxy acquiring angular momentum, and m is the mass of the object doing the torquing.

The separation $d(t)$ between nearest neighbor galaxies is $d(t) = d(R)/N_0$ where N_0 is the number of galaxies typically found in a sphere of radius $d(R)$ about a given galaxy. From the observed amplitude of the galaxy correlation function it follows that $N_0 \simeq 3$. If M_{12} is the mass of a galaxy in units of $10^{12} M_\odot$ and $R_{0.1}$ the radius of the halo in units of 0.1 Mpc, then for cosmic strings and cold dark matter we obtain

$$L \sim 10^{73} \, cm^2 \, g \, s^{-1} \, \left(\frac{\epsilon}{2} \, M_{12}^2 \, R_{0.1}^2 \, N_0^3 \right) \tag{3}$$

For hot dark matter masses and radii are typically one order of magnitude smaller. In addition, most of the mass has accreted in the last expansion time, whereas for cold dark matter t_f occurs at a redshift of about 10. This leads to a further suppression of L :

$$L \sim 10^{67} \ cm^2 \ g \ s^{-1} \left(\frac{\epsilon}{2} \ M_{11}^2 \ R_{0.01}^2 \ N_0^3 \right). \tag{4}$$

We conclude that for cosmic strings and cold dark matter tidal torquing can explain the observed angular momenta of spiral galaxies which range from $10^{73} - 10^{75} cm^2 g s^{-1}$. For hot dark matter the answer is negative and we need a different mechanism for explaining the angular momenta of spiral galaxies (maybe along the lines of the one proposed in Ref. 5).

3. Evolution of the Mass Function

If all cosmic string loops accrete mass independently up to the present time t_0, then the mass function $n(M)$ of galaxies would follow immediately from (1). For the range of R of relevance for galaxies, $n(R) \sim R^{-5/2}$. With cold dark matter the mass $M(R)$ accreted about a seed loop of radius R is proportional to R. Hence

$$n(M) \sim M^{-5/2} \, , \tag{5}$$

much too steep to match observations. $n(M)dM$ is the number density of objects in the mass interval $[M, M + dM]$. For hot dark matter, free streaming reduces accretion onto small loops, with the result[7] that $M(R) \sim R^3$ and hence

$$n(M) \sim M^{-3/2} \tag{6}$$

in rather good agreement with observations. Note that since cosmic string loops are perturbations which survive neutrino free streaming, hot dark matter with cosmic strings is not ruled out and yields an interesting cosmological model with several good features such as a reasonable mass function and flat halo velocity rotation curves.[7]

Here we shall take a closer look at galaxy formation with cosmic strings and cold dark matter. We shall show[4] that the actual mass function is quite different than (5) and in much better agreement with observations. In fact, the assumption that all string loops accrete mass independently breaks down. Already at a redshift $z(t_f) \simeq 10$ all mass has accreted onto some loop. Thereafter, competition between loops and merging will be important. Note that for cosmic strings and hot dark matter only about 10% of all matter has virialized at t_0.

We first consider the mass function at redshift $z(t_f)$ in a model with cosmic strings and cold dark matter. We must incorporate the fact that many small loops will be inside the turn–around radius of larger loops and will not seed independent structures. Let $F(R)$ denote the fraction of space inside the turn–around radius of loops with radius $\geq R$. Then the effective mass function of structures is given by $\hat{n}(R) = n(R) \left(1 - F(R)\right)$. In Fig. 1 we plot the resulting mass function at $z(t_f)$. We checked the results by numerical simulations and found agreement within the statistical error bars.

Next, we studied the evolution of the mass function for $z < 10$ using an N body code developed by E. P. S. Shellard (for more details see Ref. 4). The code is a "sticky particle" code. Two bodies merge in a given time step δt if the minimal distance between the points is smaller than the sum of the virialized radii and if the relative velocity is smaller than the escape velocity from the combined object. In Fig. 2 we plot the time dependence of the mass function and in Fig. 3 we compare the results with observational data. Between redshifts 10 and 1, the slope of $n(M)$ decreases significantly, especially at the low mass end. The final curve does not follow a single power law. The effective slope evaluated at $10^{12}M_\odot$ changes from -1.8 at $z = 10$ to -1.25 at $z = 1$. The final curve is in reasonable agreement with observations.

4. Conclusions

We have presented new results on angular momentum generation and mass function evolution in the cosmic string model. With cold dark matter, tidal torquing can easily explain the observed angular momenta of spiral galaxies, with hot dark matter a different mechanism is required. With cold dark matter, competition between loops and merging are important in analyzing the mass function. We conclude that, in contrast to naive expectations, the mass function agrees fairly well with observations.

REFERENCES

1. For reviews see e.g.,
   A. Vilenkin, *Phys. Rep.*, **121** (1985) 263;
   R. Brandenberger, *Int. J. of Modern Physics*, **A2** (1987) 77;
   N. Turok, in *"Astroparticle Physics,"* ed. A. De Rujula and P. Shaver, (World Scientific, Singapore, 1988).
2. N. Turok and R. Brandenberger, *Phys. Rev.* **D33** (1986) 2175;
   H. Sato, *Prog. Theor. Phys.* **75** (1986) 1342;
   A. Stebbins, *Ap. J. (Lett.)* **303** (1986) L1.
3. N. Turok, *Phys. Rev. Lett.* **55** (1985) 1801.
4. R. Brandenberger and E. P. S. Shellard, "Angular Momentum and Mass Function of Galaxies Seeded by Cosmic Strings," Brown preprint BROWN-HET-663, June 1988.
5. E. Bertschinger, *Ap. J.* **316** (1988) 496.
6. W. Zurek, *Phys. Rev. Lett.* **57** (1986) 2326.
7. R. Brandenberger, N. Kaiser, and N. Turok, *Phys. Rev.* **D36** (1987) 2242;
   R. Brandenberger, N. Kaiser, D. Schramm and N. Turok, *Phys. Rev. Lett.* **59** (1987) 2371;
   E. Bertschinger and P. Watts, *Ap. J.*, in press (1988).
8. B. Binggeli, in *"Nearly Normal Galaxies,"* ed. S. Faber, (Springer, New York, 1987).
9. N. Bahcall, *Ann. Rev. Astr. Astrophys.* **15** (1977) 505.

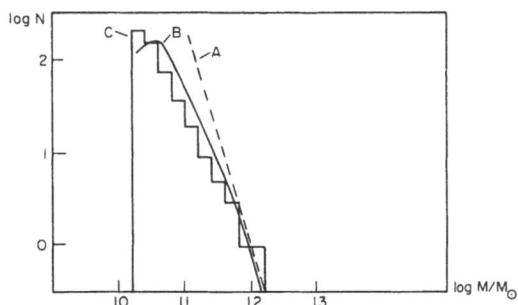

1. The initial mass function for cosmic strings and cold dark matter (redshift $z \sim 10$). Curve C presents the numerical results, curve B gives the analytical prediction, curve A shows the original mass function with a slope $-5/2$.

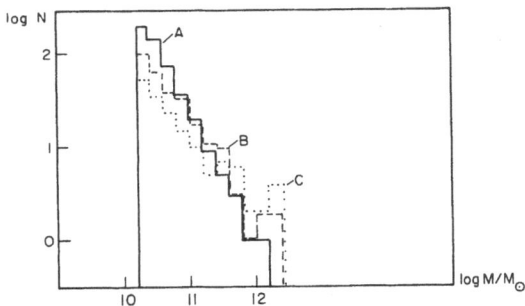

2. The evolution of the mass function. Curve A is the initial mass function at a redshift of 10. Curve B is the mass function at redshift 2.5 and curve C the mass function at $z = 1.2$.

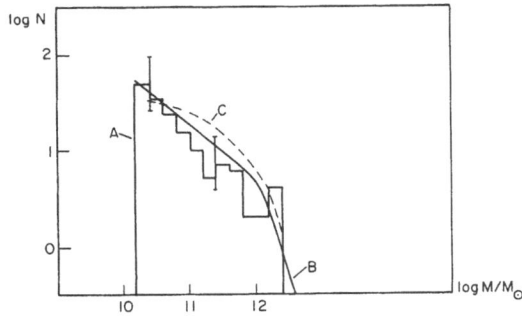

3. A comparison with observational data. Curve A gives the cosmic string mass function at $z = 1.2$, curve B is the total mass function given by Binggeli (Ref. 8) and curve C are the observational data from Bahcall (Ref. 9). Note that curves B and C can be shifted horizontally by changing the mass to light ratio. Thus, really only the slopes of the three curves should be compared. The error bars for the two points on curve A represent the statistical uncertainties based on two runs.

GALAXY FORMATION IN UNSTABLE DARK MATTER MODELS

A.A. Klypin
Space Research Institute, Moscow 117810, USSR
A.A. Doroshkevich
Institute of Applied Mathematics, Moscow 125047, USSR

ABSTRACT. Formation and evolution of the large-scale structure of galaxy distribution in **unstable dark matter** (UDM) models are discussed. Epoch of galaxy formation which for stable neutrino models (m = 30 eV) was too late (z = 0.5 - 1) for UDM models (m = 50 - 100 eV) goes to z = 3 if products of decays are relativistic at present or at least to z= 7 - 10 if the products are nonrelativistic.

1. INTRODUCTION

The model we will discuss is a modification of the pancake picture with stable (m = 30 eV) neutrinos. From our point of view the most significant objections to the stable neutrino model are as follows. (i) Epoch of **galaxy formation** is too late (z= 0.5 - 1). (We use normalization on proper slope of galaxy correlation function -1.8 at present). (ii) Structures in the model look too regular in comparison with what is observed. UDM models resolve (i): decay of a large fraction of matter essentially slows down the growth of perturbations which means that galaxy formation could start much earlier without changing the slope of correlation function at z = 0. The mass of decaying particles (50 - 100 eV) is usually 2 - 4 times larger than in stable model which reduce all scales by the same factor.

The regular appearance of structures in pancake models is due to the fact that no numerical models at present can involve very complicated but quite essential processes of galaxy formation inside superclusters. In existing numerical pancake models there should be no galaxies because of absence of perturbations on galactic scales. As a result a pancake looks very regular with smoothly varying density along its surface. In reality, quick and violent hydrodynamical and thermal instability processes must destroy the quiet picture (Doroshkevich et al. 1978).

2. THE UNSTABLE DARK MATTER MODEL

There is a variety of UDM models which differ one from the other by masses, lifetimes, types of decays, initial perturbations and so on (Doroshkevich had Khlopov, 1984a,b; Turner et al. 1984; Suto et al. 1985; Hoffman 1986). The main points of models under consideration are as follows (Doroshkevich & Khlopov 1984a,b; Doroshkevich et al. 1985, 1987).

A heavy particle ν_H ('neutrino') with mass m_H and time of decay τ (z is the redshift at t = τ) decays on two particles ν_R and ν_L with masses m_R and m_L ($m_R < m_L < m_H$). Before decays heavy particles ν_H dominated in the Universe, but there was a population of stable

C. S. Frenk et al. (eds.), The Epoch of Galaxy Formation, 327–330.

particles with small contribution Ω_{st} to the mean density. We do not specify the content of the population. In any case it should include ordinary matter ('baryons') and weakly interacting particles, which at present give missing mass on scales of galaxies and clusters of galaxies (relic stable neutrinos ν_L are nice candidates). After decays there arise a component of relativistic particles ν_R with density Ω_R and a component of massive particles ν_L with density Ω_{hot}. Because of expansion, contribution of different components to the total density changes with time, but at any moment of time $\Omega_d + \Omega_R + \Omega_{hot} + \Omega_{cold} = 1$, where Ω_{cold} is the contribution of stable particles ($\Omega_{cold} = \Omega_{st}$ at $t \ll \tau$), Ω_d is the contribution of decaying particles ($\Omega_d = 1 - \Omega_{st}$ at $t \ll \tau$).

It is very important that particles decay before the onset of the nonlinear stage in the growth of perturbations, when $\delta\rho/\rho$ is less (but not much less!) than unity. Slow growth of perturbations provides the formation of first objects at z = 3-10 (depending on particular choice of parameters) and conserves the structure up to z = 0.

3. NUMERICAL MODEL

A number of numerical models was present in our previous papers (Doroshkevich et al. 1985, 1987). The models were made for m_L=0 and showed that in this case galaxy formation could start at z≈3. The aim of present simulations is to demonstrate that there are UDM models for which galaxy formation starts much earlier - at z > 7 - 10.

Parameters of the model are as follows: $m_H = 78$ eV, $m_L = 0.15 m_H = 12$ eV, $\tau = 1.6 \cdot 10^8$ yrs, z = 11.8, Ω_{cold}=0.38, Ω_{st}=0.1. Amplitude of density perturbations estimated by the linear theory at z = 0 is $(\delta\rho/\rho) = 2.0$. The simulations were made with 64^3 particles in mesh 64^3 using standard Cloud-In-Cell method.

When selecting points which could be identified as 'galaxies' we used a two-step procedure described by Doroshkevich et al. (1987). Firstly, only points inside pancakes were selected. Secondly, it was taken into account that only first fraction of gas heated by shock could cool and may be used to form galaxies.

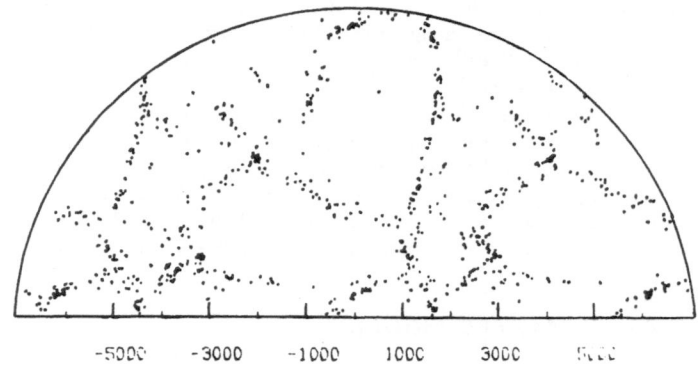

Fig. 1. Distribution of simulated galaxies within a 6 degree cone. 850 objects are shown. Recession velocities (km/s) are used as distance indicators.

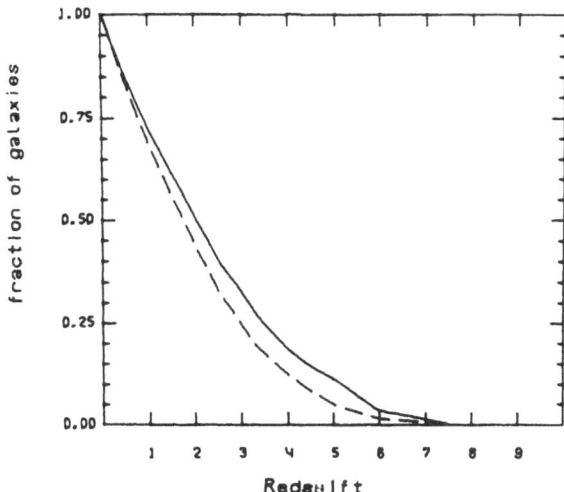

Fig. 2. Fraction of galaxies formed up to given redshift (full curve). Broken line presents fraction of matter piled up by superclusters. The functions are normalised at z=0.

Figure 1 presents wedge diagram (6 degree cone as in de Lapparent et al. (1986) for 'galaxies' in the model. The agreement with the observed distribution seems to be very encouraging. Long superclusters and giant empty voids (20-30 Mpc) are clearly seen, but the structure is not so regular as it was in the case of standard stable HDM model. Small fraction of 'galaxies' (7%) is presented in Figure 1. Random choice of points generates small-scale density fluctuations, which are absent in real density of the models, but it must be produced by galaxy formation processes.

In the range $\xi(r) > 1$ correlation function of 'galaxies' is well approximated by power law with slope γ=-1.8.

Figure 2 presents fraction of galaxies formed up to given z and fraction of points piled up into pancake. It is clearly seen that first galaxies and quasars could be formed as early as z≈7.

4. CONCLUSIONS

(1) Pancake UDM models become more attractive if main features of galaxy formation are taken into account and included in numerical simulations. (2) Galaxy formation for UDM models could start at least at z = 7 - 10.

References

DeLapparent, V., Geller, M., Huchra, J. 1986, Astrophys. J. 302, L1.
Doroshkevich, A.G., Khlopov, M.U. 1984a, Yadernaya Fizica 39, 869.
Doroshkevich, A.G., Khlopov, M.U. 1984b, M.N.R.A.S. 211, 277.
Doroshkevich, A.G., Klypin, A.A., Khlopov, M.U. 1985, Pis'ma v. Astron. Z. 11, 483.
Doroshkevich, A.G., Klypin, A.A., Khlopov, M.U. 1987, IAU Symp. No. 130. Eds. Audouze, J. et al., Reïdel, Dordrecht, Holland.

330

Doroshkevich, A.G., Shandarin, S.F., Saar, E., 1978, M.N.R.A.S. **184**, 643.
Hoffman, Y. 1986, Astrophys. J. **305**, L1.
Turner, M.S., Steigman, G., Krauss, L.M. 1984, Phys. Rev. Letters **52**, 2090.
Suto, Y., Kodama, H., Sato, K. 1985, M.N.R.A.S. 218, 637.

DISCUSSION:

SPERGEL: Your model predicts a neutrino mass that is large enough to produce an observable time of flight delay in the arrival time of neutrinos produced in a galactic supernovae. Planned detectors (e.g. Sudbury) are sensitive to neutral current interactions and if we are fortunate enough to observe a galactic supernova, we could observe your ν_H.

PEEBLES: Have you estimated the number density of rich clusters of galaxies in your model?

KLYPIN: Recently Doroshkevich estimated analytically number density of rich clusters in decaying models and found reasonable agreement with observed value. But the estimates certainly must be confirmed in numerical simulations. Unfortunately the size of simulated cube in presented models was too small to give the number density of clusters (we were mainly interested in galaxy distribution on scales 10-20 Mpc).

BONOMETTO: How would the prediction on $\Delta T/T$ change if the primeval spectrum has $n \neq 1$, as is well known to be the case within the framework of inflationary models? Is it right that constraints coming from this side become less stringent?

KLYPIN: One of our models has a power spectrum index $n = 4$. In this case the estimated value of the quadrupole moment is $Q_2 = 2\ 10^{-7}$. This should be compared with $Q_2 = 2\ 10^{-5}$ in models with an $n = 1$ spectrum and with an upper limit imposed by Relict experiment $Q_2 < 0.5\ 10^{-5}$ (valid only for $n = 4$ primordial spectrum; $Q_2 < 1.6\ 10^{-5}$ for $n = 1$ case).

JONES: What is the age of your model?

KLYPIN: 12 Gyr compared with 13 Gyr, so we lost 1 Gyr as compared with CDM model.

THE AGE OF THE GALAXY

K.C. Freeman
Mount Stromlo and Siding Spring Observatories
The Australian National University

ABSTRACT. How old are the halo globular clusters, the disk globular clusters and the oldest stars of the galactic disk ? In principle, these questions can be answered directly from observations and the theory of stellar evolution (isochrones), nuclear cosmochronology and white dwarf cooling. In practice, the various estimates do not yet agree well. 1) From isochrones, Vanden-Berg's system gives ages of about 14 Gyr for metal rich and metal poor globular clusters, while ages from the Revised Yale Isochrones range from about 12 to 18 Gyr. 2) On VandenBerg's system, the oldest disk stars have ages of about 15 Gyr; their RYI ages would be significantly younger. 3) Ages for the solar neighborhood derived from nuclear cosmochronoloy are still contentious: solar system data give an age in the range 10 to 15 Gyr, while data for nearby stars suggests an age less than about 12 Gyr. 4) The cutoff at the faint end of the white dwarf luminosity function implies an age for the local disk of 9.3 ± 2.0 Gyr, if the long estimates for white dwarf cooling times are correct. In this talk, I will review some recent results and problems in each of these topics.

1. Ages From Theoretical Isochrones

1.1. Globular Clusters

Despite great improvements in the color-magnitude diagram (CMD) data for globular clusters, the ages are still uncertain. It is not yet clear whether there are significant age differences between the metal rich disk globular clusters and the metal poor halo clusters (for a recent review, see VandenBerg 1988) and the galactic disk itself.

Most of the recent age estimates for globular clusters have been derived using the Vanden-Berg and Bell (1985: hereafter VB85) isochrones and later modifications. These isochrones were computed from stellar models with helium abundance $Y = 0.2$ and 0.3. The adopted mixing length parameter is $\alpha = 1.6$; models with this value of α fit the zero age main sequence in young open clusters, and the main sequence and lower giant branch in globular clusters, better than models with the older value $\alpha = 1.0$. Model atmospheres from Bell and Gustafsson (1978) were used: the colors of these atmospheres are normalized to Vega, and the bolometric corrections are normalized to the sun. There is a problem here: for the redder stars, $(B - V > 0.4)$ the $B - V$ colors come out too blue, by a few hundredths of a magnitude. Most fits of these isochrones to cluster CMD's require an adjustment in color of about this magnitude; without it, the cluster ages are overestimated by about $15\% \approx 2$ to 3 Gyr. This color error is believed to result from an incomplete line list for the atmospheres.

C. S. Frenk et al. (eds.), The Epoch of Galaxy Formation, 331–338.

The chemical compositions of the clusters are a major source of uncertainty. Most of the 16 to 18 Gyr ages derived for clusters in the last few years came from the VB85 isochrones, which have $Y = 0.20$ and scaled solar abundances. It now seems that: 1) $Y = 0.24$ is more appropriate for old systems, and 2) [O/Fe] is progressively enhanced in metal-poor field stars: see Sneden (1985) and Nissen et al. (1985). The derived ages *decrease* by about 1 Gyr for an increase of 0.3 in the adopted [Fe/H] or [O/Fe], or for an increase of 0.1 in Y. If the globular cluster abundance patterns are similar to those for field stars, then the appropriate [O/Fe] enhancement is 0.7 to 1.0 for the most metal-poor clusters, and about 0.3 for more metal-rich clusters like 47 Tuc. When this [O/Fe] enhancement is included in the stellar models, then the derived globular cluster ages are all about 14 Gyr, independent of [Fe/H]. The α-elements (Mg, Si, Ca...) are also enhanced in metal-poor stars (*e.g.* Tomkin et al. , 1985), but this enhancement has not yet been included in the stellar models from which the isochrones are derived.

Some of the best examples of high quality CMD's for globular clusters are those for M68, M15 and 47 Tuc. The CMD's for the metal poor ([Fe/H] < -2) clusters M68 (McClure et al. , 1986) and M15 (see VandenBerg, 1985a) are well fitted by oxygen-enhanced isochrones, with [O/Fe] = 0.7 and 1.0 respectively, for ages of about 14 Gyr. Consistent models for the zero age horizontal branch are used to constrain the distances to the clusters and also the small color shifts mentioned above. The CMD for the metal rich ([Fe/H] = -0.65) cluster 47 Tuc (Hesser et al. , 1987) gives an age of about 13.5 Gyr from isochrones with [O/Fe] = 0.3. Again, a small color correction is required.

1.2. The Oldest Disk Stars

VandenBerg (1985b) computed isochrones for disk stars, with $0 > $[Fe/H]$ > -1$, $Y = 0.25$, and $\alpha = 1.6$. For the old open cluster NGC 188, these isochrones give an age of 10 Gyr (with a small empirical correction for the error in the color-temperature relation). However, the old disk in the solar neighborhood contains non-cluster stars that are significantly older. A sample of nearby evolved F stars observed with high resolution spectroscopy and Strömgren photometry, by Nissen et al. (1985), includes stars with ages up to $\sim$ 15 Gyr on the VandenBerg scale (after correction for non-solar [O/Fe]). Butcher's (1987) sample of nearby solar-type stars with trigonometrical parallaxes also has stars with ages up to $\sim$ 15 Gyr.

A comment on the age of the thick disk: Norris et al. (1985) showed that about 25% of the nearby metal poor stars ([Fe/H] < -1.2) have low eccentricity orbits and thick disk kinematics. The implications depend on one's view about the origin of the thick disk. If the thick disk formed as part of a classical monolithic collapse, in which the most metal poor stars are also the oldest, then the presence of such metal poor stars in the thick disk may suggest that the oldest thick disk stars are as old as the globular clusters. On the other hand, in the accretion picture for the formation of the thick disk (*e.g.* see Freeman 1987), the presence of these metal poor stars may give no information about the age of the thick disk (if, for example, they are the debris of accreted metal poor satellites). The question of the age of the thick disk is discussed by Norris in this volume. For a comprehensive review of the problems in determining the age of the disk (and thick disk), see Sandage (1988).

1.3. The Revised Yale Isochrones

The Yale group were concerned about the problems in using model atmospheres to transform from temperature to color. (see §1.1). They therefore derived a semi-empirical UBVRI (color - effective temperature - gravity -[Fe/H]) calibration from well studied stars, instead of using model atmospheres (Green, 1988). The original stellar models were computed with the mixing length parameter $\alpha = 1.0$ (rather than the value $\alpha = 1.6$ used by VB85); this required an empirical correction to the effective temperature scale on the giant branch, which was derived by comparing the (model) effective temperatures along the giant branch with empirical V-K temperatures, as a function of [Fe/H]. The adopted [O/Fe] value was zero.

The Revised Yale Isochrones (Green *et al.* 1987: hereafter RYI) are a good fit to cluster CMD's over the entire abundance range. However the RYI ages show much more variation from cluster to cluster than those derived from the VandenBerg isochrones. On the RYI scale, the age for the metal rich cluster 47 Tuc is 12 to 13 Gyr; the metal poor cluster M92 has an age of 16 to 18 Gyr and the age of M15, which is yet more metal poor, is 14 Gyr (King *et al.*, 1988). To derive similar ages on the RYI scale for all globular clusters, arbitrary changes to [O/Fe] from cluster to cluster would be required.

As a consistency check, the RYI main sequence fits to cluster CMD's were used to derive an absolute magnitude - abundance relation for horizontal branch stars, which was in excellent agreement with the relation from independent synthetic horizontal branch models (Lee *et al.*, 1987). It is also interesting that RYI ages are available for the well known "second parameter" pair of clusters, NGC 362 (with a red horizontal branch) and NGC 288 (with a blue horizontal branch). Both clusters have similar [Fe/H] values (-1.4), but their RYI ages differ by about 4 Gyr, in the sense expected from horizontal branch models.

The RYI ages for disk stars are younger than those from VandenBerg. For example, the open cluster NGC 188 has an RYI age of only 6.5 Gyr, compared to its 10 Gyr age on the VandenBerg scale, and the ages of the oldest disk stars will be significantly less than 15 Gyr (see §1.2).

In conclusion, it seems clear that the VandenBerg and RYI ages for the globular clusters and for the old disk are not the same. The globular clusters have ages around 14 Gyr on VandenBerg's system, independent of [Fe/H], while their RYI ages spread from about 12 to 18 Gyr. The ages of old disk objects on the two systems are also significantly different. The two sets of ages have different implications for the epoch of galaxy formation and for the sequence of events that took place during galaxy formation. For definitive ages from isochrones, we will have to wait at least until the the detailed abundance distributions in individual globular clusters are known, and the problems of the temperature - color calibration are sorted out.

I am very grateful to Elizabeth Green for unpublished material, advice and helpful discussions on the problems discussed above.

2. Ages from Nuclear Cosmochronology

Nuclear cosmochronology uses the relative abundances and production rates of radioactive isotopes to estimate the time when nucleosynthesis began. For a recent review, see Fowler (1987). Most results are for solar system objects or nearby disk stars, and therefore pertain to the nucleosynthesis that affected the abundance distribution in the solar neighborhood: *i.e.* they give the time when *local* nucleosynthesis began (unless there has been extensive mixing). This need not be the same as the time of formation of the oldest objects in the Galaxy.

2.1. The Solar System

The age of the local disk, as defined by the beginning of nucleosynthesis, is $\Delta + t_{ss}$, where Δ is the time from the beginning of local nucleosynthesis to the formation of the solar system, and t_{ss} is the age of the solar system (4.6 ± 0.1 Gyr: see Fowler, 1987). The most used isotopes are ^{232}Th (20.27 Gyr mean lifetime), ^{235}U(1.02 Gyr) and ^{238}U(6.45 Gyr). There are others, but they are either shorter-lived or have uncertain physics in astrophysical environments. These isotopes are all formed in the rapid neutron capture process (the r-process).

To estimate Δ, the following information is needed: 1) The relative abundances (e.g. N_{232}/N_{238}, where N_{232} denotes the abundance of ^{232}Th, etc) at the time of formation of the solar system: these are calculated from the present relative abundances (measured from meteorites), using the known relative lifetimes. 2) The relative production rates (e.g. P_{232}/P_{238}), calculated from r-process networks. These relative production rates are not yet precise, because of uncertainties in nuclear data and theory, and because the site of the r-process is not precisely known. 3) A model for the time dependence of nucleosynthesis. The evidence for ongoing nucleosynthesis in the local galactic disk comes from the observed age-metallicity relation for nearby disk stars. Although there is a general trend of increasing abundance with decreasing age, the relation is clearly not one-to-one: at a given age, there can be a significant spread in abundance (see Nissen et al. , 1985; Barry, 1988). We will briefly describe two recent estimates of Δ.

Fowler (1987) used ^{238}U as the reference isotope. His two input parameters are $K_{232} = (N_{232}/N_{238})$ and K_{235}, defined similarly relative to ^{238}U. His model for the time dependence of nucleosynthesis has an initial spike of relative magnitude S_E, followed by a uniform rate of nucleosynthesis of relative magnitude $1 - S_E$ for the time Δ. This model leads to two equations for the two input parameters K_{232} and K_{235}, of the form $K_{232} = K_{232}(\Delta, S_E,$ lifetimes$)$, and similarly for K_{235}. For the input parameters, Fowler adopted $K_{232} = 1.351 \pm 0.100$ and $K_{235} = 0.246 \pm 0.035$. Solving these two equations for Δ and S_E gave $S_E = 0.17 \pm 0.02$ and $\Delta = 5.4 \pm 1.5$ Gyr. The age of the disk is then $(5.4 \pm 1.5) + (4.6 \pm 0.1) = 10.0 \pm 1.5$ Gyr. The error does not include the uncertainty in the model for the time dependence of nucleosynthesis.

Cowan et al. (1987) recalculated the production ratios P_{232}/P_{238} etc, using the best available internally consistent values for the rates of neutron capture and photodisintegration, beta decay and beta-delayed neutron emission and fission. For the site of the r-process, they adopted the He-burning shells of supernovae, with the neutron source ^{13}C$(\alpha, n)^{16}$O. Their r-process model gave the best fit to the solar system abundance distribution for an initial temperature of 4.10^8 K and density of 10^4 g cm^{-3}. They argued that, even if the adopted r-process environment is not correct, any calculation that gives the correct solar system abundance distribution for the r-process elements has correctly identified the location of the r-process capture path in the (neutron number)-(proton number) plane, and will therefore give correct relative production ratios. Their production ratios are smaller than those used by Fowler, by about 7% for P_{232}/P_{238} and 15% for P_{235}/P_{238}. For the time dependence of nucleosynthesis, they took a simple exponential model with instantaneous initial enrichment. Their derived value of Δ is between 7.8 to 10.1 Gyr, so the age of the disk is in the range 12.4 to 14.7 Gyr. The uncertainty in the production ratio of ^{232}Th/^{238}U, due to nuclear physics, adds a further spread of about 2 Gyr to the uncertainty of the galactic age. More complex chemical evolution models could increase the estimated galactic age by 2 to 3 Gyr. The range of galactic ages allowed by nuclear cosmochronology of the solar system would then be consistent with the range of ages for globular clusters inferred from theoretical isochrones.

2.2. Nearby Stars

Butcher (1987) measured the ratio of ^{232}Th (20.27 Gyr mean lifetime) to Nd (stable) in nearby solar-type dwarfs with a range of age, in order to estimate the age of the nearby galactic disk.

Assume that ^{232}Th is produced according to the production function $P_{232}p(t)$ and destroyed by radioactive decay with the constant λ_{232}; here $p(t)$ is the nucleosynthesis rate (assumed to be valid for all species) and P_{232} is the relative production rate. Now assume that nucleosynthesis began at $t = 0$ and continues to the present time $t = T_0$. For a star born at time T, its atmosphere gives a sample of the galactic composition at that time, modified only by free decay. i.e.

$$\frac{dN_{232}}{dt} = P_{232}p(t) - \lambda_{232}N_{232}$$

so the abundance seen *today* in a star born at $t = T$ is

$$N_{232,T} = P_{232} e^{-\lambda_{232}T_0} \int_0^T p(t) e^{\lambda_{232}t} dt$$

Relative to some stable species s ($\lambda_s = 0$), the abundance is

$$\frac{N_{232,T}}{N_{s,T}} = \frac{P_{232}}{P_s} e^{-\lambda_{232}T_0} \langle e^{\lambda_{232}t} \rangle_{t=0 \ to \ T}$$

where the angle brackets denote the mean value over the interval $t = 0$ to T. Now consider how the Th/Nd ratio changes for stars of different ages. If there is no observed evolution in the Th/Nd ratio, then the above equation implies that either (1) the $\langle e^{\lambda_{232}t} \rangle$ term is constant (*i.e.* a single synthesis event at the beginning) or (2) the term $\langle e^{-\lambda_{232}(T_0-t)} \rangle$ is constant ($T_0 \ll \lambda_{232}^{-1} = 20 \ Gyr$; *i.e.* no significant radioactive decay has occurred). Note that Th is an r-process element and Nd is partly an s-process element, so their relative production rates *could* vary with time. However Butcher argued that this is unlikely to be a problem, because the observed ratio of r- to s-process element abundances is approximately constant for stars with [Fe/H] $\gtrsim -1.5$.

Butcher made high resolution ($R = 100,000$), high S/N (100 to 500) observations of the spectra of the sun and 20 nearby solar type stars. Most of these stars have measured trigonometric parallaxes, so their ages (on the VandenBerg scale) can be derived directly. He used the ThII $\lambda4019.13$, NdII $\lambda4018.82$ lines: these represent the dominant ions over the region of line formation in solar-type stars and have similar ionization potentials, and will therefore have similar behaviour with gravity and temperature. He estimated the Th/Nd ratio by fitting synthetic spectra to his data: as a check, he showed that the solar Th abundance estimated in this way agrees well with the meteoritic Th abundance.

The observed Th/Nd ratio for these stars was approximately constant, over the entire age range from 1 to 15 Gyr. The implications are: 1) If nucleosynthesis was dominated by a single early event (from the equation above, this is consistent with no observed evolution in the Th/Nd ratio), then the age of the nearby disk is given by the solar system data (which Butcher took to give an upper limit of 11 Gyr: see the previous section for alternative limits). 2) At the other extreme, if the rate of nucleosynthesis was constant (*i.e.* $p(t) = $ constant), then the 2-σ limit on the slope of the observed Th/Nd–age relation gives an upper limit on the present galactic age T_0 of 9.6 Gyr. An exponential nucleosynthesis rate is also consistent with the observations, for $T_0 \lesssim 12$ Gyr.

The conclusion, of a young age for the nearby disk, depends on the assumption of the *same* p(t) for the r– and s–processes. Models with separate p(t)'s and older ages can be made consistent with the Th/Nd data; however Butcher (1988) argued that these models are then clearly inconsistent with the Eu/Ba data for his stars (Eu comes from the r-process and Ba from the s-process, and the observed Eu/Ba ratio is also approximately constant with age).

Using analytic models, Clayton (1988) has recently discussed many of the important uncertainties in nuclear cosmochronology. In particular, he showed that early infall of metal poor gas into the galactic disk can lead to significantly greater derived ages for the local disk from the U, Th chronometers, and stressed the importance of a correct model of the growth and chemical evolution of the solar neighborhood for the determination of the galactic age by nuclear chronology. He also analysed several other radioactive chronometers and concluded that, although no single method is reliable, when taken together they favor a galactic age in the range 12 to 20 Gyr.

3. The White Dwarf Luminosity Function and the Age of the Disk

White dwarfs (WD's) originate from main sequence stars with masses in the range $1 \lesssim M \lesssim 8M_\odot$. Their evolution is dominated by cooling: the heat of the degenerate core leaks out slowly through the thin insulating non-degenerate envelope. If the cooling times are long enough, then the coolest known degenerate stars are among the oldest stars of the nearby disk.

The luminosity function for WD's shows a cutoff at the faint end ($\log L/L_\odot \approx -4.5$). Comparison of this observed luminosity function with theoretical luminosity functions (calculated using theoretical cooling curves) can give an independent estimate of the age of the nearby disk, as suggested by Schmidt (1959).

The most recent derivation of the WD luminosity function is by Winget $et\ al.$ (1987) and Liebert $et\ al.$ (1988). Their data come from two sources. Stars with $M_V < 13$ are from the Palomar Green survey: this is a UB photographic survey for objects with $U - B < -0.44$, which produced 353 spectroscopically confirmed DA WD's. The M_V-color calibration was established from stars with parallaxes. The luminosity function was derived using $1/V_m$ weighting and assuming that the disk density decreases exponentially with height above the galactic plane, with a scale height of 250 pc. For the fainter stars ($M_V > 13$), the LHS catalog gave 43 confirmed WD's with declination $> -20°$ and with proper motion $\mu > 0''.8$. Most have trigonometric parallaxes. Because the LHS is a kinematically selected catalog, $1/V_m$ weighting was needed to derive the luminosity function, corrected for kinematic bias. The LHS was assumed to be complete for $V < 18$ and $0''.8 < \mu < 2''.0$. The resulting mean tangential velocity is then independent of M_V. To compare the observed luminosity function with theoretical luminosity functions (and thereby derive the age of the nearby galactic disk), bolometric corrections calculated from model atmospheres and the observed colors were used. The resulting composite luminosity function showed a marked cutoff at $\log L/L_\odot \approx -4.5$.

The next step was to compute theoretical luminosity functions. This required calculation of the WD cooling curves, and some assumptions about the history of nearby star formation. There are many theoretical calculations of WD cooling ($e.g.$ Mestel, 1952; Iben and Tutukov, 1984). The more recent ones include detailed treatment of the microscopic physical properties of the WD interior and envelope. As the WD cools and its luminosity decreases, it begins to crystallize (at $\log L/L_\odot \approx -2$), releasing its latent heat of crystallization. Thereafter the specific heat of its interior is much reduced, and more rapid (Debye) cooling sets in, to invisibility. There is some disagreement about the cooling timescales; it is not entirely clear whether the observed cutoff at the faint end of the WD luminosity function represents the age of the galactic disk ($i.e.$ there are no nearby WD's old enough to have cooled to $\log L/L_\odot \lesssim -4.5$), or is simply due to more rapid WD cooling.

Winget $et\ al.$ (1987) and Liebert $et\ al.$ (1988) argued that most of the recent cooling models do not predict that the cooling is rapid enough to produce the observed break in the luminosity function at $\log L/L_\odot \lesssim -4.5$. They constructed a model for the luminosity function, which we will describe briefly, and thereby estimated the age of the nearby disk.

They used pure carbon WD models, including convection, Coulomb interactions and crystallization. (The presence of a H/He envelope and of oxygen in the interior each affects the age at $\log L/L_\odot = -4.5$ by about 15% but with opposite sign). The age of a WD at a given $\log L$ depends on its mass, so cooling sequences were constructed for WD's in the mass range 0.4 to 1.0 $M_\odot$; these sequences were then weighted by the observed mass distribution (mean $= 0.6\ M_\odot$, $\sigma = 0.1\ M_\odot$). The epoch of formation of the galactic disk was assumed. Then the contribution to the luminosity function from each mass is zero for $\log L$ values such that the corresponding cooling times are greater than the assumed age of the disk. Matching the theoretical and observed cutoff in the luminosity function gave the cooling age for the WD population to be 9.0 ± 1.8 Gyr. The mean pre-WD evolution time of the progenitor stars is about 0.3 Gyr, so the derived age of the nearby disk is then 9.3 ± 2.0 Gyr.

An alternative view comes from Mazzitelli and d'Antona (1986), who followed the evolution of a $3M_\odot$ star from the main sequence through the WD stage, with various assumptions about the composition and atmospheric properties. They find that the cooling time to $\log L/L_\odot = -4.5$, for their most realistic models, is in the range 5.1 to 6.6 Gyr, which is significantly less than the likely age of the disk. The main difference between their models for WD cooling and those of Winget $et\ al.$ (1987) is that Mazzitelli and d'Antona used older conductive opacities which are about two times lower than those used by Winget $et\ al.$, and which therefore lead to shorter cooling times.

White dwarf cooling models constructed since 1986 give values of the cooling time to $\log L/L_\odot = -4.5$ that range from 5 to 13 Gyr. Winget and van Horn (1988) showed that these differences come almost entirely from the physics included and the parameters adopted, and not from computational differences. They tabulated the effects on the age at $\log L/L_\odot = -4.5$ of some changes in the physics and parameters:

Old to new conductive opacities:	+2.3 Gyr
He layer abundance: $z = 0.02$ to $z = 10^{-5}$	-1.8 Gyr
Pure C to pure O core:	-0.6 Gyr
Suppressing crystallization and	-0.5 Gyr (O/He WD)
release of latent heat:	-1.4 Gyr (C/He WD)

As Liebert $et\ al.$ (1988) stated, "It is obviously crucial, if the WD luminosity function is to be used as a galactic chronometer, to determine which input physics the degenerate stars use". Empirical calibration of the cooling ages is urgently needed, $e.g.$ from degenerate companions to demonstrably old stars.

References

Barry, D. 1988. Preprint.

Bell, R., Gustafsson, B. 1978. Astron.Astrophys.Suppl., **34**, 229.

Butcher, H. 1987. Nature, **328**, 127.

Butcher, H. 1988. Messenger, No. 51, 12.

Clayton, D. 1988. Mon.Not.R.Astron.Soc., **234**, 1.

Cowan, J., Thielemann, F., Truran J. 1987. Astrophys.J., **323**, 543.

Fowler, W. 1987. Q.J.R.Astron.Soc, **28**, 87.

Freeman, K.C. 1987. Ann.Rev.Astron.Astrophys., **25**, 603.

Green, E., Demarque, P., King, C. 1987. *The Revised Yale Isochrones and Luminosity Functions*, (New Haven: Yale University Observatory).

Green, E. 1988. Preprint.

Hesser, J., Harris, W., VandenBerg, D., Allwright, J., Shott, P., Stetson, P. 1987. Publ.Astron.Soc.Pa **99**, 739.

Iben, I., Tutukov, A. 1984. Astrophys.J., **282**, 615.

King, C., Demarque, P., Green, E. 1988. Preprint.

Lee, Y., Demarque P., Zinn, R. 1988. *The Second Conference on Faint Blue Stars*, ed A. Davis Philip, D. Hayes, J. Liebert (L. Davis Press).

Liebert, J., Dahn, C., Monet, D. 1988. Astrophys.J., **332**, 891.

Mazzitelli, I., D'Antona, F. 1986. Astrophys.J., **308**, 706.

Mestel, L. 1952. Mon.Not.R.Astron.Soc., **112**, 583.

McClure, R., VandenBerg, D., Bell, R., Hesser, J., Stetson, P. 1987. Astron.J., **93**, 1144.

Nissen, P., Edvardsson, B., Gustaffson, B., 1985. *Proceedings of ESO Workshop on Production and Distribution of C,N,O Elements*, p 131.

Norris, J., Bessell, M., Pickles, A. 1985. Astrophys.J.Suppl., **58**, 463.

Sandage, A. 1988. Preprint.

Schmidt, M. 1959. Astrophys.J., **129**, 243.

Sneden, C. 1985. *Proceedings of ESO Workshop on Production and Distribution of C,N,O Elements*, p 6.

Tomkin, J., Lambert, D., Balachandran, S. 1985. Astrophys.J., **290**, 289

VandenBerg, D. 1985a. *Proceedings of ESO Workshop on Production and Distribution of C,N,O Elements*, p 78.

VandenBerg, D. 1985b. Astrophys.J.Suppl., **58**, 711.

VandenBerg, D. 1988. *Globular Cluster Systems in Galaxies*, ed J. Grindlay and A. Davis Philip (Kluwer: Dordrecht), p 107.

VandenBerg, D., Bell, R. 1985. Astrophys.J.Suppl., **58**, 561.

Winget, D., Hansen, C., Liebert, J., Van Horn, H., Fontaine, G., Nather, R., Kepler, S., Lamb, D. 1987. Astrophys.J., **315**, L77.

Winget, D., Van Horn, H. 1988. *The Second Conference on Faint Blue Stars*, ed A. Davis Philip, D. Hayes, J. Liebert (L. Davis Press), p 363.

THE NATURE AND AGE OF THE GALACTIC THICK DISK

John E. Norris

Mount Stromlo and Siding Spring Observatories
Australian National University

ABSTRACT.   It is argued that the Gilmore-Reid-Wyse thick disk is the
tail of a continuous extended Galactic disk configuration. The low
abundance material in the extended disk has less extreme kinematic
parameters than advocated for the thick disk at similar metallicity.
The case is made that at the solar circle the thicker component of the
disk has an age 3-6 Gyr younger than that of the disk globular
clusters, and that the Galactic disk grew outwards on a timescale of
several Gyr.

1. INTRODUCTION

     At the end of the 1970s most astronomers were of the view that the
luminous material of the Galaxy could be well described in terms of a
thin disk structure embedded in a tenuous halo. That this was an
oversimplification became clear in the present decade as the result of
two investigations. First, Gilmore and Reid (1982) argued persuasively
from stellar number counts towards the South Galactic Pole for the
existence of a third component which they designated thick disk. They
reported that its scale height perpendicular to the Galactic plane is
1350 pc, compared to 300 pc for the thin disk, and that at the plane
the ratio of thick to thin disk components is 0.02. In further
investigations, Gilmore and Wyse suggested that the thick disk has a
velocity dispersion perpendicular to the Galactic plane σ_z ~ 60 km/s,
an asymmetric drift of ~ -100 km/s relative to the LSR, and <[Fe/H]> ~
-0.6 (Gilmore 1984; Gilmore and Wyse 1986; Wyse and Gilmore 1986). The
second result which argued strongly for the need for some component
intermediate between thin disk and halo was the demonstration by Zinn
(1985) that the globular cluster population of the Galaxy is bimodal
and comprises halo and disk components. For the latter component Zinn
reports a scale height of order 500 pc, an asymmetric drift of -70
km/s, and <[Fe/H]> ~ -0.5.
     It should be noted, however, that Bahcall and his coworkers
strongly questioned the existence of the Gilmore-Reid thick disk (see
eg Bahcall and Soneira 1984).   Thus when Rose (1985) reported that

339

C. S. Frenk et al. (eds.), The Epoch of Galaxy Formation, 339–344.

towards the North Galactic Pole there exists a well defined population of red horizontal branch (RHB) stars with parameters not unlike those of the thick disk it seemed to the present author that these objects could provide an ideal probe to investigate and say something definitive about the thick disk. Clearly if Rose were correct one would have test particles of well defined brightness (M_v ~ 1.0). Further, one would have good reason to believe that the thick disk was coeval to within a billion or so years with the disk globular clusters.

The present paper summarizes work which has been done over the past three years to utilize core-helium-burning (RHB and clump) stars as probes of the thick disk.

2. THE ROSE RHB CANDIDATES

Rose suggested that his RHB candidates belonged to a population which comprised some 10 percent of the Galactic disk, had a scale height > 500 pc, a velocity dispersion σ_z = 40 km/s, and by inference <[Fe/H]> = -0.7. To astrophysical accuracy these properties are the same as those suggested for the thick disk. As a first step in using the RHB candidates as probes of the thick disk, observations were made with the DDO intermediate band photometric system. The result of this endeavour (Norris 1987a) was to suggest that the Rose candidates were not in fact the counterpart of the RHB stars in the disk globular clusters but were similar to the so-called clump stars in the old (~5 Gyr), metal poor ([Fe/H] ~ -0.6), open clusters such as NGC 2243. One should consult the cited reference for complete details, but one way to see the effect is to compare the color distribution of the Rose RHB candidates with those of the RHB stars in the disk globular cluster 47 Tuc ([Fe/H] = -0.7) and the clump stars in the old, metal poor open clusters as is shown in the top 3 panels in Figure 1. The point to be made is that the Rose candidates have a color distribution more like that of the open clusters than of 47 Tuc. A similar conclusion was reached also by Friel (1987). The simplest explanation of these results is that the thick disk discovered by Rose is younger by a few Gyr than the disk globular cluster population. We shall return to this point in Section 4.

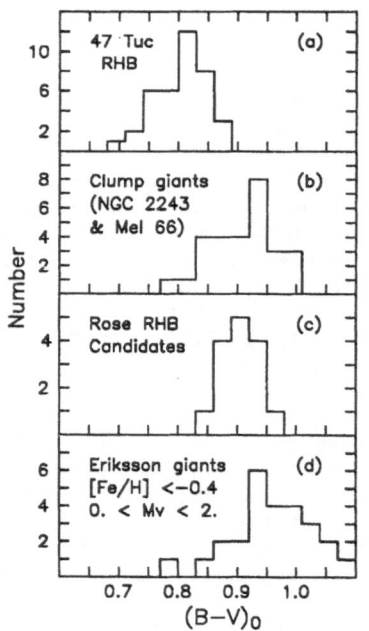

Figure 1. $(B-V)_0$ histograms for (a) the RHB of 47 Tuc, (b) clump stars in NGC 2243 and Melotte 66, (c) Rose RHB candidates, and (d) Eriksson giants.

3. THE THICK DISK AS THE TAIL OF AN EXTENDED DISK CONFIGURATION

The Rose RHB candidates have a velocity dispersion σ_z = 40 km/s, while the putative value advocated for the thick disk by Gilmore was 60 km/s. The former value was reminiscent of the result obtained by Janes (1975) for stars with [Fe/H] ~ -0.6, in his investigation of a complete sample of red giants in the solar neighborhood. His data, combined with the reinvestigation of Norris (1987b), are shown by the open circles in Figure 2. Note the increase of σ_z as [Fe/H] decreases. Clearly the lower abundance material will have a larger scale height than that having higher abundance, and will lie in a thicker disk configuration. Just what sort of a disk one might expect from this

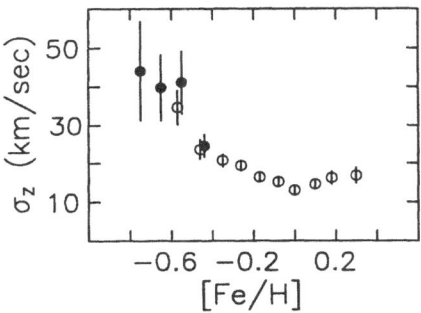

Figure 2. The dependence of σ_z on [Fe/H]. The open circles are based on the work of Janes (1975) and Norris (1987b), while the closed circles are derived from the data of Hartkopf and Yoss (1982) for stars within 1 kpc of the Galactic plane.

material? An attempt to answer this question was made by Norris (1987b), who modelled the density distribution perpendicular to the Galactic plane using the precepts laid down by van der Kruit and Searle (1981). It was found that the stellar number counts of Gilmore and Reid (1982) could be well fitted beginning with the data of Janes (1975) with only one free parameter, the scale height of the major component of the Galactic disk, which is given by z_0 = 230 pc. That is, a continuous and extended disk, having less extreme parameters than advocated by Gilmore and his coworkers, is consistent with the observed number counts. A discrete thick disk component is not necessarily required by the observations.

Gilmore (see Norris 1987c) countered that the above model did not provide a good description of his unpublished in situ velocity dispersion measurements, and that a two component thin/thick disk model in which the thick component had a velocity dispersion of 49 km/s gave a better fit. The reader should inspect Figure 9 of Norris (1987c). I take some comfort in the fact that 49 km/s is somewhat less than the 60 km/s initially suggested for the thick disk, and in the spirit of compromise suggest that the data of the sample of Janes (1975) do not permit one to make a definitive statement about the form of the σ_z versus [Fe/H] relation at low abundance - there are too few stars. In Figure 2, however, I show as filled circles the values of σ_z for [Fe/H] < -0.4 taken from the work of Hartkopf and Yoss (1982) for material within 1 kpc of the sun. The data are clearly consistent with the view that the velocity dispersion continues to about 40-45 km/s as [Fe/H] decreases to -0.8. Further work on a larger sample of stars is necessary to define the run of σ_z for [Fe/H] < -0.3. If this can be done with precision, we will be able to answer the question as to whether one has a continuous disk or a discrete thin/thick disk configuration with the attendant implications for its origin.

The question of the asymmetric drift of the thick disk is also of some interest. The preferred value of Gilmore and Wyse (1986) is -100 km/s. In contrast, most of the available observational estimates suggest a substantially smaller value. The results of Janes (1975), Sandage and Fouts (1987), Ratnatunga and Freeman (1988), and Friel (1988) all suggest a value of -30 km/s. This too is consistent with the view that the thick disk is merely the monotonic tail of an extended disk configuration.

4. ON THE AGE OF THE THICK DISK

It was recognized by Norris (1987a) that the Rose RHB candidates were spectroscopically selected and might be biased against the bluer RHB stars (with $(B-V)_0$ ~ 0.8), which might then explain their rather red color distribution seen in Figure 1. To address this problem Norris and Green (1989) undertook a survey of a sample of stars at the South Galactic Pole from the catalog of Eriksson (1978). Stars were chosen to have V < 13, and 0.70 < B-V < 1.10, and observed with the DDO photometric system with a view to isolating red giants in their core-helium-burning (RHB and clump) phase of evolution up to a distance of ~ 3 kpc from the Galactic plane. The DDO system permits one to distinguish giants from dwarfs with great efficiency, and to determine their abundances and distances. The upshot of this investigation was that among giants with M_v = 0 to 2, and further than 1 kpc from the Galactic plane, the bluer RHB stars as are found in disk globular clusters with [Fe/H] = -0.3 to -0.7 are not present in the Eriksson sample at more that the 10 percent level. The B-V histogram for the Eriksson giants with 0 < M_v < 2 and [Fe/H] < -0.4 is shown in the bottom panel of Figure 1. Here one sees a distribution which once again resembles that of the old, metal deficient open clusters such at Melotte 66 and NGC 2243, rather than that of 47 Tuc.

The case put forward by Norris and Green (1989) is that the explanation of the observations is that the thick disk is 3-6 Gyr younger that the disk globular clusters. One of the details being currently debated is just how representative 47 Tuc is of the disk globular clusters, and whether abundance uncertainties of order 0.3 dex can be responsible for the observed effects. We refer the reader to Sandage (1989) for the counter argument.

We recall four independent pieces of evidence which suggest that the bulk of the Galactic disk in the solar neighborhood is also younger than the disk globular clusters by several (~5) Gyr. They are : (1) Winget et al (1987) suggest that the absence of cool white dwarfs with M_{bol} > 16 is most simply explained if the disk has a age of 9 Gyr; (2) Schuster and Nissen (1987), using Stromgren photometry on local high velocity field stars, show that material with [Fe/H > -0.6 exhibits main sequence turnoffs consistent with ages at least 6 Gyr younger than that of the halo material ([Fe/H] < -1.0) in the sample. (One infers that they saw no turnoff corresponding to greater ages in their [Fe/H] $\geq$ -0.6 material.); (3) Barry (1988), for an unbiased sample in the solar neighbourhood, uses a calibration of calcium emission versus age to argue for a large star formation rate for ages 5-10 Gyr, with not too much star formation at greater ages; and (4) Butcher (1987, 1988), from the observed constancy of the ^{232}Th/Nd abundance ratio in a sample of local disk stars having -0.8 < [Fe/H] < 0.4, argues that either these elements were synthesized before the disk formed or the disk in the solar neighbourhood has an age less that 10 Gyr.

These results have an important implication. If the disk in the solar neighbourhood is indeed somewhat younger than the disk globular cluster population, which has a mean Galactocentric distance of order 4 kpc, one then has the possibility that galactic disks form first at the center and grow outward on a timescale of a few Gyr, as has been suggested, for example, by Larson (1976).

REFERENCES

Bahcall, J. N. and Soneira, R. M. 1984, Astrophys. J. Suppl. 55, 67.
Barry, D. C. 1988, Astrophys. J. in press.
Butcher, H. R. 1987, Nature 328, 127.
Butcher, H. 1988, The Messenger 51, 12.
Eriksson, P. -I. W. 1978, Uppsala Astron. Obs. Rep. No. 11.
Friel, E. D. 1987, Astron. J. 93, 1388.
Friel, E. D. 1988, Astron. J. 95, 1727.
Gilmore, G. 1984, Monthly Notices Roy. Astron. Soc. 207, 223.
Gilmore, G. and Reid, N. 1982, Monthly Notices Roy. Astron. Soc. 202, 1025.
Gilmore, G. and Wyse, R.F.G. 1986, Astron. J. 90, 2015.
Hartkopf, W.I. and Yoss, K. M. 1982, Astron. J. 87, 1679.
Janes, K. A. 1975, Astrophys. J. Suppl. 29, 161.
Larson, R. B. 1976, Monthly Notices Roy. Astron. Soc. 176, 31.
Norris, J. 1987a, Astron. J. 93, 616.
Norris, J. 1987b, Astrophys. J. Letters. 314, L39.
Norris, J. 1987c, in The Galaxy, G. Gilmore and R. F. Carswell, eds., Reidel, Dordrecht, p.297.
Norris, J. E. and Green, E. M. 1989, Astrophys. J. in press.
Ratnatunga, K. U. and Freeman, K. C. 1989, Astrophys. J. in press.
Rose, J. A. 1985, Astron. J. 90, 787.
Sandage, A. 1989, in Calibration of Stellar Ages, A. G. D. Philip, ed., L. Davis Press, Schenectady, in press.
Sandage, A. and Fouts, G. 1987, Astron. J. 93, 74.
Schuster, W. J. and Nissen P. E. 1987, in Stellar Evolution and Dynamics of the Outer Halo of the Galaxy, M. Azzopardi, ed., ESO, Garching, p.141.
Winget, D. E., Hansen, C. J., Liebert, J., van Horn, H. M., Fontaine, G., Nather, R. E., Kepler, S. O., Lamb, D. Q. 1987, Astrophys. J. Letters 315, L77.
van der Kruit, P. C. and Searle, L. 1981, Astron. Astrophys. 95, 116.
Wyse, R.F.G. and Gilmore, G. 1986, Astron. J. 91, 855.
Zinn, R. 1985, Astrophys. J. 293, 424.

DISCUSSION:

JONES: Is the detailed distribution of the stellar orbital parameters in the Gilmore-Wyse sample consistent with the van der Kruit-Searle potential that you fitted? It should be if your calculated scale height is to be taken as a serious criticism of their model.

NORRIS: The results of the Gilmore-Wyse survey are not yet available. It will be interesting to see how they relate to the van der Kruit-Searle formalism.

FALL: Is the diagram with numbers of stars against age that you showed a direct indication of the history of star formation in the solar neighbourhood, i.e. is the sample unbiased with respect to age?

NORRIS: No. As noted by Barry, they have to be corrected for the scale height-age correlation - depending on one's prejudices concerning the origin of that correlation.

GUNN: Might it not be that if one corrected for the kinematics/age correlation in the 25 pc sample and the resultant thickening in the old stars, that the star formation rate will be almost constant integrated over height. The numbers are so small that the existence of a cutoff is likely to be very uncertain.

NORRIS: Clearly it would be nice to extend the technique to a larger sample to investigate this question. I agree that the numbers are too small to make a strong statement.

A Simple Model for the Local Group of Galaxies

Adrian L. Melott
Department of Physics and Astronomy
University of Kansas
Lawrence, Kansas 66045 U.S.A.

ABSTRACT. The standard model for the Local Group is a two–body system originally comoving with the Universe. I generalize this model by assuming accretion onto two seed masses. The model is normalized by an assumed Hubble constant, the distance to M31, and its observed radial velocity. These assumptions result in reasonable rotation curves for the galaxy. A test of the velocity field against dwarf members of the Local Group results in excellent agreement if $H_0 = 50 \ kms^{-1} \ Mpc^{-1}$.

Kahn and Woltjer (1950) suggested that at high redshift the Local Group was a two-body system expanding with the Universe. Peebles (1971) worked out this model and verified that it produced reasonable infall velocities of the Milky Way toward M31. A modern approach should include the accretion of dark matter by these bodies, which began as small seed masses. Mishra (1985) tested this idea, but only in the approximation of isolated galaxies.

A fully self–consistent two component model (seeds and dark matter) with two galaxies will permit a check as to whether the Local Group is indeed dominated by the two large galaxies. It will help shed light on the nature of mass density fluctuations which led to galaxy formation. This model for the Local Group is especially appropriate for theories in which galaxies grew by accretion onto seed masses, such as black holes or cosmic string loops.

Our test model will be normalized by specifying three known quantities, which then "predict" other known quantities. Our seeds are embedded in a cold collisionless dark matter background, as modelled by a Particle Mesh cosmological code (Sellwood, 1987). Tests of this code show that is is capable of reproducing exact solutions for spherical infall onto a point mass and growth of a spherical void (Peebles, Melott, and Jiang 1988).

In this case, we make use of symmetry in the problem to reduce the amount of calculation. Since the masses of the Milky Way and M31 are nearly equal, we take them to be equal. Rather than placing two seed masses in one computational box, we place one in a single box of dimension $N \times N \times 2N$, and solve the Poisson equation on a system $N \times N \times 2N$ composed of the original plus a reflected image. Thus the system is periodic with period N in the x and y directions, but reflective in the z direction (but periodic with period $2N$). A particle placed off center in the box will fall toward its image and merge at the boundary. Making use of this symmetry saves considerable memory and computation time.

345

C. S. Frenk et al. (eds.), The Epoch of Galaxy Formation, 345–349.

An evolved system is shown in Fig. 1. The system contains 64^3 particles, but only those in a "slice" through the plane of the seed mass are shown. The model is normalized in the following way: A timescale is fixed by a choice of Hubble Constant, 50 $km\ s^{-1}\ Mpc^{-1}$ in this case. A lengthscale is chosen by setting the distance between the density peak and its image equal to the distance to M31. A dimensionless amplitude is chosen by requiring that the relative approach velocity of the two galaxies match the observed 120 km/s, after including the Hubble expansion. We have also studied a system with $H_0 = 100\ km\ s^{-1} Mpc^{-1}$.

The first consistency check in the model is that of the rotation curve of the galaxies, shown in Fig. 2. I do not mean the system is rotating, but rather that we can calculate the necessary speed for a stable circular orbit, which is shown.

The velocity is a bit low compared to the 220 $km\ s^{-1}$ rotational velocity generally accepted for our galaxy. However, we have not included any effects of baryonic collapse. This will, by adiabatic compression, increase the circular velocity inside the core radius and make it more closely match the velocity further out. I therefore regard that as a good fit. The velocity for the $H_0 = 100$ model is considerably higher, passing through 270 km/s.

The model has passed an important consistency check. Its severest test will be against the radial velocities (corrected for our galactic rotation) of dwarf galaxies in the Local Group. These are assumed to be test particles in the velocity field of our simulation. Knowing the distance and angular coordinates of each galaxy, we specify a position (actually a ring, due to the axial symmetry of the situation). We average over this infall velocity, and arrive at V, a predicted radial velocity for each galaxy.

In Fig. 3 we show VO, the observed radial velocity for these galaxies as a function of V. (For references on the data used, see Peebles et al. 1988). Our model appears to work very well. There is more scatter at small V, but the velocity field is extremely well mixed in that area. In general the fit is extremely good, considering the simplicity of the model. $H_0 = 100$ does not work. Predicted velocities are (in general) much too large.

This simple model for the Local Group works extremely well, suggesting that the idea of galaxies forming by accretion around black holes or cosmic string loops may not be far from the truth. Our model is probably a poor approximation to collapse of a Gaussian field, which seems to proceed as a series of one–dimensional "pancake" collapses (Melott and Shandarin 1988).

I thank Jim Peebles and Laura Jiang for their collaboration on this project. I thank NATO and the organizers of this meeting, NASA for financial support under Astrophysics Grant NAGW–1288, and NSF for supercomputer time at the John von Newmann National Supercomputer Center under NSF Grant 86–12138.

REFERENCES

Kahn, E.D. and Woltjer, L. 1959, *Ap. J.* **30**, 705.
Melott, A.L. and Shandarin, S.F. 1988 in preparation.
Mishra, R. 1985 *MNRAS* **212**, 163.
Peebles, P.J.E. 1971, *Physical Cosmology* (Princeton: Princeton University Press) p. 821.
Peebles, P.J.E., Melott, A.L. and Jiang, L.R. 1988 in preparation.
Sellwood, J.A. 1987, *Ann. Rev. Astron. Astrophys.* **25**, 151.

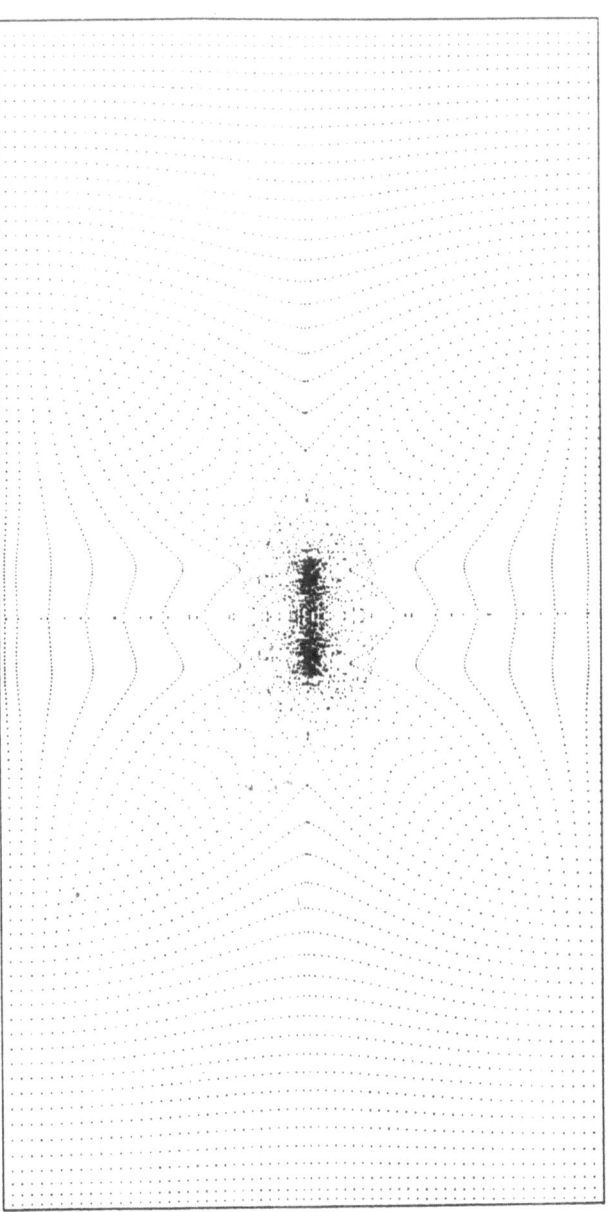

Figure 1

ROTATION CURVE

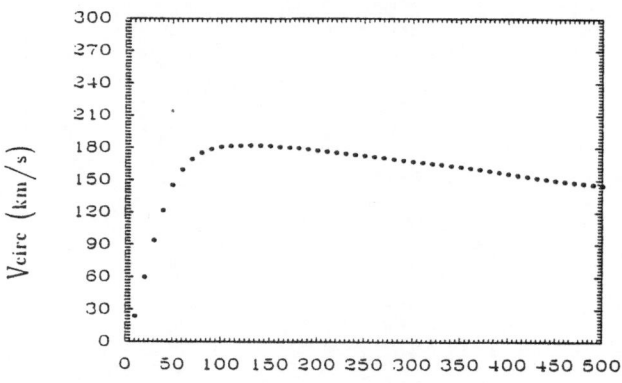

R (kpc)

Figure 2

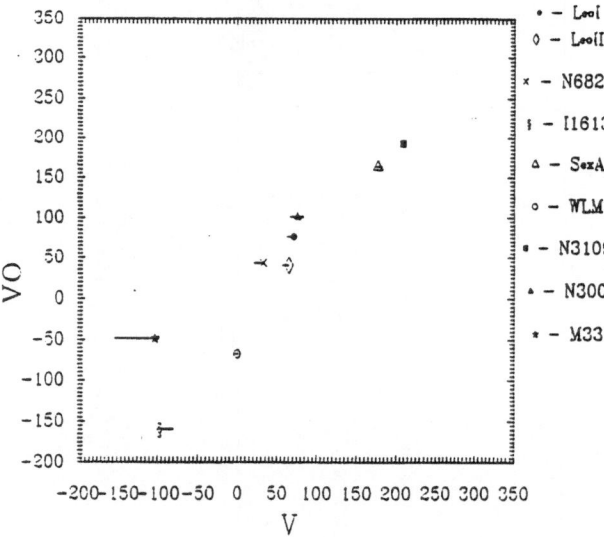

Figure 3

DISCUSSION:

CARR: Your two blobs are accreting as well as moving towards each other. How much do they accrete and when must the process begin?

MELOTT: In the case of the successful model with $H_o = 50$, the initial conditions had an overdensity of total mass about 3×10^9 $M_\odot$ in a region of diameter about 1 kpc at a redshift of 1000. Of course, it is somewhat arbitrary where we join the growing mode solution. Since the halos blend continuously into the background, the total blob mass is also a bit arbitrary. If we use the radius equal to one-half the distance between the blobs, we can estimate the mass of each blob as about 10^{12} $M_\odot$, consistent with the timing argument.

CARLBERG: A comment. Teresa Cripps and I have examined all the wide binaries in a CDM simulation to test the reliability of the timing argument masses. We find that these estimates are in remarkable agreement ($\sim 30\%$) with the total mass in half separation radius spheres around the galaxies. We therefore support a simple accretion model for the local group. Furthermore, we find that the mass within 50 or 100 kpc spheres around the galaxies severely underestimates the total mass, because the dark halos are quite fuzzy.

MELOTT: This sounds like there is general agreement with what I've found. But I think it is unlikely that any Gaussian model would reproduce the detailed agreement with the velocity field found in the two-seed model.

KOO: Have you tried to use your model to explain the apparently long persistence of compact groups, and if so, what are the results?

MELOTT: No, but it sounds like a good idea.

FUTURE OBSERVATIONAL PROSPECTS

Richard Ellis
Physics Department
Durham University, England

1. INTRODUCTION

My aim in this brief talk is to summarise some key observational issues concerning the epoch of galaxy formation, z_f, in the light of both discussions at this meeting and possible instrumental developments of importance for the 1990's. The points made serve to illustrate not only the diversity of opinion on when galaxies formed, but also the observational tasks ahead. I will concentrate mostly on uv-optical-ir opportunities because of my own bias, rather than because this wavelength region necessarily offers the most informative avenue for the future.

2. EVOLVED STELLAR POPULATIONS AT LARGE LOOKBACK TIMES

Peebles and others have claimed that "the Universe at $z \simeq 1$ was much the same as it is today" and presented this as part of the evidence for a large z_f. It is certainly true that galaxies at moderate redshifts (≤ 1) can be found with old stellar populations, either in terms of the strength of their 4000 Å break (Hamilton 1985), their optical-infrared colours (Persson 1988) or their restframe ultraviolet flux (Ellis et al 1985). However, what remains unclear is the proportion of distant galaxies sharing this property. That not all the galaxies behave in this way seems clear: colour-redshift relations for what are thought to be precursors of early-type galaxies appear to define a "red envelope" where some old galaxies are found at high z but many are significantly bluer (Djorgovski et al 1985). Whilst it is possible that the diversity in evolutionary history is environmental in origin, it highlights the need for obtaining a proper statistical understanding of the numbers of red and blue objects before jumping to major conclusions based on a few remote examples.

Again, at higher redshift, the identification of Lilly's (1988) evolved radio galaxy 0902+34 at $z > 3$ unveils the interesting possibility of galaxy formation at $z_f > 5$ but its relevance for normal galaxies remains as unclear as that of the QSO redshift decline for $z > 3$ that this same observation apparently contradicts. Ideally we should identify the most distant observable evolved population representative of the bulk of the galaxy population as well as to continue studies of individual objects that must have formed at $z > 5$. Whilst only special galaxies are currently observable beyond $z \sim 1$, we must remember that for $\Omega = 1$ the time interval from z = 1 to 3 is only a third of that back to z = 1 where controlled studies of more normal objects should soon become feasible.

For example, at redshifts $z \simeq 0.8$, corresponding to half the Hubble time, complete photometric samples can already be chosen in various restframe passbands in the rich clusters found by various groups and the proportion of old objects quantified directly. Gunn and Dressler (1988) have begun using multi-slit spectroscopy to determine the distribution of the 4000 Å break but the going is tough since few members/cluster can be accessed spectroscopically with

351

C. S. Frenk et al. (eds.), The Epoch of Galaxy Formation, 351–357.
© 1989 by Kluwer Academic Publishers.

current facilities. Furthermore successful redshifts are often obtained more easily for the members with *large* 4000 Å breaks and one can imagine biases tending to exaggerate the proportion of old galaxies at high z, although these authors actually suggest a *weakening* of the break with redshift in their samples. Because of such difficulties, the distribution of restframe *ultraviolet* ($\sim$ 2500 Å) colours for optically-red members might be a more satisfactory approach. Work in lower z clusters has shown this is a sensitive probe of residual star formation (Ellis *et al* 1985). At z $\sim$ 1 the restframe uv shifts to the peak QE of modern CCDs and at such large redshifts the colour differences between members and field galaxies (whose $\bar{z} \sim$ 0.3–0.5) using multiband techniques (c.f MacLaren *et al* 1988) are sizeable.

But how are clusters beyond z $\sim$ 1 to be found? Certainly not by optical contrast techniques which often select overlapping groups at diverse redshift at these very faint limits (c.f. Ellis 1988a). The two-colour scanning technique mentioned by Gunn at this meeting should find examples with z $>$ 1 but depending on the search criteria it may preferentially select those clusters with old stellar populations or ones dominated by star-forming galaxies. I imagine we will have to wait for X-ray or other independent techniques to construct truly well-defined samples. Nonetheless, determining the fraction of old galaxies at z $\sim$ 1 is a worthy goal as it will constrain the star formation history for normal galaxies over the previous 2-3 Gyr (for $H_o = 50$ km sec^{-1} Mpc^{-1} this probes back to z $\sim$ 3, a critical epoch).

Meanwhile, at z $\sim$ 0.3 – 0.5, *detailed* studies at intermediate spectral resolutions of $\sim$ 4 Å (e.g. Couch & Sharples 1987) already reveal significant star formation activity in marked contrast to that seen in rich environments today. This seems sufficiently prolific to be *in addition* to possible spiral-stripping effects that might produce the present day S0s. Even for those early-type galaxies lying on the "red envelope" of the colour-redshift line, many show ultraviolet excesses and/or Balmer absorption lines indicative of a recent return to the locus of an old stellar population. In short, the burst frequency observed suggests a significant amount of star formation has occurred relatively recently in quite normal cluster ellipticals (c.f. Rose *et al.* 1989).

Is the Universe really unchanging over 0 $<$ z $<$ 1? Whilst observers have been known to exaggerate the differences between present day and remote objects (presumably to justify their long integrations), it seems we are beginning to witness some theoreticians that wish to downplay the differences! Observationally, it now seems inescapable that the *range* of star formation activity seen in dense environments was significantly higher at z $>$ 0.4. Until the *physical origin* of this enhanced star formation activity is determined, it would be wise to remain open-minded about its importance for the epoch of galaxy formation.

3. FAINT BLUE PRIMAEVAL GALAXIES

Attention at this meeting has once again focussed on the faint blue galaxies first identified a decade ago by Kron (1978) and further probed by Tyson (1988) with his pioneering CCD studies. Cowie (1988) has studied those galaxies with very blue colours that dominate the count excess at B $\sim$ 24-26 and proposed that these may be high z protospheroids.

A pure star forming population with a normal IMF would present a colour B – R $>$ –0.5 for its first 10^{7-8} yr depending on its redshift and so it seems reasonable to suppose the faint blue objects are extremely active star-formers. Local low-luminosity examples are known however (c.f. NGC 4449), and we must be cautious in identifying Tyson's galaxies necessarily with a *primaeval* population. Indeed, objects with such colours could easily be low z bursting galaxies such as those identified in the brighter (B $\leq$ 22.5) redshift surveys now underway (Broadhurst *et al* 1988, Koo 1989, Colless *et al* 1989). Whilst such surveys have not yet probed to the limits where large numbers of B – R $<$ 0.5 galaxies are seen, those with spectra to B $<$ 22.5 (Colless *et al* 1988) are consistent with the bursting hypothesis.

Cowie and Tyson point to a "new" population of blue objects at B > 24. It all sounds rather familiar: such new (primaeval) populations were claimed to be present at brighter magnitudes in the photographic samples (Tinsley 1977), but turned out to be star-bursting nearby galaxies. Whilst the median colour is getting monotonically bluer with increasing magnitude, the count slope ($\gamma = dlogN/dm$) remains remarkably uniform over the entire range 21 < B <25, which is hard to understand if beyond some fundamental threshold magnitude large numbers of primaeval galaxies are waiting to be found.

First it is worth examining the only self-consistent model of the $N(m, z)$ plane to B $\sim$ 22.5 and seeing to what extent this can/cannot reproduce the gross features of the fainter CCD data. In the model proposed by Broadhurst et al (1988), the luminosity function evolves with redshift in *faint end slope* rather than *luminosity*. Only the sub-M* galaxies are bursting at z < 0.5; the luminous objects must have completed their activity well before z $\sim$ 1. The limit this places on, for example, the $\mu = 0.5$ models of Bruzual(1988), is quite severe; normal massive galaxies *cannot* have undergone a luminous phase more recently than z $\sim$ 2 otherwise high z examples would appear in the redshift surveys. The picture of a gradual depletion of activity down the luminosity function is particularly appealing since it ends today with the low-mass H II galaxies – appropriate spectral counterparts for the blue galaxies isolated in the deep redshift surveys.

Spectral analysis shows that the bursts of star formation must be brief otherwise one cannot simultaneously match the colours and Balmer absorption lines (which are a reasonably good post-burst age indicator). To isolate their cause and understand the striking similarity between the "field" and "cluster" starburst activity remain key problems. Only a moderate improvement in image quality (c.f. Thompson 1987) would determine whether such blue light is extended as compared to that seen, e.g. at 2 μm. That we are seeing gradual disk formation over 0 < z < 3 is suggested by the widespread occurrence of the phenomenon, the absence of any spectacularly luminous examples and also by Pettini's (1989) observation of a chemically-young disk at z $\sim$ 2.5.

If we extend this luminosity-dependent evolution to z $\sim$ 2-3, it is difficult to simultaneously match the count slope *and* constrain the blue end of the colour distribution. However, rather than predicting too few blue objects, such as would be the case if there were an additional primaeval population beyond B = 24, the reverse is the case. Furthermore, Majewski's (1988) U-I test limits the size of any redshift tail of optically observable galaxies beyond z$\sim$ 3 *provided* they have strong Lyman breaks. If there are primaeval luminous sources, from several accounts it seems, they must be more distant than z = 3.

A crucial question then is how many faint infrared sources have no visual counterpart. At 2 μm, one is sensitive to protogalaxies to redshifts as high as z$\sim$ 20. Preliminary work on the infrared counts suggest no new population is being found, but this depends (as it did in the early days of *optical* number counts) on accurate modelling. Mobasher et al (1988) predict the 2μm no-evolution slope to be $\gamma = 0.36$ for 18 < K < 20 on the basis of a genuine K-band luminosity function for 200 nearby field galaxies and well-understood K-corrections applicable to z $\sim$ 3-4. Any 2μm slope steeper than this would be very interesting. K $\sim$ 20 is now practical with modern infrared arrays such as UKIRT's *IRCAM* (McLean 1987) though the small areal coverage of such devices means it will be some time before statistically significant optical/infrared comparisons and K counts can be obtained.

Multicolour spectral energy distributions from the uv to the ir can isolate objects with presumed Lyman discontinuities; this is adequate for a poor man's N(z) at very faint limits. Cowie(1989) has shown the power of this approach. However, many remain suspicious of photometric redshifts whatever the tricks adopted and there is no substitute for genuine spectroscopy. Blue objects indicate emission lines and if it is simply a question of whether these are Lyα or [O II], we may not have to wait for 8 metre telescopes to know the answer. The original blue searches for objects with strong Lyα emission were done many years ago and performed at low resolution. However, multi-slit spectroscopy of candidate sources with modern detectors and a resolution optimal for *narrow* lines are worth re-considering. At the time of writing one possible z = 3.4 candidate blue galaxy with B = 24 is claimed by Cowie and Lilly (1988) and further examples are eagerly awaited.

4. INSTRUMENTATION FOR COSMOLOGY IN THE 1990s

Much has been written on this subject, mostly in the context of 8-10 metre class telescopes and Hubble Space Telescope. It is safe to assume there will be further cosmology conferences before we see those facilities in action so I will concentrate on what we can do with our existing facilities. An excellent summary of this topic was published at the meeting organised by NOAO (Davies 1988). There appear to be three growth areas:

Firstly, infrared photometry, when coupled with deep optical frames, offers the tantalising opportunity of identifying truly high z objects, both via counts and wide wavelength coverage spectral energy distributions. Larger format devices are urgently needed to match the areal coverage of optical CCDs particularly because of the need to use small pixels to minimise crowding effects at faint limits.

Secondly, redshifts are needed for those blue galaxies where emission lines are expected in a wide wavelength range from 4000 Å to 7500 Å . Are the B - R $\leq$ 0.5 objects at B > 23.5-24 a continuation of the bursting population found by Broadhurst *et al* identifiable by their strong [O II] and H II region-type spectral features? Or are a subset genuine high z galaxies undergoing spectacular star formation identifiable with Lyα emission?

Multi-slit techniques can be pushed very faint in good conditions (c.f. Ellis and Parry 1988). The Low Dispersion Survey Spectrograph (LDSS) is a multislit wide-field spectrograph built at Durham purposely for such work. At the AAT it has already pushed intermediate resolution work ($\sim$ 8 - 10 Å) a magnitude fainter (to B = 22.5) than was possible with fibres on the AAT (Colless *et al* 1988) and in searches for narrow-lined objects it could reach considerably fainter. At *lower resolution* ($\sim$ 40 - 50 Å), a preliminary survey of over 200 *compact or uv-strong* objects to B = 23.5 reveals Galactic stars, low z H II galaxies or UVX QSOs. Interestingly, virtually all B - R > 1.5 objects are cool M dwarfs; there appear to be *no* extragalactic compact red objects at very faint limits. Such a statement could probably be pushed to extraordinarily faint limits, since M stars are remarkably easy to identify, and thus it might be possible to place a stringent limit on the number of old spheroidal galaxies at z $\sim$ 2.

Thirdly, large efficient echelles are being commissioned on several 4 metre telescopes and these promise enormous gains for detailed studies of QSO absorption line systems, both in chemistry and in the spectroscopic identification of galaxies associated with deep absorption systems (Hunstead 1989). To me, the birth of this subject has been the highlight of the meeting. The techniques are proven, we just need bigger samples.

These three "key projects" are my best-buy for 4-metre cosmology in the next 3-5 years. Note that the success of each project is critically dependent on having the right detector - a problem astronomers are often powerless to solve.

Finally, although this meeting has tended to steer clear of issues related to large scale structure, clustering studies at high z will clearly tell us lots about the epoch of galaxy formation. We have already seen (Shanks 1989) the importance of establishing the QSO correlation function at various epochs and one can expect to do likewise for high z galaxies and QSO absorbers. Some of these projects are limited by telescope aperture but many are limited more by the *numbers* of objects that can be gathered in a reasonable amount of telescope time ("reasonable" for the UK time allocation committees means 3 nights a year).

An exciting proposal now at the design stage proposes to increase the prime focus field of the AAT to over 2° in diameter with a new corrector optimised for fibre spectroscopy, rather than direct imagery. This would increase 10-fold the areal coverage for fibre-optic work and a feed for several hundred objects has been proposed (Ellis 1988b, Taylor & Parry 1988). In parallel with HST and the 8-10 metre telescope studies, I believe this opportunity is a most significant step forward and one that hopefully can be completed quickly.

References

Broadhurst, T J, Ellis, R S and Shanks, T 1988 *Mon Not R astr Soc*, in press.

Bruzual, G 1983 *Astrophys J*, **273**, 105.

Colless, M M, Ellis, R S and Taylor, K 1989 these proceedings.

Couch, W J and Sharples, R M 1987 *Mon Not R astr Soc*, **229**, 423.

Cowie, L L 1989 these proceedings.

Cowie, L L and Lilly, S J, preprint.

Davies, R L 1988 *Instrumentation for Cosmology*, NOAO Publications.

Djorgovski, S, Spinrad, H and Marr, J 1985 in *New Aspects of Galaxy Photometry*, ed. Nieto, J-L, Springer, p193.

Ellis, R S 1988a in *Towards Understanding Galaxies at High Redshift*, eds. Kron, R and Renzini, A, Kluwer, p147.

Ellis, R S 1988b *A Wide Field Prime Focus for the AAT*, AAT Proposal, March 1988.

Ellis, R S and Parry, I R 1988 in *Instrumentation for Ground-Based Astronomy*, ed Robinson. L B, Springer-Verlag, p192

Ellis, R S, Couch, W J, MacLaren, I and Koo, D C 1985 *Mon Not R astr Soc*, **217**, 239.

Gunn, J E and Dressler, A 1988 in *Towards Understanding Galaxies at High Redshift*, eds. Kron, R and Renzini, A, Kluwer, p227.

Hamilton, D 1985 *Astrophys J*, **297** 31.

Hunstead, R N 1989 these proceedings.

Koo, D C 1989 these proceedings.

Lilly, S J 1988 submitted to *Astrophys J Lett*.

MacLaren, I, Ellis, R S and Couch, W J 1988 *Mon Not R astr Soc*, **230**, 249.

McLean, I 1987 in *Infrared Astronomy with Arrays*, eds. Wynn-Williams, C G & Beklin, E E, Univ of Hawaii.

Majewski, S 1988 in *Towards Understanding Galaxies at High Redshift*, eds. Kron, R and Renzini, A, Kluwer, p203

Mobasher, B, Sharples, R M and Ellis, R S 1988, in preparation.

Persson, S E 1988 in *Towards Understanding Galaxies at High Redshift*, eds. Kron, R and Renzini, A, Kluwer, p251.

Pettini, M 1989 these proceedings.

Rose, J, Sharples, R M, Ellis, R S and Bower, R G, 1989, these proceedings.

Shanks, T 198 , these proceedings

Taylor, K and Parry, I 1988 *Wide Field Prime Focus Adaption to the AAT*, AAO Technical Report.

Thompson, L A 1987 *Astrophys J*, **315**, L35.

Tinsley, B M 1977 *Astrophys J*, **211**, 621.

Tyson, J A 1988 *Astron J*, **96**, 1.

DISCUSSION:

BOWER: In your diagrams, you showed spheroids forming before discs. Is there any good reason for this?

ELLIS: Not really. As we have heard, observationally there is evidence for bursts of star formation in the interval $0 < z < 2$ which might be associated with disk formation. Furthermore as faint as we can image optically the majority of objects are apparently still visible in U and hence have $z < 3$. Yet we know *some* AGN are observed at $z > 4$. The key observation would be to search for evidence of a normal population of protospheroids at $z > 5$ in the infrared.

JONES: Is it possible, or even worthwhile, to look at the spatial and spectral structure of the extragalactic background light?

ELLIS: I'm not sure what you mean. It is possible to look at fluctuations in the sky at various wavelengths and to search spectra of blank sky for "features" such as the Lyman series. Work in this area so far has not produced any interesting results. Furthermore, the technique of comparing sky fluxes on/off a dark Galactic cloud, whilst an important experiment, has produced upper limits to the optical background light that lie well above that inferred from an extrapolation of the galaxy counts.

WOLFE: I know that the models which are popular now predict recent disk formation. However, there is growing evidence that most of the known baryonic matter at redshifts between 2 and 3 reside in layers of cool neutral hydrogen with low velocity dispersions that have a better than even chance of being protogalactic disks. If, as I suspect, these objects are also chemically evolved, then we must face the fact that disk formation evolution occurred much earlier than depicted on your viewgraph.

ELLIS: I agree the information is conflicting. Remember Pettini showed us at least one probable disk is chemically young. Quite possibly the disk "era" is very extended. With optical techniques we may only see *recent* activity $0 < z < 2$, even though a substantial portion of disk formation might have occurred at $z > 2$.

SANCISI: Is there any evidence of clustering of damped Lyα systems at $z = 2$-3? We are carrying out deep surveys at 327 MHz at Westerbork to search for HI emissions at $z = 3.3$ and could detect typical cluster masses as low as $5\ 10^{13}\ M_\odot$ and sizes 1-5 Mpc.

WOLFE: Clustering of damped Lyα systems along the line of sight is difficult to detect directly, because the probability of intervention is low compared to that of the Lyα forest population or to most metal-line systems. However, there is indirect evidence that clustering may be present. It turns out that 6 of the 16 damped systems occur in pairs along 3 lines of sight, with redshift separations, $z \leq 0.19$. The probability of finding 3 doubles by chance within this z along the entire redshift path of the survey is less than 1%. This may indicate clustering over large scales. Note however that this probability is proportional to μ^6, where μ is the mean number of systems per unit redshift interval. So a 50% error in μ could cause us to underestimate the pair probability by a factor of 10. While the formal error in μ is only $\sim 10\%$, an increase in size of the damped sample would help to strengthen this result.

KOO: Could LDSS benefit from higher dispersion to discriminate narrow emission lines? Combining several sets of data with different wavelength ranges may allow this without sacrificing information.

ELLIS: That's an interesting suggestion we are considering. At present we get ~ 30 objects in the field with $\lambda\lambda$ 3700 – 7000 at a resolution of 10 ÅThis should be adequate for detection of most emission line objects. However, at B $\sim$ 23-24 we really need good conditions for extended periods. I certainly think searching for *emission lines*, rather than gathering redshifts for statistical samples as we have been doing until now, is the only way forward to fainter regimes.

KOO: (1) Searches for blue PG were proposed in the mid 70's by Meier, who proposed both QSO like objects and faint blue fuzzy objects (i.e. faint blue galaxies), so some efforts have already been undertaken. (2) Colours have often been given a bad name, but in fairness to your statement that colours were misleading us in our models, colours did suggest as one possibility, the presence of many blue moderate redshift ($z \sim 0.7$) galaxies, but could not discriminate this picture from luminosity evolution, so spectroscopy was vital - but colours are still important diagnostics, especially when calibrated over a wide wavelength range, deep UV to near IR.

ELLIS: (1) I agree. I mentioned blue PG searches in my summary not because they hand't been tried but because technology advances and it is always worth trying new techniques on old questions. For example, so far as I know, searches for the Lyman series ($L\alpha$, $L\infty$) in slit spectra of blank sky has not been attempted in the wavelength interval 3500–5500 Å (2) My criticism of colours as a tool in studies of distant galaxies was not directed at work based itself on spectroscopic surveys (such as yours)! Rather I wanted to make 2 warnings: (i) if there are objects with composite populations (old + new) e.g. PSB galaxies it would be extremely difficult to recognise this with broad-band colours (witness the early claim that the B-O effect was caused by an excess of spiral galaxies). (ii) The precision of extracting redshifts from colours is poor for the dominant faint blue bursting population and thus this makes it hard to believe Loh's Ω test can work reliably until much more spectroscopic calibration work is done.

PEEBLES: I urge you all once again to bear in mind the fundamental distinction between galaxy formation and evolution. The former by which I mean the assembly of the bulk of the mass of the galaxy, may well be much harder to get at than the epoch z_* of formation of the bulk of the stars. But it would make no sense to base our theories of galaxy formation on the observation of z_* just because that is all that can be observed.

ELLIS: Theories of galaxy formation cannot hide behind the statement $z_g \gg z_*$, but must predict observable quantities. The difficulty I have with a high formation redshift is that it is exceedingly difficult to test observationally. In the next few years we may well get very solid clues about z_* and thus theories should concentrate on predicting that epoch rather than the more fundamental epoch when the mass accumulated.

A NEW DEEP AAT REDSHIFT SURVEY

Matthew Colless[1], Richard S Ellis[1] & Keith Taylor[2]
1. Durham University, England
2. Anglo-Australian Observatory, Australia

ABSTRACT. We present first results from a new deep redshift survey being undertaken at the AAT using a wide-field efficient multislit spectrograph, *LDSS*. This instrument has allowed us to extend the limiting magnitude of the earlier AAT fibre-optic survey by an entire magnitude to $b_J = 22.5$. At this depth the on-sky surface density of galaxies is twice the no-evolution prediction, yet our first results from LDSS indicate that, as in the earlier fibre survey to $b_J = 21.5$, the mean redshift is *not* significantly different from that for a non-evolving model. We conclude that M^* galaxies cannot have undergone significant luminosity evolution in the interval $0 < z < 1$, and that the steep galaxy count slope to faint magnitudes arises primarily because of sporadic star formation in otherwise low-luminosity galaxies.

1. INTRODUCTION

Recent photographic and CCD counts (Figure 1) show that the excess in the observed galaxy counts over a non-evolving model is as large as a factor of 2 at $b_J = 22.5$. Three equally plausible hypotheses can be suggested for this over-abundance of faint galaxies: (i) a number are primeval galaxies at redshifts $z \sim 3$ undergoing their first strong burst of star formation; (ii) there is mild luminosity evolution in most galaxies over a redshift range $0 < z < 2$.; (iii) there are far more intrinsically faint galaxies than local surveys would indicate and these dominate at faint magnitudes as they are not affected by redshift dimming.

Seeking to distinguish these possibilities, Broadhurst, Ellis and Shanks (MNRAS in press, 1988) measured the joint distribution $N(m, z)$ for a strictly magnitude-limited sample with $20.0 < b_J < 21.5$. This survey, which achieved a high degree of completeness (86%), found, surprisingly, that the mean redshift $\bar{z}$ was remarkably close to that predicted for no luminosity evolution *despite the count excess of 30% seen to this depth*. This result places a strong constraint on any evolution at the bright end of the luminosity function. Broadhurst *et al.* suggested that the excess counts arise because of *low luminosity* galaxies that suffer short bursts of star formation that increase in strength or frequency with look-back time.

To determine whether this sporadic star formation picture can be extrapolated to explain the steep count slope to much fainter magnitudes, or whether general luminosity evolution eventually becomes important, we have begun a second, deeper, redshift survey, for which we present preliminary results here.

359

C. S. Frenk et al. (eds.), The Epoch of Galaxy Formation, 359–361.

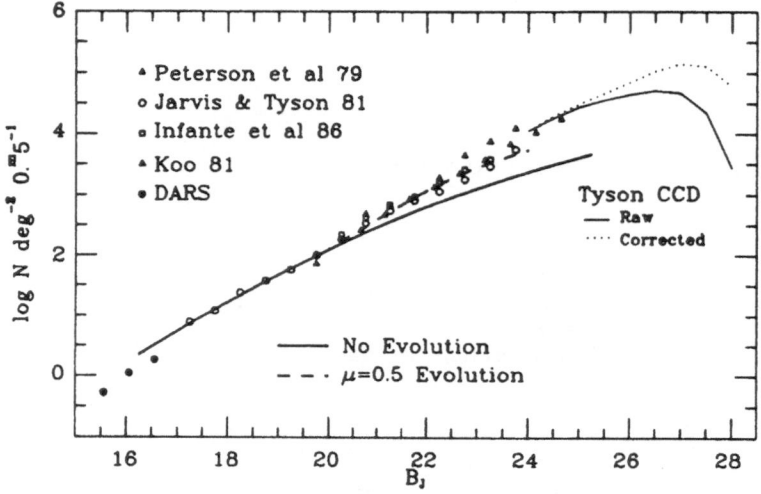

Figure 1. Recent galaxy counts in the b_J passband. The solid line is the no-evolution prediction based on local galaxy properties. The dashed line is a $\mu=0.5$ (Bruzual) luminosity evolution model (with all galaxy types undergoing identical evolution) that fits the slope of the number counts.

2. THE NEW SURVEY

In 1986–7 two of us (KT and RSE) commissioned a new *wide-field* faint object multislit spectrograph *LDSS* (Ellis & Parry 1988) which combines the sky-subtraction advantages of slits with the multiplex advantages of fibre optics. We are presently mid-way through two separate surveys using *LDSS*: (i) at a resolution of 10Å, we are extending the Broadhurst *et al.* survey a magnitude deeper to $b_J=22.5$ with the important difference that we are surveying *all* objects regardless of whether they appear to be stars or galaxies. Our multiplex advantage here is typically ×30; (ii) at a resolution of 50Å, we are surveying compact and blue objects to $b_J=23.5$ to place constraints on the numbers of active galaxies, primeval galaxies and unknown classes of object. Our multiplex advantage here is typically ×70.

In two recent observing runs we have secured high-dispersion spectra for about 170 objects in 3 fields in the range $21<b_J<22.5$. Redshifts have been derived for 123 objects (success rate 75%), of which 33 are stars. In Figure 2 we combine this new $N(m,z)$ galaxy data with that published by Broadhurst *et al.* We show the mean redshift $\bar{z}$ against apparent magnitude for $0^m.5$ intervals from $b_J=20$ to 22.5 with the error calculated from the field-to-field scatter. It is encouraging to note that in the overlap window $21<b_J<21.5$ common to both surveys, $\bar{z}$ is identical. At the fainter magnitudes of the new survey, $\bar{z}$ is continuing to rise only slowly with increasing apparent magnitude, strengthening considerably the conclusions reached by Broadhurst *et al.* The solid line represents the prediction of the no-evolution model, which fits the distribution *despite the factor 2 excess in the counts at the new magnitude limit.*

Whilst these are only preliminary results awaiting the completion of the survey, it seems that luminous galaxies cannot have evolved significantly in the interval $0<z<1$. Additionally, the combination of counts and spectroscopy of normal galaxies to the new faint limit places an even more severe constraint not only on the absence of luminous primeval galaxies at redshifts

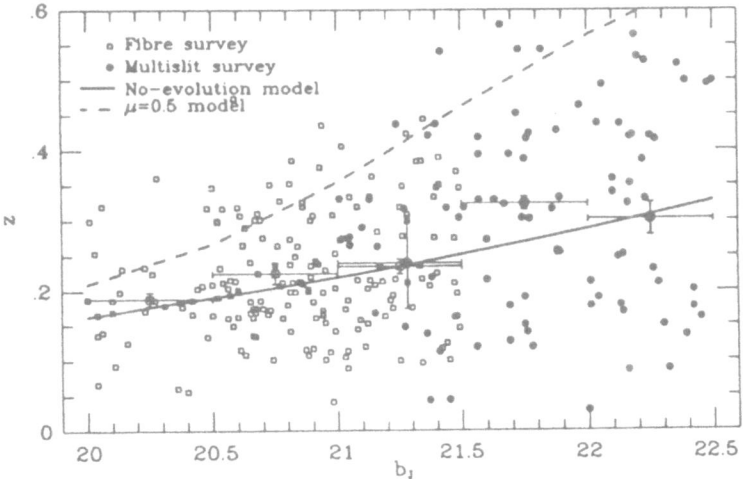

Figure 2. Joint magnitude–redshift data for 277 galaxies from two AAT redshift surveys (open circles—Broadhurst *et al.* fibre survey, filled circles—new LDSS survey). The mean redshifts in $0^m.5$ intervals (points with error bars) are compared with the predictions of the same no-evolution (solid line) and $\mu=0.5$ luminosity evolution (dashed line) models as in Figure 1.

z<4, but also on the likelihood of any luminous phase for normal galaxies to z∼2. That most of the faint galaxies to b_J∼24–26 are at moderate redshifts z<2 is consistent with the steep U counts of Majewski (1988) and others.

The absence of detectable luminosity evolution for M* galaxies means that the steep count slope can only be due to evolution at the *faint end* of the luminosity function, where the frequency and/or intensity of bursts of star formation increases with look-back time. Similarities with clusters, which also undergo sporadic bursts of star formation (Dressler & Gunn 1983), are striking.

Are we witnessing a general process associated with galaxy formation, such as the gradual forming of disks over 0<z<2? Or is the spectral evidence for bursts indicative of merging or close interactions (Tyson & Scalo 1988)? Identifying the cause and location of the star formation activity seen in the majority of faint blue field galaxies is essential if we are to understand the role that galaxy counts play in determining the epoch of galaxy formation.

References

Dressler, A., & Gunn, J.E. (1983), *Ap. J.*, **263**, 533.

Ellis, R.S., & Parry, I.R. (1988), *Instrumentation for Ground-Based Optical Astronomy*, ed. Robinson, L.B., (Springer), p192.

Majewski, S. (1988), *Towards Understanding Galaxies at Large Redshifts*, eds Kron, R.G., & Renzini, A., (Kluwer), p203.

Tyson, N.D., & Scalo, J.M. (1988), *Ap. J.*, **329**, 618.

ON THE INTERPRETATION OF GALAXY COUNTS, AND COLOR AND REDSHIFT DISTRIBUTIONS

Gustavo Bruzual A.
Centro de Investigaciones de Astronomía (C.I.D.A.)
Apartado Postal 264
Mérida 5101-A
Venezuela

ABSTRACT. The interpretation of the galaxy number counts according to traditional models requires considerable amounts of spectral evolution in early type galaxies, whereas the redshift surveys are consistent with no spectral evolution at all. This apparent contradition is examined and a possible way out is suggested.

1. INTRODUCTION

Massive surveys that provide magnitudes, colors, and red-shifts of large numbers of faint galaxies have recently become available. These data are ideally suited to probe models for the spectral evolution of galaxies. One can thus address the question of whether or not distant galaxies were more luminous and bluer in the past than nearby ones at the present epoch.

Several naive models were built in the last decade to try to understand the photometric properties of increasingly large samples of galaxies, for which no information about galaxy redshift was available. Models by Tinsley (1980), Bruzual and Kron (1980), Koo (1981), Shanks et al. (1984), and Bruzual (1987) among others, suggest that there is an excess of counts in the J band around J = 23 with respect to the no evolution prediction. In these models this excess in the number counts is interpreted as being best reproduced by high redshift early type galaxies seen at early stages of their evolution in a low q_0 universe.

The redshift distributions at faint magnitudes derived from these models predict significant numbers of galaxies at high z (z > 0.5). The Durham/AAT (Ellis 1987, Broadhurst, Ellis, and Shanks 1988) and the KPNO (Koo and Kron 1987) redshift surveys have failed to reveal but a small number of distant galaxies up to J = 21.5. In fact, the average of the z distribution derived by Broadhurst, Ellis, and Shanks

C. S. Frenk et al. (eds.), The Epoch of Galaxy Formation, 363–366.

(1988) up to J = 21.5 is z = 0.23, as expected from the no evolution models.

Thus, the interpretation of the number counts requires considerable amounts of spectral evolution in early type galaxies, whereas the redshift surveys are consistent with no spectral evolution at all.

2. ANALYSIS OF PREVIOUS MODELS

The models mentioned above are naive in the following sense. The color range observed in the Hubble sequence of galaxy types (local), is reproduced by most model builders by assigning star formation rates that decay exponentially in time, with e-folding times that increase from E/S0 to Irr galaxies. The relative number of each kind of galaxies is derived from local catalogues. Assumptions about the luminosity function of galaxies and the cosmological model lead in a straightforward way to a prediction of the number counts as a function of apparent magnitude in the passband of interest, usually one or two bands. As seen from Fig. 1, the expected z average for galaxies with J in the range from 21 to 22 is above z = 0.30 for galaxies with rest frame J-F between 0.9 and 1.1. The average z reaches values between 0.8 and 1 for J in the range from 22 to 22.5. These high values of z average are reached because the phase of rapid star formation that takes place early on in the life of these model galaxies is being sampled. When the different color classes shown in Fig. 1 are combined as indicated in Bruzual (1987) and Bruzual and Koo (1989), the values in the following table are obtained. The z distributions are broad and one expects larger numbers of galaxies at higher z's than the observations indicate. This is especially true for J > 22.

J	average z	sigma	median z
19.0 - 21.0	0.17	0.09	0.15
21.0 - 22.0	0.28	0.24	0.23
22.0 - 22.5	0.56	0.64	0.32
22.5 - 23.0	1.02	0.94	0.54
23.0 - 23.5	1.37	1.00	1.07

Figure 1 (next page). Expected average redshift derived from a model in which all galaxies evolve according to one of Bruzual (1983) μ models plotted for different apparent J ranges versus the J-F color at z = 0. μ ranges from μ = 0.01 (left) to 0.95 (right) in steps of 0.05. The Schechter (1976) luminosity function, q_o = 0, H_o = 50, and galaxy age of 16 Gyr was used in this calculation.   For comparison, the bottom frame shows the no evolution calculation.

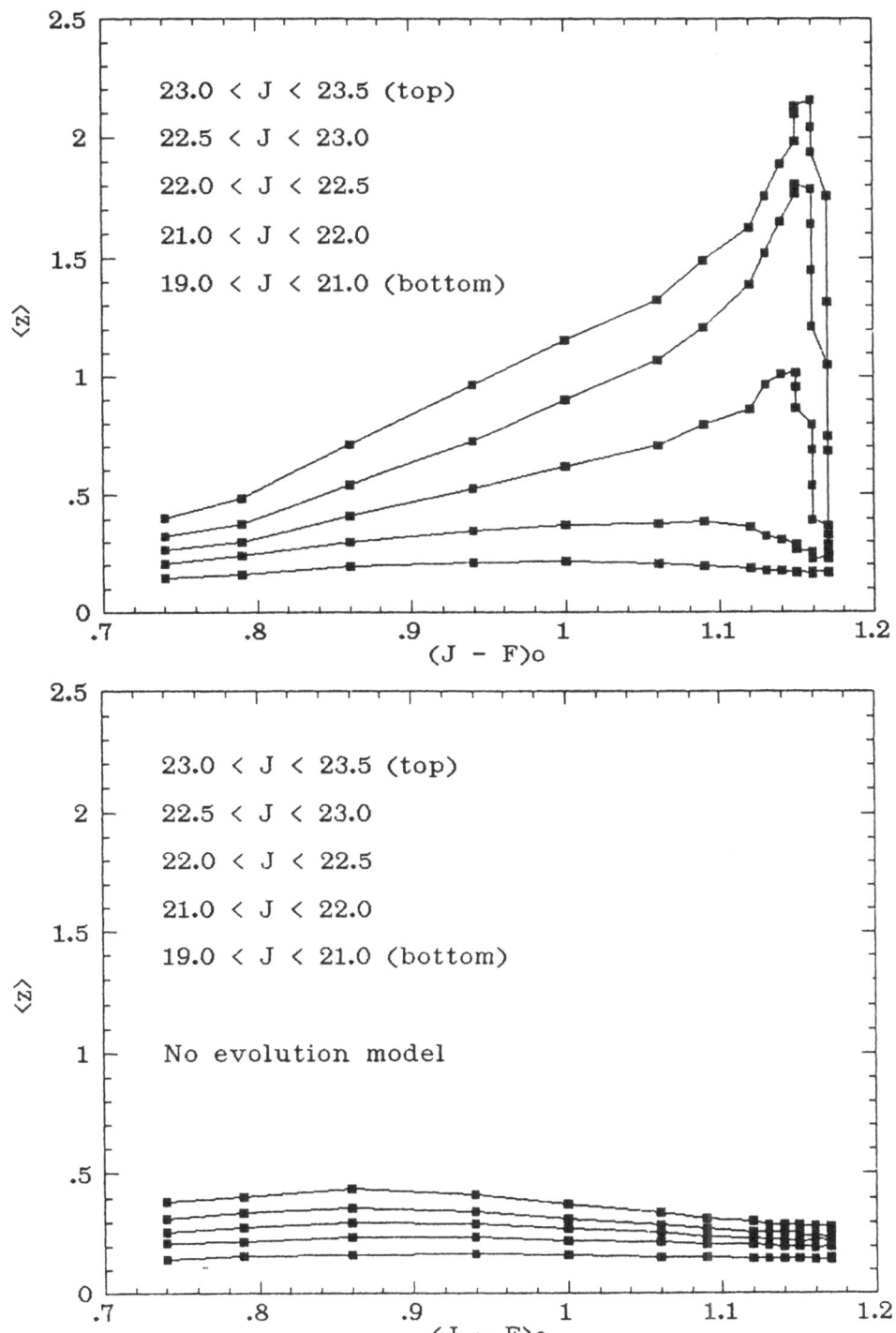

DISCUSSION

Current observations indicate that models in which galaxies evolve according to exponentially decaying SFR's are not the most realistic way to represent the color range observed in the Hubble sequence. The bright and blue early phases predicted by these models have not been detected. Bursts of star formation superimposed on an old stellar population turn a galaxy blue and luminous by amounts that depend on the strength of the burst, the age of the burst at the time of the observation, and the IMF of the stars formed during the burst. The average z of the distribution predicted for this kind of bursting galaxy is close to 0.2 and can be even lower than the no evolution prediction. Very blue galaxies are then recent bursts that disappear soon in z space, a low luminosity red galaxy remains. See Broadhurst et al. 1988 for a preliminary model along these lines.

Since color distributions change smoothly as a function of apparent magnitude, if the bursts occur at random times and in randomly chosen galaxies, the bursts must occur in such a way that the number counts and the color distributions behave as observed. The only hope to build good models to understand galaxy formation and evolution is to use all the information available: number counts, color and redshift distributions, and not just individual subsets of the data (Bruzual and Koo 1989).

REFERENCES

Broadhurst, T. J., Ellis, R. S., and Shanks, T. 1988, M.N.R.A.S., (in press).
Bruzual A., G. 1983, Ap. J., 273, 105.
Bruzual A., G. 1987, in "Towards Understanding Galaxies at Large Redshifts", eds. R. G. Kron and A. Renzini, p. 161.
Bruzual A., G., and Kron, R. G. 1980, Ap. J., 241, 25.
Bruzual A., G., and Koo, D. C. 1989, (in preparation).
Ellis, R. S. 1987, in "Towards Understanding Galaxies at Large Redshifts", eds. R. G. Kron and A. Renzini, p. 147.
Koo, D. C. 1981, Ph. D. thesis, Univ. of Calif., Berkeley.
Koo, D. C., and Kron, R. G. 1987, in "Towards Understanding Galaxies at Large Redshifts", eds. R. G. Kron and A. Renzini, p. 209.
Schechter, P. 1976, Ap. J., 203, 297.
Shanks, T., Stevenson, P. R. F., Fong, R., MacGillvray, H. T. 1984, M.N.R.A.S., 206, 767.
Tinsley, B. M. 1980, Ap. J., 241, 41.

THE INTERPRETATION OF FAINT GALAXY COUNTS

B. Guiderdoni[1] and B. Rocca–Volmerange[1,2]
[1] Institut d'Astrophysique de Paris, 98bis Boulevard Arago, F–75014 Paris
[2] Laboratoire René Bernas, Bât. 108, Université Paris XI, F–91306, Orsay

1. Introduction

Faint galaxy counts could in principle be used to constrain the value of the deceleration parameter q_0. But as in other classical cosmological tests involving galaxies, the effect of intrinsic evolution dominates the pure geometrical effect of varying q_0 between plausible values. This intrinsic evolution can be analyzed by means of a model of spectrophotometric evolution which allow to compute evolving synthetic spectra of galaxies from a minimum set of assumptions about the star formation history. Predictions of faint galaxy counts must use magnitude predictions based on these evolving spectra. We recently presented a new atlas of evolving stellar spectra (Rocca–Volmerange and Guiderdoni, 1988) based on our model (Guiderdoni and Rocca–Volmerange, 1987). In this paper, we interpret the published data of faint galaxy counts by means of this model, to constrain the intrinsic evolution (particularly z_{for}, the redshift of galaxy formation), and tentatively the value of q_0.

2. Ingredients

The $k-$ and $e-$corrections are computed from the set of synthetic spectra (Rocca–Volmerange and Guiderdoni, 1988) for a set of seven spectral types of the Hubble sequence. The luminosity function for each spectral type has a Schechter form. The values of the magnitude M_j^* of the knee (for $H_0=50$ km s^{-1} Mpc^{-1}) and of the slope α for each type are taken from King and Ellis, 1985, as well as the fraction f of each type in the "field" population. The count predictions are computed with this mix of spectral types.

3. Influence of q_0 and z_{for}

For sake of simplicity, we hereafter assume that all galaxies form at uniform redshift z_{for}. Figures 1 and 2 show the influence of varying q_0 and z_{for}. In figure 1, the various predictions without evolution are shown. There is clearly evolution at $J^+ > 22$. The observed counts have a slope steeper than the predictions without evolution, whatever the values of q_0 may be. This conclusion can be extended to any value of z_{for}.

C. S. Frenk et al. (eds.), The Epoch of Galaxy Formation, 367–368.

When evolution is taken into account, the deceleration parameter influences the counts through geometry (distance modulus and volume element) and through the time versus redshift relation $t(z)$ (in the computation of the e-corrections). Figure 1 shows the influence of varying q_0 at fixed $H_0 = 50$ and $z_{for} = 30$. The plateau seems to occur at lower levels for higher q_0, with a factor $\simeq 4$ shift in the counts between $q_0 = 0$ and $q_0 = 0.5$. Small values of q_0 are required to fit the counts. Figure 2 shows the influence of varying z_{for} for $H_0 = 50$ and $q_0 = 0.05$. z_{for} acts in the e-corrections and in the upper limit of the integral of the counts which is consequently larger for higher z_{for}. The values $z_{for} \leq 5$ are not compatible with the data. Large values of z_{for} are required to fit the counts. In both figures, Tyson, 1987, counts fainter than $J^+ \simeq 25$ are particularly constraining.

Thus it is difficult to disentangle the effects of cosmology (through q_0) and evolution (through z_{for}). **Nevertheless, the slope of the faint counts is so steep that both small q_0 and large z_{for} (i.e. large ages of galaxies) are required to fit the plateau** observed in the counts fainter than $J^+ \simeq 25$. With $H_0 = 50$, this leads to small q_0 (say, below 0.25) and large z_{for} (say, above 10), resulting into current ages of the oldest galaxies larger than 14.3 Gyr (if $q_0 = 0.25$ and $z_{for} = 10$) and possibly as high as 17 Gyr (if $q_0 = 0.05$ and $z_{for} = 30$).

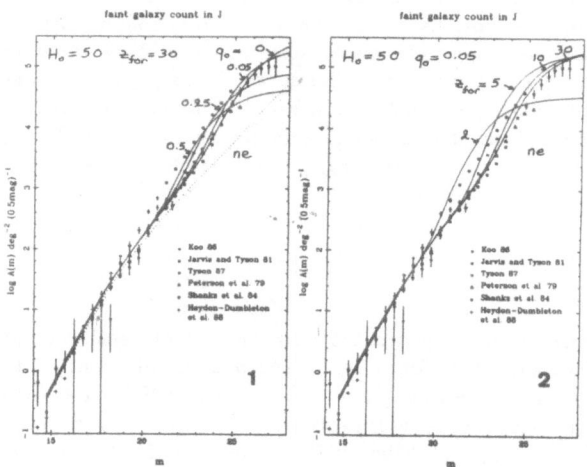

Figures 1 and 2: Influence of q_0 (resp. z_{for}) on the faint galaxy counts. $H_0 = 50$ km s^{-1} Mpc^{-1}

References

Guiderdoni, B., Rocca–Volmerange, B., 1987, Astron. Astrophys., 186, 1
King, C.R., Ellis, R.S., 1985, Ap.J., 288, 456
Rocca–Volmerange, B., Guiderdoni, B., 1988, Astron. Astrophys. Suppl. Series, 75, 93
Tyson, J.A., 1987, *preprint*

LIMITS ON DUST IN DAMPED LYMAN-ALPHA SYSTEMS AND THE OBSCURATION OF QUASARS

S. Michael Fall and Yichuan C. Pei
Space Telescope Science Institute
3700 San Martin Drive, Baltimore MD 21218
and
Department of Physics and Astronomy
The Johns Hopkins University
Homewood Campus, Baltimore MD 21218

The damped Lyα systems discovered in the spectra of quasars at high redshifts are natural places to search for dust. They have column densities greater than 10^{20} cm^{-2}, contain most of the neutral hydrogen in the Universe, and may be protogalaxies or galactic disks in an early, gas-rich phase of evolution. We compare the spectra of quasars in the Wolfe et al. (1986) survey that have damped Lyα with those that do not have damped Lyα to obtain statistical information about the reddening by dust. Our results are given in terms of the dimensionless dust-to-gas ratio $k \equiv 10^{21}(\tau_B/N_H)$ cm^{-2}, where τ_B is the optical depth in the B band in the rest frame of an absorber and N_H is the column density of neutral hydrogen. Using non-parametric tests, we find, at the 95% confidence level, $k \leq 0.41$ (GAL), $k \leq 0.29$ (LMC) and $k \leq 0.19$ (SMC), depending on whether the extinction curve is assumed to have the same shape as that in the Milky Way or the Large or Small Magellanic Clouds. Our upper limits on the dust-to-gas ratio in the damped Lyα systems are half the observed value in the Milky Way but are several times larger than the observed values in the Magellanic Clouds.

From the known redshifts and column densities of the damped Lyα systems, we set limits on the mean and the variance of the optical depth along random lines of sight. This approach is model-independent in the sense that the only assumption made about the space density, characteristic sizes and internal profiles of the absorbers or the way they evole with redshift is that in combination they match the observed properties of the damped Lyα systems. Our calculations include a correction for the effect, emphasized by Ostriker and Heisler (1984), that highly obscured quasars are less likely to be included in optically-selected samples than quasars with little dust in the foreground. Our results for the mean optical depth in the B band of the observer can be approximated by the formula $\bar{\tau}_B(z) = 0.4\tau_\star[(1+z)^{5/2} - 1]$;

369

C. S. Frenk et al. (eds.), The Epoch of Galaxy Formation, 369–370.

using the upper limits on the dust-to-gas ratio in the damped Lyα systems, we obtain $\tau_\star \leq 0.07$ (GAL), $\tau_\star \leq 0.06$ (LMC) and $\tau_\star \leq 0.05$ (SMC). For comparison, Ostriker and Heisler (1984) predict $\tau_\star = 0.4$ or 0.8 and Heisler and Ostriker (1988) predict $\tau_\star = 0.16$. All the limits presented here could be reduced significantly or a positive detection could be made by determining more accurately the spectral indices of the quasars in the Wolfe *et al.* (1986) survey.

A by-product of our study is a comparison between the true luminosity function of quasars and the one derived assuming that space is transparent. For this purpose, we fit a new and potentially useful model to the data compiled by Koo and Kron (1988). We find that the true and observed luminosity functions differ by less than 60% for $z < 3$. This is an upper limit, corresponding to our upper limits on the dust-to-gas ratio in the damped Lyα systems. Since all information about the counts of quasars, either as a function of magnitude or redshift, is also contained in the luminosity function, we conclude that any dust in the damped Lyα systems cannot be responsible for a cutoff in the counts at redshifts below 3. It is possible that dust at higher redshifts would produce an apparent cutoff but this seems unlikely on the basis of a smooth extrapolation of our limits on the mean optical depth. A complete account of our work will be published soon (Fall and Pei 1989).

REFERENCES

Fall, S.M., and Pei, Y.C. 1989, *Ap. J.*, **337**, 000.

Heisler, J., and Ostriker, J.P. 1988, *Ap. J.*, **332**, 543.

Koo, D.C., and Kron, R.G. 1988, *Ap. J.*, **325**, 92.

Ostriker, J.P., and Heisler, J. 1984, *Ap. J.*, **278**, 1.

Wolfe, A.M., Turnshek, D.A., Smith, H.E., and Cohen, R.D. 1986, *Ap. J. Suppl.*, **61**, 249.

THE STELLAR CONTENT OF EARLY-TYPE GALAXIES IN DENSE ENVIRON-MENTS

J Rose[1], R M Sharples[2], R S Ellis[3] and R G Bower[3]
1. University of North Carolina, Chapel Hill, NC, USA
2. Anglo-Australian Observatory, Australia
3. Durham University, England

ABSTRACT. We present evidence, based on an analysis of the integrated spectra of cluster and field galaxies, that the stellar content of early-type galaxies in the cores of rich clusters is different from that of early-type galaxies in low-density environments. Specifically, the large intermediate-age stellar populations inferred to exist in ellipticals in low-density environments appears to be substantially less evident in ellipticals in dense clusters. Our result is preliminary, but if supported by further work would indicate that galaxies were aware of their surroundings at an early stage of their formation.

1. INTRODUCTION

An important question concerning the development of structure in the universe is whether galaxies formed before or after larger-scale structures such as clusters of galaxies. In principle, the detection of a difference in stellar content, or other structural or dynamical characteristics, between elliptical galaxies in low-density environments versus those in dense environments can provide information concerning formation timescales of galaxies versus those for clusters of galaxies. To investigate this possibility, we have analyzed integrated spectra for elliptical and S0 galaxies in the dense clusters Abell 2670 and Coma, and compared them with spectra of galaxies in a low-density cluster (Virgo) and the field, using a system of spectral indices designed to reveal the presence of intermediate-age stars.

2. OBSERVATIONAL DATA

The new data consists of the integrated spectra of early-type galaxies in the Coma (14 galaxies) and Abell 2670 (128 galaxies) clusters, and 5 galaxies in lower density environments (including 3 in the Virgo cluster). The A2670 spectra were obtained with the multi-fiber spectrograph/IPCS combination at the 3.9m AAT and the remainder with a longslit spectrograph/IPCS combination at the 2.5m INT. In the case of A2670, the S/N ratio of the individual spectra is low, but a high S/N ratio composite spectrum was generated by co-adding the individual spectra (after adjusting them to the same redshift). This procedure was adopted for the Virgo and Coma samples also, although here the S/N of the individual spectra is much higher. In the case of Coma, morphologically-classed E's and S0's were selected from Dressler's (1984) list. In the case of A2670, we experimented with selection via morphologies (kindly provided by L

C. S. Frenk et al. (eds.), The Epoch of Galaxy Formation, 371–373.

Thompson) and via the strength of the 4000 Å break; the latter provides the largest sample used here but the conclusions are identical if the restricted morphological sample is used. In all samples, galaxies with internal velocity dispersions greater than 245 km sec^{-1} have been excluded from the analysis since excessive velocity-broadening dilutes the spectral indices used.

The new observations are compared with the photographic image-tube spectra of the isolated E galaxies M32, NGC3377, and NGC5576, and of numerous stars obtained by Rose (1985). The different spectral responses for the two data sets was allowed for using the ratio of the photographic spectrum of NGC3377 to the composite IPCS A2670 spectrum. All spectra were smoothed to a common wavelength resolution of 4.5 Å (FWHM) and rebinned to a dispersion of 2 Å /pixel.

3. ANALYSIS

Rose (1985) has developed a system of quantitative spectral indices to demonstrate that the contribution of main sequence stars to the integrated light of M32 and other nearby elliptical galaxies is substantially higher than expected from a stellar population with a mean age equal to Galactic globular clusters. He inferred that a large intermediate-age population is present in those galaxies, as advocated by O'Connell (1980). Specifically, an index (Sr II/Fe I) that measures the ratio in the residual central intensity of Sr II 4077 to Fe I 4063, when plotted against a similar index constructed using the ratio of Hδ to the same Fe I 4063 line, discriminates surface gravity in individual stars and in composite systems.

The results of our study are summarized in Figure 1, which is a diagram of Sr II/Fe I versus Hδ/Fe I. Filled squares indicate the locations of various field dwarf stars, whereas open squares indicate the locations of field giants. Note the clear separation between the two stellar sequences of different surface gravity. Also plotted (∗) are three elliptical galaxies in low-density environments studied by Rose. All three lie close to the mean dwarf relation in the diagram, indicating their light at 4000 Å is dominated by main sequence and turnoff stars. Likewise, the composite IPCS *Virgo* spectrum lies close to the dwarf sequence. However, the composite spectra of both high density clusters (A2670 and Coma - filled circles) are located approximately midway between the dwarf and giant sequences, indicating that the integrated light at 4000 Å in those galaxies is evenly divided between main sequence and evolved stars. Since Rose (1985) showed that a population as old as the globular clusters should have its light at 4000 Å evenly split between dwarfs and giants, we conclude that there is little or no indication for an intermediate-age population in early type galaxies in *dense* clusters, whereas a large intermediate-age population exists in *low-density* environments.

Errors for the composite cluster spectral indices have been estimated (1) by constructing indices for random subsamples of the individual galaxy spectra and determining the scatter in these indices, or (2) by examining the scatter in the individual spectra in Coma and Virgo, or (3) by estimating errors on the basis of the total number of counts per pixel and assuming Poisson statistics in the errors. Excellent agreement is obtained between the different methods. Using these error estimates we find that A2670 deviates by 2.0 σ from the relation defined by the galaxies in low-density environments. The effect for Coma is less significant - 1.3 σ, but *in the same sense as for A2670.*

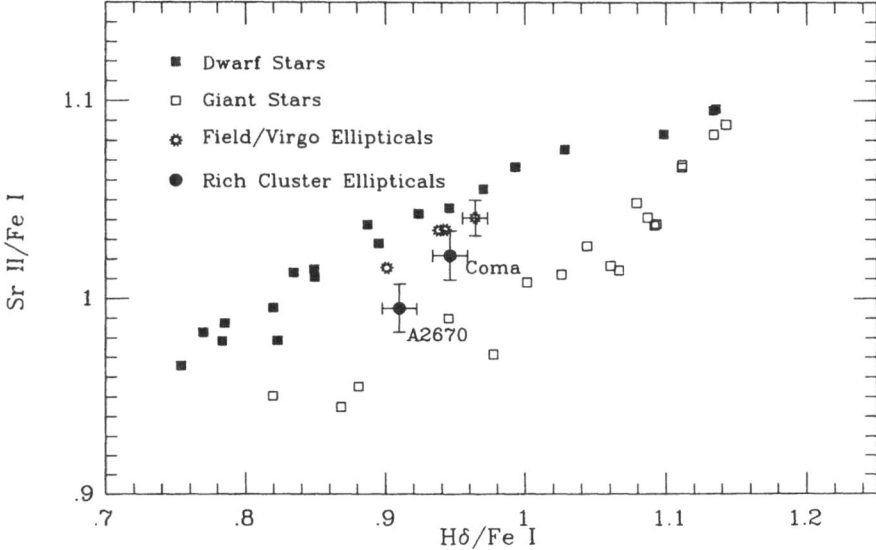

Figure 1: Line ratios (Sr II/Fe I) and (Hδ/Fe I) derived for the composite and individual spectra of ellipticals in various environments as compared to the sequence for stellar dwarfs and giants.

4. CONCLUSION

The stellar content of early-type galaxies appears to be correlated with environment - ellipticals in dense clusters have a less prominent intermediate-age population than those in low-density environments. The measurements presented here are, however, very demanding in terms of S/N ratio, and it would thus be highly desirable to repeat the analysis with a number of other clusters of various richnesses.

References

Dressler, A 1984 *Astrophys. J.*, **281**, 512.
O'Connell, R 1980 *Astrophys. J.*, **236**, 430.
Rose, J 1985 *Astron. J.*, **90**, 1927.

A NEW STUDY OF ABSORPTION LINE DENSITY IN QSO's $Ly\alpha$ FOREST

BIAN YULIN*
Royal Observatory, Blackford Hill, Edinburgh EH9 3HJ, U.K.

1. Introduction

It has been shown by many authors that $N(z_{abs})$, the number of $Ly\alpha$ absorption lines in unit redshift interval at the absorption redshift z_{abs} in QSOs' $Ly\alpha$ forest is given by the fomula $N(z_{abs}) = N_0(1 + z_{abs})^\gamma$, with γ near or somewhat larger than 2.

On the other hand, the possible correlation between the absorption line density $N(z_{abs})$ and the emission redshift z_{em} of the QSO itself has been found by Bian et al.(1986) at first, supported by Bian et al.(1987) and re-found independently by Tytler(1987).

In this paper the dependence of line density on both the emission and absorption redshifts is re-examined as well as discussed if the "inverse effect" is surely of existance.

2. Sample

The sample used in this paper consists of 20 QSOs' $Ly\alpha$ forests with 527 absorption lines altogether, the range of emission redshift from 1.715 to 3.780, the range of absorption redshift from 1.501 to 3.780, and 8.4787 as the sum of covered ranges of absorption redshifts. The 20 member QSOs are Q2000-330, Q1442+101, Q2126-158, Q0805+046, Q0528-250, Q0002-422, Q0100+130, Q0453-423, Q1623+268, Q1623+269, Q0237-233, Q1448-232, Q1225+317, Q0122-380, Q1101-264, Q0421+019, Q0119-046, Q0002+051, Q1115+080, Q0215+015.

3. Probable Correlation Between Absorption Line Density And Emission Redshift

(1) In order to compare absorption line density $N(z_{abs})$ of different quasars at the same absorption redshift z_{abs}, we divided the entire redshift range covered by the 20 QSOs into 2,3,4,5, and 6 intervals of equal length, grouped the QSOs with neighbouring z_{abs} together, and made the statistics on $N(z_{abs})$ in relation to z_{em} for the grouped QSOs in every interval. As a result, it is shown once more that not only the larger z_{em} QSOs have larger mean density, but more importantly at a rather high confidence level, that statistically at the same values of z_{abs}, QSOs with larger z_{em} possess a greater number of $Ly\alpha$ absorption lines.

(2) As pointed out in both Bian et al.(1986) and Bian et al.(1987), the above result necessarily leads to the following: over the entire wavelength range covered by our sample, the mean density N of $Ly\alpha$ absorption lines of each quasar also increases with emission redshift z_{em}. The linear regression equation relating N and z_{em} for the 20 QSOs is $N = 26.9 + 35.8(z_{em} - 1.7)$, with correlation coefficient $r = 0.861$, which is far above the fiducial value $r_{0.05} = 0.444$ or $r_{0.01} = 0.561$.

---

* Visitor from the Beijing Astronomical Observatory, Beijing, China

C. S. Frenk et al. (eds.), The Epoch of Galaxy Formation, 375–376.
© 1989 by Kluwer Academic Publishers.

(3) Similar to the procedures in Bian *et al.*(1986), 11 QSOs with line numbers $n \geq 20$ are used to examine their separate behaviour of $N(z_{abs})$ with z_{abs} by dividing each $Ly\alpha$ forest into 3 sections with equal lengths and evaluating $\bar{z}_{abs}$ and $N(\bar{z}_{abs})$ for each section separately. There seems nothing in common in the behaviour of $N(\bar{z}_{abs})$ with $\bar{z}_{abs}$ among difference quasars. This result is unfavourable to the suggested existence of the "inverse effect" (see Murdoch *et al.*,1986 and Hunstead *et al.*,1988).

4. The Global Estimate of γ by ML Method

(1) By applying Murdoch *et al.*(1986)'s procedures to our sample, using the maximum likelihood (ML) method, I have got the best estimate of γ and its standard deviation 2.20 ± 0.28, which is nearly same as the adopted value 2.17 ± 0.36 in Murdoch *et al.*(1986).

(2) The effect of extreme z_{em} QSOs on the estimated γ is investigated and it is worked out that both the quasars with highest and lowest z_{em} do not individually have a significant effect on the global estimate of γ.

(3) It may be expected that the γ value estimated by using ML method would not change significantly even if the QSOs with $z_{em} > 4.0$ are added to the statistical sample.

5. The γ Estimated from the $Ly\alpha$ Forest of Individual QSOs

(1) The γ values are estimated respectively by using ML method to the $Ly\alpha$ forests of individual QSOs.

(2) Based on a non-parametric test, it is shown that there is no significant difference between the numbers of positive and negative γ values. This result, together with the item (3) in section 3, suggests that it is still hard to say whether the "inverse effect" exists.

Thanks to both Dr.R.W.Hunstead's and Dr.R.F.Carswell's suggestion, the calculations in this paper were repeated after the Workshop with different wavelength cutoff λ_c applied for different QSOs (Murdoch *et al.*,1986) and the original points were supported once again.

References

Bian,Y.L., Chen,J.S., and Zou,Z.L. 1986, *Acta Astrophysica Sinica*, 6, 101. = *Chin. Astron. Astrophys.* 1986, 10, 173.
Bian,Y.L., Zou,Z.L., and Chen,J.S. 1987, in *Proc. of the Fourth IAU Asian-Pacific Regional Meeting* (Beijing: Science Press), in press.
Hunstead,R.W., Murdoch,H.S., Pettini,M., and Blades,J.C. 1988, *Ap.J.*, **329**, 527.
Murdoch,H.S., Hunstead,R.W., Pettini,M., and Blades,J.C. 1986, *Ap.J.*, **309**, 19.
Tytler,D. 1987, *Ap.J.*, **321**, 69.

THE EDINBURGH/DURHAM SOUTHERN GALAXY CATALOGUE

C.A. COLLINS and N.H. HEYDON-DUMBLETON
University of Edinburgh, Dept of Astronomy, Blackford Hill, Edinburgh EH9 3HJ
H.T. MACGILLIVRAY
Royal Observatory, Blackford Hill, Edinburgh EH9 3HJ

ABSTRACT. We describe the construction of the Edinburgh/Durham Southern Galaxy Catalogue. Results on the number–magnitude counts and angular two-point correlation function are presented. Galaxy formation models involving CDM with extra large-scale power are severely constrained.

1. The Construction of the Catalogue.

The Edinburgh/Durham Southern Galaxy Catalogue (EDSGC) is a large-scale machine based optical galaxy catalogue produced from COSMOS scans of SERC(J) survey plates. COSMOS Threshold Mode data has been obtained for all images on ~ 200 plates. Image classification and photometric calibration has been carried out on a mosaic of 60 plates centred on the SGP. The COSMOS deblending software has been applied to these datasets to overcome problems associated with image merging in crowded fields.

Heydon-Dumbleton et al. (1988b) details the fully automated image classification technique that is applied over the full magnitude range of the survey. The limiting magnitude of the EDSGC ($b_j = 20$) is set so that the final catalogue is $> 95\%$ complete with $< 10\%$ stellar contamination. We have shown that the residual rms variation in the number of objects classed as galaxies is only $\simeq 3\%$. To ensure the accuracy of the zero-point photometric calibration we have obtained CCD sequences (in B & V) for every second field. Those plates without sequences can be calibrated using the 4 overlaps to adjacent plates with CCD sequences (Heydon-Dumbleton et al. 1988a). We estimate that the rms variation in the limiting magnitude of the EDSGC will be 0.04 mag.

The image classification and photometric accuracy of the EDSGC, coupled with our control of systematic errors and careful data reduction have produced an objective catalogue that is deeper and more uniform than any existing optical galaxy survey. The EDSGC has a wide range of applications to many areas of astronomy and will be made generally available in the near future.

2. Preliminary Results

Galaxy number–magnitude counts for a 100 deg^2 region of our survey are compared with the counts of previous authors in Fig. 1. This region is considerably larger than any previously

C. S. Frenk et al. (eds.), The Epoch of Galaxy Formation, 377–378.

378

surveyed and provides the most reliable estimate of galaxy counts at bright magnitudes. There is significant deviation between our counts and a no-evolution model for $b_j > 18.75$.

Fig. 2 (Collins $et\ al$ 1988) shows the angular two-point correlation function ($b_j \leq 20$) for two regions of our survey, scaled to the depth of the Shane & Wirtanen counts (1967) (Groth & Peebles 1977). The correlation function for a region ~ 100 deg^2 calibrated to high accuracy confirms a break from a power-law corresponding to a physical scale $\sim 7h^{-1}$ Mpc. A partially calibrated region of ~ 500 deg^2 gives rise to a break feature on larger scales, indicating that systematic number density variations of $\simeq 20\%$ may affect the position of the break. In the context of current theories of galaxy formation, models involving a standard cold dark matter power spectrum with additional large-scale power (Bond & Couchman 1988) appear to be excluded.

References

Bond, J.R. & Couchman, H., 1987. Proceedings of the Second Canadian Conference on General Relativity and Relativistic Astrophysics, ed. A. Coly & C. Dyer (Singapore: World Scientific).
Collins, C.A., Heydon-Dumbleton, N.H. & MacGillivray, H.T., 1988. $Mon.\ Not.\ R.\ astr.\ Soc.,$ in press.
Groth, E.J. & Peebles, P.J.E., 1977. $Astrophys.\ J.,$ **217**, 385.
Heydon-Dumbleton, N.H., Collins, C.A. & MacGillivray, H.T., 1988a. $Large\text{-}Scale\ Structure\ in$ $the\ Universe\ -\ Observational\ and\ Analytical\ Methods,$ Springer-Verlag, in press.
Heydon-Dumbleton, N.H., Collins, C.A. & MacGillivray, H.T., 1988b. $Mon.\ Not.\ R.\ astr.\ Soc.,$ submitted.
Shane, C.D. & Wirtanen, C.A., 1967. $Publs\ Lick\ Obs.$ **Vol 22**, Part 1.

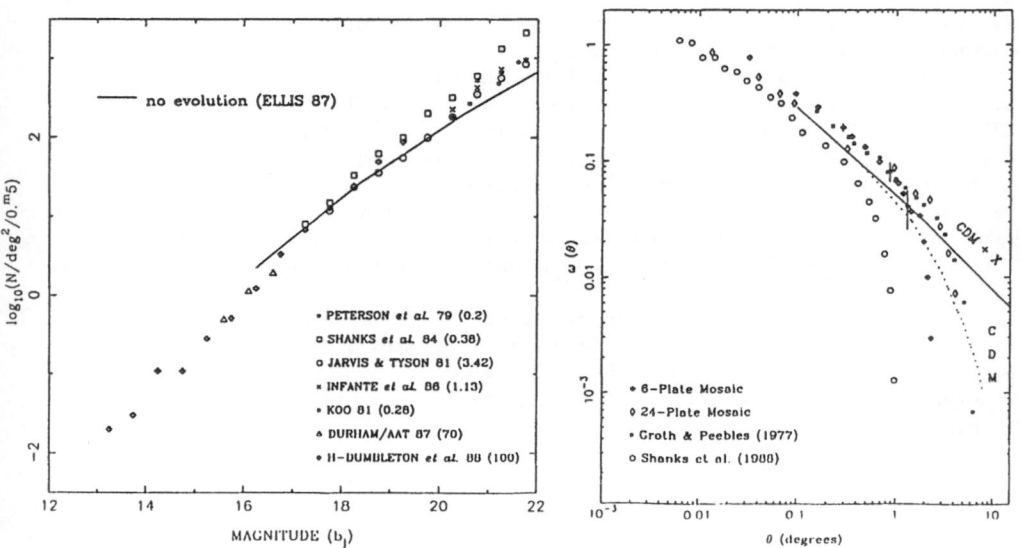

Fig.1 Galaxy Number-Magnitude Counts.   Fig.2 Angular Correlation Function.

MASS-TO-LIGHT RATIOS AND THE AGE OF GALAXIES

Tom Broadhurst and Paolo Salucci
Physics Department, University of Durham, UK

ABSTRACT. We present a method for constraining the present age of spiral galaxies or equivalently the value of the Hubble constant H_o, for a given cosmological model. This is achieved by calculating a *dynamical* measure of the disc mass-to-light ratio, M_D/L_B, for a sample of spiral galaxies, (with accurate rotation curves and colours) and by comparing the resulting tight relation between M_{disc}/L_B and $(B-V)_o^T$, with theoretical predictions of galaxy evolutionary models.

The observed mass-to-light ratios are proportional to H_o, but the model predictions are proportional to galaxy age, t_*, which is *inversely* proportional to H_o (for the Friedmann models). Consistency between theory and observations can only be produced, for a given value of Ω_o, within a narrow range of combinations of H_o and t_*.

For the case of $\Omega_o = 1$, we find consistency for $t_* = 9 \pm 1$ Gyr and $H_o = 75 \pm 8 km s^{-1} Mpc^{-1}$.

1. DYNAMICAL DISC MASS-TO-LIGHT RATIOS

The dynamical mass of a spiral galaxy, $M_{tot} \equiv G^{-1} V_{25}^2 R_{25}$ is a poor measure of the luminous (disc) mass, given the presence of possibly large amounts of dark matter. However it is possible to estimate the fraction of total mass in the disc component M_{disc}/M_{tot} from the value of the mean slope of the outer rotation curve, $< dlogV/dlogR >$, thus inferring dynamically the true disc mass (Persic and Salucci, 1988; see also Salucci and Frenk 1988. The contribution of the dark halo mass to the total mass within R_{25} is found to vary as

$$log \frac{M_{disc}}{M_{tot}} = -0.08 - 1.16 < dlogV/dlogR >$$

Thus, the contribution from dark halo matter is more important for lower luminosity galaxies, which is evident from their steeper rotation curves (Rubin *et al.* 1980). The resultant M_{disc}/L_B ($\propto H_o$) are plotted versus $(B-V)_o^T$ in Figure (1) which shows a tight relation between mass-to-light ratios and colour.

2. EVOLUTIONARY PREDICTIONS FOR THE MASS-TO-LIGHT VS. COLOUR RELATION

Now we compare the above corrected observational relation with the predictions of standard evolutionary models, based on the Bruzual stellar synthesis code,

379

C. S. Frenk et al. (eds.), The Epoch of Galaxy Formation, 379–381.

(Bruzual, 1983), assuming various histories of star formation. These model predictions are very insensitive to the details of the star formation history as first recognised by Tinsley (1981), but are sensitive to the low mass limit of the initial mass function adopted (here we use the standard Miller-Scalo IMF, Miller and Scalo, 1979) and in particular to the galaxy age assumed. In fact $(M_{lum}/L_B)_{pred} \propto t_*$ (Tinsley 1981, Broadhurst and Salucci, 1988), or $t_* \simeq H_o^{-1}$ (assuming galaxy age is $\approx$ age of universe). Note that $M_{lum} \cong M_{disc}$ for the sample of spiral considered here since the bulge contributions to L_B are $< 20\%$, which is negligible in terms of the scatter in Figure 1.

This dependency on H_o is in the opposite sense to the *observed* mass-to-light ratios, i.e. the ratio

$$\frac{M_{disc}/L_B}{(M_{lum}/L_B)_{pred}} \propto H_o^2$$

and provides a means of tightly constraining t_* and H_o.

To parameterise the star formation histories we take for convenience the simple model of an initial 1 Gyr burst with a constant star formation thereafter, so that for any adopted age $\sim 5\%$ gas is left today. The fraction of stars formed during the initial burst is varied from 0 to 1 producing the bluest and reddest extremes in colour respectively. Other star formation histories were considered (Broadhurst and Salucci 1988) but we find the slope of the relation between $(M_{lum}/L_B)_{pred}$ and $(B\text{-}V)_o^T$ is independent of star formation history, galaxy age and IMF, and is in excellent agreement with the corrected observations (Figure 1).

Note that if no correction is made for the presence dark matter, the observed relation (mass- to-light ratio *vs* colour) is flat and inconsistent with the predictions of evolutionary models as concluded by Tinsley (1981). Thus it would seem we are able to clear up an important discrepancy for these evolutionary models, which are used widely in other contexts.

3. RESULTS AND DISCUSSION

The theoretical and observational relations agree in their zero-points only for a very small range in galaxy's age and H_o. For example, in the case $\Omega_o = 1$ considered here (and assuming the timescale of galaxy formation short compared to the Hubble time), we can strongly constrain the value of H_o. This is demonstrated in Figures 1a) to 1c): Figure 1a) shows that, if we chose $H_o = 50$, the implied cosmological age of 12 Gyr is inconsistent with the data. In fact only the age ~ 5 Gyr is compatible with the observations for this value of H_o. Conversely for $H_o = 100$, the best fit age (11 Gyrs) implies $\Omega_o \sim 0$. To achieve the consistency with $\Omega_o = 1.0$, $H_o = 75 \pm 8$ must be adopted, so that the age implied is 9 ± 1.0 Gyrs.

Note that changes in the low mass limit to the IMF can affect our results, but if this limit is lowered to produce more low mass stars the outcome is a *higher* value of H_o for a given Ω_o. Further support for a high value of H_o is indicated by comparison of the observed range in colour with the range produced by the evolutionary model: high values for H_o are preferred but note that a few objects may be tollerated (5%) blueward of (B-V)=0.5 due to non smooth star formation in low (disc) mass systems (see Broadhurst, 1988).

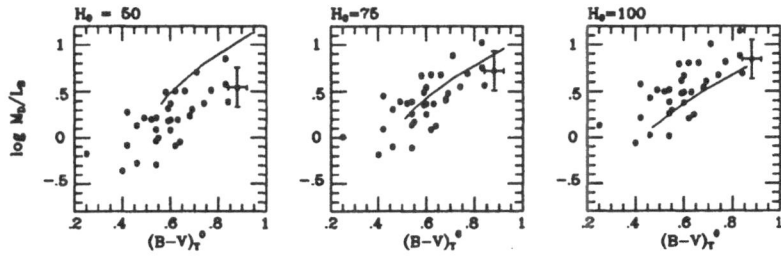

Figure 1. Data points are disc mass-to-light ratios M_{disc}/L_B versus colour (B-V)$_o^T$ values derived from observations of non-local spiral galaxies (Sample C of Persic and Salucci, 1988), for different values of H_o (indicated top left). The theoretical curves refer to the predictions of evolutionary models described in the text. The present day ages of these model predictions are chosen to be consistent with the H_o used to calculate the data values with $\Omega_o = 1$. Also shown is the averaged dynamical M/L_B and colour (transformed into the B_T system) for a sample of elliptical galaxies (square point with error bars) of Djorgovski and Davis (1987).

The scatter in this relation, 0.7 mag, is addressed in greater detail in Broadhurst and Salucci (1988) and may be expected to reduce to only $\sim$ 0.4 mag with good photometry, with the remaining scatter that of the Tully-Fisher relation.

REFERENCES

Bruzual, G. 1983. *Astrophys. J.*, **273, 105**
Broadhurst, T.J. and Salucci, P. *in preparation*
Broadhurst, T.J. 1988, *PhD Thesis, University of Durham*
Djorgovski, S. and Davis, M. 1987, *Astrophys. J.*, **313, 59**
Miller, G.E. and Scalo, J.M 1979, *Astrophys. J. Suppl.*, **41, 513**
Persic, M. and Salucci, P. 1988. *Mon. Not. R. astr. Soc.*, **234**, 131.
Rubin, V.C., Ford, W.K., Thonnard, N., 1980 *Astrophys. J.*, **238**, 471
Salucci, P. and Frenk, C.S. 1988, *Mon. Not. R. astr. Soc.*, in press
Tinsley, B.M. 1981 *Mon. Not. R. astr. Soc.*, **194**, 63.

EMISSION LINES AND STAR FORMATION IN RADIO GALAXIES

J.R. Allington-Smith
Physics Department, Durham University
South Road, Durham DH1 3LE, UK

ABSTRACT. A comparison of spectroscopic data from two independent complete samples of radio sources shows that there is a correlation between radio luminosity and the equivalent width of [OII]λ3727 for radio galaxies with $z \sim 0.5$. Other data, although incomplete, suggests that this relationship may also hold at higher redshift. The previously reported correlation between the 'blueness' parameter (an indicator of continuing star formation) and [OII] equivalent width is confirmed. It is suggested that the latter correlation is not fundamental but arises as a result of fundamental relationships between line strength and radio power and between the blueness parameter and redshift. This implies that the stellar populations of radio galaxies at *moderate* redshift are relatively unaffected by interactions with the radio emitting plasma.

Recent work on the 1-Jy sample (Allington-Smith 1988 and references therein) has resulted in a large body of spectroscopic data which can be directly compared with that of the 3C sample (Spinrad *et al.* 1985). Since the 1-Jy sample is a factor ~ 5 fainter than 3C, this allows us to determine the variation of properties with radio luminosity independently from redshift — something which cannot be done with a single sample because of the very strong correlation between radio luminosity and redshift in flux density limited samples.

Figure 1 shows a histogram of rest-frame [OII]λ3727 equivalent width (W_λ^o) for the 3C and 1-Jy samples for radio galaxies with $0.40 \leq z < 0.55$. It is very likely that both samples are complete in this redshift interval. There is a significant difference between the two histograms which indicates the presence of a correlation between radio luminosity at 408 MHz (P_{408}) and equivalent width which may be parameterised in the form $W_\lambda^o \propto P_{408}^\gamma$ with $\gamma = 0.37 \pm 0.01$.

Figure 2 shows a plot of equivalent width against radio luminosity for all the data available in the literature including the redshift-limited sub-sample described above. Although the data in this plot are not complete and are therefore subject to selection bias, it is interesting to note that the relation derived above (dashed line) follows the data quite well suggesting that this correlation may apply over a wide range of redshift. This picture is consistent with the known weakness of narrow emission lines in fainter radio-selected samples (e.g. the 5C surveys). This relationship may be compared to the correlation between radio and [OIII] line luminosity for Seyfert galaxies noted by Whittle (1985).

For 3C galaxies, Lilly and Longair (1984) showed that there was a correlation between [OII] equivalent width and a 'blueness' parameter defined as $(r - K) - (r - K)_C$ where $(r - K)_C$ is the colour predicted by the 'C-model' of Bruzual (1981) in which there is no star formation after an initial burst lasting $1 Gyr$ which occurred $16 Gyr$ before the present. They noted that there is a similar relationship between [OII] line strength and (B-V) colour in nearby ordinary

C. S. Frenk et al. (eds.), The Epoch of Galaxy Formation, 383–385.

spiral galaxies (Dressler and Gunn, 1982) although with equivalent widths typically an order of magnitude smaller.

Figure 3 shows a plot of equivalent width against blueness for all available 3C and 1-Jy data. There are two 1-Jy radio galaxies which are significantly redder than the rest (negative blueness). Both are unusual since they lie close to the line of sight of other galaxies that may have contaminated the K photometry. The 1-Jy radio galaxies appear to follow the same trend defined by the 3C data but the numbers are too small to be certain. Note also that the 3C galaxies tend to have stronger lines than their 1-Jy counterparts for the same value of the blueness parameter.

How can these two correlations be explained? The simplest explanation is that there is a *fundamental* relationship between radio luminosity and equivalent width (due to some physical process in the Narrow Line Region) which produces an *apparent* blueness – equivalent width correlation. This would arise quite naturally if we interpret the known tendency for distant radio galaxies to be bluer than expected as being due to a *fundamental* relationship between the amount of continuing star formation and cosmic time. To complete the picture, we need only remember that redshift and radio luminosity are strongly correlated in flux density limited samples.

Support for this simple picture comes from the fact that the 1-Jy and 3C data points follow the *same* trend in Fig. 2 but are *separated* in Fig. 3 in the sense that the 3C galaxies have the stronger lines for constant value of the blueness parameter. Since the relationship between redshift and radio power is a function of the limiting flux density of the radio sample, it would be expected that there would be a displacement of the 1-Jy and 3C data points in the blueness – equivalent width plane in precisely the same sense as seen in Fig. 3.

This picture suggests that there is no interaction between the radio source and the stellar populations. Indeed any link of this type would require the 1-Jy and 3C data points in Fig. 3 to lie along the same locus. However, there are some problems with this scheme. Firstly, it is based on a small dataset which is incomplete outside $0.40 \leq z < 0.55$. Secondly, as redshift increases, there is a tendency for the morphology of the radio galaxies to change and become aligned with the axis of the radio structure (e.g. McCarthy *et al.* 1987) and for the narrow emission line activity to become increasingly extra-nuclear (e.g. 3C368; Djorgovski *et al.* 1987). This implies that, at very high redshift, other processes may be operating which could provide a direct link between the radio source and star formation in the host galaxy in contrast to the situation suggested here for lower redshift.

References

Allington-Smith, J.R., Spinrad, H., Djorgovski, S. & Liebert, J., 1988. *M.N.R.A.S.*, **234**, 1091.

Bruzual, G.A., 1981. Ph.D thesis, University of California.

Djorgovski, S., Spinrad., H., Pedelty, J., Rudnick, L. & Stockton, A., 1987. *A.J.*, **93**, 6.

Dressler A. & Gunn, J.E., 1982. *Ap.J.*, **263**, 533.

Lilly, S.J. & Longair, M.S., 1984, *M.N.R.A.S.*, **211**, 833.

McCarthy, P., van Breugel, W., Spinrad, H. & Djorgovski, S. 1987. *Ap.J.*, **321**, L29.

Spinrad, H., Djorgovski, S., Marr, J. & Aguilar, L., 1985. *P.A.S.P.*, **97**, 932.

Whittle, M., 1985. *M.N.R.A.S.*, **213**, 33.

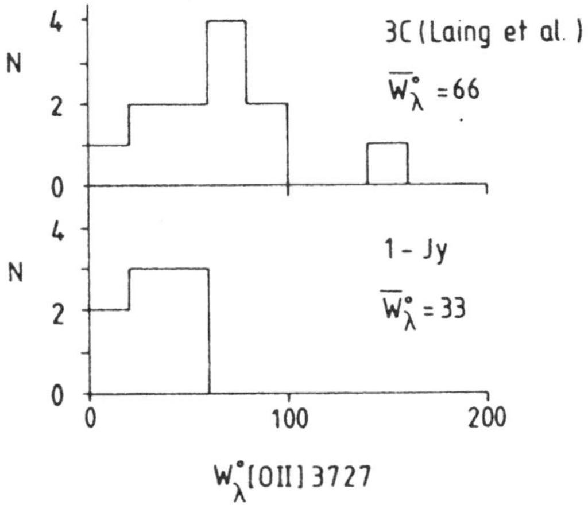

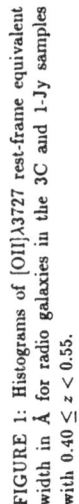

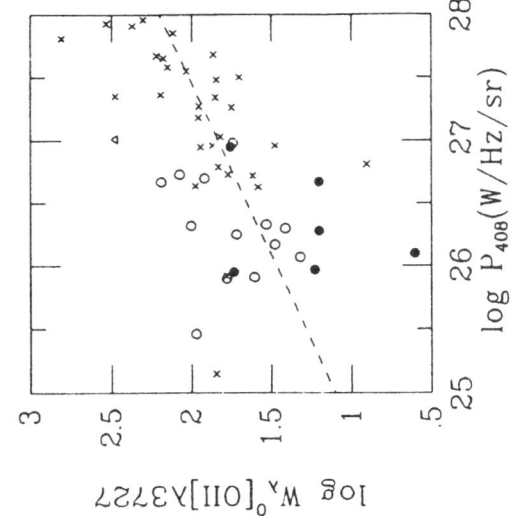

FIGURE 1: Histograms of [OII]λ3727 rest-frame equivalent width in Å for radio galaxies in the 3C and 1-Jy samples with $0.40 \leq z < 0.55$.

FIGURE 2: A plot of [OII] equivalent width against radio luminosity for all data from the literature for the 1-Jy sample (circles) and the 3C sample (crosses). The filled circles indicate 1-Jy galaxies believed to be in clusters.

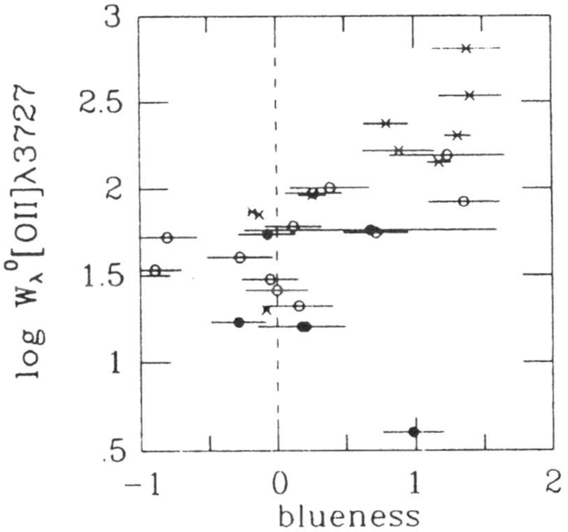

FIGURE 3: A plot of [OII] equivalent width against the blueness parameter described in the text using the same symbols as Fig. 2.

THE AGE OF THE RADIOGALAXY 0902+34 AT THE REDSHIFT z=3.395

B. Rocca–Volmerange[1,2] and B. Guiderdoni[1]
1 : Institut d'Astrophysique de Paris, CNRS, F–75014 Paris, France
2 : Université de PARIS XI, F–91405 Orsay-Campus, France

1. Abstract

Primeval galaxies would give informations on processes of galaxy formation and are excellent constraints for observational cosmology. The recent discovery of the extremely distant (z=3.395) radiogalaxy 0902+34 by Lilly, 1988, which does not appear as a primeval galaxy, can help us to research the epoch of galaxy formation and to predict appearances of the most remote galaxies only if we are able to estimate the present and past star formation activities. With the help of our atlas of synthetic spectra by Rocca-Volmerange and Guiderdoni, 1988, (RVG), we may estimate the star formation history of the radiogalaxy 0902+34. The best fit of the observations (VIJK) gives two bursts of star formation initiated 0.1 Gyr and 3 Gyr earlier. Consequences of such an age on the epoch of galaxy formation z_{for} and then on the age of the Universe are given. From these results and fits of other independant observations, we conclude to remote epochs of galaxy formation (high z_{for}) in a very old Universe, inducing low q_0 and H_0.

2. Spectral Signatures of a Young Galaxy

Predictions of the main spectral features of a young galaxy are made with a burst of star formation calculated in the Atlas of Synthetic Spectra of Galaxies (RVG), based on our spectrophotometric model of galactic evolution (Guiderdoni and Rocca-Volmerange, 1987). This model brings several improvements relative to our previous models (Rocca-Volmerange et al, 1981). It is caracterised by a better set of stellar input data. The observational UV spectra from the IUE Atlas (Wu et al, 1983) are connected to theoretical predictions between 1200Å down to 200 Å from Mihalas, 1972 models extended by Borsenberger and Gros, 1978 and to the observational stellar library from Gunn and Stryker, 1983. The horizontal branch and asymptotic giant-branch evolutionary phases are taken into account along the evolutionary tracks. The model also estimates evolution

C. S. Frenk et al. (eds.), The Epoch of Galaxy Formation, 387–390.

of neutral gas, nebular emission from ionised gas and extinction by dust. The Initial Mass Function (IMF) based on Scalo, 1986, is standard and the efficiency 1 $M_\odot$ Gyr^{-1} × Mass of the bursting star formation law corresponds to a maximum consumption of gas for 1 Gyr. The nebular emission is calculated from the ionising photon number issued from the Lyman continuum and absorbed by the gas at any time. A comparison of the Lyα emission lines of the distant radiogalaxies (Spinrad, 1987, Djorgovski, 1987) with emission calculated from our evolving stellar population of massive stars and photoionisation models with typical parameters of an HII region (Stasińska, 1984) shows that in such distant radiogalaxies, this last process mainly originates the Lyα emission line (Spinrad, 1987, Rocca-Volmerange, 1988). This is also the case of 0902+34 (see also Spinrad, this conference) which shows in the Lyman line a flux similar to the Lyα galaxies (Lilly, 1988). Spectra of active burst (0.1 Gyr old) or passive burst with respective ages 1.1 Gyr, 2.1 Gyr,.. and an activity stopped at 1 Gyr are shown on figure 1 from 400Å to 2.1μm. Evolution of continuum as well as absorption lines typical from the stellar population and the 4000 Å discontinuity clearly appears.

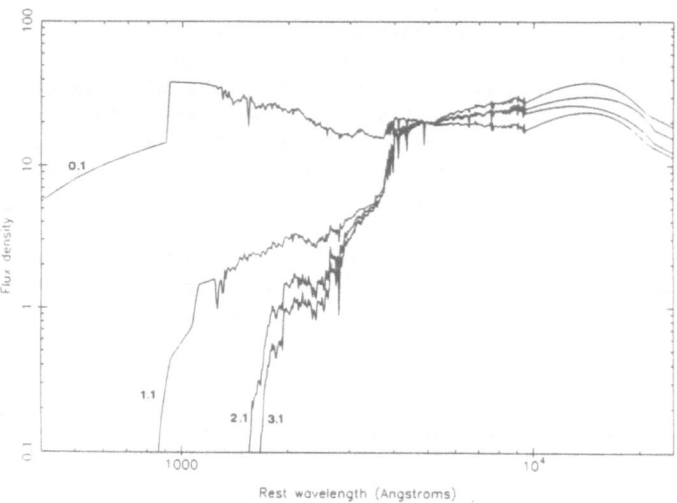

FIGURE 1

Spectra of 1Gyr bursts at various ages in relative units (Normalisation at 4990Å). See text for details

3. An Age of the Radiogalaxy 0902+34

This radiogalaxy has been discovered by Lilly, 1988. After the Lyα and CIV emission lines, the redshift is $z= 3.395$. It is important to note that, at such a redshift, the Lyα and Hβ lines coincide with the V and K broad-bands, increasing the possibility of detection. It has been selected on the basis of a faint infrared emissivity and a very red J-K color ≥ 2.75. Its emissivity in Lyα line ($\simeq 2.1 \times 10^{-18}$ Wm^{-2}) and its equivalent width ($\simeq 1000$ kms^{-1}) are similar to those of the 3CR radiogalaxies observed by Spinrad, Djorgovski and their team. Lilly suggested from essentially the emissivities in K and V bands, that this galaxy does not appear to be primeval since an old population appears to be emitting in the K band. With our model, we can give a more precise estimate of age for this galaxy. According to the different images through the three filters VIK (Lilly, 1988), bursts appear as the best scenarios to simultaneously reproduce the emission in the far–UV and visible light. If a present (galaxy-frame) burst fits the far-UV emission, the epoch of galaxy formation will be given by the age of the most recent passive burst which fits the stellar emission in the K band. Figure 1 is the best fit of VIJK colours when the nebular component is excluded according to Lilly,1988. It is obtained with a sum of two bursts: a present (galaxy-frame) one with an age 10^8 yr and a burst initiated 2, 3, 4 and 5 Gyr earlier with a duration 1Gyr. Details of the respective contributions of 0.1 Gyr and 3 Gyr bursts are also plotted. The Lyα line has been substracted in observations as in models from the total flux. The contribution of the line is about 50% of the total flux.

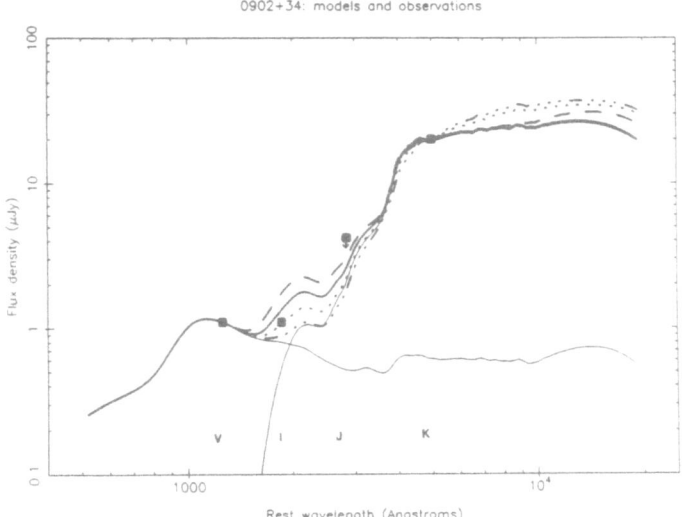

FIGURE 2

A fit of the VIJK photometry of the radiogalaxy 0902+34 with the sum of a burst of age 0.1 Gyr and a passive 1 Gyr burst of age 2 Gyr (dashed), 3 Gyr (full), 4 Gyr (dots), 5 Gyr (dot-dashed). Detailed spectra of burst 0.1 and 3 Gyr are given (thin lines).

4. Consequences on Cosmological parameters

An age 3 Gyr is extremely constraintful for the epoch of galaxy formation. The following table gives the corresponding values of z_{for} for various values of the deceleration parameter q_0. and the constant of Hubble $H_0 = 50$ km s^{-1} Mpc^{-1}.

	$q_0 = 0$	$q_0 = 0.05$	$q_0 = 0.5$
Universe Age in Gyr	19.56	17.57	13.04
$t(z)$ in Gyr	4.45	3.13	1.41
$t(z_{for})$ in Gyr for an age 3Gyr	1.45	0.13	-
z_{for}	12	15	-

As a conclusion, the values of $H_0 = 50$ and $q_0 \leq 0.05$ needed by our model to fit the most distant galaxy 0902+34 are very low. This is essentially depending on the age of the oldest stellar population (3 Gyr) infering high values of z_{for} (≥ 15) and an old Universe (≥ 17Gyr). These results are in agreement with those deduced from the faint galaxy counts (Guiderdoni and Rocca-Volmerange, this conference) which are based on independant observational data. The common point of all these results is the stellar ages used in the evolutionary models essentially realted to the ages of the globular clusters. A more complete analysis will be published in a forthcoming paper.

References

Borsenberger, J., Gros, M., 1978, *Astron. Astrophys. Suppl. Series*, 31, 291

Djorgovski, S.G., 1987, in *Towards Understanding Galaxies at Large Redshift*, Kron, R.G., Renzini, A. eds., Kluwer, p.259

Guiderdoni, B., Rocca–Volmerange, B., 1987,(GRV), *Astron. Astrophys.*, 186, 1

Gunn, J.E., Stryker, L.L., 1983, *Astrophys. J. Suppl. Series*, 52, 121

Lilly, S., *Astrophys. J.*, 333, 161

Mihalas, D., 1972, *NCAR Technical Note*, STR 76

Rocca–Volmerange, B., 1988a, *M.N.R.A.S.*, in press

Rocca–Volmerange, B., Guiderdoni, B., (RVG), 1988b, *Astron. Astrophys., Sup. Ser.*, 75, 93

Rocca-Volmerange, B., Lequeux, J., Maucherat-Joubert, M., 1981, *Astron. Astrophys.*, 104, 177

Scalo, J.M., 1986, *Fundamental of Cosmic Physics*, 11, 1

Spinrad, H., 1987, *High Redshift and Primeval Galaxies*, Ed. J. Bergeron, D. Kunth, B. Rocca-Volmerange, Tran Thanh Van, Ed Frontieres,p. 59

Stasińska, G., 1984, *Astron. Astrophys. Sup. Ser.*, 55, 15

Wu et al., 1983, *IUE Ultraviolet Spectral Atlas*, NASA No22

STATISTICS OF RADIO GALAXY POPULATIONS AND GALAXY FORMATION

J. A. Peacock
Royal Observatory, Blackford Hill, Edinburgh EH9 3HJ

ABSTRACT. The radio luminosity functions of radio-loud quasars and elliptical radio galaxies appear identical above $P_{2.7GHz} \simeq 10^{25}$ WHz^{-1}sr^{-1} – both quasars & galaxies displaying a 'redshift cutoff' at $z \gtrsim 2$. This result may well reflect the collapse epoch of massive galaxies, rather than that of the groups/clusters in which these objects reside.

1. The radio luminosity function at high redshift

It has been known for some time that the degree of cosmological evolution displayed by radio galaxies & quasars at $z \lesssim 1$ is similar, and evidence is now becoming available that this is also the case at higher redshifts, based on a study of the Parkes Selected Regions (Downes *et al.* 1986; Dunlop 1987; Dunlop *et al.* 1989). These regions are the deepest in the Parkes survey ($S_{2.7GHz} \geq 0.1$ Jy) and provide a total of 178 radio sources over 216 deg^2. CCD frames have been taken to B $\simeq$ 25, R $\simeq$ 24 and infrared photometry to K $\simeq$ 19, yielding detections of all but 4 of the sources. Spectroscopy is largely complete to R $\lesssim$ 22, providing redshifts for about 50% of the sample, including essentially all the quasar candidates. For the remaining very faint galaxies, it is necessary to rely initially on redshift estimation using the infrared photometry. The 'standard candle' relation is well defined to K $\simeq$ 18 (*cf.* the recent detection of a radio galaxy with z=3.395 and K=18.8 by Lilly 1988). In fact very nearly all the Selected Region galaxies were detected at K $\leq$ 17.5. The distribution of K magnitudes is shown in Fig. 1, from which it is clear that very few of the galaxies can lie at $z > 2$, unless they are implausibly luminous.

This result parallels that from the quasars in this sample: the highest redshift in the Selected Regions is $z = 3.71$ (Dunlop *et al.* 1986), but the next highest is $z = 2.9$ and there are only 3 quasars with $z > 2.5$ (out of a total of 49). These results confirm a trend noted by Peacock (1985) in the data on flat-spectrum quasars: radio samples at intermediate flux levels contain maximum redshifts no higher than those seen in brighter samples. In short, there exist objects with $P_{2.7GHz} \simeq 10^{27}$ WHz^{-1}sr^{-1} at $z \simeq 1$ in bright complete samples which should have high-redshift counterparts in the fainter samples. Since these do not appear, we can deduce that the RLF must have fallen at high redshifts (at least for radio powers in this region); this can be done in a variety of ways. The complete-sample data can be combined with other information in the free-form expansion method used by Peacock (1985); alternatively, one may simply bin the data and deduce an RLF directly (with poissonian error bars). The results of both these procedures are shown in Fig. 2.

The evidence for a reduction in the RLF at $z \gtrsim 2$ is quite clear from this figure; a very similar diagram results for $\Omega_o = 0$. The result can be backed up by a 'banded' V/V_{max} analysis.

C. S. Frenk et al. (eds.), The Epoch of Galaxy Formation, 391–396.
© *1989 by Kluwer Academic Publishers.*

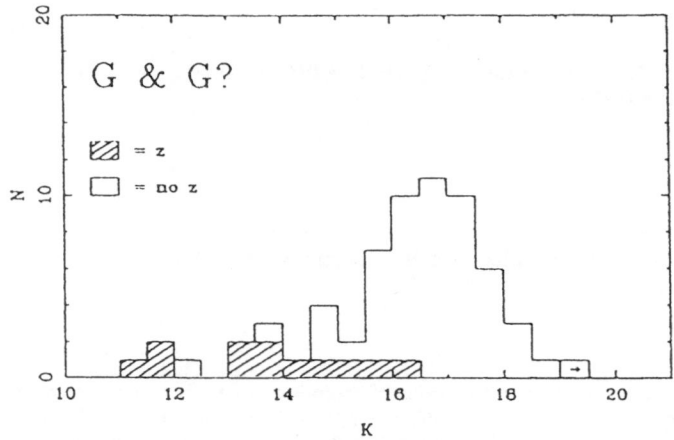

Figure 1 *The distribution of K magnitudes for objects in the Parkes Selected Regions. Note the relative lack of objects with $K \gtrsim 17.5$.*

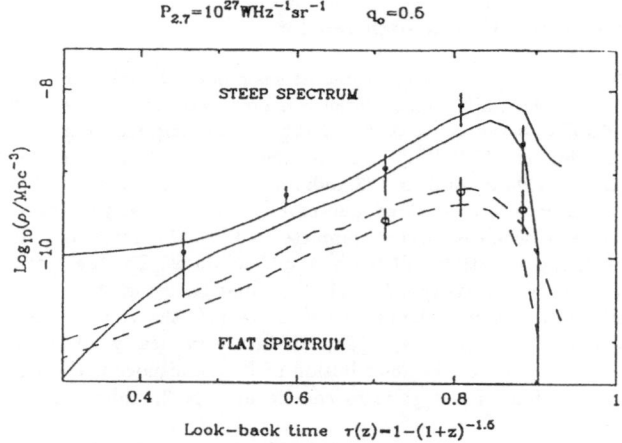

Figure 2 *A cut through the RLF at constant power. Note the reduction in comoving density for both radio spectral types (steep-spectrum: largely galaxies; flat-spectrum: largely quasars) at $z \gtrsim 2$. $\Omega = 1$ and $h = 0.5$ are assumed; a similar diagram results for lower values of Ω.*

Table 1 shows the values of $\langle V_e/V_a \rangle$ resulting from a combination of the Parkes Selected Region data with brighter samples, considering $z > 2$ only.

2. Caveats

Could this result be wrong? For this to be the case, we must have missed some objects at very high redshift; we can see how large this number would have to be by taking the (well-determined) RLF at $z = 2$ and assuming that it stayed constant at higher redshifts. On this

	n	$\langle V_e/V_a \rangle$	
		$\Omega_0 = 1$	$\Omega_0 = 0$
Flat-Spectrum (all)	25	0.328	0.359
Flat-Spectrum Quasars	24	0.342	0.374
Steep-Spectrum (all)	23	0.395	0.423
Steep-Spectrum Quasars	10	0.417	0.431

Table 1 *Results of the banded V_e/V_a test for 2.7-GHz complete samples with flux-density limits in the range 0.1 to 2 Jy. Redshifts $z > 2$ only are considered.*

hypothesis, the Selected Regions should contain 18 steep-spectrum sources with $z > 2.5$, which seems highly unlikely in the light of the above data.

The next major step in this area will be to determine the high-redshift evolution at lower luminosities. The ideal database for this will probably be the Leiden-Berkeley Deep Survey (Windhorst 1984), which now has optical data of approximately the same degree of completeness as the Parkes Selected Regions. Indeed, Windhorst (1984) claimed that there was evidence for a deficit of low-power galaxies at $z \gtrsim 1$. In fact, the highest redshift seen to date in the LBDS is $z = 2.4$ (Koo, private communication), so it may be that the behaviour described here applies approximately at all radio powers.

In any case, it is fascinating that the recent results for radio-quiet quasars seem to be in qualitative agreement about a deficit for $z \gtrsim 2$ (with the possible exception of the very highest luminosities; Warren 1988). This suggests that there was some critical feature of the Universe at that particular epoch which was favourable to the development of active nuclei.

3. Radio galaxy environments

The determination of the local galaxy density about active galaxies has long been of interest for what it may tell us about the triggering of the AGN phenomenon. The recent results of Yee & Green (1987) are especially interesting, as they claim to have discovered a change of quasar environment with redshift. Their result is however based on an approximately flux-limited sample, so those objects at high redshift are also the most radio-luminous. To see whether this matters, consider our knowledge of the correlation of environment with radio power for radio galaxies. In all cases, we will be measuring the density of the environment in terms of the spatial cross-correlation amplitude $B_{gr} = \xi(r = 0.5h^{-1}Mpc)$; for orientation, the values of this parameter corresponding to Abell richnesses 0,1 & 2 are approximately 110, 250 & 380. It was established by Longair & Seldner (1979) that B_{gr} showed a strong dependence on radio power, with $\langle B_{gr} \rangle$ changing by a factor ~ 2 on crossing the Fanaroff-Riley division at $P_{2.7GHz} \simeq 10^{23.5}h^{-2}$ WHz^{-1}sr^{-1}, in the sense that the weaker sources occupied the denser environments; these results were amplified by Prestage & Peacock (1988).

The above investigations were unable to study the environments of the most powerful radio galaxies as they were limited to $z \lesssim 0.2$ by the use of photographic material. It then became clear that there was indirect evidence that these more luminous sources (of which the archetypes at lower redshifts are Cygnus A & 3C295) might indeed lie in rich environments (Yates *et al.* 1986). These suggestions have been followed up by deep CCD imaging of very powerful radio galaxies at $0.4 \lesssim z \lesssim 0.7$ (Yates *et al.* 1988). The interpretation of the results is not entirely straightforward, as the visibility of high-z clusters depends strongly on one's assumptions about the evolution of the optical cluster luminosity function. Nevertheless, the data appear to be consistent with a single B_{gr}–P relation for both FRII radio galaxies & quasars.

This would make good physical sense: Prestage & Peacock (1988) have argued that the only reason that low-power FRII sources avoid dense environments is that they cannot supply enough ram pressure to overcome the static IGM pressure – a constraint that poses no difficulty for the more luminous galaxies and quasars. In fact, there is some evidence (G. Hill, private communication) that the 'gap' in environmental richness may close up at higher redshift ($z \simeq$ 0.5). This could plausibly be attributed to a small amount of evolution of the intracluster medium. In any case, there seems no doubt that radio galaxies well above the FR borderline are in environments at least as dense as an Abell $R = 0$ system at $z \simeq 0.5$.

Finally, note that any changes of B_{gr} with redshift may have an interesting implication for theories of biased galaxy formation. The calculation of B_{gr} assumes virialized clusters, so that $\xi(r) \propto (1 + z)^{-3}$ at fixed *proper* r. If galaxy clustering were in fact mainly the result of initial bias (and therefore approximately fixed in *comoving* terms) we would expect $(1 + z)^{-1.8}$. It is in principle possible to infer an increase in B_{gr} at high z, simply because we may have built in incorrect assumptions about the evolution of clustering.

4. Clustering of radio galaxies

It is interesting to consider the relation of the above environmental results to data on the large-scale clustering of quasars and radio galaxies. The reason this is important is that we know that there is a trend between the richness of a galaxy system and the amplitude of its spatial cross-correlation function (Bahcall & Soneira 1983), so clustering and environmental studies may well have a common interpretation.

Work is currently in progress on an all-sky sample of galaxies which approximate the selection criteria $S_{1.4GHz} > 0.5$ Jy; $0.01 < z < 0.1$; $|b| > 15°$. Figure 3 shows the correlation function for the luminous FRI galaxies only (i.e. $10^{23.5} < P_{2.7GHz} < 10^{24.5}$ WHz^{-1}sr^{-1}); there is no detectable clustering from other power ranges.

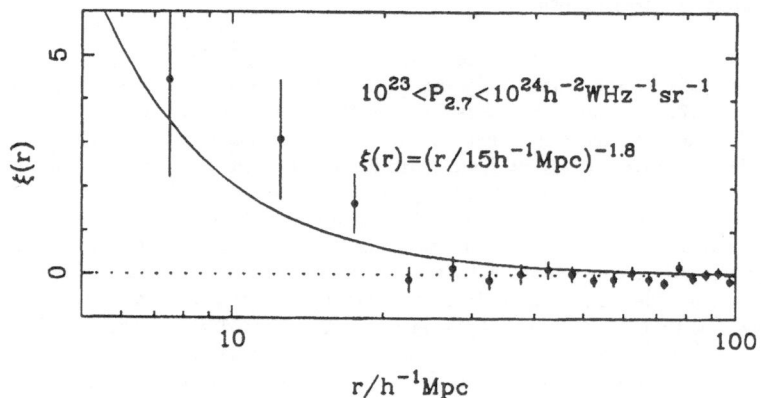

Figure 3 *The two-point correlation function estimated from 120 galaxies with spectroscopic redshifts and radio powers in the range $10^{23} \lesssim P_{2.7} \lesssim 10^{24}$ $h^{-2} WHz^{-1}sr^{-1}$. No significant clustering is seen in other power ranges.*

This result is especially exciting in view of our knowledge both of the environments of radio galaxies and of the spatial correlations of rich clusters. From Prestage & Peacock (1988), we know that the above radio power range should pick out the objects of richest environments at z $<$ 0.1 – having average Abell richnesses of about R = 0. The work of Bahcall & Soneira (1983) would suggest a scale-length of $s_0 \simeq 25$ h^{-1} Mpc for R $\geq$ 1 clusters, with a correlation of s_0 and richness. Sutherland (1988) conversely finds $s_0 \simeq 14$ h^{-1} Mpc with no richness correlation above R = 0. In either case, our results are consistent with the optical determinations for these intermediate-richness systems, but without the problematic selection effects caused by cluster selection. These results should be regarded as rather preliminary, as they are based on only $\sim$ 120 galaxies with redshifts out of a total sample of about 400. Spectra of a further $\sim$ 100 galaxies are being analyzed; this will reduce the error bars in Fig. 3 by a factor $\sim$ 2.

5. Gravitational collapse and triggering of radio galaxies

It is an interesting challenge to see how these results might be interpreted in the 'standard' Cold Dark Matter model. Efstathiou & Rees (1988) have pointed the way by calculating the distribution of the masses of object which may have collapsed by a given redshift in the CDM clustering hierarchy. At any epoch, there is a characteristic largest mass-scale which has just collapsed; this corresponds today to the richest Abell clusters. The obvious temptation, in dealing with an apparently special epoch at z $\simeq$ 2, is to ask what was special about the critical mass at that time.

The answer is model-dependent in that it depends on the normalization of the power spectrum, via the bias parameter b. We are looking for the radius of spheres which have $\delta\rho/\rho \simeq 1.7$ in order to collapse at that redshift. According to Narayan & White (1988), this is $\simeq$ 0.15 h^{-1}Mpc $(b = 2.5)/0.5$ h^{-1}Mpc $(b = 1.5)$. The mass in such a sphere is relatively small: $\lesssim 10^{11.5}$ $h^{-1}M_\odot$. For the lower value of b, this number is significant in two ways: it is close both to the mass inferred for radio galaxies, and to the maximum galaxy mass allowed by dissipation in a CDM universe (Blumenthal et al. 1984). This is suggestive of a picture in which the central engine in active galaxies is triggered when such galaxies are dynamically young, and declines rapidly thereafter. The main alternative to this would involve triggering by mergers at a time significantly after the formation of the main galaxy. To see the sort of difficulty this would pose in CDM, consider the space densities involved. For $\gtrsim L^*$ galaxies, the comoving density is $\simeq 0.3\phi^*$, for the lowest mass galaxies expected in CDM ($\sim 10^6 M_\odot$) it is $\simeq 50\phi^*$. The radii containing even one such object are respectively 3.6 h^{-1}/1.0 h^{-1} Mpc, which will not have collapsed by z = 2.

We can make this argument more empirical. The above figures for Abell R = 0 clusters with $\xi \propto r^{-1.8}$ suggests an integrated overdensity within 0.5 h^{-1} Mpc of $\sim$ 300. Now, the spherical collapse model suggests that objects which collapse at redshift z_c virialize at a present overdensity of $176(1+z_c)^3$ (Narayan & White 1988). In this case, even if mass were to trace light, such clusters could not have existed much before z = 0.5. If one inserts a bias in overdensity of a factor $\sim$ 10 necessary to reconcile observed M/L ratios, the limit becomes considerably more severe.

It therefore seems unlikely that activity peaking at z $\simeq$ 2 can be accounted for in terms of collapse of older galaxies into a group or cluster. A further argument which supports the idea that radio galaxies at z $\simeq$ 2 may be dynamically young is their angular extent. Taking the above radii would suggest observed extents of 30–90 h^{-1} kpc, in good agreement with many of the 3C galaxies (McCarthy et al. 1988). Finally, the *abundance* of objects is roughly explicable in this picture. The total density of luminous radio sources at z = 2 is $\sim 10^{-6}h^3$ Mpc^{-3}, which compares with a value for ϕ^* of about $10^{-2.3}h^3$ for elliptical galaxies (King & Ellis 1985). The optical luminosity for powerful radio galaxies is of course rather high: a median total B luminosity of $\sim 10^{10.8} \, h^{-2} L_\odot$ – *i.e.* about $7L^*$ (*e.g.* Lilly & Prestage 1988). The integrated density of such objects is therefore about $10^{-6.4}$ – consistent with the idea that the most luminous galaxies produce powerful radio sources with near-unit probability. This fact remains unexplained, but empirically it is quite satisfying that the massive galaxies which power radio sources are expected to form at about the redshift where the radio-source population displays peak activity.

6. Acknowledgements

I am grateful to my collaborators in the projects described here: Chris Collins, James Dunlop, Simon Lilly, Lance Miller, David Nicholson & Mark Yates.

References

Bahcall, N.A. & Soneira, R.M. 1983. *Astrophys. J.*, **270**, 20.

Blumenthal, G.R., Faber, S.M., Primak, J.R. & Rees, M.J., 1984. *Nature*, **311**, 517.

Downes, A.J.B., Peacock, J.A., Savage, A. & Carrie, D.R. 1986. *Mon. Not. R. astr. Soc.*, **218**, 31.

Dunlop, J.S. 1987. PhD thesis, Univ. of Edinburgh.

Dunlop, J.S., Downes, A.J.B., Peacock, J.A., Savage, A., Lilly, S.J., Watson, F.G. & Longair, M.S. 1986. Nature, **319**, 564.

Efstathiou, G. & Rees, M.J., 1988. *Mon. Not. R. astr. Soc.*, **230**, 5P.

Green, R.F. & Yee, H.K.C., 1984. *Astrophys. J.*, **54**, 495.

King, C.R. & Ellis, R.S., 1985. *Astrophys. J.*, **288**, 456.

Lilly, S.J. 1988. *Astrophys. J.*, in press.

Lilly, S.J. & Prestage, R.M., 1988. *Mon. Not. R. astr. Soc.*, **225**, 531.

Longair, M.S. & Seldner, M. 1979. *Mon. Not. R. astr. Soc.*, **189**, 433.

McCarthy, P.J., van Breugel, W., Spinrad, H. & Djorgovski, S., 1987. *Astrophys. J.*, **321**, L29.

Narayan, R. & White, S.D.M., 1988. *Mon. Not. R. astr. Soc.*, **231**, 97P.

Peacock, J.A. 1985. *Mon. Not. R. astr. Soc.*, **217**, 601.

Prestage, R.M. & Peacock, J.A. 1988. *Mon. Not. R. astr. Soc.*, **230**, 131.

Sutherland, W. 1988. *Mon. Not. R. astr. Soc.*, in press.

Yates, M.G., Miller, L. & Peacock, J.A. 1986. *Mon. Not. R. astr. Soc.*, **221**, 311.

Yates, M.G., Miller, L. & Peacock, J.A. 1988. *Mon. Not. R. astr. Soc.*, in press.

Yee, H.K.C. & Green, R.F. 1987. *Astrophys. J.*, **319**, 28.

Warren, S.J., this volume.

Windhorst, R.A. 1984. PhD thesis, Univ. of Leiden.

SIMULATIONS OF THE VISUAL APPEARANCE OF GALAXY CLUSTERS AT HIGH REDSHIFT

H. K. C. Yee
Département de Physique, Université de Montréal
C. P. 6128, Succursale A
Montréal, PQ, H3C 3J7, Canada

1. Introduction

Observing high redshift galaxy clusters is the most efficient method for studying galaxies at large redshift. Systematic searches for high z clusters include those of Gunn, Oke and Hoessel (1986), Ellis *et al.* (1987), and Yee and Green (1987) who use quasars as possible markers for distant clusters. However, almost all clusters discovered so far have $z < 1$. Can clusters at very high redshift be identified visually? In deep images required to detect clusters at high redshifts, the background counts dominate, and the contrast in galaxy density decreases drastically.

To address the question of the visibility of clusters at high z, I have created an algorithm, as part of the image processing system PPP (Yee, 1989), to simulate CCD observations of galaxy clusters situated in realistic background fields.

2. Simulation Algorithm

The simulation consists of two parts: the background field and the cluster field. Observing conditions are input through parameters for the CCD detector, filter, telescope, seeing, integration time, and sky brightness.

Background field: A randomly generated background field is created according to observed number-magnitude count relations for galaxies and stars (e.g. Yee, Green, and Stockman 1986). Stars have a point-spread function (PSF) parametrized using the fit by Moffat (1969). For all galaxies, a redshift is assigned according to random draws from $z - m$ distributions computed using the luminosity function (LF) models in Yee and Green (1987). Using properties of local galaxies from Boroson (1981), Dressler (1980) and Strom and Strom (1978), parameters such as morphological type, bulge/disk ratio, and scale length are generated. The galaxies are then inclined and rotated randomly in space and convolved with the PSF profile. Galaxies and stars are generated down to 1 mag fainter than the expected detection limit (5σ). The file of object parameters can be saved, so that frames with different observing conditions for the same patch of the 'sky' can be made.

Cluster field: Galaxy cluster members are created using an empirical King-law radial distribution with the absolute magnitudes drawn from galaxy LFs of different morphological types (e. g. King and Ellis, 1985). The distribution of morphological types as a function of galaxy density follows that of Dressler (1981). A redshift and the amount of luminosity evolution in M^* for the cluster can then-be chosen by the observer. The parameters of the observed member galaxies

C. S. Frenk et al. (eds.), The Epoch of Galaxy Formation, 397–398.

are generated according to these assigned values. The member galaxies are then added onto the background field.

3. Examples of Simulations and Discussion

A cluster of moderate richness (Abell class 1) with a relatively compact core radius of 300 kpc was created and 'observed' at different redshifts and evolutions. The computations simulate observing at the CFHT 3.6 m using a CCD of 300×490 pixels with a scale of 0."4/pixel with a seeing of FWHM of 1."1. Simulated observations with integration times of 1200 sec to 3 hours for clusters at $z = 0.55$ to 1.25 were made. Also, simple luminosity evolutions in M^* from 0 to 2.0 mag were applied. Figure 1 shows the resultant simulation of a cluster at $z = 1.0$ with a large evolution of 2.0 mag in M^*.

From the visual impression of the simulations, one can conclude that moderately rich clusters with a small amount of luminosity evolution (< 0.8 mag) can be readily identified up to $z \approx 0.8$. At $z > 1.$, even with an evolution of 1.5 mag, the contrast in galaxy density is difficult to pick up visually; although, statistically, a significant enhancement in galaxy density may be present. Increasing the depth of observation alone will not alleviate the problem of high background density. This is because the increase in background counts is steeper than the slope of the luminosity function of the cluster past M^*. However, if there has been significant evolution of M^* (> 1.5 mag), clusters at $z > 1$ can be much more readily detectable.

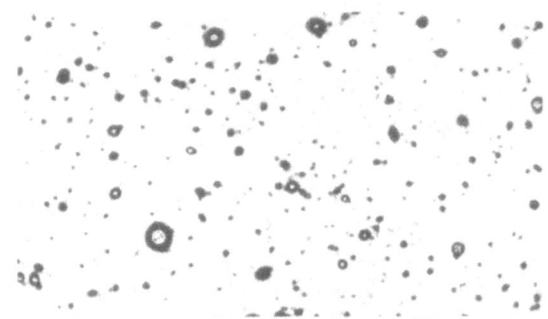

Figure 1. Simulation of a 3 hour CCD observation, using a 3.6 m telescope, of a cluster of galaxies at $z = 1.0$ with an evolution of 2 mag in M^*. The cluster is centered at the center of the frame.

References

Boroson, T. 1981, *Ap. J. Suppl.* **46**, 177.

Dressler, A. 1980, *Ap. J.* **236**, 351.

Ellis, R. S., Couch, W. J., MacLaren, I., and Koo, D. 1988, in preparation.

Gunn, J. E., Oke, J. B., and Hoessel, J. G. 1986, *Ap. J.* **306**, 80.

King, C .R. and Ellis, R. S. 1985, *Ap. J.* **288**, 456.

Moffat, A. F. J. 1969, *A. A.* **3**, 455.

Strom, S. E., and Strom, K. M. 1978, *A. J.* **83**, 732.

Yee, H. K. C. 1989, preprint to be submitted to *P. A. S. P.*

Yee, H. K. C. and Green, R. F. 1987, *Ap. J.* **319**, 28.

Yee, H. K. C., Green, R. F., and Stockman, H. S. 1987, *Ap. J. Suppl.* **62**, 681.

CENTRAL GALAXY FORMATION BY COOLING FLOWS

R.M. Johnstone and A.C. Fabian
Institute of Astronomy, Madingley Road, Cambridge CB3 0HA, U.K.

ABSTRACT. A brief review of star formation in cluster cooling flows is given. Preliminary results from a new analysis of the surface brightness of central galaxies in cluster cooling flows, in which the profile of the mass deposition is taken into account, can allow more normal star formation than had previously been thought.

1. Introduction

It is now well known that in a large fraction of galaxy clusters the hot X-ray emitting intracluster gas is the site of mass deposition in the form of cooling flows (see e.g. Fabian, Nulsen & Canizares 1984). Measured mass deposition rates span the range 1-500 $M_\odot$ yr^{-1} for different objects, with values of $\sim 100\,M_\odot$ yr^{-1} being common. Whatever forms from the cooled gas is deposited over a region with a radius of ~ 200 kpc, around the central cluster galaxy.

The consistency of mass deposition rates measured by different X-ray diagnostics characteristic of different temperature regions in the cooling phase provides strong evidence that the flows are in a steady state and hence are of a comparable age to the cluster. If a typical age for a flow is 10^{10} years then around $10^{12}\,M_\odot$ of material will be deposited over the lifetime of the flow. This means that the central galaxies are still forming from the intracluster gas.

Over the past ten years or so many attempts have been made to search for this cooled matter. Obvious candidates for the residue are massive regions of ionized, neutral or molecular gas. What form the deposited mass ends up as will depend on how far the X-ray emitting gas is able to cool. Searches have indeed revealed the presence of all three forms of hydrogen, but the implied mass of gas is several orders of magnitude lower than predicted above. The failure to find cool gas in cooling flows has fuelled the idea that the gas forms stars.

2. Colours of Central Galaxies

The luminosity and colour of the cooling flow galaxies can be a sensitive diagnostic of the stellar population present. Fabian *et al.* (1982) showed that the entire mass deposition rate cannot form stars with a disk-galaxy-like mass function otherwise extremely blue, over-luminous galaxies would result. Most cooling flow galaxies have colours typical of giant elliptical galaxies although some cooling flow galaxies do have some excess blue light in their centres.

Sarazin & O'Connell (1983) made the first in-depth study of the effect of an accretion population on the colour of cooling flow galaxies, and this work was followed up by Romanishin (1986). Until now, however, the mass deposition profile within the cooling flow has not been taken into account. In general, all of the cooled gas has been assumed to be deposited within

C. S. Frenk et al. (eds.), The Epoch of Galaxy Formation, 399–400.

a relatively small region near the centre of the cooling flow galaxy. The X-ray data (Thomas *et al.* 1987) show that the profile of mass deposition varies like $\dot{M}(r) \propto r$ within the cooling radius. If the mass function does not change with radius the resulting surface brightness profile follows an r^{-1} form within the cooling radius.

Schombert (1986) has presented surface brightness profiles for a large number of brightest cluster member galaxies, many of which are cooling flow galaxies. These are represented well by the de Vaucouleurs $r^{\frac{1}{4}}$ law only over a relatively narrow range in surface brightness. In general a simple power-law fit to the profiles yields an index $\sim 1.5 - 2$. The flatness of the cooling flow profile means that the accretion population cannot dominate the V-band light except close to the cooling radius.

We have just started modelling the spectra of an accretion population using the techniques of Bruzual (1981) and the library of stellar spectra from Pickles (1985). This allows us to normalize the cooling flow profile to give the correct V-band luminosity. Comparison with observed surface brightness profiles shows that previous work by others has set too tight a constraint on the amount of normal, disk-galaxy like star formation that can take place. Preliminary results for a trial model of the Perseus cluster are shown in Figure 1. The filled circles represent the observed surface brightness profile from Schombert (1986) whereas the solid line is the computed profile for a 10^{10} yr old cooling flow forming 300 $M_\odot$ yr^{-1} of stars with a power-law mass function (of number index 3) ranging between 0.13 and 1.0 $M_\odot$. The V magnitude of the accretion population is 11.7. It can be seen that in this model, near the centre of the galaxy the accretion population contributes only 10 per cent of the V light. Near the cooling radius (~ 200 kpc) the accretion population dominates the light. If reliable colour maps can be made far out in cooling flow galaxies this is the place to look for the effects of an accretion population. Indeed, the sizes of cooling flow galaxies are generally of the same order as the cooling radius.

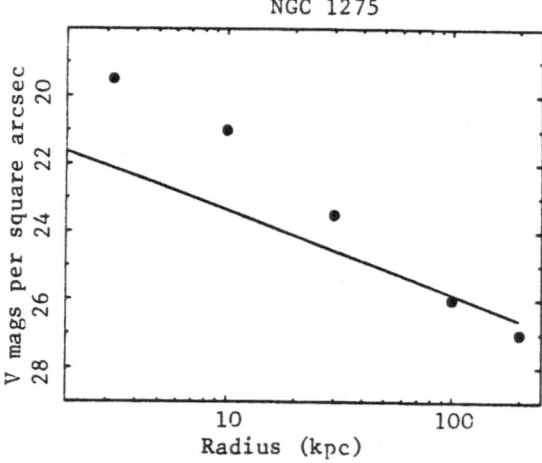

Figure 1. V-band surface brightness profile of NGC 1275 (filled circles) compared with the predicted profile (solid line) from a model cooling flow in which stars between 0.13 and 1.0 $M_\odot$ form at 300 $M_\odot$ yr^{-1} within 150 kpc.

References

Bruzual, G.A., 1981. *PhD Thesis*, University of California, Berkeley, CA.
Fabian, A.C., Nulsen, P.E.J. & Canizares, C.R., 1982. *Mon. Not. R. astr. Soc*, **201**, 933.
Fabian, A.C., Nulsen, P.E.J. & Canizares, C.R., 1984. *Nature*, **310**, 733.
Pickles, A.J., 1985. *Astrophys. J. Suppl. Ser.*, **59**, 33.
Romanishin, W., 1986. *Astrophys. J.*, **301**, 675.
Sarazin, C.L. & O'Connell, R.W., 1983. *Astrophys. J.*, **268**, 552.
Shombert, J.M., 1986. *Astrophys. J. Suppl. Ser.*, **60**, 603.
Thomas, P.A., Fabian, A.C. & Nulsen, P.E.J., 1987. *Mon. Not. R. astr. Soc.*, **228**, 973.

Constraints on the amplitude of primordial density fluctuations

Roberto Scaramella[*]   &   Nicola Vittorio[†]

[*] *International School for Advanced Studies (S.I.S.S.A.), Trieste, Italy.*
[†] *Dipartimento di Fisica, Università dell'Aquila, L'Aquila, Italy.*

We report work in progress for properly constraining the amplitude of the initial density fluctuations. We use Monte-Carlo simulations for obtaining the distribution functions for various observables (such as, e.g., the Cosmic Microwave Background [CMB] temperature auto-correlation function [a.c.f]) that would be recorded by an ensemble of different observers. We then discuss how to use *local* measurements of CMB temperature distribution in limiting the amplitude of primordial density fluctuations.

The statistics of the CMB temperature field is completely descibed by the two point angular a.c.f. $C(\alpha) \equiv \langle \Delta T/T(\hat{n}_1) \cdot \Delta T/T(\hat{n}_2) \rangle_{\Omega}$, where $\alpha = \cos^{-1}(\hat{n}_1 \cdot \hat{n}_2)$, averaged over a *single* sky (e.g. a *given* realization of last scattering surface). Under the assumption of adiabatic, scale-free, gaussian density perturbations of primordial index n and amplitude A, and smoothing by a gaussian beam of dispersion σ, we can generalize the method given in Ref. 1 as follows, and therefore have the general expression for the a.c.f. by writing $C(\alpha, \sigma) = \mathcal{A} \cdot S(\alpha, \sigma)$

$$\mathcal{A} \equiv \frac{2^{n-1} A}{4\pi \, r_0^{(n+3)}} \frac{\Gamma(3-n)}{\left[\Gamma\left(\frac{4-n}{2}\right)\right]^2} \quad ; \quad S(\alpha, \sigma) \equiv \sum_{\ell=2}^{\infty} X_\ell^2 \cdot W_\ell(\alpha, \sigma) \tag{1}$$

where $r_0 \equiv 2c/H_0$, $\mathcal{A}$ is an overall adimensional normalization factor ($\mathcal{A} = A/(\pi^2 \, r_o^4)$ if $n = 1$), $X_\ell^2 \equiv c_\ell / \langle c_\ell \rangle$ is a stochastic variable with a normalized $\chi_{(2\ell+1)}^2$ d.f. (Ref. 2, [AW]), and finally

$$W_\ell(\alpha, \sigma) \equiv (2\ell+1) \frac{\Gamma\left(\frac{2\ell+n-1}{2}\right)}{\Gamma\left(\frac{2\ell+5-n}{2}\right)} P_\ell(\cos \alpha) \exp\left\{ -[(\ell+\tfrac{1}{2})\sigma]^2 \right\} \tag{2}$$

Experiments with large sky coverage give upper limits on the quadrupole anisotropy (we define $Q_2^2 \equiv \sum_{m=-2}^{m=2} |a_2^m|^2$). The Berkeley and Princeton groups (Ref. 3; [BPG]) set $Q_2 < 2.5 \cdot 10^{-4}$ at the 90 % confidence level. The Princeton Group (Ref. 4; [F]) placed also an upper limit on the a.c.f. : $C(\alpha, 2^\circ.9) < 1.4 \cdot 10^{-9}$ for $10^\circ < \alpha < 180^\circ$. More recently, the RELIC satellite borne experiment (Ref. 5; [R]) set $C(20^\circ, 2^\circ.4) < 5.5 \cdot 10^{-10}$ and $Q_2 < 1.1 \cdot 10^{-4}$ at the 95 % confidence level (there is a factor 4π beetwen our and R definition of Q_2^2). Assuming $n = 1$ in the fitting procedure makes the latter limit even more severe: $Q_2 < 5.7 \cdot 10^{-5}$.

C. S. Frenk et al. (eds.), The Epoch of Galaxy Formation, 401–402.
© *1989 by Kluwer Academic Publishers.*

The observational upper limits are of course obtained by observing *our* realization of the last scattering surface. Ensemble averages can then be constrained by knowing the d.f. of different observables (e.g. quadrupole, a.c.f., $\Delta T/T$). AW pioneered and applied this kind of arguments to the quadrupole anisotropy. They found that the upper limit on the expected (i.e. ensemble averaged) quadrupole is less stringent than that obtained comparing the expected value with the experimental upper limit taken at face value. Let us now consider a given observable, y, for wich an observational upper limit y_{lim} is available. Let us also assume that this observable is theoretically distributed according to a probability function $P_{th}(y)$ and with an expected value $\langle y \rangle$. Is then possible to define a parameter γ_ϵ such that $P_{th}(y \leq \gamma_\epsilon \cdot \langle y \rangle) = \epsilon$. For $y = y_{lim}$ we obtain $P_{th}(y_{lim} \leq \gamma_\epsilon \cdot \langle y \rangle) = \epsilon$ or, equivalently, $P_{th}(y_{lim} \geq \gamma_\epsilon \cdot \langle y \rangle) = 1 - \epsilon$. Then, at the $(1 - \epsilon)$ confidence level (hereafter c.l.), one has $y_{lim} \geq \gamma_\epsilon \cdot \langle y \rangle$. This implies at the same c.l. an upper limit on the ensemble average of the observable y : $\langle y \rangle \leq \gamma_\epsilon^{-1} \cdot y_{lim}$. For ϵ small, $0 < \gamma_\epsilon \ll 1$. Because of this the theoretical upper limit on $\langle y \rangle$ obtained in this way is weaker than that obtained by limiting $\langle y \rangle$ directly with y_{lim}. For example, as shown by AW, one finds that the upper limit on the expected squared quadrupole is almost one order of magnitude greater than the observational upper limit at the 99% c.l.

The c.l. considered so far take into account only the theoretical probability distribution. On the other hand, the quoted upper limits y_{lim} have a confidence level by themselves, a fact which should also be taken in account.

Hence a more realistic approach consists in considering the joint probability with c.l. $1 - \epsilon(\langle y \rangle) = erf\left[y_{lim}/(\sqrt{2}\sigma_{err})\right] \cdot P_{th}(\langle y \rangle \leq \gamma_\epsilon^{-1} \cdot y_{lim})$, where a good approximation for experiments with large sky coverage is to assume the d.f. for y_{lim} to be gaussian with dispersion σ_{err}. A preliminary application of the above discussion can be done to results obtained for the $n = 1$ case. Results on the a.c.f. obtained by Monte-Carlo simulating an ensemble of ten thousands different observers are presented in Table I. We note that the limit obtained from F is also valid for open universes, $\Omega_0 \gtrsim 0.2$ (Ref. 1).

Table I

Observable y	Experimental limit y_{lim} quoted	adopted (96%)	$\gamma_{0.01}^{-1}$	Ensemble average upper limit (95%)	Upper limit on fluctuations amplitude $\mathcal{A}_{max}$ (95%)
Q_2 [BPG]	$2.5 \cdot 10^{-4}$ (90%)	$3.1 \cdot 10^{-4}$	$\sqrt{9.01}$	$9.3 \cdot 10^{-4}$	$8.3 \cdot 10^{-8}$
Q_2 [R]	$1.1 \cdot 10^{-4}$ (95%)	$1.1 \cdot 10^{-4}$	$\sqrt{9.01}$	$3.3 \cdot 10^{-4}$	$1.0 \cdot 10^{-8}$
$C(10°, 2°.9)$	$1.4 \cdot 10^{-9}$ (90%)	$1.8 \cdot 10^{-9}$	1.85	$3.3 \cdot 10^{-9}$	$1.4 \cdot 10^{-9}$
$C(20°, 2°.4)$	$5.5 \cdot 10^{-10}$ (95%)	$5.8 \cdot 10^{-10}$	4.0	$2.3 \cdot 10^{-9}$	$2.1 \cdot 10^{-9}$

References

[1] Scaramella, R., and Vittorio, N., 1988, *Ap.J. (Letters)* **331**, L53.

[2] Abbott, L.F., and Wise, M.B., 1984 *Ap.J. (Letters)* **282**, L47.

[3] Lubin, P., and Villela, T., 1986, in *Galaxy Distances and Deviations from Universal Expansion*, Madore, BF., and Tully, R.B., eds., D.Reidel Pub. Co.

[4] Fixen, D.J., Cheng, E.S., and Wilkinson, D.T., 1980, *Phys.Rev.Lett.* **44**, 1563.

[5] Lukash, V.N., and Novikov, I.D., 1987, I.A.U. Symp. **127**, Hewitt *et al.* (eds.), D.Reidel Pub. Co.

SPOTTINESS IN THE STRUCTURE OF THE MICROWAVE BACKGROUND RADIATION

J.L. Sanz & E. Martínez–González
Departamento de Física Moderna
Universidad de Cantabria
39005 Santander, Spain

1. Summary

Most of the observations on the microwave background radiation (MBR) have concentrated on the analysis of temperature dispersion and amplitud of dipole and quadrupole components. However, other statistical features can be stracted from the available maps of this background (Lubin et al. 1985; Klypin et al. 1987). The distribution of pips and spots is closely related to the density fluctuations present at the recombination time, so that such distributios can be used to test the primordial spectrum.

Let us assume a Gausssian random field on the sphere, characterised by a correlation $C(\alpha)$, to represent the temperature fluctuations. Then, the mean number of spots N_s over the whole celestial sphere, above the level $\nu C(0)^{1/2}$ can be stimated in the form (Vanmarcke 1983)

$$N_s = 2\pi^{-1}\lambda\left[\mathrm{erfc}(2^{-1/2}\nu)\right]^{-1}e^{-\nu^2/2}$$

where $\lambda = -C(0)^{-1}C''(0)$. We have represented $N_s(\nu)$ in figure 1 for three types of observations: a) single beam experiment with $\sigma = 5°$ (Klypin et al. 1987), b) beam switching experiment with beamthrow angle $\alpha = 6°$ and beamwidth $\sigma = 2°.2$ (Melchiorri et al. 1981) and c) double beam switching experiment with $\alpha = 8°.2$ and $\sigma = 2°.4$ (Watson et al. 1988). In all cases a scale-invariant primordial spectrum and a Gaussian response for the receiver, with dispersion σ, have been assumed.

C. S. Frenk et al. (eds.), The Epoch of Galaxy Formation, 403–404.

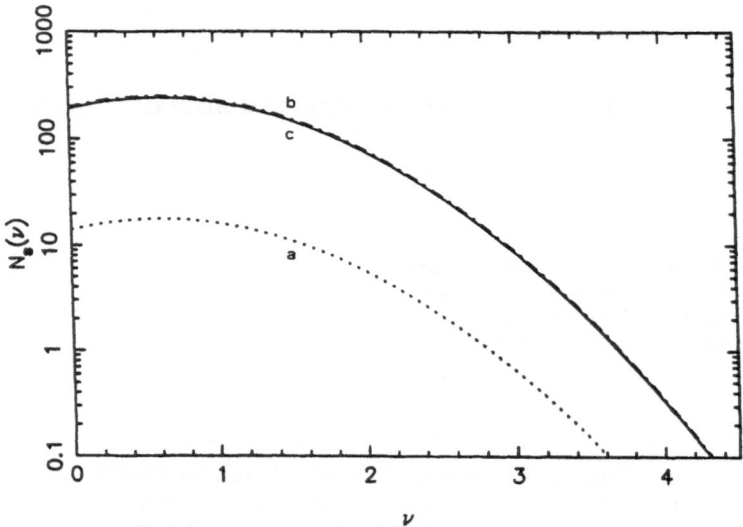

Figure 1 The number of spots in the microwave background radiation expected from a scale-invariant spectrum of primordial fluctuations.

On the other hand, the probability to have the hottest pip above the level ν is given by the double exponential distribution (Leadbetter *et al.* 1983)

$$P = 1 - \exp\{-\exp[-a(\nu - a)]\}$$

where $a = [2\log(\lambda^{1/2}\cos\delta)]^{1/2}$ and δ is the declination. Regarding the Watson *et al.* (1988) experiment, the probabilities to find the hottest pip above the levels $\nu = 2, 2.5$ and 3 are 38% , 20% and 10% , respectively. Thus, if the hotpip centred on the Bootes void is a real feature and the level of brightness is above 2 rms, then we must appeal either to non-intrinsic fluctuations or to non-Gaussian random fields to represent the density fluctuations at the recombination time. The same conclusion can be obtained by analysing the mean pip lenght (Martínez–González and Sanz 1988).

References

Klypin, A.A., Sazhin, M.V., Strukov, I.A. & Skulachev, D.P. 1987, *Sov. astron. Lett.*, **13**, 259
Leadbetter, M.R., Lindgren, G. & Rootzen, H. 1983, *Extremes and Related Properties of Random Secuencies and Processes*, Springer-Verlag, New York
Lubin, P., Villela,T., Epstein, G. & Smoot, G. 1985, *Ap.J.*, **298**, L1
Martínez–González, E. & Sanz, J.L., 1988, *M.N.R.A.S.* (Submitted)
Melchiorri, F., Melchiorri, B. O., Ceccarelli, C. & Pietranera, L. 1981, *Ap.J.*, **250**, L1
Vanmarke, E.H. 1983, *Random Fields: Analysis and Synthesis*, MIT Press, Cambridge, Massachusetts
Watson, R.A., Rebolo, R., Beckman, J.E., Davies, R.D. & Lasenby, A.N. 1988, In: *The Large Scale Structure and Motions in the Universe*, eds. Giuricin, G., Mardirossian, F., Mezzetti, M. & Ramella, M., Kluwer, Dordrecht, Holland

ENERGETIC CONSTRAINTS ON SPECTRAL DISTORTIONS OF THE MICROWAVE BACKGROUND

Cedric G. Lacey & George B. Field
Harvard-Smithsonian Center for Astrophysics
Cambridge, MA, U.S.A.

1. Introduction

Matsumoto *et al.* (1988) have claimed detection of a distortion from blackbody form in the spectrum of the microwave background: relative to a $T = 2.74\,\mathrm{K}$ black-body, there is an excess in the background which peaks at $\lambda \approx 600\,\mu\mathrm{m}$ and contains roughly 10% of the total energy density. Matsumoto *et al.* suggest that this excess may be produced by thermal emission from dust heated by stars at large redshift, or by Compton scattering of microwave background photons by a hot intergalactic medium. We investigate the energetic constraints on these mechanisms, assuming that the required energy is generated by a population of stars formed with the same Initial Mass Function (IMF) as found in the solar neighborhood, and assuming bounds on the cosmological baryon density derived from the standard model of primordial nucleosynthesis. A recent determination of the latter by Kawano *et al.* (1988) gives $\Omega_b h^2 < 0.022$, where $h = H_0/100\,\mathrm{km\,s^{-1}\,Mpc^{-1}}$, and we assume $0.4 < h < 1$. We use Scalo's (1986) determination of the solar neighborhood IMF, which for $m \gtrsim m_1 = 2\,\mathrm{M_\odot}$ can be approximated by a power law, $\phi(m) = (x-1)(\xi_1/m_1^2)(m/m_1)^{-(x+1)}$, with $x \approx 1.7$ and $\xi_1 \approx 0.4$, and an upper mass limit $m_U = 100\,\mathrm{M_\odot}$.

2. Emission by Dust Grains

In the case of thermal emission from dust grains heated by starlight, we assume that the dust has the same properties and same abundance relative to the gas as in the local interstellar medium. The dust absorption cross-section (per hydrogen atom) for $\lambda \gtrsim \lambda_0 = 10^{-3}\,\mathrm{cm}$ is represented by a power law, $\sigma_\lambda = \sigma_0(\lambda_0/\lambda)^\alpha$, where we take either $\alpha = 2$, $\sigma_0 = 7 \times 10^{-23}\,\mathrm{cm^2}$ (Draine & Lee 1984), or $\alpha = 1$, $\sigma_0 = 2 \times 10^{-23}\,\mathrm{cm^2}$ (Rowan-Robinson 1986). A dust grain at temperature T_d then radiates power $P_d \propto T_d^{4+\alpha}$. The spectrum of the excess radiation can be fit by a single temperature dust emission model, with observed temperature $T_{d0} = 4.4\,\mathrm{K}$ for $\alpha = 1$ and $T_{d0} = 3.7\,\mathrm{K}$ for $\alpha = 2$ (Matsumoto *et al.*). To reproduce this, we assume that the dust is heated by starlight to a temperature $T_d(z) = T_{d0}(1+z)$ for $0 < z < z_h$. In practice, this means that most of the energy is emitted near redshift z_h. For a given dust abundance, we can then calculate the required z_h and total stellar energy input (we assume that 100% of the starlight is absorbed by dust). Assuming that all the baryons are either in stars or in gas containing dust, we find that the efficiency ϵ_s with which stellar mass needs to be converted into energy is minimized when they are in the ratio $\Omega_g/\Omega_s = 1/(\alpha + 3/2)$. Assuming an

C. S. Frenk et al. (eds.), The Epoch of Galaxy Formation, 405–406.

$\Omega = 1$ cosmology, and substituting the limit on Ω_b, we obtain limits on the conversion efficiency $\epsilon_s \gtrsim (1.5 \times 10^{-3} h^{2/5}, 2.7 \times 10^{-3} h^{2/7})$, and on the redshift of energy injection $(1 + z_h) \gtrsim (11 h^{2/5}, 20 h^{2/7})$, for $\alpha = (1, 2)$ respectively. We compare the requirement on ϵ_s with the efficiency available from nuclear burning in a population of stars having our standard IMF, based on stellar evolution models, and taking account of finite stellar lifetimes. The most favourable case for reconciling theory with observations is to have small H_0 ($h = 0.4$). Even then, nuclear burning in stars provides insufficient energy by at least a factor 2 or 5, for $\alpha = 1$ or $\alpha = 2$ respectively.

3. Compton Scattering by Hot Intergalactic Medium

We suppose that the IGM is heated to temperature T_h at redshift z_h, and subsequently cools by Compton scattering and by adiabatic expansion. For non-relativistic scattering, the distortion in the microwave background spectrum is given by the parameter $y = \int (kT_e/m_e c^2) n_e \sigma_T c \, dt$, which depends only on the thermal energy density of the plasma as a function of z. The observed distortion is fit with $y = 0.028$. The energy input required to reproduce this is minimized if the energy is injected at $(1 + z_h) \approx 17 h^{2/5}$. The required efficiency for converting baryonic mass into energy is then $\epsilon_C \gtrsim 3.0 \times 10^{-3} h^{2/5}$, substituting the limit on Ω_b. We compare this with the energy obtainable from supernova explosions in a stellar population (since the IGM could not be heated by photo-ionization to the high temperatures required). Assuming that each star with $m \geq 10 M_\odot$ explodes with energy $E_{SN} = 10^{51}$ erg and integrating over our standard IMF gives an efficiency $\epsilon_{SN} \approx 2.9 \times 10^{-6}$, which is too small by at least a factor of 700.

4. Conclusions

Explanations for the spectral distortion which require the energy to be generated by a population of stars formed with a normal IMF thus appear to fail, assuming that the dust (where relevant) has the same properties as found locally, and that the standard theory of primordial nucleosynthesis is correct. Among the possibilities for reconciling theory with observations are: (1) Non-standard models of primordial nucleosynthesis may allow larger Ω_b. (2) The IMF may contain a larger fraction of massive stars (but one must be careful not to overproduce heavy elements). (3) The first generation of stars might be Very Massive Objects, which generate energy more efficiently than ordinary stars and then collapse to black holes. (4) The dust might have different properties. (5) The IGM might be heated by active galactic nuclei. (6) Some more exotic mechanism such as superconducting cosmic strings or decaying particles may be involved.

A full account of this work has been published in *Astrophys. J.* **330**, L1 (1988).

References

Draine, B.T., & Lee, H.M. 1984. *Astrophys. J.*, **285**, 89.

Kawano, L., Schramm, D., & Steigman, G. 1988. *Astrophys. J.*, **327**, 750.

Matsumoto, T., Hayakawa, S., Matsuo, H., Murakami, H., Sato, S., Lange, A.E., & Richards, P.L. 1988. *Astrophys. J.*, **329**, 567.

Rowan-Robinson, M. 1986. *Mon. Not. Roy. Astron. Soc.*, **219**, 737.

Scalo, J.M. 1986. *Fund. Cos. Phys.*, **11**, 1.

CBR POLARIZATION BY COSMIC DUST AT HIGH REDSHIFTS

Bronisław Rudak[1], Mirosław Panek[2]
Copernicus Astronomical Center
[1] 87-100 Toruń, Chopina 12/18
[2] 00-716 Warszawa, Bartycka 18

Observations of Matsumoto et al.(1988) revealed an excess over the $2.75 K$ black-body spectrum at submillimeter wavelengths. This excessive radiation can be explained as coming from hypothetical cosmic dust present at high redshifts ($z_d = 10 - 30$) (Hayakawa et al. 1987, Hogan and Bond 1987). Another observational effect of this dust can be the CBR polarization in the submillimeter region. This is possible only if the dust grains are nonspherical and aligned over sufficiently large scales. The most efficient alignment mechanism involves magnetic fields. Therefore, the detection of the CBR polarization could support both, the cosmic dust hypothesis and the existence of the primordial magnetic fields. First we have to answer the question whether the intensity of the magnetic field, $B \simeq 10^{-9} (1 + z)^2 \, Gs$, is sufficient to obtain significant alignment.

We consider paramagnetic spheroidal grains with material temperature $T_s = 3.55 (1 + z) K$ embedded in the gaseous component of the density $n = 2.5 \cdot 10^{-7} (1 + z)^3 \, cm^{-3}$ and temperature T_{gas} which is left as a free parameter. Grains have typical size $a \simeq 10^{-5} cm$ and density $\varrho_g \simeq 2 \, g \, cm^{-3}$, and rotate with thermal or suprathermal velocities. The alignment of maximum inertia axis $\mathbf{J}$ with $\mathbf{B}$ is assumed to proceed via the Davis - Greenstein process.

The alignment parameter $Q_A = \frac{3}{2} \langle \cos^2 (\mathbf{J}, \mathbf{B}) \rangle - \frac{1}{2}$ is

$$
Q_A = \begin{cases}
Min \left(4 \frac{B_9^2}{T_{3.55}} T_{gas}^{-1/2} \, \eta, \, 1 \right), & \text{thermal rotation} \\
Min \left(0.04 \frac{B_9^2}{T_{3.55}} (a/10^{-5})^{-2} (\varrho_g/2)^{-1} \, \eta, \, 1 \right), & \text{suprathermal}
\end{cases}
$$

where $\eta = 1$ for ordinary paramagnetics, and $\eta = 10 - 10^6$ for superparamagnetics. One can see that Q_A can reach relatively high values, even if $\eta = 1$.

Let us assume that the magnetic field has a domain structure. The value of $| \mathbf{B}|$ is uniform over the whole space and $\mathbf{B}$ is uniform within a domain, but orientations of $\mathbf{B}$ in different domains are random. The mean value of polarization is a statistical sum of effects from all domains along the line of sight (also some fluctuations of the CBR total intensity are expected). The characteristic angular scale θ to observe this polarization is given by the typical comoving size of a domain l_c: $\theta = l_c / (1.3 h^{-1} Mpc) \, arc\,min$. If the dust epoch lasted for Δz at z_d

C. S. Frenk et al. (eds.), The Epoch of Galaxy Formation, 407–408.
© *1989 by Kluwer Academic Publishers.*

then the CBR now observed crossed N_D dusty domains:

$$N_D \approx \frac{70}{\theta(arc\,min)} \left(\frac{\Delta z}{z_d}/0.1\right) \left(\frac{10}{z_d}\right)^{1/2}.$$

We have found the polarization fraction $P = (I_{max} - I_{min})/(I_{max} + I_{min})$ assuming the dust optical depth $\tau = 0.2 \left(\frac{700\,\mu m}{\lambda}\right)^{\alpha}$, $\alpha = 1$ or 2, and spheroidal grain axes ratio $b/a = 2$ (oblate) or $\frac{1}{2}$ (prolate). Results for $N_D = 10$ are shown on Figure 1.

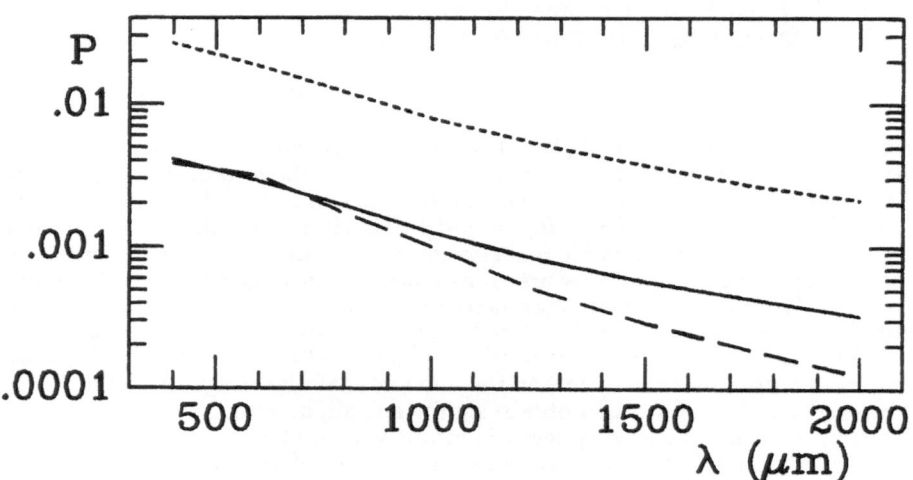

Figure 1. The polarization fraction P vs. the wavelength λ for oblate grains:
$Q_A = 0.1$, $\alpha = 1$ – continuous line,
$Q_A = 0.1$, $\alpha = 2$ – dashed line,
$Q_A = 0.7$, $\alpha = 1$ – dotted line.

To extrapolate these results for another values of N_D or Q_A we can use the scaling $P \propto Q_A/\sqrt{N_D}$.

We can see that P grows with frequency and therefore short wavelengths are most interesting for observations. Unfortunately, for $\lambda < 400\,\mu m$ the CBR spectrum is dominated by interstellar dust emission component.

If we expect the primordial magnetic fields to exist on scales of galaxies then the few arcminute angular scale is preferred for observations.

References

Matsumoto, T., et al., 1988. *Astrophys. J.*, **329**, 567.
Hayakawa, S., et al., 1987. *Publ. Astr. Soc. Japan*, **39**, 941.
Hogan, C.J., Bond, J.R., 1987. *NATO ASI "The Post-Recombination Universe"*, in press.

GROWTH OF PERTURBATIONS IN A COLLAPSING PROTOGALAXY

Cedric G. Lacey
Harvard-Smithsonian Center for Astrophysics
Cambridge, MA, U.S.A.

ABSTRACT. A method is presented for analyzing the evolution of short-wavelength perturbations inside a gas cloud which is undergoing inhomogeneous and/or anisotropic collapse. The method is applied to background collapses having spherical or planar symmetry.

1. General Method

Consider an inhomogeneous fluid flow with density $\rho(\mathbf{x}, t)$ and velocity $\mathbf{u}(\mathbf{x}, t)$, acted on by pressure $p(\rho)$ and by gravity. The trajectory of a fluid particle in the background unperturbed flow can be expressed as $\mathbf{x} = \mathbf{x}(\mathbf{X}, t)$, where $\mathbf{X}$ is a comoving coordinate. We expand this around the trajectory $\mathbf{x} = \mathbf{x}_0(t)$ of a particular fluid particle, giving

$$(\mathbf{x} - \mathbf{x}_0)_i = A_{ij}(t)(\mathbf{X} - \mathbf{X}_0)_j + O(\mathbf{X} - \mathbf{X}_0)^2$$

For studying *local* behavior of short wavelength perturbations, it is sufficient to retain only the first term in this expansion. In the same approximation, the local velocity field is linear and the local density uniform for the background flow. An arbitrary *linear* perturbation can then be decomposed into locally plane waves having spatial dependence $\exp[i\mathbf{q} \cdot (\mathbf{X} - \mathbf{X}_0)] = \exp[i\mathbf{k} \cdot (\mathbf{x} - \mathbf{x}_0)]$, where the *comoving* wavevector $\mathbf{q}$ has components fixed in time, while the *proper* wavevector $\mathbf{k}$ evolves in time as $k_i(t) = A_{ji}^{-1}(t)q_j$.

In the case that the background flow has zero vorticity *i.e.* a *pure shear flow*, we can derive a second-order equation for the evolution of the relative overdensity $\epsilon = \delta\rho/\rho$ in a plane wave. A shear flow is particularly simple if the principal axes of the local shear do not rotate with time. Then we can write $A_{ij}(t) = \mathrm{diag}(a_1(t), a_2(t), a_3(t))$ and $\mathbf{k} = (q_1/a_1, q_2/a_2, q_3/a_3)$, and the evolution equation is

$$\frac{d^2\epsilon}{dt^2} + 2\left[\frac{\sum_i (q_i/a_i)^2(\dot{a}_i/a_i)}{\sum_i (q_i/a_i)^2}\right]\frac{d\epsilon}{dt} + \left[c_s^2\sum_i (q_i/a_i)^2 - 4\pi G\rho_0\right]\epsilon$$

$$= 2i\left[\sum_i \frac{q_i V_{\perp i}\dot{a}_i}{a_i^3} - \left(\frac{\sum_i (q_i/a_i)^2(\dot{a}_i/a_i)}{\sum_i (q_i/a_i)^2}\right)\sum_i \frac{q_i V_{\perp i}}{a_i^2}\right]$$

In the above $c_s = (dp/d\rho)^{1/2}$ is the sound speed, and the quantity $\mathbf{V}_\perp$ is a measure of the perturbation in vorticity, which is constant in time as a result of Kelvin's circulation

409

C. S. Frenk et al. (eds.), The Epoch of Galaxy Formation, 409–410.

theorem. We notice several things from the above equation : (1) The effects of the perturbed gravity and perturbed pressure are contained entirely in a term $[4\pi G\rho_0 - c_s^2 k^2]\epsilon$. The effects of gravity and pressure exactly balance at the Jeans wavenumber $k_J \equiv (4\pi G\rho_0)^{1/2}/c_s$. This is the same result as one derives for perturbations in a static medium. For $k < k_J$, gravity dominates and this term tends to drive growth of ϵ; for $k > k_J$, pressure dominates, and tends to drive oscillations of ϵ (i.e. sound waves). (2) The expansion/contraction of the background flow results in the appearance of a $d\epsilon/dt$ term with coefficient depending on $\dot{a}_i/a_i$, as in the analysis of perturbations to a homogeneous isotropic cosmological model, although the coefficient has a more complicated form because the expansion factors are different along different directions. In the cosmological case, one considers an expanding background ($\dot{a}_i/a_i > 0$), so that this term tends to cause decay of ϵ, but for a background flow which is collapsing ($\dot{a}_i/a_i < 0$), this term can drive a *growth* of ϵ. This turns out to be an important effect in a typical background collapse. (3) *Unlike* the case of a homogeneous isotropic background flow, perturbations in vorticity ($\mathbf{V}_\perp \neq 0$) in general drive perturbations in density. Thus, there are pure density perturbations ($\epsilon \neq 0$, $\mathbf{V}_\perp = 0$), but in general no pure vorticity perturbations ($\epsilon = 0$, $\mathbf{V}_\perp \neq 0$).

2. Applications

I have applied this method to the analysis of pressure-free inhomogeneous background collapses having either *spherical* or *planar* symmetry. In this case, one can derive an analytic solution for the background flow, valid up to the time at which trajectories intersect. The specific background solutions I consider are derived from the evolution into the *non-linear* regime of initially small density perturbations in an expanding cosmological background. This "background perturbation" could represent a collapsing protogalaxy. I then investigate the evolution of small-scale, small amplitude perturbations on top of this background flow. The small-scale perturbations are not assumed to have any particular symmetry.

In both cases, the background flow is initially nearly homogeneous, and the local velocity field around any point nearly isotropic. At this stage, growth of small-scale perturbations is driven mainly by *local self-gravity*. However, as the collapse proceeds, gravitational tidal fields cause the local velocity field to become increasingly anisotropic. Corresponding to the growth in anisotropy, there is a change in perturbation behaviour: growth of density perturbations is now driven mainly by the *kinematics of the background flow*, with local self-gravity becoming negligible. For plane-wave perturbations, perturbation growth is fastest when the wavevector $\mathbf{k}$ is aligned with the principal axis of the local flow which is contracting the fastest ($\dot{a}_i/a_i$ the most negative). Perturbation growth is therefore anisotropic.

This result that background kinematics come to dominate over self-gravity in driving small-scale perturbation growth seems likely to apply to the late stages of a generic low-pressure background collapse. The anisotropic nature of the perturbation growth in the linear regime when kinematics are important suggests that the non-linear evolution of perturbations takes the form of collapse to planar structures or "pancakes". These pancakes are unlikely to be gravitationally bound in their transverse dimensions, but may themselves undergo fragmentation on small scales into self-gravitating objects, if the gas can cool sufficiently.

A full account of this work will be published in *Astrophys. J.* **336** (1989).

THE TURNAROUND EPOCH OF CLUSTERS OF GALAXIES

Jacob D. Bekenstein and Eyal Maoz
Physics Department, Ben Gurion University, Beersheva 84105, Israel

ABSTRACT. As described elsewhere in these proceedings, galaxies are now seen at $z \approx 5$. In the hierarchical scenario present day clusters of galaxies should have turned around later. We verify this point by developing a new (Hubble constant independent) estimator of the turnaround redshift z_t of a cluster in terms of its velocity dispersion, angular size, and recession velocity.

We assume: (i) spherically symmetric protoclusters, (ii) negligible intracluster gas dynamics, (iii) clusters are relaxed today, and (iv) the universe is open (Friedmann $\Omega_o < 1$ model). The turnaround and present cosmological mean densities, ρ_t and ρ_o, are related by

$$\rho_t = \rho_o(1 + z_t)^3 = (3H_o^2/8\pi G)\Omega_o(1 + z_t)^3 \qquad (1)$$

Write $\rho_{cl} = C_1 C_2 \rho_t$ for the mean cluster density, where C_1 is the density contrast between protocluster perturbation and the surrounding universe at turnaround, while C_2 is the factor by which the density grows in the collapse and virialization. If M is the mass contained within some reasonable cluster radius R_{cl}, then $\rho_{cl} = 3M/(4\pi R_{cl}^3)$. M may be estimated via the virial theorem,

$$M = 3\gamma R_{cl}\sigma^2 G^{-1} \qquad (2)$$

where σ is the one-dimensional velocity dispersion in the cluster, and γ is a structure factor of order unity. Combining all this we get

$$1 + z_t = (6\gamma/\Omega_o C_1 C_2)^{1/3}\alpha^{2/3} \qquad (3)$$

where $\alpha \equiv \sigma/(R_{cl}H_0) = \sigma/(\Theta V_r)$, Θ is the angular radius corresponding to R_{cl}, and V_r is the observed cluster recession velocity, assumed identical with its Hubble velocity. Expression (3) for z_t is independent of the Hubble scale. Present cluster parameters enter only through the combination α.

In the Einstein-De Sitter model $C_1 = 9\pi^2/16$ (Peebles 1980). The traditional wisdom that an initially cold perturbation collapses by a factor of two in radius to virialization ($C_2 = 8$) gives (with $\gamma = 1$) the rule of thumb

$$1 + z_t = 0.51(\sigma/\Theta V_r)^{2/3}\Omega_o^{-1/3} \qquad (4)$$

When Ω_o is not near unity, the naive estimate of C_1 can be improved by comparing the parametric law of change of the radius of a recollapsing pressureless protocluster with that of a Friedmann universe. The result is

$$C_1 = (\pi^2/8)(cosh\eta_t - 1)^3(sinh\eta_t - \eta_t)^{-2} \qquad (5)$$

C. S. Frenk et al. (eds.), The Epoch of Galaxy Formation, 411–412.

where $cosh\eta_t = 1 + 2(\Omega_o^{-1} - 1)(1 + z_t)^{-1}$. Since z_t itself enters in (5), an iterative procedure starting with the zeroth order z_t in (4) is required. To improve the estimate of C_2 from virialization imagine that the protocluster turns around as a homogeneous cold sphere of mass M and radius R_m, and collapses to a singular isothermal sphere cutoff at radius $R_{cl} = GM/3\sigma^2$ with general 1-D velocity dispersion σ. Equating the initial and final energies we get

$$C_2 = (R_m/R_{cl})^3 = (6/5)^3. \tag{6}$$

Putting (5) and (6) into (3) and starting the iteration with (4) we obtain:

Turnaround Redshifts

	$\alpha=$	2	3	4	5	6
$\Omega_o=$	0.1	.94	1.7	2.5	3.2	3.8
	0.5	.54	1.1	1.5	2.0	2.4
	0.9	.36	.78	1.2	1.5	1.8

The range $\alpha= 2 - 6$ spans the observed clusters. It is evident that protoclusters turned around after the epoch at which galaxies turned on, even in a low Ω_o universe. In a dense universe the turnaround epoch is so late that the sparser clusters (small α) may not have had time to relax.

Reference

Peebles, P.J.E. (1980), The Large Scale Structure of the Universe (Princeton University Press, Princeton).

THE CORRELATIONS OF PEAKS IN RANDOM NOISE

Peter Coles
Astronomy Centre
University of Sussex,
Falmer,
Brighton BN1 9QH,
U.K.

ABSTRACT. Calculations of the clustering properties of peaks in random noise (both gaussian and non-gaussian) have been performed. The results indicate that it is difficult to reconcile the observed correlations of rich clusters with a model where they form at the peaks of a field of CDM perturbations.

The observational evidence that rich (Abell) clusters of galaxies are more strongly clustered than galaxies themselves (Bahcall & Soneira 1983) led Kaiser (1984) to suggest that such enhanced correlations might be due to clusters forming at high peaks of a gaussian field of density perturbations. The statistical properties of peaks of such fields are different to those of the underlying fields and one can arrange for peaks to be more strongly clustered than typical field points. Other authors, notably Politzer & Wise (1984), have performed more accurate calculations based on Kaiser's original idea.

These studies, however, assume that clusters form from high-level regions of the density field, rather than local maxima, because it is easier to calculate the region-region correlation function than the peak-peak correlation function. In Coles (1986) I pointed out the limitations of this approach and showed that the easier calculation systematically overestimates the peak-peak correlation function, $\xi_{pk-pk}(r)$, at distances less than $\sim 25h^{-1}$ Mpc. To see what the form of $\xi_{pk-pk}(r)$ is really like we have to consider local maxima of random fields rather than just high regions.

Unfortunately, calculations of $\xi_{pk-pk}(r)$ are rather difficult for 3-d random fields so I have concentrated on the behaviour of peaks of 1-d random noise (whose behaviour is qualitatively identical to those in the 3-d case). The full details of this work can be found in Coles (1988), but it is worth making a few points here about the results obtained. Firstly, the zero-crossings of $\xi_{pk-pk}(r)$ are not necessarily the same as the underlying correlation function $\xi(r)$. Secondly, $\xi_{pk-pk}(r) \rightarrow -1$ as $r \rightarrow 0$ (peaks cannot lie on top of each other). Finally, the enhancement of $\xi_{pk-pk}(r)$ with respect to $\xi(r)$ is extremely dependent on the shape of $\xi(r)$ (i.e. on high order derivatives) so a simple expression like that of Politzer & Wise (1984) cannot be derived. One manifestation of this last point is that the enhancement predicted in CDM models (where this picture of cluster formation is usually discussed) is lower than the Politzer-Wise estimate. Furthermore, the zero-crossing of $\xi_{pk-pk}(r)$ occurs at even smaller r ($\sim 25h^{-1}$ Mpc) than the zero of $\xi(r)$ for the CDM perturbations ($\sim 34h^{-1}$ Mpc). The observations suggest that $\xi_{cc}(r)$ is positive out to greater distances than this and thus it is difficult to reconcile the observations with the CDM model. It has to be said, however, that observational uncertainties in the cluster-

413

C. S. Frenk et al. (eds.), The Epoch of Galaxy Formation, 413–414.

cluster correlation function (Sutherland 1988) and theoretical difficulties concerning the correct normalisation of the CDM spectrum (Bardeen *et al.* 1986) make it impossible at the moment to rule out the CDM model using these calculations but the model does appear to be on rather shaky ground (see also Lumsden, Peacock & Heavens in this proceedings).

In addition to these studies of peaks in gaussian noise, some non-gaussian models have also been considered. Such studies are important because we know that $(\delta\rho/\rho)_{rms} \sim 1$ on cluster scales in the CDM scenario which shows immediately that the distribution of mass is non-gaussian (a gaussian with mean and variance unity would predict negative densities). Although we expect that rich clusters should trace peaks of the primordial mass distribution fairly accurately, the number of peaks above a fixed threshold and their correlations will not be exactly as predicted using a gaussian distribution for the overdensities. One can attempt to include this effect in the calculations by using second order perturbation theory to calculate the expected skewness of the matter distribution (Peebles 1980) and then using a lognormal distribution with this skewness as a model for the nonlinear mass distribution. A lognormal distribution was chosen because this is a simple transformation of a gaussian and therefore if one can calculate $\xi_{pk-pk}(r)$ for a gaussian, one can do the same for a lognormal (for an application of this technique in a different context see Coles & Barrow 1987). In addition it is interesting that counts of galaxies in cells seem to follow a lognormal distribution to reasonable accuracy (Peebles 1980). This calculation (details are in Coles 1988) shows that the form of $\xi_{pk-pk}(r)$ for lognormal noise is very similar to the gaussian case but that the enhancement is slightly greater ($\sim 20\%$) for the same threshold. This would tend to ease one of the problems facing CDM. However, there are are difficulties with this escape route because the shape of $\xi(r)$ is not preserved during nonlinear evolution. It has been shown (Juszkiewicz, Sonoda & Barrow 1984) that one expects the zero-crossing of $\xi(r)$ to move inwards as evolution becomes non-linear thus exacerbating the zero crossing problem for CDM models. These calculations should provide a useful consistency check on studies of nonlinear clustering using N-body experiments.

The basic conclusion of this work is that the Politzer-Wise expression for the two-point correlation function of peaks of noise does not fit the accurate evaluation of $\xi_{pk-pk}(r)$ at any sensible distance. It remains to be seen whether the same problems exist with higher-order correlation functions but until rigorous calculations have been performed one should be very doubtful about the applicability of any calculations based on the high region approximation.

References

Bahcall, N.A. & Soneira, R.M., 1983. Astrophys. J., 270, 20.
Bardeen, J.M., Bond, J.R., Kaiser, N. & Szalay, A.S., 1986. Astrophys. J., 304, 15.
Coles, P., 1986. Mon. Not. R. astr. Soc., 222, 9P.
Coles, P., 1988. *D. Phil. Thesis*, University of Sussex.
Coles, P. & Barrow, J.D., 1987. Mon. Not. R. astr. Soc, 228, 407.
Juszkiewicz, R., Sonoda, D.H. & Barrow, J.D., 1984. Mon. Not. R. astr. Soc., 209, 139.
Kaiser, N., 1984. Astrophys. J., 284, L9.
Peebles, P.J.E., 1980. *The Large Scale Structure of the Universe*, Princeton University Press, Princeton.
Politzer, H.D. & Wise, M.B., 1984. Astrophys. J., 285, L1.
Sutherland, W., 1988. Mon. Not. R. astr. Soc., *submitted*.

DENSITY MAXIMA AS SITES FOR GALAXY FORMATION

A. F. HEAVENS
University of Edinburgh, Blackford Hill, Edinburgh, EH9 3HJ
J. A. PEACOCK
Royal Observatory, Blackford Hill, Edinburgh, EH9 3HJ
S. L. LUMSDEN
University of Edinburgh, Blackford Hill, Edinburgh, EH9 3HJ

ABSTRACT. In this paper we briefly review the successes and failures of the theory which identifies local maxima in the primordial density field as sites for structure formation. We present more fully the results of recent calculations on the clustering properties and mass function predicted by the theory.

1. Introduction

In confronting theory with observation, one of the principal difficulties is in comparing like with like. For example, the theory of the growth of gravitational instability usually makes specific predictions of the properties of the mass distribution, whereas observation is normally concerned with luminous (or at least detectable) matter. By identifying local maxima in the mass density as sites for structure formation, we can analyse the properties of regions which we believe will be associated with observable objects. This theory has developed via analysis of electrical noise (Rice 1954), through water waves (Longuet-Higgins 1957) to the three-dimensional problem of cosmological structure formation (Doroshkevich 1970, Peacock & Heavens 1985 [PH], Bardeen *et al.* 1986 [BBKS], Couchman 1987; see also Adler 1981, Vanmarcke 1983, Bertschinger 1987). The method is an analytic complement to N-body simulations, powerful in its ability to predict the properties of rare events, for example.

2. General Method

This section outlines the method of calculating the properties of peaks in gaussian noise. If the reader is interested principally in the results of this sort of analysis, he may proceed directly to section 2.

We wish to find the probability distribution for a quantity subject to the constraint that the point in space at which the quantity is measured is a local maximum of the (fractional over-)density field, $\delta(\mathbf{x})$. If we wish to consider objects of a certain mass, a low-pass filter of appropriate size is applied to the density field to erase all substructure.

C. S. Frenk et al. (eds.), The Epoch of Galaxy Formation, 415–420.

The method is to write down the joint probability distribution for this smoothed δ and its first two derivatives (to specify that the point is a maximum) and any other quantity whose distribution is to be calculated. The joint distribution of this system of N variables V_i is then, under rather weak conditions, a multivariate gaussian:

$$f(V_i)d^N V_i = \frac{1}{(2\pi)^{N/2}\|\mathbf{M}\|^{1/2}} \exp(V_i M_{ij}^{-1} V_j) d^N V_i \qquad (1)$$

where $M_{ij} \equiv < (V_i - \overline{V_i})(V_j - \overline{V_j}) >$, and the angle brackets and superior bars indicate mean values. If the density field is described by a Fourier power spectrum $|\delta_{\mathbf{k}}|^2$, with random phases for the components, then the density field will be gaussian (in the linear regime), and the V_i and M_{ij} can be calculated from the power spectrum.

Near a maximum, $\partial\delta/\partial x_i = \partial^2\delta/\partial x_i\partial x_j\, dx_j$, and the elements of the derivative of δ are replaced by a volume element, so we end up with a number density of peaks with the desired properties.

The constraint that the point is a maximum is then applied, by setting $\partial\delta/\partial x_i = 0$, and the second derivatives are integrated out (the region of integration is chosen such that all the principle values of $\partial^2\delta/\partial x_i\partial x_j$ are negative; see BBKS). In the simplest cases, this leaves a joint distribution for the overdensity of the peak, δ (usually expressed in terms of the r.m.s. overdensity variation σ_0: $\delta \equiv \nu\sigma_0$), and the variable of interest. More complicated quantities such as angular momentum are treated in a similar way.

3. Overdensities and Collapse Redshifts

The distribution of overdensities of peaks was calculated by PH and BBKS, and shows that peaks have mean overdensities $\nu \sim 1.5-2$, depending on power spectrum. The distribution (Fig.1) also gives the distribution of collapse redshifts, if we take the prescription that an object forms when δ reaches a constant value of order unity. Note that the collapse redshift normalisation is somewhat arbitrary, and will depend on the normalisation of the power spectrum, and the mass of the objects under consideration. However, the relative values of $(1+z)$ for the different height peaks are as shown in the figure.

3.1. SHAPES

PH and BBKS also investigated the shapes of the peaks, defined by the values of the principal axes of the second derivative matrix. Both found that peaks are triaxial, with an asymptotic result that peaks are spherical in the high-ν limit. For heights of practical interest $\nu \lesssim 4$, the asymptotic limit is inappropriate.

4. Streaming motions

The streaming velocities of peaks are systematically smaller than points in the field (BBKS). This is understandable, as the principal small-scale movement of matter will be towards peaks. The joint probability distribution of the streaming motions on two scales can be used (Peacock, Lumsden & Heavens 1987) to constrain large-scale damping models. The $\Omega_0 = 1$ adiabatic neutrino and baryon models fit the streaming data rather

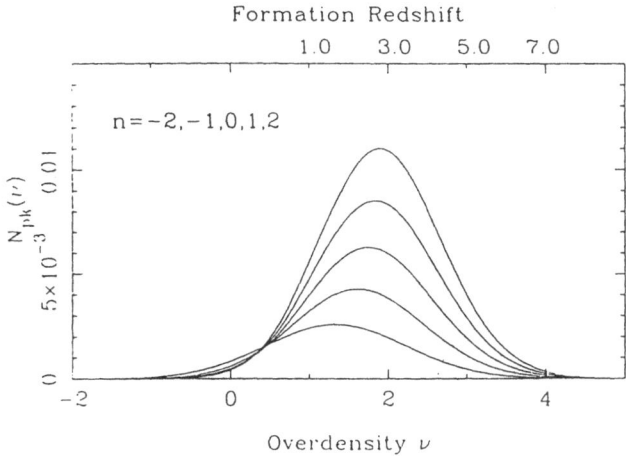

Figure 1. *The distribution of peak heights and collapse redshifts for a smoothed gaussian field. The different curves correspond to power spectra $|\delta_k|^2 \propto k^n \exp(-k^2 R_f^2)$. The number density in the range $d\nu$ is $N_{pk}(\nu) R_f^{-3} d\nu$. The collapse redshifts have been chosen such that a $\nu = 2$ peak collapses at a redshift 3. For normalisation to other redshifts, see text.*

well, but the latter with some difficulty: we must both allow $H_0 \simeq 25\,\mathrm{kms}^{-1}\mathrm{Mpc}^{-1}$ for consistency with nucleosynthesis, and find an alternative mechanism for galaxy formation, since most of the baryon pancakes remain uncollapsed at the present.

5. Angular Momentum

Heavens & Peacock (1988) calculated the angular momentum of matter in the vicinity of peaks. This calculation showed that high, early-forming peaks are subjected to systematically higher torques than low peaks. This is offset by the shorter collapse time during which the torques act. The angular momentum acquired is more-or-less independent of the peak height, but with a large dispersion. The most interesting result is perhaps that the spin parameter anticorrelates with peak height, with a median value of about 0.05, in good agreement with N-body simulations and observations of elliptical galaxies. The scaling is $\lambda \sim 0.08/\nu$, which leads to a weak anticorrelation with mass ($\lambda \propto M^{-1/6}$ for CDM on galaxy scales). Note however that the distributions (Fig.2) are very broad. Whilst spin parameters characteristic of spiral galaxies could in principle be generated entirely by this method, the galaxies would have to be associated with very low peaks, $\nu \sim 0.1$, which have little chance of survival (see later). A more plausible explanation must surely lie in dissipative effects outside the scope of this method.

418

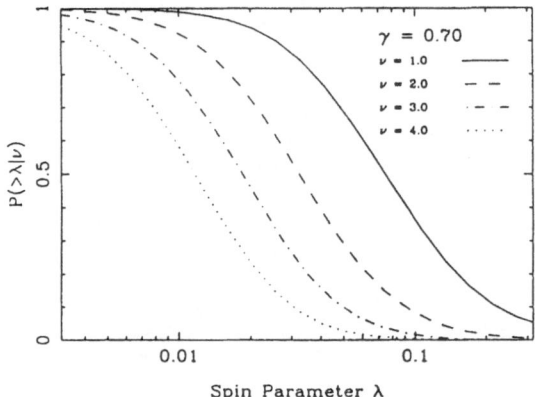

Cumulative Spin Parameter Distribution

Figure 2. *The cumulative spin parameter distributions for peaks of different heights. When the peak height distribution is folded in, the median is about $\lambda \sim 0.05$, but with a large dispersion.*

6. Mass Function

In order to calculate this, we need to smooth the power spectrum on all mass scales simultaneously, and identify those peaks which will exist at any given epoch. A straightforward method is to argue that only those peaks which are just collapsing (in the sense that $\delta_{pk} \simeq 1$) will exist. The justification for this is that if a peak on a certain mass scale has collapsed in the recent past, it will have accreted mass and will be part of a slightly larger system which is just collapsing. This argument (Bond 1987) is essentially a more physically justifiable variant of the Press & Schechter (1974) analysis. The agreement with N-body simulations is not bad, but it predicts too few objects with masses somewhat less than M_*, the mass for which $\sigma_0 \simeq 1$. Bond (1987) showed that there was good agreement with the meagre data on groups and clusters, provided one allowed oneself the freedom to scale all the peak masses down from the closure value. There are at least two ways in which this simple prescription could be modified: firstly, low mass peaks ($M \ll M_*$) in this picture must come from very low peaks $\nu \ll 1$. Almost all of these will actually be incorporated into larger mass systems which have already collapsed (Peacock & Heavens 1989). Secondly, with this prescription, systems of a given mass exist only for an infinitessimal time. If we relax this strict criterion, then we modify the mass spectrum, keeping more objects with M a little below M_*.

7. Clustering Properties

The main difficulty in confronting theory with observation has been that theory usually describes the properties of the mass, whereas observation is more concerned with luminous (or at least detectable) matter. One of the attractive aspects of identifying peaks as

Mass Function for CDM

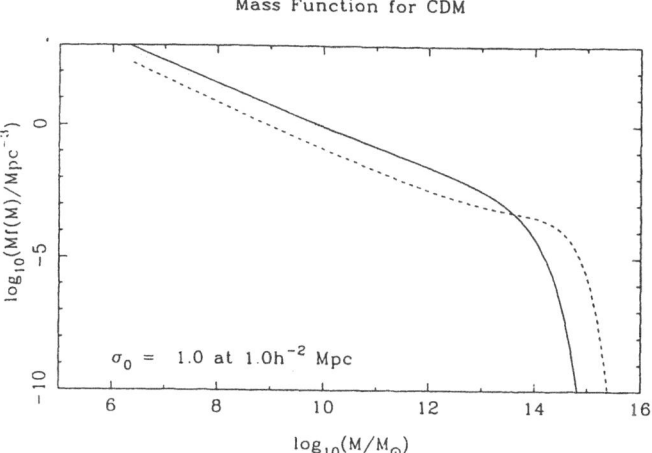

Figure 3. *The mass spectrum predictions for the Press-Schechter theory (solid), with the survival correction included (dotted; no mass renormalisation) and the peaks theory (dashed). The survival correction is very small for CDM on these scales because the spectrum is very flat. The scalings with Ω_0 and h are $R \propto (\Omega_0 h^2)^{-1}$, $M \propto (\Omega_0 h^2)^{-2}$, $M f(M) \propto (\Omega_0 h^2)^3$.*

sites of structure formation is that it allows a more direct comparison with observation to be made. Kaiser (1984) demonstrated that if clusters of galaxies could be identified with high *regions* of the density field, then their large correlation amplitude could be partly explained. One difficulty is that the zeroes of the mass autocorrelation function ξ_m and the high region correlation function ξ coincide (Politzer & Wise 1984: PW, Jensen & Szalay 1986: JS) which presents difficulties for models such as 'canonical' ($\Omega_0 = 1$, biased) CDM where ξ_m crosses zero at a small separation. Not surprisingly, high *peaks* show the same behaviour, but it is difficult to quantify the effect exactly, owing to the large number of variables (N=20). Various approximation schemes have been tried for this and related problems (BBKS, PW, JS, Lumsden, Heavens & Peacock 1989: LHP), but none is accurate in the region of interest, where $\xi \gtrsim 1$. In one dimension, it can be shown that ξ_{pk} does not cross zero at the same radii as ξ_m (LHP, Coles 1988), and static simulations of the 3D case also show this behaviour. Fig. 4 shows the peak-peak and mass correlation functions for clusters within CCDM, along with a hatched region bounded by Sutherland's (1988) and Bahcall & Soniera's (1983) estimates of the correlation function of Abell clusters. The peaks have $\nu \geq 2.7$ and the gaussian smoothing length is 10 Mpc ($H_0 = 50$). Notice that the zero-crossing of ξ_{pk} is at a smaller r that that of ξ_m. (It is worth noting that the use of a gaussian smoothing function eliminates the 'ringing' which is apparent if a sharp cut-off in k-space is employed (e.g. White et al. 1987). This can distort the clustering length and zero-crossing points). Clearly CCDM is in difficulties, if the clustering data are reliable. CDM with $\Omega_0 \simeq 0.2$ fits well, as the characteristic length scales as Ω_0^{-1}.

420

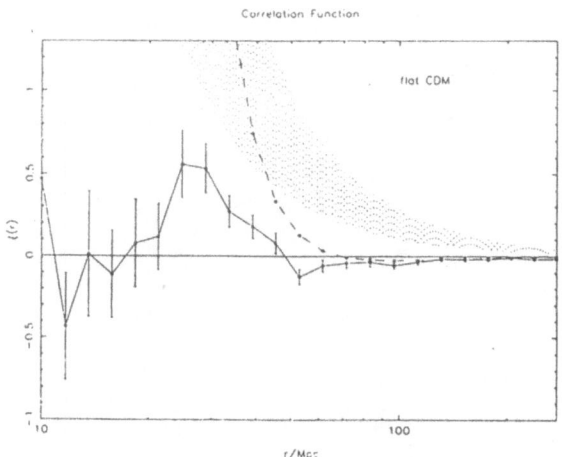

Figure 4. *The correlation function (solid) of peaks of the CCDM density field, with $\nu > 2.7$. The smoothing length is 10 Mpc, and $H_0 = 50\,\mathrm{kms^{-1}Mpc^{-1}}$. The hatched area is bounded by the Abell cluster correlation function determinations of Sutherland (1988) and Bahcall & Soniera (1983). The broken line is the best analytic approximation to date for the 3D peak correlation function. It is constrained to fit at large distances, but illustrates the difficulties.*

References

Adler, R.J., 1981. *The Geometry of Random Fields*, Wiley, Chichester.
Bahcall, N.A. & Soniera, R.M., 1983. *Astrophys. J.*, **270**, 20.
Bardeen, J.M., Bond, J.R., Kaiser, N. & Szalay, A.S., 1986. *Astrophys. J.*, **304**, 15 [BBKS].
Bertschinger, E., 1987. *Astrophys. J.*, **323**, L103.
Bond, J.R., 1987. Proc. Pont. Acad. of Sciences *"Large Scale Motions in the Universe"*.
Coles, P., 1988. *Ph.D. Thesis*, University of Sussex.
Couchman, H.M.P., 1987. *Mon. Not. R. astr. Soc.*, **225**, 777.
Doroshkevich, A.G., 1970. *Astrophysica*, **6**, 320.
Heavens, A.F. & Peacock, J.A., 1988. *Mon. Not. R. astr. Soc.*, **232**, 339.
Jensen, L.G. & Szalay, A.S., 1986. *Astrophys. J.*, **305**, L5 [JS].
Kaiser, N., 1984. *Astrophys. J.*, **284**, L9.
Longuet-Higgins, M.S., 1957. *Phil. Trans. R. Soc. Lond.*, **A249**, 321.
Lumsden, S.L., Heavens, A.F. & Peacock, J.A., 1989. *Mon. Not. R. astr. Soc.*, submitted [LHP].
Peacock, J.A. & Heavens, A.F., 1985. *Mon. Not. R. astr. Soc.*, **217**, 805 [PH].
Peacock, J.A. & Heavens, A.F., 1989. In preparation.
Peacock, J.A., Lumsden, S.L. & Heavens, A.F., 1987. *Mon. Not. R. astr. Soc.*, **229**, 469.
Politzer, H.D. & Wise, M.B., 1984. *Astrophys. J.*, **285**, L1 [PW].
Press, W.H. & Schechter, P., 1974. *Astrophys. J.*, **187**, 425.
Rice, S.O., 1954. *Selected Papers on Noise and Stochastic Processes*, p.133. Ed. Wax, N., Dover.
Sutherland, W., 1988. *Mon. Not. R. astr. Soc.*, **234**, 159.
Vanmarcke, E.H., 1983. *Random Fields: Analysis and Synthesis*, MIT Press, Cambridge, Mass.
White, S.D.M., Frenk, C.S., Davis, M. & Efstathiou, G., 1987. *Astrophys. J.*, **313**, 505.

A CROSS-CORRELATION METHOD TO TEST THE DEPENDENCE
OF THE CLUSTERING OF GALAXIES ON LUMINOSITY

A. BLANCHARD[1], J.-M. ALIMI[1], D. VALLS-GABAUD[2]
[1] *Observatoire de Meudon, D.A.E.C.*
   Place Janssen, 92190 Meudon, France
[2] *Institut d'Astrophysique de Paris*
   98bis, Bd Arago, 75014 Paris, France

ABSTRACT. In order to compare the clustering of two populations of galaxies, we present a simple method based on a cross–correlation analysis. This method is free of the main uncertainties in the estimation of the correlation function, and therefore allows us to set reliable limits on any possible luminosity segregation.

Strong efforts have been done recently to examine in detail whether or not the clustering of galaxies depends on luminosity as expected in biased models. Two main limitations in the determination of the correlation function make such a study difficult. Firstly, the estimated correlation function varies when different volumes are used. This implies that comparative studies of the clustering must be done within the same volume. The second limitation is the normalisation problem, as the determination of the correlation function needs the knowledge of the luminosity function. This last problem can be eliminated by using the cross–correlation. We have computed the following quantity within the CfA catalogue:

$$\alpha(L_2, L_1, R) = \frac{\int_0^R dr \xi_{L_1,L_1}(r)}{\int_0^R dr \xi_{L_2,L_1}(r)} = \frac{\int_0^R dr \frac{N_{11}(r)}{N_{1p}(r)} \frac{n_p}{\bar{n}(L_1)} - 1}{\int_0^R dr \frac{N_{21}(r)}{N_{2p}(r)} \frac{n_p}{\bar{n}(L_1)} - 1} \tag{1}$$

α is clearly unsensitive to the normalisation $\bar{n}(L_1)$. It is interesting to note that if α is equal to 1 for (L_1, L_2) and (L_2, L_1) then ξ_{L_1,L_1} is necessary equal to ξ_{L_2,L_2}. Therefore, the use of this method does not reduce the generality of the result. In practice, the variations due to different Virgocentric models are larger than the uncertainty due to the normalisation, as illustrated in Figure 1. We have therefore estimated a possible range for α from this variations. For instance, we deduce the following range $0.98 \leq \alpha \leq 1.03$ from Figure 1 (the small–scale pattern was not taken into account). This was done for a large range of luminosities (from M = $-15.$ to M = -19.5 with h = 1). The comparison with simple biased models is done in Figures 2a and 2b. The details of the prediction are given in the contribution by D. Valls–Gabaud in this volume. Completely independent pairs were also used, and the conclusions were the same, namely, no significant segregation is found in the CfA catalogue. Any possible luminosity segregation effect greater than 5% per unit magnitude interval is unlikely.

C. S. Frenk et al. (eds.), The Epoch of Galaxy Formation, 421–422.

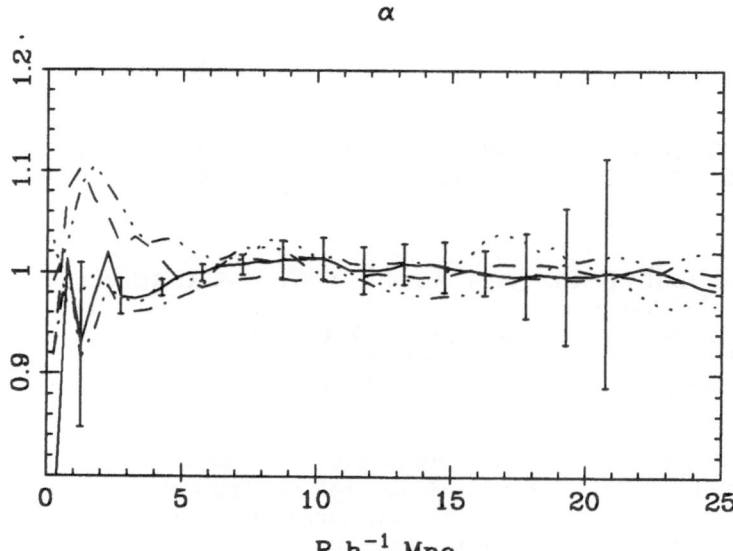

Figure 1. The coefficient α versus R. The luminosities L_1 and L_2 correspond respectively to M=-17.5, M=-18.5. The errors bars reflect the uncertainty due to counting, and the lines correspond to different Virgocentric flow corrections. From this figure, it is concluded that $.98 \leq \alpha \leq 1.03$ in the spatial range $4-17h^{-1}Mpc$.

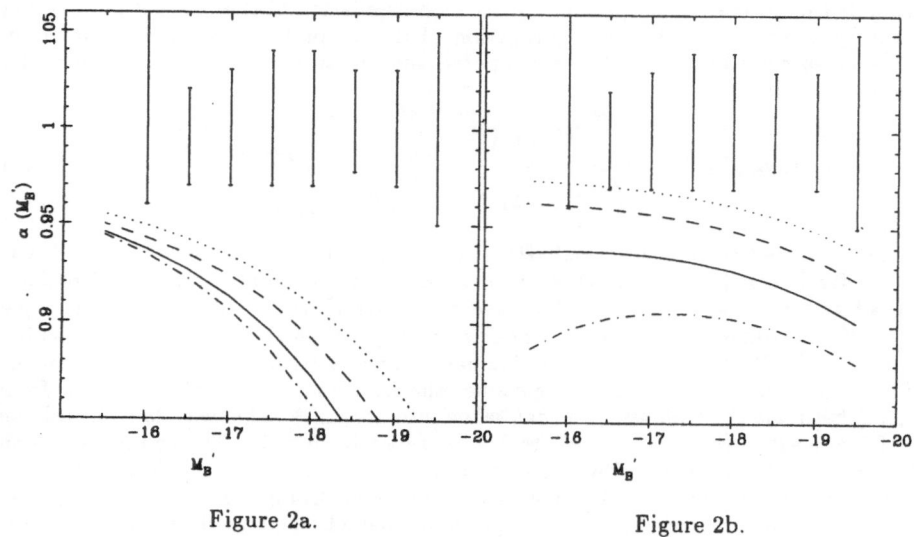

Figure 2a.

Figure 2b.

Figure 2. The comparison with models using the B.B.K.S. formalism. The models in Figure 2a are with a smoothing radius R_S scaling as $L^{1/3}$. In the second case R_S is kept constant.

CONSTRAINTS ON THE FORMATION REDSHIFT
OF BRIGHT GALAXIES IN BIASED SCENARIOS

D. VALLS-GABAUD[1], A. BLANCHARD[2], J.-M. ALIMI[2]
[1] *Institut d'Astrophysique de Paris*
98 bis, Bld. Arago, 75014 Paris, France
[2] *D.A.E.C., Observatoire de Meudon*
Place Janssen, 92190 Meudon, France

ABSTRACT. The cross–correlation analysis of luminosity segregation effects in the clustering of galaxies gives stringent constraints on the bias parameter of bright galaxies. This, in turn, limits the range on the epoch of their formation. Two simple prescriptions are used to relate the peak threshold to the luminosity. It is found that the bias parameter should be smaller than 1.7 for galaxies brigther than $M = -18.5 + 5 \log h$, and therefore their formation redshift greater than about 8.

We have investigated simple biased models using the BBKS technique[1] and made predictions concerning the expected luminosity segregation effect in the clustering of galaxies. Using the gaussian random field formalism, the cross–correlation function of galaxies brighter than L_1 with galaxies brighter than L_2 can be written as

$$\xi_{L_1 L_2}(r) = b(L_1) \, b(L_2) \, \xi(r) \tag{1}$$

where b is the bias parameter, given by $b = (<\tilde{\nu}>/\sigma_o) + 1$ (Ref. 1). Our measure of the luminosity segregation effect[2,3] is given by $\alpha(L_2, L_1) = b(L_1)/b(L_2)$. For L_2 greater than L_1, the coefficient is expected to be smaller than one in the presence of segregation.

Two simple prescriptions are used to relate the peak treshold ν to the luminosity[4]. The first one fixes the smoothing radius R_s and makes the number density of peaks above ν_1 to match the number density of galaxies brighter than L_1

$$n_{pk}(\nu \geq \nu_1, R_s) = n_g(L \geq L_1) = \int_{L_1}^{\infty} \Phi(L) \, dL \tag{2}$$

The second prescription, following the Press and Schechter technique, is to match the total mass in the fluctuation to that in galaxies

$$n_{pk}(\nu \geq \nu_1, R_s) \, R_s^3 \propto \int_{L_1}^{\infty} L \, \Phi(L) \, dL \tag{3}$$

assuming a constant M/L ratio. These simple prescriptions give the $\nu(L)$ relationship, which is used to derive the relation between the bias parameter and the luminosity, and then the α coefficient[4].

423

C. S. Frenk et al. (eds.), The Epoch of Galaxy Formation, 423–424.
© *1989 by Kluwer Academic Publishers.*

424

The comparison of the observed limits on this coefficient[2,3] to the above predictions (Fig. 2, ref. 3) gives severe constraints on the bias parameter[4] and therefore on the epoch of non–linear growth, defined as

$$z_{for} = \frac{\nu\,\sigma_o(R_s)}{f_c} - 1 \tag{4}$$

where σ_o is the r.m.s. value of the density fluctuation field, ν the peak height in σ_o units, $f_c \sim 1.69$ the collapse factor, and R_s the smoothing scale. In the first prescription (Fig. 1a), the brighest galaxies collapse earlier than the fainter ones, as expected. At fixed R_s, the higher the luminosity, the higher the ν value, in order to decrease the number density of peaks. Therefore, z_{for} increases with luminosity. In the second one, the higher the luminosity, the higher the smoothing scale, decreasing the r.m.s. value and the number density of peaks. This gives the behaviour illustrated in Fig. 1b, where z_{for} decreases with luminosity. In both Figures, the increase of z_{for} for small R_s is easy to understand, as the r.m.s. value increases with larger smoothing scales. The comparison of Fig. 1 with Fig. 2 in Ref. 3 shows that reliable limits may be obtained. Note, however, that smaller smoothing scales ($R_s \leq 0.1 h^{-1}$Mpc) might agree with our limits, but such low values would imply a very small baryonic fraction enclosed in these fluctuations ($2\ 10^{-4}\ \Omega_o$) and an important secondary infall, which seems unlikely.

The limits on the luminosity segregation effect in the clustering of galaxies that we have obtained[2,3] give severe constraints on the bias parameter[4] ($b \leq 1.6$ for galaxies brighter than $M = -15 + 5\log h$). This translates into $z_{for} \geq 8$ in the (z_{for}, M) diagram. These limits are consistent with other independent estimations.

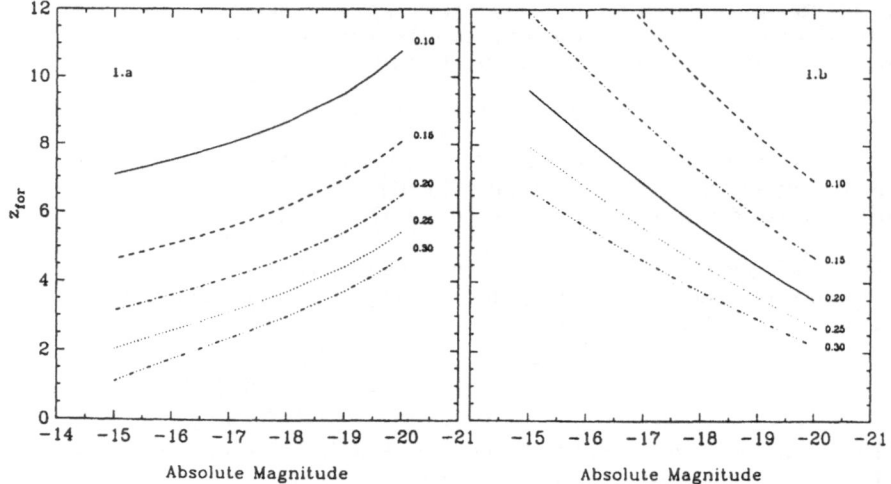

References
[1] Bardeen, J.M., Bond, J.R., Kaiser, N., Szalay, A. (1986) *Astrophys. J.* **304**, 15
[2] Alimi, J.M., Valls–Gabaud, D., Blanchard, A. (1988) *Astron. Astrophys.* (in press).
[3] Blanchard, A., Alimi, J.M., Valls–Gabaud, D. (1989) this volume.
[4] Valls–Gabaud, D., Alimi, J.M., Blanchard, A. (1988) *submitted.*

BIASED THEORIES WITH NON-ϑ THRESHOLD

Silvio A. Bonometto
Stefano Borgani

Dept. of Physics of the University of Perugia, Via Pascoli, 06112 Perugia, Italy
I.N.F.N. – Gruppo collegato di Perugia

Abstract. *The effects of selecting peaks with a non-ϑ threshold are analysed, both within Gaussian and non-Gaussian contexts. Results concerning the galaxy mass function are outlined.*

In the frame of a biased theory of galaxy formation, let $\rho(x)$ be the background matter density field, $\rho = <\rho(x)>$ its average and $\epsilon_R(x)$ its smoothed fluctuations after cutting–off the high frequency components from $\epsilon(x) = \rho(x)/\rho - 1$. Then $\xi(r) =<\epsilon(x)\epsilon(x+r)>$ and $\xi_R(r) =<\epsilon_R(x)\epsilon_R(x+r)>$ (unsmoothed and smoothed 2–point background functions) coincide for $r > R$ while, for $r < R$, $\xi_R(r)$ tends to σ_R^2 (mass–variance over the scale R). In this frame the biased density field is defined according to

$$\rho_{\nu,R} = f[\epsilon_R(x) - \nu\sigma_R] \tag{1}$$

the usual choice for the *threshold function* f being the ϑ step distribution, providing a net selection of peaks. It is also often assumed that the fluctuation field $\epsilon(x)$ is Gaussian.

However both $f \neq \vartheta$ (see, e.g., Bardeen *et al.*, 1986; Szalay, 1987) and non–Gaussian fluctuations (Matarrese *et al.* 1986, Grinstein and Wise, 1986; Fry, 1986; Bonometto *et al.* 1987, etc.) have been considered. Here we report some results on the joint effects of these assumptions; then we discuss some thresholds; finally we present an application to the multiplicity function.

Let us start from the partition integral

$$Z[J_x] = \int D[\epsilon_x]P[\epsilon_x]exp\{iJ_x\epsilon_x\} \tag{2}$$

(here function arguments are indicated as indices). The background *connected* n–point functions

$$\xi_{con}^{(n)}(x_1,..,x_n) = (\delta/\delta J_{x_1})..(\delta/\delta J_{x_n})lnZ[J_x]|_{J=0} \tag{3}$$

vanish, for $n > 2$, in the Gaussian case, for which $P[\epsilon] = exp\{-(1/2)\epsilon_x K_{x,x'}\epsilon_{x'}\}$. The peak *disconnected* n–point functions

$$\Pi_{\nu,R}^{(n)}(x_1,..,x_n) = <\rho_{\nu,R}(x_1)..\rho_{\nu,R}(x_n)> = \int D[\epsilon_x]P[\epsilon_x]\rho_{\nu,R}(x_1)..\rho_{\nu,R}(x_n) \tag{4}$$

easily furnish non–vanishing connected functions at all orders. They can be built starting from $h(\phi)$ and l defined according to $(i\phi)^l e^{h(\phi)} = (2\pi)^{-1/2} \int_{-\infty}^{+\infty} d\alpha f'(\alpha)exp(-i\alpha\phi)$, which convey the information about the threshold, and from

$$w_{R,[r_n]}^{(n)} = \xi_{R,con}^{(n)}(x_1..x_n)/\sigma_R^n \qquad (for\ n > 2\ and\ for\ n = 2\ if\ x_1 \neq x_2)$$

C. S. Frenk et al. (eds.), The Epoch of Galaxy Formation, 425–426.

$$w_R^{(n)} = \xi_{R,con}^{(n)}(x,..,x)/\sigma_R^n \qquad (for\ n > 2) \qquad (5)$$

which convey the information on non–Gaussianety. The whole information will then reach

$$\Pi_{\nu,R}^{(n)}(x_1..x_n) = \sigma_R^{ln} \sum_{L=0}^{\infty} \sum_{[m_L]} [\prod_{N=0}^{L} \prod_{[r_N]=1}^{n} (W_{R,[r_N]}^{(N)}/N!)^{m_{N,[r_N]}}/m_{N,[r_N]}!] \prod_{r=1}^{n} a_{m(r)}^{(l)}(2^{-1/2}\nu) \qquad (6)$$

through combinations like $W_{R,[r_n]}^{(n)} = w_{R,[r_n]}^{(n)} + [h^{(n)}(0)/(i\sigma_R)^n] \prod_{k=1}^{n-1} \delta_{r_k,r_{k+1}}$. Various indeces in (6) bear a clear meaning (see Borgani and Bonometto, 1988). In turn $a_m^{(l)}(z) = 2^{-\frac{m+l-1}{2}} e^{-z^2} H_{m+l-1}(z)$ (for $m + l > 1$) and $a_o^{(o)}(z) = (\pi/2)^{1/2} erfc(z)$.

From (6) it can be seen that, in the large ν limit, the n–point connected functions do not depend on the peculiar f. This result had been already seen, but just for $f(\alpha) = exp(\nu\alpha/\sigma_R)$ (Szalay, 1987). The expression of $\Pi_{\nu,R}^{(1)}$, instead, leads to the expected number densities, which are however different from the ϑ–threshold results. Here we shall consider two particular thresholds and relate the corresponding number densities to those obtainable for a ϑ threshold, both in the Gaussian and in the non–Gaussian case. For $f(\alpha) = (2\pi)^{-1/2} exp\{-(\mu^2/2\sigma_R^2)\alpha^2\}$ (Bardeen et al., 1986) and $f(\alpha) = (2\pi\delta^2)^{-1/2} \int_{-\infty}^{\alpha} dy\ exp(-y^2/2\delta^2)$ (giving back ϑ for small δ's) we have

$$< \rho_{\nu,R} > = < \rho_{\nu,R} >_\theta (\nu/\mu) exp(\nu/2^{1/2}\mu)^2 \quad and \quad < \rho_{\nu,R} > = < \rho_{\nu,R} >_\theta exp(\nu\delta/2^{1/2}\sigma_R)^2 \quad (7)$$

respectively.

The above thresholds can be given a direct meaning in the frame of theories of galaxy formation. The former one assumes that only those peaks whose size ranges around a given level can be seen now as actual galactic systems. E.g., systems related to very high peaks might form too early being soon absorbed in greater structures of later formation. The latter one, instead, may be related to non sphericity in primeval collapses. If t_o is the present time and we assume, e.g., that systems have just recollapsed by now, the turning around took place exactly at $t_o/2$ only for spheres. Allowing a spread around the level $\nu\sigma_R$, takes other possible geometries into account.

We fit the data on multiplicity function, taking the latter effect into account. It is known that the slope of such data, in the range between individual objects and Abell clusters, cannot be fit within a canonical CDM model. Here we assume that non–sphericity has a lower impact over greater mass scales. In effects, peaks of encreasing height are known to be encreasingly spherical.

Besides $\Omega = 1$, a primeval spectrum with $n = 1$, a biasing parameter $b = 1$ and a CDM transmission factor, we assumed systems to exist at *recollapse* time. In (7), then, we took $\delta = K(M/M_\odot)^{-\beta}$. We obtained a satisfactory fit for $M/L = 500$, $K = 62.4$, $\beta = 0.36$. A curve obtained with the above parameters meets all error bars of data on Abell clusters and Turner and Gott (1976) groups. Moreover the corresponding curve with $K = 0$, while unable to meet most data concerning groups, passes across error bars concerning individual galactic systems (Bahcall, 1979).

References

Bahcall, N.A. 1979, Ap.J. **232**, 689.

Bardeen, J.M., Bond, J.R., Kaiser, N., and Szalay, A.S., 1986, Ap.J. **304**, 15.

Bonometto, S.A., Lucchin, F., and Matarrese, S., 1987, Ap.J. **323**, 19.

Borgani, S., and Bonometto, S.A., 1988, preprint.

Fry, J.N., 1986, Ap. J. **308**, L71.

Grinstein, B., and Wise, M.B., 1986, Ap.J. **310**, 19.

Matarrese, S., Lucchin, F., and Bonometto, S.A., 1986, Ap.J.Lett. **310**, L21.

Szalay, A.S., 1987, preprint.

Turner, E.L., and Gott, J.R., 1976, Ap.J. **209**, 6.

SINKING SATELLITES AND DISK HEATING OF SPIRAL GALAXIES

Lars Hernquist[1] and P.J. Quinn[2]
[1]The Institute for Advanced Study, Princeton, NJ
[2]Space Telescope Science Institute, Baltimore, MD

Most large galaxies possess a system of smaller satellite companions, such as the Magellanic Clouds around our own galaxy. Dynamical friction on the luminous body of the primary galaxy and its dark halo will lead to the decay of satellite orbits and some companions may eventually be assimilated by the large galaxy. For disk primaries, the kinetic energy available from a satellite accretion can easily be comparable to the total vertical kinetic energy of the disk. In such cases, the primary will be significantly perturbed by an accretion and perhaps destroyed by continued bombardment.

The response of a disk to a decaying satellite was examined in some detail by [1], using an approximate numerical technique which ignored the self–gravity of the primary. A self–consistent treatment has subsequently been considered by [2], but only in the limit that the disk remains perfectly flat and the satellite is confined to the disk plane. We are currently investigating changes in the three–dimensional structure of self–gravitating disks resulting from tidal interactions with less massive companions using a hierarchical tree code [3,4]. Owing to their inherent efficiency and lack of geometrical restrictions, tree codes are ideal for simulating highly distorted systems at high spatial resolution.

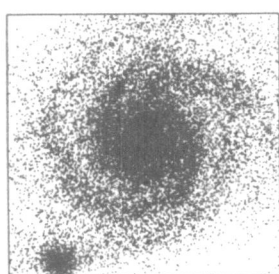 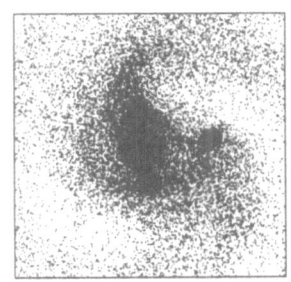

Figure 1   Figure 2

Figures 1 and 2 show face–on views of the response of a stellar disk to a perturbing satellite at two intermediate times during the latter's orbital decay. The disk, consisting of $N = 32,768$ particles, initially had an exponential density profile in radius with scale–length $h = 3.5$ kpc and a $\mathrm{sech}^2 z/z_0$ distribution vertically, with scale–height $z_0/2 = 350$ pc. The disk is embedded in a non–singular isothermal halo with scale–length $\gamma = 3.5$ kpc and asymptotic rotation velocity $v_\infty = 220$ km/sec. For these parameters, the halo mass out to the edge of the disk, $R_{disk} = 21$

C. S. Frenk et al. (eds.), The Epoch of Galaxy Formation, 427–428.

kpc, is approximately $M_{halo} \approx 3.25 M_{disk}$. The halo is rigid but allowed to move in response to the satellite according to a simple "unpinning" procedure [1]. The satellite has a mass 10% of the disk, is modeled by a spherical Jaffe profile, and consists of 4096 particles. Initially the orbit was circular at the edge of the disk, inclined by 30 degrees with respect to the disk plane.

The response to the perturbation is significant, even though the satellite is but a small fraction of the total mass of the system and the disk is strongly stabilized by its dark halo. At early times, as in Figure 1, large amplitude spiral patterns can be seen. Later, in Figure 2, a central bar is induced in the disk, generalizing a result obtained by [5] for more massive companions. This bar is transient and phase mixes out within a rotation period after the satellite orbit has decayed completely. The total sinking time in this case is roughly 3×10^9 years.

Perhaps more significantly in the context of galaxy formation, the vertical structure of the disk has been severely altered. Figure 3 shows an edge–on view of the disk initially (left panel) and at the completion of the interaction (right panel). (The scale here is expanded by a factor of two relative to Figures 1 and 2.) A detailed analysis shows that the average vertical scale-height of the disk has been increased by a factor ≈ 2.2 as the result of the satellite accretion. Furthermore, since the satellite tends to settle into the disk plane before sinking radially [1], the disk in the right panel is flared towards the edges and slightly warped.

 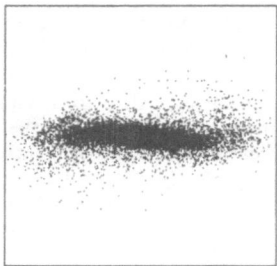

Figure 3

The damage that can be induced in spiral galaxies by even small potential perturbations underscores the fragile nature of disks. Our results suggest that either most disks have not participated in interactions such as that depicted here, or have been reformed by secondary processes. Both possibilities indicate that disks formed late, perhaps via infall, during a relatively quiescent phase of galaxy evolution. Furthermore, if disks are often reformed, interactions with less massive companions may account for some thick disks and warps. We caution, however, that this interpretation is not unique. Less violent processes appear to be capable of producing morphologically similar structures [6].

References

[1] Quinn, P.J. and Goodman, J. 1986, *Ap. J.*, **309**, 472.
[2] Valtonen, M.J., Valtaoja, L., Sundelius, B., Donner, K.J., and Byrd, G.G. 1988, *Ap. J.*, submitted.
[3] Barnes, J. and Hut, P. 1986, *Nature*, **324**, 446.
[4] Hernquist, L. 1987, *Ap. J. Suppl.*, **64**, 715.
[5] Noguchi, M. 1987, *M.N.R.A.S.*, **228**, 635.
[6] Spergel, D. 1989, this volume.

GALAXY FORMATION IN ASYMMETRIC DARK HALOES

Kandaswamy Subramanian
Astronomy Centre
University of Sussex
Falmer. Brighton
BN1 9QH, UK

The possible consequences of asymmetry in dark galactic haloes for galaxy formation are examined, by adopting an idealized model of protogalactic collapse. The protogalaxy can lose a significant fraction of its initial angular momentum to the dark halo if it initially rotates about an axis close to the halo's middle axis. We speculate that elliptical galaxies form when a protogalaxy loses most of its initial angular momentum to the dark halo. Asymmetric halo cores may also lead to triaxial bulges and assist the formation of central bars and compact objects.

Luminous galaxies are thought to be surrounded by large massive haloes of dark matter, which dominate gravitationally outside a few galactic core-radii. The halo is expected to crucially affect protogalactic collapse. Most studies of protogalactic collapse in a dark halo assume the halo to be spherically or axially symmetric. However, it is far from clear whether this assumption is valid. The initial asymmetry of proto-halo density peaks, instabilities which enhance this asymmetry during the gravitational collapse of the halo and tidal forces of neighbouring clumps are all likely to lead to haloes being asymmetric. Also N-body simulations of halo formation by Frenk et al. (1988) show that the short to long axis ratio is $\frac{1}{2}$ for $\sim\frac{1}{2}$ of the haloes. Clearly it is important to consider the effect of such halo asymmetry in determining the properties of the galaxy forming within it.

We study the effect of halo asymmetry in the context of the White and Rees (1978) picture of galaxy formation. In order to make a preliminary but analytical study, we idealize the protogalactic collapse as a pressureless (cooling timescale < dynamical timescale), test particle collapse of gas rotating initially as a solid body, in a rigid uniform density triaxial halo. Also, the initial

C. S. Frenk et al. (eds.), The Epoch of Galaxy Formation, 429–430.
© *1989 by Kluwer Academic Publishers.*

angular momentum, $\tilde{L}_0$, is not assumed to be aligned with any particular axis of the halo.

We show that the protogalaxy first forms a caustic sheet essentially normal to the halo's short axis, and then collapses along the middle axis to form a near filament before the gas self-gravity becomes important. One important question is; how much angular momentum $\tilde{L}$, does the protogalaxy lose to the halo during collapse? We show that there is negligible loss of $\tilde{L}$ prior to the formation of a caustic surface; but a significant loss is possible thereafter if the protogalaxy remains as a pancake (due to efficient cooling of gas), rotating about an axis in its plane. When $\tilde{L}_0$ lies in the plane of the longest and middle axes of the halo, this configuration of large $\tilde{L}$ loss is achieved. Further, the greatest loss of $\tilde{L}$ (~50-100 per cent $\tilde{L}_0$ for a halo short to long axis ratio $\lesssim$ $^1/_3$) occurs for $\tilde{L}_0$ parallel to the middle axis, because then the pancake rotates in the plane in which the halo's potential is most anisotropic. The loss of $\tilde{L}$ is also greater for collapse within more oblate haloes.

These results tempt us to speculate that dark haloes are typically quite asymmetric and elliptical galaxies (with little angular momentum support) are those protogalaxies whose initial rotation axis is close to the halo's middle axis and hence have lost most of their angular momentum to the dark halo. Further, in an asymmetric halo the gas falls on to preferred planes which need not coincide with the plane normal to $\tilde{L}$. This raises the question of how disk galaxies form at all in highly asymmetric haloes. Finally, in an asymmetric halo core even the details of the protogalactic collapse could be as obtained in our idealized model. Efficient star formation in the dense caustic and filament stages of the protogalaxy collapse, and subsequent relaxation may lead to a triaxial bulge and a central bar, respectively. If significant $\tilde{L}$ has been lost by this stage the gas left over in the bar could collapse further to form a central compact object. A detailed account of this work can be found in Subramanian (1988).

References

Frenk, C.F., White, S.D.M., Davis, M. & Efstathiou, G., 1988. Astrophys. J., 327, 507.
Subramanian, K., 1988. Mon. Not. R. astr. Soc., (in Press).
White, S.D.M. & Rees, M.J., 1978. Mon. Not. R. astr. Soc., 183, 341.

GAS IN A COSMOLOGICAL N-BODY SIMULATION

P. A. Thomas & R. G. Carlberg
C.I.T.A.,
60, St. George St.,
Toronto, Ontario,
M5S 1A1, Canada.

ABSTRACT. We look at the behaviour of gas in a cosmological n-nody simulation which contains cold dark matter and dissipative gas. The cooling algorithm is chosen so that galaxy formation occurs as in the scheme of Rees & Ostriker (1977). Perturbations of the microwave background radiation are small except through the cores of clusters of galaxies.

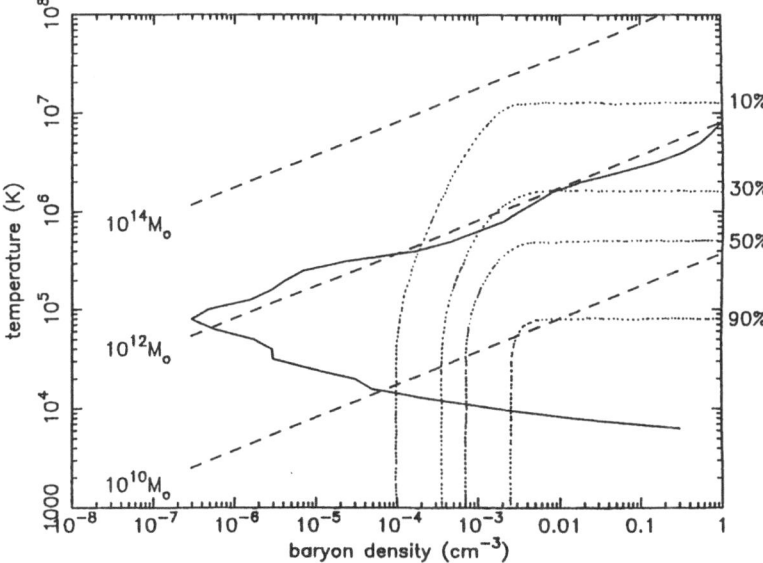

Figure 1. The cooling algorithm for the gas is based on a simple kinetic theory for particle interactions. The dotted lines show the fraction of gas particles which will cool at a given mean density and temperature. Cooling and star formation principally occur in that region of the diagram (to the right of the solid line) for which the cooling time is shorter than the dynamical time.

C. S. Frenk et al. (eds.), The Epoch of Galaxy Formation, 431–432.

432

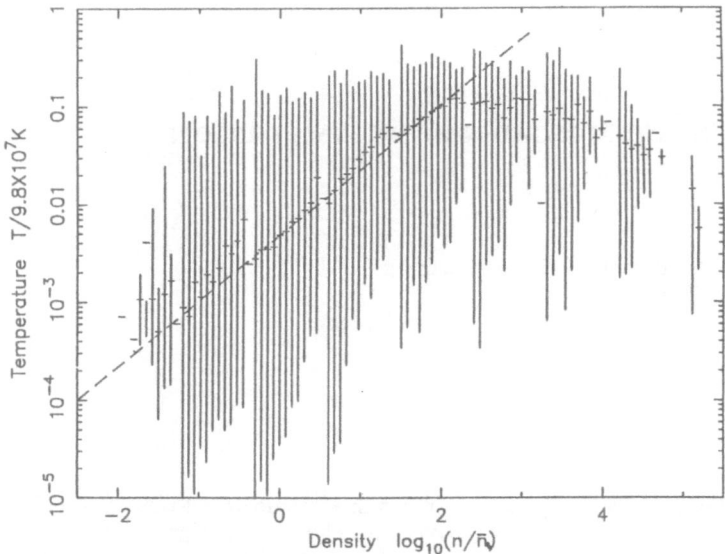

Figure 2. The 'phase diagram' of the gas particles at $z = 0$ showing the full range of measured temperatures. Low density gas has expanded adiabatically and so follows the dashed line, $T \propto n^{2/3}$. Virialised gas in clusters is approximately isothermal with the highest density gas being slightly cooler.

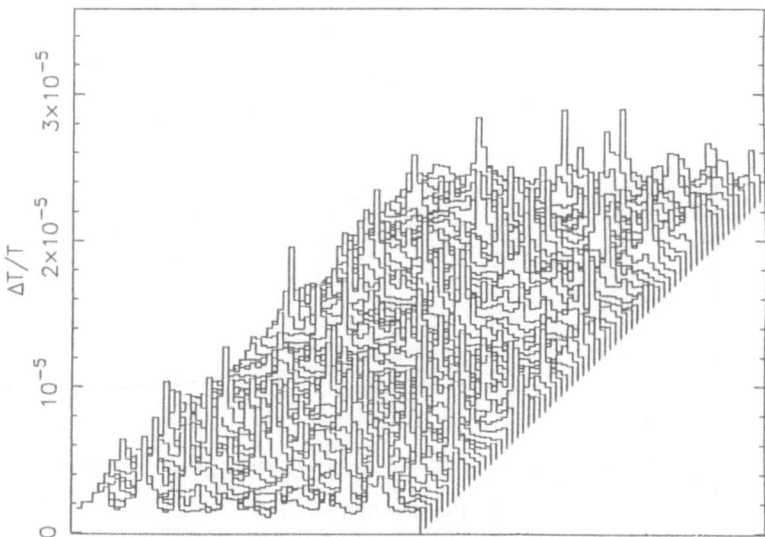

Figure 3. The Sunyaev-Zeldovich fluctuations in 1' cells on the sky. The mean dispersion is small ($\Delta T/T \approx 2 \times 10^{-6}$) but with occasional larger fluctuations ($\Delta T/T \lesssim 10^{-5}$) for lines of sight which pass through clusters back to $z \approx 1$; This could rise to 3–4×10^{-5} for Abell richness class 1 clusters.

DISSIPATIONAL GALAXY FORMATION

Neal Katz[1] and Lars Hernquist[2]
[1]Princeton University Observatory
[2]The Institute for Advanced Study

We have begun to study the formation of galaxies by gravitational collapse in different cosmological scenarios, including both dark and baryonic matter. The numerical experiments are being conducted using a new, general–purpose code for evolving self–gravitating fluids in three dimensions [1]. Hydrodynamic properties are determined using a Monte Carlo–like approach known as smoothed particle hydrodynamics (SPH). However, unlike most previous implementations of SPH, gravitational forces are computed with a hierarchical tree algorithm.

Here we present simulations of the collapse of constant density perturbations, initially rotating and in Hubble–flow expansion, ignoring the influence of external tidal fields. These results demonstrate that the morphology of the resultant galaxy is sensitive to a number of factors, including the perturbation spectrum on scales smaller than those of individual galaxies.

Figure 1 shows the final distributions of gas (left panel) and dark matter (right panel) in the collapse of a white noise density perturbation in a universe with $\Omega = 1$, ignoring the external tidal field. Initially, the matter is in solid body rotation, with angular momentum parameter $\lambda = 0.07$, and expanding with the Hubble flow at a redshift $Z = 17.9$. The total mass of the system is $7 \times 10^{11} M_\odot$ and the ratio of dark:baryonic matter is 10:1. The initial velocity dispersion of the dark matter is $\sigma = 0.2$, in natural units, corresponding to 44 km/sec. The initial radius of the system is 62.5 kpc. In this calculation star formation is ignored. Dissipation is included through Compton and radiative cooling appropriate for a pure hydrogen plasma.

 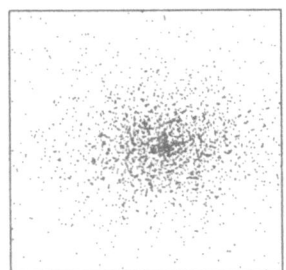

Figure 1

The figure shows a view perpendicular to initial axis of rotation and measures 125×125 kpc. The time corresponds to a redshift $z = 0.85$.

C. S. Frenk et al. (eds.), The Epoch of Galaxy Formation, 433–434.

As the system expands ($z = 17.8$ to $z = 2.7$), the dark matter fragments into clumps as the result of gravitational instability. Since the gas represents a negligible fraction of the total mass of the system, it is dragged along by the dark matter. Following turnaround, the clumps merge, shock heating the gas and dispersing the dark matter. The gas then rapidly cools and begins to form a disk in the smooth background potential of the dark matter from $z = 1.9$ to $z = 1.4$. The disk is bar unstable since it is dynamically cold and star formation has been ignored. At the final time, the disk is quite thin with a thickness approximately 0.1 of its diameter, showing some evidence for warping near its edges. The dark matter is more extended, as expected for a collisionless collapse.

The dark matter is slightly oblate and its spin axis is misaligned with that of the gas, warping the disk. Note that although the total angular momentum is conserved, the angular momentum of each species is not separately conserved, owing to tidal torquing during the fragmentation and collapse. Within statistical uncertainties the rotation curve is flat from $r \sim 6-20$ kpc. The halo has a core radius ~ 5 kpc, and the density of dark matter drops rapidly beyond 20 kpc. This latter feature is likely an artifact of the finite size of the top–hat perturbation.

We are currently conducting experiments similar to those discussed previously, but allowing for the conversion of gas into stars. The star formation rate, which is somewhat uncertain, is parameterized as a simple power–law function of the local cooling time, τ_{cool}, as follows

$$\frac{d\rho_{stars}}{dt} = C\tau_{cool}^{-q},$$

where $q \sim 1 - 2$ and C is a constant. Gas is locally converted into stars probabilistically if the flow is converging, the cooling time is less than the dynamical time, and the sound crossing time is greater than the dynamical time. These conditions give a crude approximation to a Jeans criterion, which is adequate given our somewhat coarse resolution in mass scales.

Comparisons with the simulation above indicate that the primary effect of star formation is, in this case, to seed the development of a central bulge. The disk, which forms through infall, consists mainly of gas, owing to the low densities present there. Future simulations, with varying star formation efficiency, will determine if disk galaxies with realistic stellar profiles can be formed through collapse.

The principal factors determining Hubble type are not well-known. The results presented here are still preliminary; consequently they do not provide a definitive answer to this puzzle. We note, however, that our simulations suggest a strong dependence on the form of the perturbation spectrum at mass scales smaller than those of individual galaxies. In the models presented here, structure develops in the expanding and collapsing perturbation owing to gravitational instability. The detailed form of this structure is partly determined by the initial power spectrum.

We have tested the severity of this dependence by evolving models differing only in the way that particles were distributed initially. In one case, particles were laid down at random, corresponding to a white noise spectrum. In the other, particles were distributed uniformly on a grid, giving a somewhat quieter start. There is a much stronger disk component in the run beginning with lattice initial conditions. In retrospect this result is perhaps not too surprising. Less large–scale power is present if particles are distributed on a grid, hence the fragmentation and collapse will be relatively smooth and a disk can be formed. If the fragmentation proceeds rapidly and the collapse is clumpy, as for Poisson initial conditions, rather large potential perturbations will exist in the dark matter and disks will be less likely to survive.

Reference

[1] Hernquist, L. and Katz, N. 1988, *SIAM J. Sci. Stat. Comput.*, submitted.

TIDAL ORIGIN OF STARBURSTS AND ACTIVE GALACTIC NUCLEI

Lars Hernquist
The Institute for Advanced Study, Princeton, NJ

Many observations suggest that starbursting and nuclear activity in galaxies are strongly correlated with the presence of tidally interacting companions. Furthermore, simple theoretical arguments indicate that perturbations in the potential resulting from such encounters could effectively couple to the nuclear regions, perhaps fueling a central source.

I have begun to study these effects in detail using a new, general–purpose code for evolving self–gravitating fluids in three dimensions [1]. Figure 1 shows a face–on view of the structure arising in the gaseous component of an isolated disk consisting of both stars and gas after five rotation periods. A total of 32,768 particles were used, of which 8192 represent gas, and the mass ratio stars:gas is 10:1. The stars and gas were initially distributed as in [2], and the disk is stabilized by a rigid halo with mass $M_{halo} \approx 3.25 M_{disk}$ out to 21 kpc. The detailed structure in this case is partly an artifact of the initial conditions, which consisted of an isothermal gas sheet at $T = 10^4 K$, and the fact that cooling was suppressed for $T \lesssim 10^4$ to inhibit the formation of dense clumps. However, subsequent studies indicate that the global distribution of the gas is not sensitive to these assumptions.

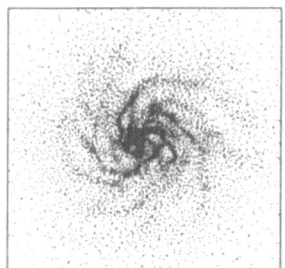

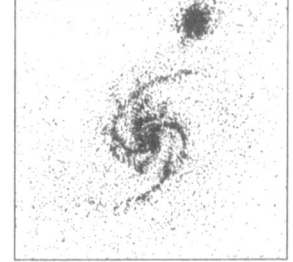

Figure 1   Figure 2

An example of the response of the gaseous component of a disk identical to that in Figure 1 to a decaying satellite is shown in Figure 2. The satellite has a mass 10% of the disk and consists of 4096 particles. Initially the orbit was circular at the edge of the disk, inclined by 30 degrees with respect to the disk plane. A strong two–armed spiral pattern is clearly visible, owing to the dipole nature of the perturbation. The gas, though but a small fraction of the total mass, causes the companion orbit to decay nearly 50% more rapidly than in the purely stellar dynamical case [2]. This is presumably because the disk remains colder dynamically when

435

C. S. Frenk et al. (eds.), The Epoch of Galaxy Formation, 435–436.

436

dissipation is included. At the conclusion of the simulation, the total radiated energy amounts to nearly 80% of the initial kinetic energy of the companion.

Of even greater interest is the accumulation of a dense central concentration of gas in the disk during the final stages of the interaction. The dramatic nature of this event can be most clearly seen in the radial distribution of gas at late times. Figure 3 shows the number of gas particles in radial bins as a function of radius, measured in kpc. The distribution just prior to these times is only slightly more concentrated than that initially. The final catastrophe does not occur until a time $1.9 \times 10^9 < t < 2.0 \times 10^9$ years after the beginning of the simulation; in other words the central gas mass has formed on a time–scale $< 10^8$ years. This gas is mostly self–gravitating, has a mass $\approx 2 \times 10^9 M_\odot$, and contains approximately 40% of the gas initially distributed throughout the disk. Furthermore, it occupies a region of order 200 parsecs in radius.

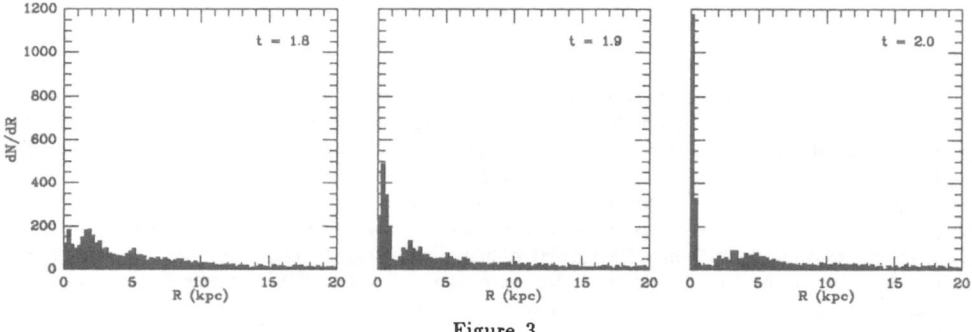

Figure 3

Unlike results obtained earlier by [3], the instability shown here is a consequence of the self–gravitating response of both the gas and stars. Initially, a bar induced in the disk stars [2] triggers a relatively slow funneling of gas towards the center. Once a critical mass of gas accumulates in the inner regions such that it becomes mildly self–gravitating, the rate of radial flow is accelerated as it becomes bar unstable. During these latter stages a mass $\gtrsim 10^9 M_\odot$ is deposited into the central regions on a time–scale $\sim 10^7$ years. This instability is purely a gasdynamical effect; there is no tendency for a nuclear concentration of stars to form.

The time–scales and central gas densities are in excellent agreement with those inferred for luminous nuclear starburst galaxies. It is also not unreasonable to suppose that some of this gas will be transported to the very center of the disk, perhaps powering a central source and thereby initiating nuclear activity.

It is likely that gas in disks would respond strongly to any large non–axisymmetric perturbation, including bars, oval distortions, global asymmetries in the halo, or fluctuations in the potential during the formation of the galaxy. Events such as that discussed here would probably be more common at high redshifts where galaxies are presumably gas–rich and encounters are frequent, perhaps explaining some properties of quasars. Finally, if ellipticals and/or bulges are formed through a series of encounters with less massive companions, starbursting and nuclear activity may well be intrinsic to galaxy formation itself.

References

[1] Hernquist, L. and Katz, N. 1988, *SIAM J. Sci. Stat. Comput.*, submitted.

[2] Hernquist, L. and Quinn, P.J. 1989, this volume.

[3] Noguchi, M. 1988, *Astron. Astrophys.*, in press.

THE TOPOLOGY OF LARGE SCALE STRUCTURE: OBSERVATIONS

David H. Weinberg
Department of Astrophysical Sciences
Princeton University

ABSTRACT. The genus-threshold density relation provides a quantitative measure of the topology of large scale structure. This measure has been applied to several large galaxy redshift surveys. On scales greater than about twice the galaxy correlation length, the topology of the galaxy distribution is consistent with Gaussian density fluctuations. On a scale ~ 600 km/s, the observed genus curves display a shift in the direction of a "meatball" topology. We find no evidence for "bubbles" in the galaxy distribution, although simulated data sets suggest that we could have detected them if they were present.

1. Introduction

A variety of qualitative models have been proposed to describe the large-scale structure of the galaxy distribution: a hierarchy of more or less isolated clusters residing in a low density background, a cellular or "bubble" structure where walls of galaxies surround large voids, and a network of filaments. Gott, Melott, and Dickinson (1986, [GMD]) pointed out that a wide class of theoretical models, in which large scale structure grows from primordial density fluctuations that are Gaussian with random phases, predicts a spongelike topology for structure that remains in the linear regime. In a series of papers (Weinberg, Gott, and Melott 1987 and references therein), we have extended the approach of GMD into a quantitative method for measuring the topology of large scale structure. We smooth the galaxy distribution by convolving with a Gaussian window, $\exp(-r^2/\lambda^2)$, and define density contours that surround specified fractions of the total volume. We measure the genus of these contour surfaces,

$$G_s \equiv \text{(number of holes) - (number of isolated regions)},$$

by calculating the integrated curvature of the contour and applying the Gauss-Bonnet theorem. Multiply connected, spongelike contours have many holes and a positive genus. Contours that surround isolated high or low density regions have negative genus. The dependence of contour genus on threshold density provides a quantitative measure of the underlying topology. For Gaussian density fluctuations, the genus-threshold density relation ("genus curve") has a universal form:

$$G_s = \frac{V}{4\pi^2} \left(\frac{<k^2>}{3} \right)^{3/2} (1 - \nu^2) e^{-\nu^2/2}, \tag{1}$$

where V is the survey volume and ν is the threshold density in standard deviations from the mean. Bubble distributions, filamentary nets, and distributions dominated by isolated clusters

437

C. S. Frenk et al. (eds.), The Epoch of Galaxy Formation, 437–438.

("meatball topology") produce asymmetric genus curves that can be distinguished easily from random phase curves.

We have applied this technique to several large galaxy redshift surveys: the Arecibo redshift survey of Giovanelli and Haynes, the CfA redshift survey, the dwarf galaxy redshift survey of Thuan and Schneider, and the nearby galaxy survey of Tully. We have also used a sample of Abell clusters to study the topology on very large scales. This paper presents a brief summary of these results — a more detailed overview is given by Gott *et al.* 1988.

2. Results and Implications

We have two data samples that probe large scales, the Abell cluster sample smoothed at $\lambda = 5000$ km/s and a deep subset of the Giovanelli and Haynes survey smoothed at $\lambda = 1200$ km/s. The genus curves for both of these samples indicate a Gaussian topology (equation [1]), in agreement with the standard models in which superclusters and voids grow from random phase, primordial density fluctuations. On smaller scales, $\lambda \sim 600$ km/s, the observed genus curves shift away from the random phase curve in the direction of a "meatball" topology. The shift is quite striking in the Giovanelli and Haynes sample, which is dominated by the Perseus-Pisces supercluster, and it is present to a lesser degree in the CfA sample.

To assess the theoretical implications of our results, we have created simulated data sets with the same survey volume, galaxy density, and selection criteria as the observational data. Genus curves from cold dark matter (CDM) realizations are either random phase or meatball shifted. The CDM model is overall the most successful in reproducing the observations, although none of our simulations show a shift as large as that in the Giovanelli and Haynes data. Some biased neutrino realizations show a meatball shift, but others yield genus curves that are shifted in the opposite direction or have very low amplitude. Bubble models, in which galaxies reside on walls surrounding empty voids, do not describe the observations well at all. Models with typical bubble diameters ranging from 2500 to 4000 km/s consistently produce genus curves that are shifted in the opposite direction from the data. Apparently the voids in the galaxy distribution are not completely surrounded by walls of galaxies; instead they connect to one another through low density tunnels, creating a multiply connected, spongelike topology. Our results suggest that biased CDM models (e.g. White *et al.* 1987) offer a more promising explanation than simple bubble models for the "frothy" appearance of large scale structure.

The observational situation should improve greatly in the next few years. We have plans to analyze the IRAS redshift survey and the southern sky redshift survey, each comparable to the best data sets studied to date. The extended CfA survey should provide redshifts for $\sim 12,000$ galaxies, and there are even proposals for dedicated telescopes that would take 100,000 or 1,000,000 redshifts. With such data samples, we will be able to study topology on a variety of scales with good statistical accuracy, allowing a detailed test of theories for the formation of galaxies and large scale structure.

References

Gott, J. R., Melott, A. L., and Dickinson, M. 1986, Ap. J. **306**, 341 (GMD).

Gott, J. R., Miller, J., Thuan, T. X., Schneider, S. E., Weinberg, D. H., Gammie, C., Polk, K., Vogeley, M., Jeffrey, S., Bhavsar, S. P., Melott, A. L., Giovanelli, R., Haynes, M. P., Tully, R. B., and Hamilton, A. J. S. 1988, Ap. J. , submitted.

Weinberg, D. H., Gott, J. R., and Melott, A. L. 1987, Ap. J. **321**, 2 (WGM).

White, S. D. M., Frenk, C., Davis, M., and Efstathiou, G. 1987, Ap. J. **313**, 505.

GALAXY FORMATION: THE BOARD GAME

Lisa Florman Weinberg[1]
David H. Weinberg[2]
[1] Art History Department, Columbia University
[2] Astrophysical Sciences Department, Princeton University

ABSTRACT. We summarize the rules of "Galaxy Formation!", a board game, and describe the course of the game played at the end of the conference.

1. Overview of Rules

The game action takes place a few years in the future, beginning with the flight of Space Telescope. Each player (or collaboration thereof) chooses a theory for the formation of galaxies, large-scale structure, and everything else that matters. Currently, the available choices are: cold dark matter (biased, with $\Omega = 1$, and open, with $\Omega = 0.2$), explosions, cosmic strings (non-superconducting), cosmic strings (superconducting), isocurvature baryons, massive neutrinos, and primordial magnetic fields. A player may represent himself or may choose to assume another "persona", whose style, mannerisms, and theoretical prejudices should guide his play. Persona Icons (caricatures) of a number of eminent astrophysicists are provided with the game. The object of the game is to acquire enough Plausibility Points to have one's theory declared the one true and correct theory of galaxy formation.

The game board is divided into five Observational Areas: galaxy properties (masses, morphologies, rotation curves), galaxy clustering (covariance function and other clustering statistics), very large scale structure (cluster correlations, large-scale velocity flows, giant voids and superclusters), high redshift observations (galaxy evolution, high-z galaxies, quasars, Lyman-α clouds), and the microwave background. Each Observational Area contains Paper squares, and a player who lands on one may submit a paper explaining observations in that area in the context of her theory. If accepted (rapid refereeing is ensured by rolling the die), the paper earns plausibility points, the exact number depending on whether the Observational Area is a strength or weakness of the theory in question. Moving about the game board, players must avoid such perils as computer errors that swallow six months of work (and one turn), appointment to presidential blue-ribbon panels (one turn), and triplets (three turns), while angling to acquire superb graduate students (extra turn), or Macarthur fellowships (five years of funding).

The quest for plausibility points may be advanced or hindered by Observational Developments. When a player lands on an Observational Development space, he draws an Observational Development card. Each card has two possibilities (e.g. microwave background fluctuations are detected or new limits are set on fluctuations), and the one that appears right side up in the shuffled deck applies. The effects on the plausibility of different theories are listed on the card, although a player may attempt to persuade the others that an observational development

439

C. S. Frenk et al. (eds.), The Epoch of Galaxy Formation, 439–440.

should affect his theory in some other way. In addition, each player may choose to ignore **one** observational development during the course of the game.

Other essential aspects of the scientific endeavor have been included in the game. Players who exhaust their funding must spend their turn writing grant proposals and spin the Funding Spinner to determine the term (if any) of the grant awarded. Whenever two players land on the same space, they have the option of forming a Scientific Organizing Committee and calling a conference. This requires all other players to join them on the board, but the conference organizers lose their next turn while they edit the proceedings. A Hubble Spinner is provided for observational determinations of the Hubble constant.

A player may acquire enough plausibility points to win the game outright — winning in this way requires many papers and a remarkable string of good luck with the observations. Alternatively, after acquiring a lesser number of plausibility points a player may attempt a Consensus Win. She is given two minutes to explain why her theory should be declared the one true and correct theory of galaxy formation. Other players have one minute to discuss the question, after which they vote secretly. If there are at least two no votes, the attempted win fails. If all votes are yes, the attempted win succeeds. If there is only one no vote, the win also succeeds. The lone dissenter is branded as a quack, doomed to a life of writing papers that no one else believes and hearing whispers about "what a good theorist he used to be."

Short Game rules are provided for dilettantes who lack the energy and perseverance to resolve the issue of galaxy formation in the proper way. A Short Game lasts for a fixed period of six fiscal years, after which the two players with the highest plausibility scores are eligible to win. Each has two minutes to argue why he should be declared the victor. The winner is then decided by a majority vote.

2. The Durham Game

Fortified by a week of scientific discussion, a fine Indian dinner, and fuel from the college pub, several conference participants gathered for a Friday night game. Personae and theories were chosen up as follows (we identify the actual players by initials only, in order to maintain a false illusion of anonymity): Ostriker (CAT) and Zel'dovich (JEG) defended explosions. Burbidge (DNS) and Schmidt (AB) united behind primordial magnetic fields. White (NRT) and Geller (ELZ) adopted cold dark matter. Peebles (PJEP) took isocurvature baryons. Both flavors of cosmic strings were advocated, non-superconducting by Rees (RHB), and superconducting by Witten (JEB). Gott (TRL) and Silk (CGL) supported massive neutrinos. Efstathiou (NSK) played the Skeptic, an irritating character who defends no theory of his own but writes papers that negate other player's papers.

Gott and Silk quickly won lucrative TV and book contracts, so they retired early from astronomy (and the game) to a life of fame and fortune. Peebles, once his term as Ap J editor was up and his papers had to be refereed again, followed suit. Efstathiou was a generally ineffective skeptic, but he did manage to savage cold dark matter on several occasions. Rees lost three turns finding a job, after his department was shut down.

Cold dark matter acolytes were cheered by several Observational Developments: dwarf galaxies were found in voids, disk galaxies were shown to be rare at $z > 1$, the Great Attractor "faded away", and, most dramatically, photinos were discovered. Despite the troubling observation of two $z > 6$ quasars, cold dark matter seemed well poised to win the game. In the end, though, it was Witten and superconducting cosmic strings that carried the day, largely on the basis of the gravitational wave background (detected by timing studies of millisecond pulars), some creative paper titles, and a better performance in the final conference. However, since this game was played under the Short Rules instead of the Knock-Down Drag-Out Rules, the outcome cannot be considered definitive.

Author Index

Subject Index

4000 Å break 21, 36, 175, 351, 371.

Absorption line systems (see also Lyman alpha absorption) 107.
Active galactic nuclei (AGN) (see also Quasars, Radio galaxies) 435.
Angular momentum 4, 321, 417, 429.

Biased galaxy formation 6, 15, 211, 213, 243, 246, 257, 265, 275, 281, 413, 415, 421, 423, 425.
Black holes 130, 147, 232, 309.
Bulges (see Galaxies: morphology)

Chemical abundances (see metals)
Clustering (see Quasars: clustering, Correlation functions, etc.)
Clusters of galaxies
  Abell 259, 413, 415.
  elliptical galaxies in 167, 179, 371.
  high redshift 57, 167, 185, 196, 352, 391, 397.
  velocity dispersions 170, 185, 257.
  X-ray properties 261, 310.
Cold dark matter (CDM) 3, 15, 22, 147, 216, 227, 243, 257, 275, 311, 415, 419.
Cooling flows 2, 23, 163, 309, 399, 431.
Cooling times 5, 23, 265, 431.
Correlation functions
  cluster 413, 418.
  cross 421.
  galaxy 7, 377.
  quasar 141.
  radio galaxy 394.
Cosmic strings 220, 228, 315, 321.
Cosmological constant 216.
Cosmological evolution (see Galaxy evolution, Quasars, etc.)

Damped Lyman alpha systems (see Lyman alpha absorption)
Dark haloes 22, 148, 243.
Dark matter
  cold (see Cold dark matter)
  hot 7, 217.
  Jupiters 309.
  unstable (see Decaying elementary particles)
Decaying elementary particles 159, 233, 327.
Density fluctuations (see Fluctuation spectrum)

Density maxima (see also Gaussian random fields) 265, 271, 281, 413, 415.
Disk formation 1, 33, 265, 433.
Dust (see also Obscuration, Microwave background)
  microwave background from grains 407.
  dust-to-gas ratios 111, 369.
  redshifted emission from early epochs 405.

Eddington limit 130, 147, 232, 238.
Emission line regions 39, 57, 63, 163, 383.
Environment 167, 185, 371, 391.
Epoch of galaxy formation 1, 18, 36, 153, 315, 416.
Explosion models 4, 159, 311, 315.
Evolution of galaxies (see Galaxy evolution)

Faber-Jackson relationship 243.
Fisher-Tully relationship 243.
Fluctuation spectrum
  cold dark matter (see Cold dark matter)
  isocurvature 10, 217.
  normalisation 22, 149, 245, 257, 391, 401.

Galaxies
  age 63, 191, 342, 379, 387.
  blue 80, 187, 238, 352.
  clustering (see Clustering)
  colours 33, 39, 64, 75, 295, 363.
  dwarf 76, 115.
  E+A 169.
  elliptical 167, 179, 371.
  field 71.
  flat spectrum 16, 31.
  groups 179.
  interactions (see also Mergers) 167.
  luminosity function (see Luminosity function)
  mass function (see Mass function)
  morphology (Hubble sequence) 57, 265.
  normal 191.
  number counts (see Number counts)
  primæval (see Primæval galaxies)
  Seyfert (see Seyfert galaxies)
  spectra 15, 31, 39, 63, 167, 191.
  starburst (see Starburst galaxies)
  young 15, 107.
Galaxy catalogues (see Surveys)
Galaxy counts (see Number Counts)
Galaxy evolution 1, 17, 72, 85, 359, 363, 367, 387.

443